11/70

PURE AND APPLIED MATHEMATICS

A Series of Texts and Monographs

Edited by: R. COURANT · L. BERS · J. J. STOKER

PURE AND APPLIED MATHEMATICS

A Series of Texts and Monographs

Edited by: R. COURANT · L. BERS · J. J. STOKER

VOLUME XX

DIFFERENTIAL GEOMETRY

J. J. STOKER

WILEY-INTERSCIENCE
a Division of John Wiley & Sons
New York · London · Sydney · Toronto

Library of Congress Catalog Card Number: 69–16131
SBN 471 82825 4

Printed in the United States of America

To Heinz Hopf

PREFACE

More than thirty-five years ago I was introduced to the subject of this book by my friend and teacher Heinz Hopf through his lectures at the Technische Hochschule in Zürich. I had expected to take my degree there in applied mathematics and mechanics, but Heinz Hopf made such an impression on me, and created such an interest for the subject in me, that I wrote my thesis in differential geometry in the large on a topic suggested by him. My professional career afterwards turned in the main to fields concerned with mathematics in relation to problems in mechanics and mathematical physics generally. However, differential geometry has continued to fascinate me and to cause my thoughts to return again and again to various problems in the large—particularly during the rather frequent occasions when I happened to be teaching a course on the subject. Unfortunately, my efforts in this direction have had rather meagre results, so that I feel myself to be an amateur in the field. However, I am an amateur in the etymological sense of that word, and hope that something of my love for differential geometry will be infectious and will carry over to readers of my book.

In the introduction which follows this preface I outline the contents of the book and indicate the ways in which it differs from others in its attitudes and in its selection of material. In brief, it is stated that the book is intended for students and readers with a minimum of mathematical training, but still has the intention to deal with much that is relatively new in the field, particularly in differential geometry in the large. It also has as one of its purposes the introduction and use of three different notations: vector algebra and calculus; tensor calculus; and the notation devised by Cartan, which employs invariant differential forms as elements in an algebra due to Grassman, combined with an operation called exterior differentiation.

It is now my pleasant duty to thank a number of my friends and colleagues for the help and advice they have given me. Louis Nirenberg and Eugene Isaacson used the manuscript in courses, read it in detail, and spent much time and effort in making specific corrections as well as suggestions of a general character. K. O. Friedrichs also made some use of the manuscript in a course, and I benefited from a number of discussions with him about a variety of matters of principle and logic which aroused his interest. H. Karcher gave me a number of valuable suggestions about some parts of Chapter VIII, which deals with problems in the large; his help is acknowledged in that chapter at the appropriate places.

I owe much to Miss Helen Samoraj, who typed the manuscript in several versions, uncovered many errors and mistakes, and prodded me from time to time to get on with the job. I wish also to thank Carl Bass for drawing the figures.

In 1964 the Guggenheim Foundation gave me a Fellowship; during that time this book was finally organized and carried far toward completion in some of its major portions.

Finally, I am very happy to acknowledge the help given to me by the Mathematics Branch of the Office of Naval Research. I do this with particular pleasure because I have felt for years that the Office of Naval Research has had a very remarkable and beneficial effect on the progress of science in this country.

JAMES J. STOKER

Professor of Mathematics
Courant Institute of
Mathematical Sciences
New York University

INTRODUCTION

Differential geometry is a subject of basic importance for all mathematicians, regardless of their special interests, and it also furnishes essential ideas and tools needed by physicists and engineers.   But important as these considerations are, the value of the subject, for the author at least, arises rather from the great variety and beauty of the material itself, and for the close ties it has with important portions of algebra, topology, non-euclidean geometry, analysis generally (in particular with the theory of partial differential equations), and in mechanics and the general theory of relativity.   Beside all that it furnishes a great variety of fascinating unsolved problems of its own that are of a particularly challenging nature.

In writing this book the author had in mind these different points of view, and the corresponding classes of potential readers with their various interests.   The intention, therefore, is not to present a treatise for advanced students and specialists, but rather to present an introductory book which assumes no more at the outset than a knowledge of linear algebra and of the basic elements of analysis—in other words such preparation as an advanced undergraduate student of mathematics could be expected to have, and the kind of preparation to be expected in the early years of graduate study for the other classes of readers indicated above.   It turns out, happily, that even quite recent and interesting advances in the subject can be dealt with on the basis of such relatively scanty foreknowledge.

Since there are quite a number of books about differential geometry in print the author feels it his duty to say in what ways his book differs from others in its attitudes and its selection of material.   A brief outline of the contents of the book, chapter by chapter, is therefore given here.

Chapter I gives a brief summary of the basic facts and notations of vector algebra and calculus that are used in the book.

Chapter II deals with the theory of regular curves in the plane.   Most books, if they deal with plane curves at all, consider them as a special case of space curves.   In this book a relatively long chapter is devoted to them because they are of great interest in their own right and their theory is not in all respects the same as it is for space curves.   In addition, it is possible to present the theory of plane curves in such a way as to give the basic general motivations once for all for the underlying concepts of differential geometry so that the concepts can be introduced without much motivation in the more complicated cases of space curves and surfaces. Included in Chapter II is a discussion of some problems in differential geometry in the large. Among them is a proof of the Jordan theorem for smooth plane curves having a uniquely determined tangent vector. This belongs in differential geometry, whereas the theorem for merely continuous curves properly belongs in topology. The author found no proof in the literature of the Jordan theorem including a proof of the fact that the interior domain is simply connected, for the simpler case of smooth curves, in spite of the fact that the theorem in this form is the most widely used—for example in the integration theory for analytic functions of a complex variable, and in mechanics.

Chapter III deals with the theory of twisted curves in three space. The concepts of arc length s, curvature κ, and torsion τ are introduced. The Frenet equations are derived, and on the basis of the existence and uniqueness theorems for ordinary differential equations it is shown that the three invariants s, κ, and τ form a complete set of invariants in the sense that any two curves for which these quantities are the same differ at most by a rigid motion. Important connections of this theory with the kinematics of rigid body motion, and of the motion of a particle under given forces, are discussed.

Chapter IV deals with the basic elements of the theory of regular surfaces in three dimensional space. This revolves to a large extent around the two fundamental quadratic differential forms which serve to define the length of curves on the surface and the various curvatures that can be defined on it. Interesting special curves such as the asymptotic lines and lines of curvature, and their properties, are studied. The solution of many problems in differential geometry (and in other disciplines as well) can often be made very simple once an appropriate special system of curvilinear coordinates is introduced. The author thought it reasonable to justify such procedures in a number of important cases by an appeal to the existence and uniqueness theorems for ordinary differential equations.

Chapter V is concerned with two special classes of surfaces that are very interesting in their own right and that also serve to illustrate how the theory of Chapter IV can be used. These are the surfaces of revolution and the developable surfaces. The old-fashioned classification of the developables as cylinders, cones, or tangent surfaces of space curves is given up since this classification rules out many valid developables, i.e. many easily defined surfaces that are not composed entirely of parabolic points. Instead these surfaces are defined as those for which the Gaussian curvature is everywhere zero, and various properties of them in the large are treated on this basis.

Chapter VI treats the fundamental partial differential equations of the theory of surfaces in three-space. These come about by expressing the *first derivatives* of the two tangent vectors of the coordinate curves on the surface, and of its unit normal vector, as linear combinations of these vectors themselves. These equations are given the names of Gauss and Weingarten. They form an over-determined system, i.e. there are many more equations than there are dependent functions to be determined. Solutions thus exist only when certain compatibility conditions are satisfied, and these conditions are equations due to Gauss and to Codazzi and Mainardi. The equation due to Gauss embodies what is perhaps the most striking theorem in the whole subject : it says that the Gaussian curvature, defined originally for a surface in three-space, is really independent of the form of the surface in three-space so long as the lengths of all curves on the surface remain unchanged in any deformation of it. This theorem gave rise in Gauss's mind to the fruitful idea—later on developed in full generality by Riemann—of dealing with inner differential geometry, i.e. to geometrical questions that concern only geometry in the surface as evidenced by the nature of the length measurements on it. In this kind of geometry all geometric notions arise from the functions which, as its coefficients, serve to define the first fundamental form ; much of the later portions of the book are concerned with such inner, or intrinsic, geometries. In Chapter VI it is shown that a surface exists and is uniquely determined within rigid motions once the coefficients of the two fundamental forms are given, and if these functions satisfy the compatability conditions ; this is done by integration of the basic partial differential equations. The theorem follows from a basic theorem concerning over-determined systems (a theorem proved in Appendix B).

Chapter VII has as its purpose a treatment of *inner* differential geometry of surfaces, but it is done nevertheless by considering the surfaces to lie in three-space. In this way an intuitive geometric motivation for the concepts of the inner, or intrinsic, geometry of surfaces is made direct and simple. The concept chosen as basic for the whole chapter is the beautiful one due to Levi-Civita, of the parallel transport of a vector along a given curve on the surface. From this the notion of the geodetic curvature, denoted by κ_g, of a given curve is derived, and the special curves called geodetic lines are defined as those for which κ_g vanishes. All of these concepts, though derived for surfaces in three-space, are seen to belong to the intrinsic geometry of surfaces since they make use in the end only of quantities that are completely determined by the coefficients of the first fundamental form. Nevertheless it is quite interesting to know their relation to surfaces embedded in three-space. The geodetic lines, though defined initially as those curves along which $\kappa_g = 0$, can also be defined through studying curves of shortest length between pairs of points on a surface. This important problem is treated partially in Chapter VII. It is shown that the condition $\kappa_g = 0$ (which really is a second order ordinary differential equation) is in general only a *necessary* condition in order that a geodetic line should be a curve of shortest length. On the other hand it is shown that a small enough neighborhood of any point p can always be found such that any point q of it can be joined to p by a uniquely determined geodetic line of shortest length, when compared with the length of any other curve joining p and q. Since the differential equation determining the geodesics is of second order it follows that a uniquely determined geodesic exists through a given point p in every direction. In fact, a certain neighborhood of p is covered simply by these arcs, which can be taken, together with their orthogonal trajectories, as a regular parameter system— in complete analogy with polar coordinates in the plane or spherical coordinates on the sphere. This is one of those special coordinate systems referred to earlier that have the effect of simplifying the solutions of particular problems. One such problem concerns the surfaces of constant Gaussian curvature K, which are seen to furnish models for the three classical geometries, i.e. the Euclidean for $K = 0$, the elliptic geometry for $K > 0$, and the hyperbolic or Lobachefsky geometry for $K < 0$ when the straight lines are defined as geodesics in their whole extent. The Lobachefsky geometry is treated in some detail. The Gauss-Bonnet formula is derived (in the small) in this chapter. This formula relates the integral of the Gaussian curvature over a simply connected domain to the integral of the geodetic curvature over the boundary curve of the domain. A tool used to accomplish this is also derived; it is the beautiful result that the integral of the Gaussian curvature over a domain is equal to the angle change that results when a vector is transported parallel to itself around the boundary of the domain.

Chapter VIII is probably the chapter that makes the book most different from others because it deals with a considerable variety of the fascinating theorems of differential geometry in the large, especially for two-dimensional manifolds. For this purpose an introduction to the concept of a manifold in n dimensions is given intrinsically. This leads to the special case of a Riemannian manifold. Since most of the material of the chapter is then specialized to two-dimensional manifolds—in fact in large part to the compact two-dimensional manifolds—it was thought reasonable to interpolate a brief description of the facts from topology about them that are needed later on. Except for the last two sections of the chapter the theorems in the large are all concerned with inner differential geometry, thus indicating that this kind of geometry is very rich in content. Once abstract surfaces or manifolds have been given a metric it is possible to consider them in a natural way as metric spaces by defining a distance function in them, and to introduce the concept of *completeness*. This means, roughly speaking, that the manifold contains no boundary points at finite distance from any given point; thus this condition is a restriction only for open, or non-compact, manifolds. The theorem of Hopf and Rinow, which establishes the equivalence of four different characterizations of completeness, is proved as a by-product of the proof of one of the most important single theorems in differential geometry in the large, i.e. the theorem that a curve of shortest length exists joining any pair of points on a complete manifold, and that this curve is a geodetic line. A section is devoted to angle comparison theorems, of rather recent date, for geodetic triangles on surfaces. Geodetically convex domains are studied; in particular it is shown that sufficiently small geodetic circles and geodetic triangles are geodetically convex. The Gauss-Bonnet formula is used to prove the beautiful theorem that the integral of the Gaussian curvature over the area of a two-dimensional compact surface is not only an isometric invariant, but is also a topological invariant with a value fixed by the Euler characteristic. Vector fields on surfaces are considered and an index is assigned to their isolated singularities, i.e. to points where the field vector is the zero vector. This makes it possible to prove a theorem due to Poincaré, with the aid of the theorem on the change of angle resulting by parallel transport of a vector around a simple closed curve, that determines the sum of the indices in question on a compact surface in terms of its Euler characteristic. The theorem on the existence of shortest arcs as geodesics referred to earlier was a nonconstructive existence theorem. It is of great interest to approach this problem more directly as a two-point boundary value problem for the second order ordinary differential equation that characterizes geodetic lines. This leads to Jacobi's theory of the second variation and to a sufficient condition, based on the notion of a conjugate point, for the existence of the shortest join when the comparison curves are restricted to a neighborhood of a geodesic joining two points. This theory in turn makes it possible to prove the generalization of a famous theorem of Bonnet given by Hopf and Rinow, i.e. that a complete two-dimensional surface with Gaussian curvature above a certain positive bound is of necessity compact, because it has a diameter that can be estimated in terms of the bound on the curvature, and consequently is readily seen to be topologically a sphere. (Bonnet *assumed* the surface to be topologically a sphere lying in three-space, and then gave a bound for its diameter in terms of the bound on the Gaussian curvature.) The theorem of Synge is next dealt with; this theorem states that a compact manifold with an even number of dimensions and with positive Gaussian curvature is simply connected. The consideration of problems in intrinsic geometry in the large ends with a discussion of covering surfaces of complete two-dimensional surfaces with nonpositive Gaussian curvature—they are obtained by expanding geodetic polar coordinates over the surface. The final two sections of the chapter are concerned with complete surfaces lying in three-dimensional space. The first of these deals with Hilbert's famous theorem on the non-existence in three-space of a complete regular surface with constant negative Gaussian curvature. Two proofs of the theorem are given. One of them is a version of Hilbert's original proof, but it makes use of the covering surface just mentioned above; the other is a version of a proof due to Holmgren. The final section treats a generalization of a theorem due to Hadamard. This theorem states that a compact surface in three-space with positive Gaussian curvature is the full boundary of a convex body, or, in other words it is an ovaloid. Thus local convexity of the surface combined with its assumed closure guarantees that no double points or self-intersections can occur—in contrast with what can occur for locally convex closed curves in the plane. The theorem is generalized to the case of complete surfaces in three-space, with the result that the open, that is, noncompact, surfaces are the full boundary of an unbounded convex body.

Chapter IX treats the elements of Riemannian geometry on the basis of a systematic, though brief, introduction to tensor calculus. The point of view of Cartan is taken in doing this. Some applications to problems in the large are made, e.g. the extensions of the theorem of Hopf and Rinow, and of Synge's theorem, to n-dimensional manifolds are treated. Although it might be thought to fall out of the scope of a book on differential geometry to treat the general theory of relativity, the author nevertheless thought it good to do that. The reason is simply that this application of Riemannian geometry is so striking and beautiful, and it lends itself to a not too lengthy treatment, on a somewhat intuitive basis, even when the special theory of relativity and the relativistic dynamics of particle motion are first explained.

Chapter X has two purposes in view. One of them is to introduce still another notation to those already used. It is a notation due to Cartan which applies an algebra introduced by Grassman, and which employs an alternating product, to elements that are invariant differential forms. (They are invariant by virtue of the fact that their coefficients are the components of alternating covariant tensors.) In addition, an operation called exterior differentiation is introduced. This leads to the construction of new invariant differential forms, of higher degree, from any given one. It turns out that this notation is particularly effective in dealing with compatibility conditions and in converting volume integrals into surface integrals with the use of Green's theorem—both of which are basic operations in differential geometry. The notation is then applied here to vector differential forms in order to formulate the geometry of two-dimensional surfaces in three-space. The compactness of the notation is rather remarkable. In particular, the derivation of such a basic theorem as that of the isometric invariance of the Gaussian curvature, is very elegant. Minimal surfaces are treated here. However, most of the applications treated in this chapter are concerned with differential geometry in the large. These include various characterizations of the sphere (Chern's theorem), and three classic theorems concerning the uniqueness within motions of closed convex surfaces in three-space. The three theorems prove the uniqueness of the surface when (1) the line element is prescribed, or (2) the sum of the principal radii of curvature is prescribed as a function of the direction of the surface normal (Christoffel's theorem), or (3) the same as (2) but the Gaussian curvature is prescribed (Minkowski's theorem). These problems are all solved with the aid of an appropriately chosen invariant scalar differential form which results by taking a scalar triple product of three vector differential forms that involve vectors from both of two examples of the surfaces satisfying the given conditions.

A number of problems are formulated at the end of the chapters. The author tried to invent some new problems to serve as exercises; it is hoped that they will be found interesting and instructive without being too difficult.

The book has two appendices. Appendix A summarizes the main facts and formulas needed from linear algebra in a form suitable for ready reference in the book, together with brief discussions of geometry in affine, Euclidean, and Minkowskian spaces. Appendix B gives brief formulations, without proofs, of the basic existence and uniqueness theorems for ordinary differential equations, and a proof of the existence and uniqueness theorems—of such vital importance in differential geometry—for the solutions of over-determined systems of partial differential equations when appropriate compatibility conditions are satisfied.

This outline of the contents of the book should support the earlier statement concerning the author's intentions, i.e. to write (1) a thorough but elementary treatment of differential geometry for young students, that (2) includes a treatment of a rather large number of problems of differential geometry in the large, and that (3) makes a point of introducing and using three different notations employing vectors, then tensors, and finally invariant differential forms. In addition, it is hoped that all of these things can be done successfully on the basis of a minimum of preparation in other mathematical disciplines.

CONTENTS

Chapter III   Space Curves

Chapter IV   The Basic Elements of Surface Theory

CHAPTER I

Operations with Vectors

1 The Vector Notation

This chapter presents briefly the principal rules for operating with vectors, and a collection of those formulas which are useful in differential geometry. No attempt at completeness nor at an axiomatic treatment of vector algebra is made—for that, the student should consult the books about linear algebra (e.g., the book of Gelfand [G.2]), however, a summary of those parts of linear algebra that are most relevant to differential geometry is included as Appendix A of this book. In any case, only vector algebra and the elements of vector calculus are needed in the first eight chapters. Later on in Chapter IX the tensor calculus, and in Chapter X the notation based on invariant forms and their exterior derivatives, will be introduced and applied.

Vectors are denoted by Latin letters in bold-faced type, usually as capital letters, except for the case of unit coordinate vectors, which will be denoted by small letters. The rectangular components of a given vector, which are, of course, scalars, will be represented by the corresponding small letter with a subscript:

$$(1.1) \qquad \mathbf{X} = (x_1, x_2, x_3).$$

It is often convenient to work with the representation in terms of components; in general, as (1.1) indicates, the coordinate axes will be denoted by x_1, x_2, x_3, as in Fig. 1.1, and they will be chosen so as to form a right-handed coordinate system. The components of the vector are also the coordinates of its end point, the initial point being the origin. By the length $|\mathbf{X}|$, or magnitude, of a vector we mean the length of the straight-line segment from the origin to the point with the coordinates x_1, x_2, x_3; thus we have

$$(1.2) \qquad |\mathbf{X}| = \sqrt{x_1^2 + x_2^2 + x_3^2}$$

as the definition for the magnitude of $\mathbf{X}$.

1

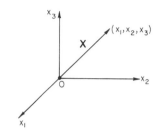

Fig. 1.1   A vector and its components.

2. Addition of Vectors

The characteristic property of vectors that distinguishes them from scalars is embodied in the law of addition, which is the familiar parallelogram law, as indicated in Fig. 1.2. We write

$$(1.3) \qquad\qquad \mathbf{Z} = \mathbf{X} + \mathbf{Y} = \mathbf{Y} + \mathbf{X}.$$

The order in which the vectors are added is immaterial. Also, the ordinary plus sign is used to denote vector addition. It should be stated explicitly that *vectors can be added in general only when they are attached to the same point.* (In the kinematics and mechanics of rigid bodies certain special types of vectors are not thus restricted, but that is a very exceptional state of affairs.)

Fig. 1.2   Addition of vectors.

In a sum of several vectors parentheses may be introduced or taken away at will:

$$(1.4) \qquad \mathbf{X} + (\mathbf{Y} + \mathbf{Z}) = (\mathbf{X} + \mathbf{Y}) + \mathbf{Z} = \mathbf{X} + \mathbf{Y} + \mathbf{Z}.$$

In terms of the representation using components, the rule (1.3) reads

$$(1.5) \quad \mathbf{Z} = (x_1 + y_1, x_2 + y_2, x_3 + y_3) = (y_1 + x_1, y_2 + x_2, y_3 + x_3).$$

3. Multiplication by Scalars

Various different sorts of products occur in vector algebra. Consider first the product $\alpha\mathbf{X}$ of a scalar and a vector; this means geometrically that the length, or magnitude, of $\mathbf{X}$ is multiplied by α, but the direction is either left unaltered (if $\alpha > 0$) or reversed (if $\alpha < 0$). If $\alpha = 0$, the result is the vector zero, which is, however, not printed in bold-faced type since no confusion will result in this exceptional case. In terms of the components of $\mathbf{X}$ the product $\alpha\mathbf{X}$ is given by

(1.6) $\alpha\mathbf{X} = (\alpha x_1, \alpha x_2, \alpha x_3)$.

In this notation the fact that $\alpha\mathbf{X}$ is opposite in direction to $\mathbf{X}$ for α negative is clear. It is also clear that a difference of two vectors is to be interpreted as the sum of $\mathbf{X}$ and of -1 times $\mathbf{Y}$, or, as it is also put as the sum of $\mathbf{X}$ and of the vector obtained by reversing the direction of $\mathbf{Y}$.

The following rules hold for the product of a scalar and a vector:

$$(\alpha + \beta)\mathbf{X} = \alpha\mathbf{X} + \beta\mathbf{X}, \qquad \alpha(\mathbf{X} + \mathbf{Y}) = \alpha\mathbf{X} + \alpha\mathbf{Y}$$

(1.7)

$$\alpha(\beta\mathbf{X}) = (\alpha\beta)\mathbf{X} = \alpha\beta\mathbf{X}$$

4. Representation of a Vector by Means of Linearly Independent Vectors

An important fact about vectors in the three-dimensional Euclidean space is that any vector $\mathbf{V}$ can be expressed in one, and only one, way as a linear combination of any three vectors $\mathbf{X}, \mathbf{Y}, \mathbf{Z}$ which do not lie in the same plane; that is, uniquely determined scalars α, β, γ exist under these circumstances such that

(1.8) $\mathbf{V} = \alpha\mathbf{X} + \beta\mathbf{Y} + \gamma\mathbf{Z}$.

Three vectors $\mathbf{X}, \mathbf{Y}, \mathbf{Z}$ that are not at all in the same plane are said to be linearly independent.

In two dimensions, that is, in the plane, any vector can be expressed as a linear combination of any two others which are not in the same straight line; again it is said that the vector is expressed as a linear combination of linearly independent vectors.

5. Scalar Product

Another kind of product, called a scalar product, involves the multiplication of two vectors, but in such a manner as to yield a scalar quantity. The notation for this product is $\mathbf{X \cdot Y}$; it is in fact sometimes called the dot product of the vectors. It is defined as follows:

(1.9) $$\mathbf{X \cdot Y} = |\mathbf{X}|\,|\mathbf{Y}|\,\cos\theta,$$

in which θ is the angle, $0 \le \theta \le \pi$, between the two vectors, as shown in Fig. 1.3. It is the product of the lengths of the two vectors and the cosine of the

Fig. 1.3   The scalar product.

angle between them. It is also the product of the length of either one of the vectors and the length of the projection of the other vector on it. The following rules for operating with this product hold:

(1.10) $$\mathbf{X \cdot Y} = \mathbf{Y \cdot X},$$
$$\mathbf{X \cdot (Y + Z)} = \mathbf{X \cdot Y} + \mathbf{X \cdot Z},$$
$$(\alpha\mathbf{X}) \cdot \mathbf{Y} = \alpha(\mathbf{X \cdot Y}) = \alpha\mathbf{X \cdot Y}.$$

The special case in which

(1.11) $$\mathbf{X \cdot Y} = 0$$

is quite important; this equation holds not only if $\mathbf{X}$ or $\mathbf{Y}$ is zero but also if neither $\mathbf{X}$ nor $\mathbf{Y}$ is the zero vector but the two are *orthogonal*. We observe also that

(1.12) $$\mathbf{X \cdot X} = |\mathbf{X}|^2.$$

The scalar product of a vector with itself thus gives the square of the magnitude of the vector. Sometimes $\mathbf{X \cdot X} = \mathbf{X}^2$ is written if there is no danger of ambiguity.

Consider an orthogonal right-handed coordinate system with vectors $\mathbf{u}_1, \mathbf{u}_2, \mathbf{u}_3$ along the coordinate axes, with $|\mathbf{u}_1| = 1$ (i.e., these are so-called *unit* vectors). Any vector can be represented in the form [cf. (1.8)]

$$\mathbf{X} = x_1\mathbf{u}_1 + x_2\mathbf{u}_2 + x_3\mathbf{u}_3.$$

In this case the scalars x_i are at once seen to be the components of $\mathbf{X}$. Take also another vector $\mathbf{Y}$ expressed in the same form:

$$\mathbf{Y} = y_1\mathbf{u}_1 + y_2\mathbf{u}_2 + y_3\mathbf{u}_3.$$

The scalar product of the two vectors can be expressed in terms of the components x_i and y_i simply by using the rules given in (1.10); the result is

$$(1.13) \qquad \mathbf{X} \cdot \mathbf{Y} = x_1 y_1 + x_2 y_2 + x_3 y_3,$$

since

$$(1.14) \qquad \mathbf{u}_i \cdot \mathbf{u}_j = \delta_{ij} = \begin{cases} 0, & i \neq j, \\ 1, & i = j, \end{cases}$$

these last being relations which hold for any system of mutually orthogonal unit vectors. The convenient and much used symbol δ_{ij}, called the Kronecker delta, is introduced in (1.14). A special case furnishes the well-known relation for the square of the magnitude of a vector:

$$(1.15) \qquad \mathbf{X} \cdot \mathbf{X} = |\mathbf{X}|^2 = x_1^2 + x_2^2 + x_3^2.$$

6. Vector Product

Another type of product involving two vectors will be much used. It is a product which yields a new *vector*, and not a scalar, in contrast with the above defined scalar product. The vector product of $\mathbf{X}$ and $\mathbf{Y}$ is a *vector* $\mathbf{Z}$ defined as follows (cf. Fig. 1.4):

$$(1.16) \qquad \mathbf{X} \times \mathbf{Y} = \mathbf{Z} = (|\mathbf{X}|\,|\mathbf{Y}|\sin\theta)\mathbf{u},$$

in which $\mathbf{u}$ is a unit vector perpendicular to both $\mathbf{X}$ and $\mathbf{Y}$ and so taken that the vectors $\mathbf{X}, \mathbf{Y}, \mathbf{u}$, in that order, form a right-handed system. It is important to observe that $\mathbf{X} \times \mathbf{Y} = -\mathbf{Y} \times \mathbf{X}$, i.e., the vector product is not commutative. Note also that the vector $\mathbf{X} \times (\mathbf{Y} \times \mathbf{Z})$ is not in general the

Fig. 1.4 The vector product.

same as the vector $(\mathbf{X} \times \mathbf{Y}) \times \mathbf{Z}$, since the first is in the plane of $\mathbf{Y}$ and $\mathbf{Z}$, the second in the plane of $\mathbf{X}$ and $\mathbf{Y}$. The following rules involving this product can be established with no great difficulty:

$$(1.17) \qquad \begin{aligned} \mathbf{X} \times (\mathbf{Y} + \mathbf{Z}) &= \mathbf{X} \times \mathbf{Y} + \mathbf{X} \times \mathbf{Z}, \\ (\alpha \mathbf{X} \times \mathbf{Y}) &= \alpha(\mathbf{X} \times \mathbf{Y}) = \alpha \mathbf{X} \times \mathbf{Y}. \end{aligned}$$

Note that $\mathbf{X} \times \mathbf{Y}$ furnishes the area, with a certain orientation, of the parallelogram determined by $\mathbf{X}$ and $\mathbf{Y}$. In speaking, this product is read "$\mathbf{X}$ cross $\mathbf{Y}$," and, indeed, it is often referred to as the cross product.

As with the scalar product, the special case

(1.18) $$\mathbf{X} \times \mathbf{Y} = 0,$$

in which the vector product vanishes, is important. It occurs, clearly, if either $\mathbf{X}$ or $\mathbf{Y}$ is the zero vector, but also if $\mathbf{X}$ and $\mathbf{Y}$ fall in the same straight line, that is, if $\mathbf{X}$ and $\mathbf{Y}$ are linearly dependent. In particular, it is always true that

(1.19) $$\mathbf{X} \times \mathbf{X} = 0,$$

a formula which comes into play rather often.

The vector product of two vectors $\mathbf{X}$ and $\mathbf{Y}$, when each is represented as a linear combination of a set $\mathbf{u}_i$ of orthogonal unit vectors forming a right-handed system, is readily calculated. The rules in (1.17) can be used to obtain this product in the form

(1.20) $$\mathbf{X} \times \mathbf{Y} = \mathbf{u}_1(x_2 y_3 - x_3 y_2) + \mathbf{u}_2(x_3 y_1 - x_1 y_3) + \mathbf{u}_3(x_1 y_2 - x_2 y_1),$$

when it is observed that $\mathbf{u}_i \times \mathbf{u}_i = 0$ and that $\mathbf{u}_i \times \mathbf{u}_j = \pm \mathbf{u}_k$, the sign depending upon whether or not j follows i in the order $1 - 2 - 3 - 1$. A useful way to remember the formula (1.20) is to put it in the form

(1.21) $$\mathbf{X} \times \mathbf{Y} = \begin{vmatrix} \mathbf{u}_1 & \mathbf{u}_2 & \mathbf{u}_3 \\ x_1 & x_2 & x_3 \\ y_1 & y_2 & y_3 \end{vmatrix},$$

which, if developed as though it were an ordinary determinant, leads to (1.20).

The vector product, unlike the scalar product, is not invariant under all orthogonal transformations of the coordinates, but rather is seen to change sign if the orientation of the coordinate axes is changed.

7. Scalar Triple Product

Finally, it is useful to introduce and discuss a special type of product involving three vectors that is defined by the formula $(\mathbf{X} \times \mathbf{Y}) \cdot \mathbf{Z}$. That is, the scalar product of $\mathbf{Z}$ is taken with the vector product of $\mathbf{X}$ and $\mathbf{Y}$; it is called the mixed product, or scalar triple product. As can be read from Fig. 1.5, it represents the volume (with a definite sign) of the parallelepiped, with the three vectors determining its edges. The sign of the product is positive if $\mathbf{X}, \mathbf{Y}, \mathbf{Z}$, in that order, form a right-handed system of vectors; otherwise the sign is negative.

The following formulas hold:

(1.22) $$(\mathbf{X} \times \mathbf{Y}) \cdot \mathbf{Z} = (\mathbf{Y} \times \mathbf{Z}) \cdot \mathbf{X} = (\mathbf{Z} \times \mathbf{X}) \cdot \mathbf{Y},$$

but

$$(\mathbf{Y} \times \mathbf{X}) \cdot \mathbf{Z} = -(\mathbf{X} \times \mathbf{Y}) \cdot \mathbf{Z}.$$

From the second expression in the first line and the fact that $\mathbf{X \cdot V} = \mathbf{V \cdot X}$ it is clear that $\mathbf{X \cdot (Y \times Z)} = \mathbf{(X \times Y) \cdot Z}$, so that dot and cross may be interchanged. In fact, there can be no ambiguity in omitting the parentheses altogether, since the vector product is defined only for two vectors. Thus $\mathbf{X \cdot Y \times Z}$ must mean $\mathbf{X \cdot (Y \times Z)}$.

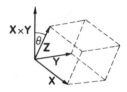

Fig. 1.5   The scalar triple product as a volume.

A useful fact can now be stated: three vectors are linearly independent (and thus span the space) if, and only if, their scalar triple product does not vanish. Or, phrased differently, a necessary and sufficient condition that three vectors should lie in a plane, and thus be linearly dependent, is that the scalar triple product of them should vanish. These and other statements about the scalar triple product can be verified by expressing the three vectors in terms of a system of orthogonal unit coordinate vectors. It is found easily that the triple product is given by the following determinant, the elements of which are the components of the vectors in a right-handed coordinate system:

$$(1.23) \qquad \mathbf{X \cdot Y \times Z} = \begin{vmatrix} x_1 & x_2 & x_3 \\ y_1 & y_2 & y_3 \\ z_1 & z_2 & z_3 \end{vmatrix}.$$

8.   Invariance Under Orthogonal Transformations

A large part of this book is concerned with the geometry of curves and surfaces which are located in the Euclidean plane or in Euclidean three-space. It is clear that a property of a curve or surface which is entitled to be called a geometrical property must be independent of the special choice of a coordinate system in the space; or, expressed in a different way, such a property should be an invariant under orthogonal linear transformations of the coordinates. The vector notation is well suited for the detection of such properties, since a vector by definition is such an invariant. The scalar product defined above is also an invariant under orthogonal transformations, as one could easily check by a calculation, but which is also obvious from its geometrical interpretation. The vector product is an invariant only under those orthogonal transformations which preserve the orientation of the axes; it changes sign if the orientation is changed. These facts are again easily verified by a calculation, and they are also obvious from the geometrical interpretation of the vector product. The scalar triple product also is an invariant only if the orientation of the coordinate axis is preserved. In general the geometrical properties of curves and surfaces will be defined in terms of vectors, together with the various products of them; thus the invariant character of these properties will be evident.

It might be added that the course pursued in this book eventually leads, in a quite natural way, through the study of the inner geometry of surfaces, to the consideration of geometrical properties that are invariant with respect to more general transformations. At that time the introduction of a more general notation than the vector notation—the tensor notation, for example —becomes a necessity.

When dealing with curves and surfaces in Euclidean space it is natural to speak of *invariance with respect to rigid* motions, and this will sometimes be done. This notion of invariance is conceptually different from that of invariance with respect to transformations of coordinates in the space. By a rigid motion is meant a change of position of an object in the space that preserves the distance between each pair of its points. However, as is well known (see, for example, Appendix A for a discussion of various matters of this kind), such a motion can be described in Euclidean geometry by a mapping of the whole space on itself that preserves distances, and this in turn is achieved by an appropriate orthogonal transformation. Thus, in the end, the two conceptually different notions of invariance both refer to invariance with respect to orthogonal transformations: in the one case with respect to a linear transformation of the whole space into itself, in the other to a transformation of the coordinate system of the space regarded as fixed.

9. Vector Calculus

The vectors dealt with in differential geometry will depend in general upon one or more real scalar parameters. This means simply that the components x_i of the vectors are functions of the parameters. For example, the end point of the vector

$$\mathbf{X}(t) = (x_1(t), x_2(t), x_3(t))$$

will in general fill out a segment of a curve in three-dimensional space when the parameter t varies, as indicated in Fig. 1.6; evidently this is nothing but a short-hand notation which gives the equations of the curve segment in the parametric form $x_i = x_i(t)$, $i = 1, 2, 3$. The vector function $\mathbf{X}(t)$ is said to be *continuous* in $\alpha \le t \le \beta$ if the functions $x_i(t)$ are defined and continuous over the interval. The vector $\mathbf{X}(t)$ is said to be *differentiable* if that is true of the coordinates $x_i(t)$, and the derivative of it is defined by the expression

$$(1.24) \qquad \frac{d\mathbf{X}(t)}{dt} = \mathbf{X}'(t) = \left(\frac{dx_1}{dt}, \frac{dx_2}{dt}, \frac{dx_3}{dt}\right),$$

or, in terms of a $\mathbf{u}_i$-system of orthogonal unit coordinate vectors, by

$$(1.25) \qquad \mathbf{X}'(t) = x_1'\mathbf{u}_1 + x_2'\mathbf{u}_2 + x_3'\mathbf{u}_3.$$

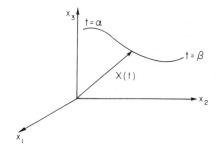

Fig. 1.6   Space curve given by a vector $\mathbf{X}(t)$.

Later on it will become clear why this definition for the derivative of a vector with respect to a scalar parameter is reasonable and appropriate. Here it perhaps suffices to notice that the vector $\mathbf{X}'(t)$ is in the direction which it is customary to define as the direction of the tangent to the curve represented by $\mathbf{X}(t)$.

It is easy to verify that the following rules for differentiation of the various products hold:

$$\frac{d}{dt}\left(\alpha(t)\,\mathbf{X}(t)\right) = \alpha'\mathbf{X} + \alpha\mathbf{X}',$$

$$\frac{d}{dt}\left(\mathbf{X}\cdot\mathbf{Y}\right) = \mathbf{X}'\cdot\mathbf{Y} + \mathbf{X}\cdot\mathbf{Y}',$$

$$(1.26) \qquad \frac{d}{dt}\left(\mathbf{X} \times \mathbf{Y}\right) = \mathbf{X}' \times \mathbf{Y} + \mathbf{X} \times \mathbf{Y}',$$

$$\frac{d}{dt}\left(\mathbf{X}\cdot\mathbf{Y}\right) \times \mathbf{Z} = \mathbf{X}'\cdot\mathbf{Y} \times \mathbf{Z} + \mathbf{X}\cdot\mathbf{Y}' \times \mathbf{Z} + \mathbf{X}\cdot\mathbf{Y} \times \mathbf{Z}'.$$

One caution should be given: the order of the factors must be strictly observed whenever the vector product is involved.

Integration of a vector with respect to a parameter over the range $\alpha \leq t \leq \beta$ is defined, as might be expected, as follows:

$$(1.27) \qquad \int_\alpha^\beta \mathbf{X}(t) \, dt = \mathbf{u}_1 \int_\alpha^\beta x_1(t) \, dt + \mathbf{u}_2 \int_\alpha^\beta x_2(t) \, dt + \mathbf{u}_3 \int_\alpha^\beta x_3(t) \, dt.$$

Also, if the upper limit is variable so that the process of integration yields a vector $\mathbf{Y}(t)$ given by

$$(1.28) \qquad\qquad \mathbf{Y}(t) = \int_\alpha^t \mathbf{X}(\tau) \, d\tau,$$

it follows immediately from (1.27) that

$$(1.29) \qquad\qquad \frac{d\mathbf{Y}(t)}{dt} = \mathbf{Y}'(t) = \mathbf{X}(t).$$

In other words the analog of the so-called fundamental theorem of the calculus holds for vectors once the above definitions are given for the derivative and the definite integral.

It is frequently useful to employ an analog of the mean value theorem for differentiable scalar functions to a vector function $\mathbf{X}(t)$. Consider, for example, the difference $\mathbf{X}(t_1) - \mathbf{X}(t_0)$, which is given, from (1.5), in terms of the components of $\mathbf{X}(t)$ by

$$\mathbf{X}(t_1) - \mathbf{X}(t_0) = (x_1(t_1) - x_1(t_0), x_2(t_1) - x_2(t_0), x_3(t_1) - x_3(t_0))$$
$$= (x_1'(\xi_1), x_2'(\xi_2), x_3'(\xi_3)) \cdot (t_1 - t_0).$$

The second line is a consequence of the mean value theorem which says that $x_i(t_1) - x_i(t_0) = x_i'(\xi_i)(t_1 - t_0)$ for some value ξ_i between t_0 and t_1. It is often convenient to write this expression in the form

$$(1.30) \qquad\qquad \mathbf{X}(t_1) - \mathbf{X}(t_0) = \overset{*}{\mathbf{X}}{}' \cdot (t_1 - t_0),$$

with $\overset{*}{\mathbf{X}}{}'$ a vector having the property

$$(1.31) \qquad\qquad \lim_{t_1 \to t_0} \overset{*}{\mathbf{X}}{}' = \mathbf{X}'(t_0),$$

which holds since the three quantities ξ_i all lie between t_0 and t_1 and the derivatives $x_i'(t)$ are assumed to be continuous. Thus while there is no mean value theorem for vector functions in the same sense as there is for scalar functions of a real variable, the application of that theorem to the components separately leads formally to the relations (1.30) and (1.31) and they can be used in analysis, as will be seen, in much the same fashion as the corresponding formulas are used for scalar functions. The earmark of this procedure in what follows in this book is the star over a derivative of a vector function.

PROBLEMS

1. In the equations $\mathbf{X}\cdot\mathbf{Y} = \alpha$, $\mathbf{X} \times \mathbf{Y} = \mathbf{Z}$ is $\mathbf{Y}$ uniquely determined if $\mathbf{X}$ and α or $\mathbf{X}$ and $\mathbf{Z}$ are given? (Impossibility of defining an inverse of these multiplications.)

2. Give a proof of the rule for differentiating $\mathbf{X}(t)\cdot\mathbf{Y}(t)$.

3. It is given that $\mathbf{X}(t)$ is differentiable and that $|\mathbf{X}(t)| = 1$. Show that $\mathbf{X}(t)$ is orthogonal to $\mathbf{X}'(t)$.

4. Prove that $\mathbf{X}(t) = \mathbf{A} \cos t + \mathbf{B} \sin t$ represents an ellipse. ($\mathbf{A}$ and $\mathbf{B}$ are linearly independent constant vectors.)

5. If $\mathbf{X} = \alpha_1\mathbf{u}_1 + \alpha_2\mathbf{u}_2 + \alpha_3\mathbf{u}_3$ is a unit vector, show that the constants α_i are the direction cosines of the line containing $\mathbf{X}$ which is directed in the same sense as $\mathbf{X}$.

6. Verify the identity of Lagrange:

$$(\mathbf{X} \times \mathbf{Y})\cdot(\mathbf{U} \times \mathbf{V}) = \begin{vmatrix} \mathbf{X}\cdot\mathbf{U} & \mathbf{Y}\cdot\mathbf{U} \\ \mathbf{X}\cdot\mathbf{V} & \mathbf{Y}\cdot\mathbf{V} \end{vmatrix}.$$

7. It is given that $\mathbf{X}(t)$ is differentiable. Show that

$$\mathbf{X}'(t_0) = \lim_{t \to t_0} \frac{\mathbf{X}(t) - \mathbf{X}(t_0)}{t - t_0}$$

by using (1.24).

CHAPTER II

Plane Curves

1. Introduction

From one point of view this chapter could be regarded as unnecessary, since the theory of twisted curves in three-dimensional space is treated in the next chapter, and that theory could be specialized for the case of plane curves. However, there are a number of good reasons for dealing with plane curves separately, quite aside from their specific interest for their own sake. To begin with, plane curves are the simplest objects dealt with in differential geometry, but for all that their study reveals something of the general attitudes and points of view that prevail in differential geometry, even in surface theory. Thus some quite simple developments regarding plane curves foreshadow a good many things to be taken up later in which the circumstances are more complicated. In addition, there are some specific differences in principle worth pointing out with respect to the possible methods of treating plane curves as contrasted with those for space curves.

2. Regular Curves

Experience has shown that it is useful and reasonable to deal in differential geometry (and in other disciplines as well, such as the theory of analytic functions of a complex variable) with a *class* of plane curves called *regular curves*. A regular curve is defined as the locus of points (cf. Fig. 2.1) traced out by the end point of a vector $\mathbf{X}(t)$ in an x_1,x_2-plane:

$$(2.1) \qquad \mathbf{X}(t) = (x_1(t), x_2(t)), \qquad \alpha \leq t \leq \beta,$$

and such that $\mathbf{X}(t)$ satisfies the following conditions:

(a) $\mathbf{X}(t)$ has continuous second derivatives in the interval $\alpha \leq t \leq \beta$, and

(b) $\mathbf{X}'$, the derivative of $\mathbf{X}(t)$, is nowhere zero.

These conditions merit some discussion.   First of all, it might be noted
in connection with the condition (a) that for a good deal of the discussion to
follow it would be sufficient to require the existence of a continuous first
derivative.   In general in this book the existence of a certain finite number
of derivatives of the functions employed will be assumed, but the minimum
number of derivatives needed from case to case will not always be stated.
On the other hand, it is *not* desirable to require the functions to be analytic,
as is commonly done in the older literature.   It is necessary to operate carefully with the tools of analysis, but it is nevertheless geometry rather than
analysis that is the subject of this book.

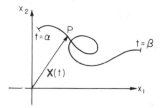

Fig. 2.1   Plane curve defined by a vector.

The condition (b) is more trenchant than condition (a), and it is important to understand why such a condition (which at first sight might seem unnecessarily restrictive) should be imposed.   The purpose of the condition—
as is also the purpose of analogous conditions for curves and surfaces in
three-dimensional space—is *to ensure that the mapping of the t-interval into the*
x_1,x_2-*plane is topological in the small.*   By a topological mapping of one object on another is meant here, as in general in mathematics, that the mapping
sets up a one-to-one point correspondence that is continuous in both directions.   In Fig. 2.1 the curve shown is not the topological image of a *t*-inter-
val, since it has a double point, and hence it is not in one-to-one correspondence with that interval.   However, in the small (that is, in a sufficiently
small neighborhood of any point) the mapping is one-to-one if the curve is
regular, as is now to be shown.   Consider two points $\mathbf{X}(t_0)$, $\mathbf{X}(t_1)$ with
$\alpha \le t_0, t_1 \le \beta$ and write

$$\mathbf{X}(t_1) - \mathbf{X}(t_0) = (x_1(t_1) - x_1(t_0), x_2(t_1) - x_2(t_0))$$

$$= (x_1'(\xi_1), x_2'(\xi_2)) \cdot (t_1 - t_0),$$

the second line being a result of the mean value theorem; i.e.,

$$x_i(t_0) - x_i(t_1) = x_i'(\xi_i)(t_1 - t_0), \qquad t_0 < \xi_i < t_1.$$

In accordance with some remarks made at the end of the previous chapter, the above equation is put in the form

$$\mathbf{X}(t_1) - \mathbf{X}(t_0) = (t_1 - t_0)\overset{*}{\mathbf{X}}',$$

with $\lim_{t_1 \to t_0} \overset{*}{\mathbf{X}}' = \mathbf{X}'(t_0)$. Since $\mathbf{X}'(t) \neq 0$ anywhere, and this derivative is continuous, it is clear that $\mathbf{X}(t_1) \neq \mathbf{X}(t_0)$ for t_1 near to t_0 but not equal to it, since $\overset{*}{\mathbf{X}}'$ is not the zero vector. In other words, any pair t_0, t_1 of distinct points of the t-interval with $|t_1 - t_0|$ sufficiently small corresponds always to a pair of distinct points in the x_1,x_2-plane. The mapping is therefore locally one-to-one. As a rule in this book, as was stated in the introduction, interest is focused, at least initially, on properties of curves and surfaces in the small, and consequently a double point such as P in Fig. 2.1 is regarded as two distinct points, each of which lies on a different curve segment.

3. Change of Parameters

It is often convenient to shift from one parameter representation of a curve to another. Quite generally, many important results in differential geometry can often be made direct and easy to achieve once a special parametric representation has been tactfully chosen.

In the present case, if $\mathbf{X}(t)$, $\alpha \leq t \leq \beta$, represents a regular curve, the introduction of a new parameter τ, $\gamma \leq \tau \leq \delta$, is brought about by mapping the t-interval in a one-to-one way on a τ-interval by a suitable function $t = \psi(\tau)$. The locus in the plane is of course assumed not to be changed; i.e., $\mathbf{X}(t)$ and $\mathbf{X}(\psi(\tau)) \equiv \mathbf{X}(\tau)$ are assumed to yield the same point for the corresponding values of t and τ. A suitable function $t = \psi(\tau)$ is obtained if it is assumed that

$$(2.2) \qquad\qquad \psi'(\tau) \neq 0,$$

and in this case the curve will be a regular curve with τ as parameter if $\psi(\tau)$ has a continuous second derivative. This follows because the condition (2.2) means that t is a monotonic function of τ; hence the t-interval and the τ-interval are in one-to-one correspondence. Thus the inverse function $\tau = \phi(t)$ exists and $\phi(t)$ would, like $\psi(\tau)$, have a continuous second derivative; hence $\mathbf{X}(\psi(\tau))$ has a continuous second derivative and is such that

$$(2.3) \qquad\qquad \frac{d\mathbf{X}}{d\tau} = \frac{d\mathbf{X}}{d\psi}\frac{d\psi}{d\tau}$$

is not the zero vector since $d\mathbf{X}/d\psi \equiv d\mathbf{X}/dt \neq 0$ holds because the curve was assumed to be regular with t as parameter. (It is convenient on occasion to use the term *regular parameter* in such cases.) Thus the conditions required of $\mathbf{X}(\tau)$ in order that it should represent a regular curve are satisfied.

In particular, it is always possible to introduce as a local parameter one or the other of the coordinates x_1, x_2, for the following reasons. Since $\mathbf{X}'(t) \neq 0$ holds, it follows that $x_1'(t_0)$ and $x_2'(t_0)$ cannot both be zero. Suppose that $x_1'(t_0) \neq 0$. In that case $t = t(x_1)$ is defined as a function of x_1 in a certain neighborhood of the value $x_1(t_0)$ of x_1, hence that

$$\mathbf{X}(x_1) = (x_1,\, x_2(x_1))$$

is a valid representation of the curve with x_1 as parameter. The curve can then be represented in the more usual way, which dispenses with vectors and defines the curve points by giving one coordinate as a function of the other:

$$(2.4) \qquad\qquad x_2 = x_2(x_1).$$

This also provides, incidentally, another way of seeing that the condition $\mathbf{X}'(t) \neq 0$ ensures that the mapping into the x_1,x_2-plane is one-to-one in the small: the desired mapping is achieved by a one-to-one orthogonal projection on the x_1-axis.

It is worth pointing out that while the condition (b) is an appropriate sufficient condition for regularity of a curve, it may be violated in spite of the fact that the curve itself has no points that could reasonably be called singular from the geometric point of view. For example, the straight line defined by

$$\mathbf{X}(t) = (t,\, t),$$

is a regular curve in the sense of the above definition for $-\infty < t < \infty$. The vector

$$\mathbf{X}(t) = (t^3,\, t^3)$$

clearly yields the same *locus* of points for $-\infty < t < \infty$, but the condition $\mathbf{X}'(t) \neq 0$ is violated for $t = 0$. Evidently, this comes about merely because of an inappropriate choice of the parameter representation. On the other hand, the curve given by

$$\mathbf{X}(t) = (t^2,\, t^3)$$

has what should properly be called a singularity for $t = 0$ (for reasons that will be pointed out later) in the form of a cusp (cf. Fig. 2.2), which is not due simply to a bad choice of parameter representation. This is a type of ques-

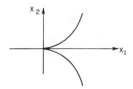

Fig. 2.2   The curve $\mathbf{X} = (t^2, t^3)$.

tion which is considered only in passing in this book. It might be added that such questions, including the classification of types of singularities, can be handled in a rather simple and complete way if the curve is assumed to be analytic (see Pogorelov [P. 5]).

4. Invariance Under Changes of Parameter

It is clear that a regular curve as defined above is a geometrical object which is invariant under transformations of the coordinate axes (for reasons discussed in the preceding chapter). In addition, in differential geometry it is usually regarded as necessary that a geometrical object should be invariant under parameter transformations as well. Here, it is made a matter of definition that two curves $\mathbf{X}(t)$ and $\mathbf{X}(\tau)$ are regarded as the same if $\tau = \phi(t)$ is a transformation of the sort discussed above. However, such an attitude, which means that the plane curves are invariants in this sense under parameter transformations, would be quite inappropriate in mechanics, for example, since *motions* of a particle along a given curve in the time t are not the same if the curve points are traversed at a different rate, although the shape of the trajectory remains the same: different forces are, in fact, required if the motions are different, and hence the two "mechanical objects" consisting of a fixed trajectory and a particle traversing it in different ways are different. Thus it is interesting and important to point out from time to time whether a given entity is, or is not, invariant under parameter transformations.

5. Tangent Lines and Tangent Vectors of a Curve

A tangent line at a point P_0 of a regular curve is commonly defined as a straight line at the point that has as its direction the limiting direction of chords obtained by joining P_0 to points P of the curve near to it and then allowing P to approach P_0. Such chords are evidently line segments joining the end points of vectors $\mathbf{X}(t_0)$ and $\mathbf{X}(t)$, with t_0 and t parameter values corresponding to P_0 and P. The difference quotient $[\mathbf{X}(t) - \mathbf{X}(t_0)]/(t - t_0)$, for $t \neq t_0$, is evidently a vector in a line through the origin that is parallel to the chord joining P_0 and P (see Fig. 2.3). It is assumed that $\mathbf{X}(t)$ is a regular curve; consequently the derivative $\mathbf{X}'(t)$ exists and is given by $\lim_{t \to t_0} [\mathbf{X}(t) - \mathbf{X}(t_0)]/(t - t_0)$. It follows that the vector $\mathbf{X}'(t)$ lies in a line parallel to the limiting direction of the chords under consideration. *The tangent line at point P_0 is now defined as the straight line through P_0 parallel to the direction fixed by the derivative $\mathbf{X}'(t_0)$ of $\mathbf{X}(t)$.* Since $\mathbf{X}'(t) \neq 0$ holds for regular curves,

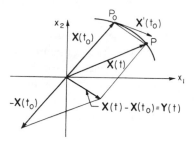

Fig. 2.3   Tangent vector of a curve.

this definition yields a *uniquely determined* line: in fact, one of the main reasons for imposing the condition $\mathbf{X}'(t) \neq 0$ was to achieve that.

The tangent line is a regular curve given by the vector equation

$$(2.5) \qquad \mathbf{T}(r) = \mathbf{X}(t_0) + r\mathbf{X}'(t_0), \qquad -\infty < r < \infty,$$

in which (cf. Fig. 2.4) the point of tangency is the point fixed by $t = t_0$, and r is the parameter on the tangent line. [Problem 1 at the end of the chapter requires a proof that $\mathbf{T}(r)$ represents a regular curve.]

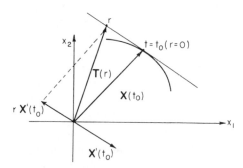

Fig. 2.4   Tangent line of a curve.

In the differential geometry of curves it is advantageous to define tangent *vectors* localized at the points of the curves. The derivative $\mathbf{X}'(t)$ could be, and by some writers is, defined as a tangent vector, but this definition would have the disadvantage of placing all of the tangent vectors at the origin rather than at the appropriate points on the curve. A reasonable way out is to use the possibility afforded by Euclidean geometry (but which is not available in other geometries to be studied later in this book) of moving vectors parallel to themselves. Thus the tangent vector at a point $\mathbf{X}(t_0)$ of the curve is *defined* as *the vector obtained by translating* $\mathbf{X}'(t_0)$ *parallel to itself to this point*. It is then still denoted by $\mathbf{X}'(t_0)$. It would also be possible to proceed in another way to achieve the same end result by first translating the origin of the coordinate system in the plane to a point on the curve and then taking the derivative $\mathbf{X}'(t)$. It is clear, and in any case easy to show, that both procedures would lead to the same tangent vector.[1]

It is to be noted that the tangent vector $\mathbf{X}'(t)$ is invariant under coordinate transformations but not under parameter transformations, since $d\mathbf{X}/d\tau = (d\mathbf{X}/dt)(dt/d\tau)$. However, the notion of a *tangent line* is seen to be independent of the choice of a parameter representation of $\mathbf{X}(t)$.

6. Orientation of a Curve

The choice of a particular parameter representation fixes a direction of travel along the curve, which in turn has a relation to the tangent vector defined above. To study this matter, the vector $\mathbf{Y}(t)$ is defined as follows:

$$(2.6) \qquad \mathbf{Y}(t) = \mathbf{X}(t) - \mathbf{X}(t_0) = (t - t_0)\overset{*}{\mathbf{X}}{}',$$

with $\overset{*}{\mathbf{X}}{}'$ a vector that tends to the derivative $\mathbf{X}'(t_0)$ as $t \to t_0$ (again a derivative with a * is used). The vector $\mathbf{Y}$ is parallel to the secant $\overline{P_0 P}$, as shown in Fig. 2.3. Hence, if $|t - t_0|$ is small enough, $\mathbf{Y}(t)$ has nearly the direction of the tangent vector $\mathbf{X}'(t_0)$ if $(t - t_0)$ is positive and nearly the opposite direction if. $(t - t_0)$ is negative. In other words, $\mathbf{X}'(t_0)$ *points into that half-plane bounded by a line normal to* $\mathbf{X}'(t_0)$ *in which the curve points lie for* $t > t_0$, *i.e., for increasing values of* t. It is thus reasonable to say that the direction of the tangent vector fixes the direction of travel along the curve when the parameter values increase. If the parameter transformation from t into $-t$ is made, it is clear that the direction of $\mathbf{X}'(t)$ is reversed; in fact, that is the case for any transformation $\tau = \tau(t)$ if τ' is negative, as can be seen from (2.3). The orientation of a curve is thus not in general an invariant property with respect to all parameter transformations, but only for those that satisfy the inequality $dt/d\tau > 0$.

One of the reasons for calling the point $t = 0$ of the curve $\mathbf{X}(t) = (t^2, t^3)$, $-\infty < t < \infty$ (see Fig. 2.2) a singular point can now be given. Since t changes sign on passing through $t = 0$, an abrupt reversal in the direction of travel takes place, although t changes in a monotonic way. A still better reason, perhaps, for calling the point a singularity is that the limiting direction of chords could be any direction if the chords were to be drawn through

---

[1] Some stress is put on this matter because it is often a sore point in mechanics, and much confusion can arise there because of a failure to recognize that it is only very exceptionally that vectors can be moved along their lines, and still less parallel to themselves, while continuing to represent a given physical entity correctly.

points chosen appropriately on each side of $t = 0$, followed by a passage to the limit.

7.　Length of a Curve

The length L_α^β of a curve is defined by the definite integral

$$(2.7) \qquad L_\alpha^\beta = \int_\alpha^\beta \sqrt{\mathbf{X}'(t)\cdot\mathbf{X}'(t)}\, dt = \int_\alpha^\beta \sqrt{x_1'^2 + x_2'^2}\, dt.$$

That the integral exists is clear, since $\mathbf{X}'(t)$ is a continuous function, so that $|\mathbf{X}'(t)|$ has this property also.　The geometric motivation for this definition is that L_α^β is the limit of the lengths of a sequence of polygons inscribed in the arc when the length of the longest side in each member of the sequence tends to zero.　The proof of this fact can be found in books on calculus, and will not be given here.

It is of interest to consider the length $s(t)$ of the curve from a fixed point to a variable point:

$$(2.8) \qquad s(t) = L_\alpha^t = \int_\alpha^t \sqrt{\mathbf{X}'(\tau)\cdot\mathbf{X}'(\tau)}\, d\tau.$$

From this the formula

$$(2.9) \qquad \frac{ds}{dt} = \sqrt{\mathbf{X}'^2(t)}$$

results.　The arc length s of a *regular* curve can therefore always be introduced as a regular parameter, since $\mathbf{X}'(t) \neq 0$ holds on such a curve, and $s(t)$ has the same number of continuous derivatives as the curve.　From (2.7) it is clear that the length of an arc of a curve is always positive.　However, $s(t)$ is evidently a positive quantity only if $t > \alpha$ holds in (2.8), i.e., when the length of an arc is measured in the direction of its orientation.　Furthermore, if s is introduced as parameter the following relation holds:

$$s = \int_\alpha^s \sqrt{\mathbf{X}'(\tau)^2}\, d\tau,$$

from which it follows that

$$(2.10) \qquad |\dot{\mathbf{X}}(s)| = \frac{ds}{ds} = 1.$$

In other words, *the tangent vector is a unit vector when the arc length is chosen as parameter.*　Note also that a dot was used to denote differentiation with respect to this special parameter—a notation which will be used frequently in this book from now on in dealing with curves for which the arc length is the parameter. It is worth remarking also that if $|\mathbf{X}'(t)| = 1$, then t measures the arc length within an additive constant; this follows at once from (2.8).

8. Arc Length as an Invariant

Since the length of a given arc of a curve is defined with the aid of vectors and an invariant operation on them, it is clear that it is a number that has a value independent of the choice of a coordinate system. This quantity is, as was pointed out in the previous chapter, therefore also invariant under rigid motions in the plane. It is perhaps reasonable, however, to verify this fact exceptionally by a direct calculation.

Thus the points $(x_1(t), x_2(t))$ of the curve are assumed to go into the points $(y_1(t), y_2(t))$ by a transformation that preserves the distances between all pairs of points. This in turn is achieved, as is well known, by a linear transformation of the form

$$y_1 = \alpha_{11}x_1 + \alpha_{12}x_2 + \beta_1,$$

$$y_2 = \alpha_{21}x_1 + \alpha_{22}x_2 + \beta_2.$$

The matrix (α_{ij}) is an orthogonal matrix defining a rotation, and the constants β_1, β_2 evidently yield a translation. The following relations hold for the coefficients α_{ij} of the orthogonal matrix:

$$\alpha_{11}^2 + \alpha_{21}^2 = \alpha_{12}^2 + \alpha_{22}^2 = 1, \qquad \alpha_{11}\alpha_{12} + \alpha_{21}\alpha_{22} = 0.$$

The following equations for y_1' and y_2' are obtained at once:

$$y_1' = \alpha_{11}x_1' + \alpha_{12}x_2',$$

$$y_2' = \alpha_{21}x_1' + \alpha_{22}x_2',$$

and an easy calculation shows that the following identity holds:

$$y_1'^2 + y_2'^2 = x_1'^2 + x_2'^2.$$

From (2.7) the invariance of the length L_α^β under rigid motions is thus seen to follow.

It is interesting to observe in addition that the arc length of a regular curve is an invariant with respect to *transformations of the parameter* used to define the curve. This is an immediate consequence of the fact that the arc length is defined by the definite integral (2.7), since definite integrals are invariant with respect to changes of the variable of integration of the sort permitted for regular curves.

9. Curvature of Plane Curves

In differential geometry the notion of curvature of curves and surfaces dominates large parts of the theory, and it is indeed a notion of great importance. In the case of plane curves only one concept of curvature comes into play, but for space curves and surfaces it is necessary to introduce various different notions of curvature. The concept of curvature for plane curves will be introduced here not in the simplest and briefest way, but in such a way as to foreshadow some later developments.

It is useful to begin by introducing the concept of the *total curvature* κ_T of a plane curve, with reference to Fig. 2.5. The angle $\theta(s)$, the inclination

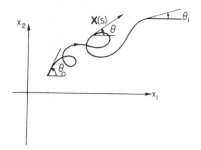

Fig. 2.5   Total curvature of a plane curve.

of the tangent to the curve with the x_1-axis, is assumed to be introduced in such a way that $\theta(s)$ is a continuous function of the arc length s. Once this is done, the total curvature is defined rather naturally as the difference in the values of θ at the end points of the curve:

$$(2.11) \qquad \kappa_T = \theta_1 - \theta_0.$$

This quantity, in effect, measures in a reasonable way the total angle through which the tangent has turned. However, the phrase "introduced in such a way that $\theta(s)$ is a continuous function" requires a little discussion. The tangent *direction* is fixed by the vector $\dot{\mathbf{X}}(s)$ at every point, but the *angle* θ at any point is determined only within additive integral multiples of 2π, so that to define $\theta(s)$ as a continuous function presents a slight problem. However, the matter is readily disposed of by employing an argument from analysis, which often comes into play in situations of this sort that involve considerations in the large (and this is such a case)—in fact, the discussion of curvature was started here in this fashion partly in order to provide an opportunity to introduce these ideas. First of all $\theta(s)$ can be uniquely defined without difficulty as a continuous function in an open *neighborhood* of any point of the interval $s_0 \leq s \leq s_1$ since $\dot{\mathbf{X}}(s)$ is a continuous unit vector and $\theta(s)$ can be defined with the aid either of $\cos \theta(s) = \dot{x}_1(s)$ or of $\sin \theta(s) = \dot{x}_2(s)$ by going over to the inverse sine or cosine function.   But by the Heine-Borel covering theorem the bounded closed interval $s_0 \leq s \leq s_1$ can be covered by choosing a finite number of the open neighborhoods within each of which $\theta(s)$ is continuous.   It is then clear that the definition of θ in each of the neighborhoods can be chosen so that θ is continuous throughout the interval.

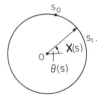

Fig. 2.6   The circular image.

There is a useful way of interpreting (2.11) which will permit a generalization to space curves (and even to surfaces) whereas (2.11) itself cannot be so generalized.   The unit tangent vectors $\dot{\mathbf{X}}(s)$ are carried to the fixed point 0, so that their end points trace out an arc of the unit circle centered at 0 (cf. Fig. 2.6) when s traverses the interval $s_0 \leq s \leq s_1$.   The result of this mapping of the curve into the circle is what is called the *circular image* of the curve. The total curvature κ_T is defined as the length of the arc covered on the circle, but with the understanding that the arc length on the circular image must be taken with an appropriate sign [fixed by the sign of $\dot{\theta}(s)$], and that portions of the circle may be covered more than once by the circular image.

It is quite natural to define the average curvature over the arc by $\kappa_T/(s_1 - s_0)$ and from that to arrive at the following *definition* for the curvature at a point:

$$(2.12) \qquad\qquad \kappa(s) = \lim_{\Delta s \to 0} \frac{\theta(s + \Delta s) - \theta(s)}{\Delta s} = \dot{\theta}(s).$$

It is readily seen that $\dot{\theta}(s)$ is continuous if $\ddot{\mathbf{X}}(s)$ is continuous, since from $\cos \theta(s) = \dot{x}_1(s)$ it follows that $\dot{\theta} \sin \theta(s) = -\ddot{x}_1(s)$, or, in case $\theta = 0$, from $\dot{\theta} \cos \theta(s) = \ddot{x}_1(s)$.

It is clear from the definition that κ is invariant under transformations of coordinates which preserve the orientation of the axes, but that it changes its sign if the orientation of the axes is reversed.   It is also clear that κ changes its sign, but retains the same numerical value if the orientation along the curve is changed.

The curvature κ of a curve given by a vector $\mathbf{X}(s)$ is now to be calculated.

Since $\dot{\mathbf{X}}(s)$ is a unit vector, it may, upon using a $\mathbf{u}_1,\mathbf{u}_2$-system of orthogonal unit coordinate vectors, be written in the form

$$\dot{\mathbf{X}}(s) = \cos\theta(s)\mathbf{u}_1 + \sin\theta(s)\mathbf{u}_2.$$

A differentiation of this formula yields

$$\ddot{\mathbf{X}}(s) = (-\sin\theta(s)\mathbf{u}_1 + \cos\theta(s)\mathbf{u}_2)\frac{d\theta}{ds},$$

or

$$\ddot{\mathbf{X}}(s) = (-\sin\theta(s)\mathbf{u}_1 + \cos\theta(s)\mathbf{u}_2)\kappa,$$

in view of (2.12). The vector in parentheses on the right is at once seen to be a unit vector; it follows that

$$(2.13) \qquad\qquad |\kappa| = \sqrt{\ddot{\mathbf{X}}\cdot\ddot{\mathbf{X}}} = |\ddot{\mathbf{X}}|,$$

or, in other words, the numerical value $|\kappa|$ of the curvature κ is the length of the vector $\ddot{\mathbf{X}}(s)$. The following relations are easily verified:

$$(2.14) \qquad \dot{\mathbf{X}} \times \ddot{\mathbf{X}} = \kappa(\cos^2\theta + \sin^2\theta)(\mathbf{u}_1 \times \mathbf{u}_2) = \kappa\mathbf{u}_1 \times \mathbf{u}_2$$

and

$$(2.15) \qquad\qquad\qquad \dot{\mathbf{X}}\cdot\ddot{\mathbf{X}} = 0.$$

The last equation states that $\dot{\mathbf{X}}$ and $\ddot{\mathbf{X}}$ are orthogonal; this conclusion might have been drawn at once from $\dot{\mathbf{X}}\cdot\dot{\mathbf{X}} = 1$ by differentiation. The equation (2.14) implies that the plane curve is thought of as placed in a three-dimensional space, since $\dot{\mathbf{X}} \times \ddot{\mathbf{X}}$ is a vector orthogonal to the x_1,x_2-plane, and that would seem unnecessary and rather unaesthetic; this procedure could have been avoided but it seems hardly worthwhile to do so here.

10. The Normal Vector and the Sign of κ

If $\kappa = 0$ holds along an entire arc, the arc is a portion of a straight line. In fact, if $\kappa \equiv 0$ holds, it follows from (2.13) that $\ddot{\mathbf{X}} = 0$, or $\dot{\mathbf{X}}(s) = \mathbf{b}$ (a constant vector). Another integration yields $\mathbf{X}(s) = \mathbf{a} + \mathbf{b}s$, which is the equation of a straight line. The quantity κ, if not equal to zero, thus has at least something to do with the way in which a curve deviates from a straight-line course.

In studying the significance of the curvature κ for $\kappa \neq 0$ it is convenient to introduce at each point of the curve a pair of orthogonal unit vectors $\mathbf{v}_1, \mathbf{v}_2$. The vector $\mathbf{v}_1$ is defined as the tangent vector of the curve, and thus is the same as $\dot{\mathbf{X}}(s)$, and therefore is the tangent vector as defined in Section 5, in case the arc length s is chosen as the curve parameter:

(2.16) $$\mathbf{v}_1(s) = \dot{\mathbf{X}}(s).$$

In Section 5 these vectors were, by convention, located on the curve. Thus the vectors $\mathbf{v}_1(s)$ define a *field* of vectors along the curve. For the first time (but it will not be the last in this book) it is desired to differentiate such a field of vectors. Since the vectors are not attached to the same point, that can be done only by introducing an appropriate, and very natural, convention, i.e., that the vectors $\mathbf{v}_1(s)$ in a neighborhood of a point P fixed by $s = s_0$ (see Fig. 2.7) are moved parallel to themselves to P and then differentiated

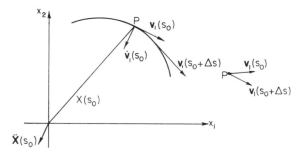

Fig. 2.7   Differentiation of the field of tangent vectors.

with respect to s to define $\dot{\mathbf{v}}_1(s)$ at $s = s_0$. This amounts to determining $\ddot{\mathbf{X}}(s)$ at $x = s_0$, from (2.16), except that the vector $\ddot{\mathbf{X}}(s_0)$ is attached to the origin, and not to the point P. However, as in Section 5, it would be possible to define $\dot{\mathbf{v}}_1(s)$ as the vector obtained by translating $\ddot{\mathbf{X}}(s)$ parallel to itself to point P. In differential geometry the first procedure is the more natural, and also it is the only procedure available in situations to be discussed in some of the final chapters of this book, when one of the principal concepts *to be introduced by definition* is a concept of local parallelism.

It is useful to introduce a field of unit vectors $\mathbf{v}_2(s)$ along the curve normal to the vectors $\mathbf{v}_1(s)$. This field is uniquely defined by requiring that the vectors $\mathbf{v}_1, \mathbf{v}_2$, in that order, have the same orientation as the coordinate axes (see Fig. 2.8). Under these circumstances the following equation will be shown to hold:

(2.17) $$\kappa\mathbf{v}_2 = \ddot{\mathbf{X}}.$$

First of all it is clear that $\ddot{\mathbf{X}}$ and $\mathbf{v}_2$ are linearly dependent, since $\mathbf{v}_1 = \dot{\mathbf{X}}$. From (2.13) it is known that the length of the vector $\ddot{\mathbf{X}}$ is $|\kappa|$, and therefore only the choice of the plus sign on κ in equation (2.17) requires justification

in order to establish the correctness of (2.17).   From (2.14) it is known that
$\dot{\mathbf{X}} \times \ddot{\mathbf{X}} = \kappa(\mathbf{u}_1 \times \mathbf{u}_2)$, with $\mathbf{u}_1$ and $\mathbf{u}_2$ orthogonal unit coordinate vectors; but
$\dot{\mathbf{X}} \times \ddot{\mathbf{X}} = \mathbf{v}_1 \times (\kappa\mathbf{v}_2) = \kappa(\mathbf{v}_1 \times \mathbf{v}_2)$ from (2.17), and hence the plus sign must
be chosen in (2.17), since $\mathbf{u}_1$, $\mathbf{u}_2$ and $\mathbf{v}_1$, $\mathbf{v}_2$ are two pairs of vectors with the

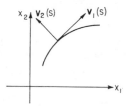

Fig. 2.8   A moving system of coordinate vectors.

same orientation.   In effect, (2.17) means that κ will be positive or negative
depending on whether $\mathbf{v}_2$ and $\ddot{\mathbf{X}}$ are directed in the same or the opposite
sense.

The sign of κ has geometric significance.   Assuming $\mathbf{X}(s)$ to possess a
continuous second derivative, the following relation holds:

$$\mathbf{X}(s) = \mathbf{X}(s_0) + (s - s_0)\,\dot{\mathbf{X}}(s_0) + \tfrac{1}{2}(s - s_0)^2\,\ddot{\mathbf{X}}^*,$$

with $\ddot{\mathbf{X}}^* \to \ddot{\mathbf{X}}(s_0)$ as $s \to s_0$.   (For this the remarks at the end of Chapter 1
need to be extended in an obvious way with the aid of Taylor's theorem with
a remainder.)   The equation of the tangent line at s_0 is (cf. Fig. 2.9)

$$\mathbf{T}(s) = \mathbf{X}(s_0) + (s - s_0)\,\dot{\mathbf{X}}(s_0),$$

and the vector $\mathbf{Z}(s)$ defined by

$$\mathbf{Z}(s) = \mathbf{X}(s) - \mathbf{T}(s) = \tfrac{1}{2}(s - s_0)^2\,\ddot{\mathbf{X}}^*$$

is thus a vector pointing in the direction from the tangent line to the curve
points.   [In Fig. 2.9 the vector $\mathbf{Z}(s)$ is drawn from the end point of the vector

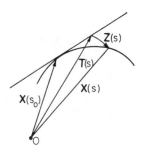

Fig. 2.9   Deviation of a curve from its tangent.

$T(s)$ instead of from the origin, where it belongs in principle; however, it is very convenient to break the rule and do such things on occasion, as was done earlier. It would be pedantic and heavy-handed to observe the rule meticulously and to make explanations over and over again because of it; no harm can result once the basic rule, and the convention of using a parallel transport on occasion, are properly understood.] Since $(s - s_0)^2$ is positive, it follows that $\ddot{X}^*$, and hence also $\ddot{X}(s_0)$ for s near enough to s_0, points to that side of the tangent line in which the curve points lie. [Note that this reasoning breaks down if $\kappa(s_0) = 0$, hence $\ddot{X}(s_0) = 0$.] The side of the tangent line to which $\ddot{X}$ points is called the positive side, and the following is then seen to hold in view of (2.17):

(a) if $\kappa > 0$, v_2 points to the $+$ side and
(b) if $\kappa < 0$, v_2 points to the $-$ side.

The curvature κ is thus negative in the case shown in Fig. 2.8.

11. Formulas for κ

It is of interest to calculate κ using the components of the vector $X(s)$ that defines the curve. The equation (2.14) can be written as follows:

$$(2.18) \qquad \dot{X} \times \ddot{X} = (\dot{x}_1 \ddot{x}_2 - \dot{x}_2 \ddot{x}_1)(u_1 \times u_2) = \kappa(u_1 \times u_2).$$

Hence κ is given by

$$(2.19) \qquad \kappa = \dot{x}_1 \ddot{x}_2 - \dot{x}_2 \ddot{x}_1 = \begin{vmatrix} \dot{x}_1 & \dot{x}_2 \\ \ddot{x}_1 & \ddot{x}_2 \end{vmatrix}.$$

Formula (2.19) brings out the fact that κ changes sign when the orientation of the coordinate axes is changed.

It is of interest to determine κ when the curve is given by a parameter t other than s. For $dt/ds \neq 0$, the following formulas are obtained:

$$\dot{X} = X' \frac{dt}{ds},$$

$$\ddot{X} = X'' \left(\frac{dt}{ds}\right)^2 + X' \left(\frac{d^2t}{ds^2}\right).$$

Since $ds/dt = \sqrt{(X')^2}$ [cf. (2.9)], the following relation is found from (2.18):

$$\kappa(u_1 \times u_2) = \frac{(X' \times X'')}{(X' \cdot X')^{3/2}} = \frac{x_1' x_2'' - x_2' x_1''}{(x_1'^2 + x_2'^2)^{3/2}} (u_1 \times u_2),$$

so that

$$(2.20) \qquad \kappa = \frac{x_1' x_2'' - x_2' x_1''}{(x_1'^2 + x_2'^2)^{3/2}}.$$

This is the well-known formula for the curvature of a plane curve given in the parametric form $x_1 = x_1(t)$, $x_2 = x_2(t)$.

From (2.20) it is readily checked that the curve $X(t) = (t^2, t^3)$ shown in Fig. 2.2 is singular at the origin in the sense that $κ$ becomes infinite there.

12.   Existence of a Plane Curve for Given Curvature $κ$

The following important theorem holds:

If $κ(s)$, $s_0 \leq s \leq s_1$, is an arbitrary continuous function, there exists one and only one regular curve (within a rigid motion) for which $κ(s)$ is the curvature and s the arc length.

Once that is shown, it is clear that the two invariants $κ$ and s form what can reasonably be called a complete set of invariants for a plane curve, since they determine it uniquely within rigid motions. The proof of the theorem is quite straightforward. Since $κ = \dot{θ}(s)$ [from (2.12)], it is natural to set

$$(2.21) \qquad\qquad θ(s) = θ(0) + \int_0^s κ(σ)\, dσ.$$

Once $θ(s)$ has been determined in this way a unit vector $\dot{X}(s)$ can be defined as follows:

$$\dot{X}(s) = \cos θ(s)u_1 + \sin θ(s)u_2.$$

From this a vector $X(s)$ is defined by the integral

$$(2.22) \qquad X(s) = X(0) + u_1 \int_0^s \cos θ(σ)\, dσ + u_2 \int_0^s \sin θ(σ)\, dσ.$$

It is now to be shown that $X(s)$, as defined by (2.22), is a regular curve for which s is the arc length and $κ(s)$ is the curvature. To this end it is sufficient, and quite simple, to verify by a direct calculation that

(a) $\dot{X}(s) \cdot \dot{X}(s) = 1$, and
(b) $\ddot{X}(s) = (-\sin θ(s)u_1 + \cos θ(s)u_2)\dot{θ}$.

From (a) it follows that the curve is a regular curve and that the parameter s is the arc length. From (b) it follows that $\dot{θ} = κ(s)$ is the curvature of the curve, in view of the developments leading up to equation (2.17). It is also readily seen that $X(s)$ has the required differentiability properties.

Finally, it is readily seen that any two curves $X^1(s)$, $X^2(s)$ that have the same arc length s and curvature $κ(s)$ differ at most by a rigid motion, as follows. By a rigid motion the two curves can be brought together at the initial point so that $X^1(0) = X^2(0)$ and $\dot{X}^1(0) = \dot{X}^2(0)$: all that is required is an appropriate translation followed by a rotation. From $\dot{\mathbf{X}}^1(0) = \dot{\mathbf{X}}^2(0)$ it follows from (2.21) that $\theta_1(s) = \theta_2(s)$ holds for all s since $\kappa(s)$ is the same for both curves and θ_1 and θ_2 can be assumed to coincide for $s = 0$. Hence for the tangent vectors $\dot{\mathbf{X}}^1(s) \equiv \dot{\mathbf{X}}^2(s)$ holds and finally [see (2.22)]

$$\mathbf{X}^1(s) = \int_0^s \dot{\mathbf{X}}^1(\sigma)\, d\sigma + \mathbf{X}^1(0) = \mathbf{X}^2(s),$$

and this proves the theorem.

13. Frenet Equations for Plane Curves

The arc length s of a regular curve is chosen as parameter, and the fields $\mathbf{v}_1(s)$, $\mathbf{v}_2(s)$ of tangent and normal vectors are defined as in the above section 10. Consequently every vector can be expressed as a linear combination of $\mathbf{v}_1$ and $\mathbf{v}_2$. In particular, the derivatives $\dot{\mathbf{v}}_1$ and $\dot{\mathbf{v}}_2$ of the vectors $\mathbf{v}_1$ and $\mathbf{v}_2$ themselves can be so expressed. Thus the following equations must hold in terms of certain appropriate, and uniquely determined, scalar functions $\alpha_i(s)$, $\beta_i(s)$:

(2.23) $\dot{\mathbf{v}}_1 = \alpha_1 \mathbf{v} + \alpha_2 \mathbf{v}_2,$

(2.24) $\dot{\mathbf{v}}_2 = \beta_1 \mathbf{v}_1 + \beta_2 \mathbf{v}_2.$

Since the relations $\mathbf{v}_1 \cdot \mathbf{v}_1 = \mathbf{v}_2 \cdot \mathbf{v}_2 = 1$ and $\mathbf{v}_1 \cdot \mathbf{v}_2 = 0$ hold, differentiations with respect to s lead to the equations

(2.25) $\dot{\mathbf{v}}_1 \cdot \mathbf{v}_2 + \mathbf{v}_1 \cdot \dot{\mathbf{v}}_2 = 0,$

(2.26) $\mathbf{v}_1 \cdot \dot{\mathbf{v}}_1 = \mathbf{v}_2 \cdot \dot{\mathbf{v}}_2 = 0.$

Scalar multiplication of (2.23) and (2.24) by $\mathbf{v}_1$ and $\mathbf{v}_2$, respectively, and use of (2.26) and $\mathbf{v}_1 \cdot \mathbf{v}_2 = 0$, yield $\alpha_1 = \beta_2 = 0$. Scalar multiplication of (2.23) and (2.24) by $\mathbf{v}_2$ and $\mathbf{v}_1$ and use of (2.25) yield $\beta_1 + \alpha_2 = 0$. Equations (2.23) and (2.24) thus become

$$\dot{\mathbf{v}}_1 = \alpha_2 \mathbf{v}_2, \quad \text{and} \quad \dot{\mathbf{v}}_2 = -\alpha_2 \mathbf{v}_1.$$

It is known that $\dot{\mathbf{v}}_1 = \kappa \mathbf{v}_2$ [cf. (2.16) and (2.17)]. Hence the equations desired take the following form:

(2.27)
$$\dot{\mathbf{v}}_1 = * \quad \kappa \mathbf{v}_2,$$
$$\dot{\mathbf{v}}_2 = -\kappa \mathbf{v}_1 \quad *,$$

in which the *'s have been put in purely to emphasize the skew-symmetric character of the right-hand sides. These differential equations are called the equations of Frenet for the special case of plane curves.

These equations form a system of linear homogeneous differential equations for the determination of the vectors $\mathbf{v}_1(s)$, $\mathbf{v}_2(s)$. They have a uniquely determined solution for these vectors as soon as the initial values $\mathbf{v}_1(0)$, $\mathbf{v}_2(0)$ are prescribed, provided that $\kappa(s)$ is any arbitrarily given continuous function of s (see Appendix B, for example). Once $\mathbf{v}_1(s) = \dot{\mathbf{X}}(s)$ is thus determined, $\mathbf{X}(s)$ can be found by an integration. Thus the Frenet equations (2.27) could be used to establish the results of the preceding section, evidently in an unnecessarily complicated way.

The generalization of the Frenet equations to space curves will be derived along much the same lines in the next chapter. Even for surfaces there are differential equations somewhat analogous to the equations (2.27). In fact, the analogous equations for space curves and surfaces are the basis for the discussion of existence and uniqueness theorems of the kind proved in the preceding section for plane curves; however, in these more complicated cases a direct procedure using explicit integral representations is not available. Even so, the theory for plane curves just discussed points the way to be followed.

14.   Evolute and Involute of a Plane Curve

Interesting applications of the theory of plane curves are furnished by curves called *evolutes* and *involutes* of a given curve $\mathbf{X}(s)$. It is assumed that the arc length is the parameter, and that the curvature κ is not zero on the given curve. The evolute is defined as the locus of the *centers of curvature* of the curve $\mathbf{X}(s)$, the center of curvature being the point on the normal to the curve, in the direction of the normal vector $\mathbf{v}_2$, that is at the distance $\rho(s) = \kappa^{-1}(s) > 0$ from the curve; the quantity $\rho(s)$ is called the *radius* of *curvature* of the curve $\mathbf{X}(s)$. (The curvature can be assumed to be positive since a proper choice of the orientation of the curve will accomplish that.) The equation fixing the points of the evolute is therefore

$$(2.28) \qquad\qquad \mathbf{Y}(s) = \mathbf{X}(s) + \rho\mathbf{v}_2(s).$$

However, s will not in general be the arc length on $\mathbf{Y}(s)$; in fact, the derivative $\mathbf{Y}'$ is given by

$$\mathbf{Y}'(s) = \dot{\mathbf{X}}(s) + \dot{\rho}\mathbf{v}_2(s) + \rho\dot{\mathbf{v}}_2(s),$$

but since $\dot{\mathbf{v}}_2 = -\kappa\mathbf{v}_1$ [see (2.27)], hence $\rho\dot{\mathbf{v}}_2 = -\mathbf{v}_1 = -\dot{\mathbf{X}}$, it follows that

$$(2.29) \qquad\qquad \mathbf{Y}'(s) = \dot{\rho}\mathbf{v}_2(s).$$

Thus the evolute is a regular curve if $\dot{\rho} \neq 0$, which, since $\kappa = 1/\rho$ is assumed not to vanish, is equivalent to the requirement $\dot{\kappa} \neq 0$. In addition it is clear from (2.29) that if the evolute is a regular curve its tangent falls along the normal to the original curve $\mathbf{X}(s)$, and that the element of arc length $|\mathbf{Y}'(s)|\ ds$ of the evolute is $|\dot{\rho}|\ ds$, since $\mathbf{v}_2$ is a unit vector. As a consequence, the length $L_{s_0}^{s_1}$ of a regular arc of the evolute corresponding to the interval between s_0 and s_1 is given by

$$(2.30) \qquad\qquad L_{s_0}^{s_1} = |\rho(s_1) - \rho(s_0)|,$$

since $\dot{\rho}$ does not change sign on such an arc. Thus the length of an arc of the evolute $\mathbf{Y}(s)$ is equal to the difference in length of the radii of curvature of $\mathbf{X}(s)$ at the two end points of the arc.

These remarks give rise to the following kinematical interpretation of the relation between the evolute $\mathbf{Y}(s)$ and the original curve $\mathbf{X}(s)$, which is called rather naturally an *involute* of $\mathbf{Y}(s)$. In Fig. 2.10 imagine a flexible string to

Fig. 2.10   Evolute and involute

be fixed at a point $Y(s_1)$ on the evolute E, and then wrapped around it to point $Y(s_0)$ from which it is held stretched tight to the point P given by $X(s_0)$ on the involute I, with ρ_0 the distance between these two points; the straight portion of the string between the two curves falls, of course, along the tangent to the evolute. Now suppose the string to be unwrapped from the evolute while being kept tightly stretched. The above discussion indicates that the locus of point P on the string is an involute I of E since P would be expected to move orthogonally to the string (because s_0 is the instantaneous center of rotation of the straight segment of the string), and the length of string unwrapped from any arc just equals the difference in length of the straight pieces at its two ends. In fact, it is not difficult to prove (it is assigned as a problem at the end of this chapter) that the locus of any point such as P on the string is an involute of the curve E. Thus while a given curve has only one evolute, a given evolute has infinitely many involutes.

It is of interest to define the *osculating circle* or *circle of curvature* of a regular curve $\mathbf{X}(s)$ at any of its points. This curve is defined as the circle with its center at the center of curvature of $\mathbf{X}(s)$ and with a radius equal to $\rho(s)$. (One of the problems at the end of this chapter requires a proof that this circle is the one which fits the curve best in the sense that it is the only circle tangent to the curve which has also the same second derivative as the curve at the point of tangency.) The locus of the centers thus is the evolute $\mathbf{Y}(s)$ of $\mathbf{X}(s)$, and if the curve $\mathbf{Y}(s)$ is regular along a certain arc, that is if $\dot{\rho} \neq 0$ on $\mathbf{X}(s)$, the one-parameter family of circles of curvature has an interesting geometrical property, i.e., that no two of those circles have a point in common, or, in other words, the circles are all nested one within another. The proof is simple. First of all, the evolute E contains no straight-line segments; this follows from (2.29) and $\dot{\rho} \neq 0$ because of the fact that $\mathbf{v}_2$ is not a constant vector since $\kappa \neq 0$ on the involute I. Thus the centers of any two circles of curvature that are on E are at a distance apart that is smaller than the length of the arc of E between the two points.[1] But the length of this arc is the difference in length of the radii of the two circles, and hence one of them lies inside the other.

15. Envelopes of Families of Curves

The preceding paragraph was introduced in part because it offers an amusing way to begin the discussion of *envelopes of a one-parameter family of plane curves*. The family is supposed given by a vector $\mathbf{X}(t, \alpha)$ that furnishes for each value of the parameter α, $\alpha_0 \leq \alpha \leq \alpha_1$, a regular curve with t as parameter on it. A very good example of an envelope is furnished by an arc of any regular curve, on which $\dot{\kappa} \neq 0$ is assumed, together with its own family of circles of curvature as the one-parameter family of curves in question. The given curve is *a regular curve which is everywhere tangent to a member of the curve family without being itself a member of it*—and that is basically the definition of an envelope of a family of plane curves. In older books one sometimes finds (cf. Graustein [G.5], p. 64, for example) a different way of looking at the problem of defining an envelope, as follows. Consider any two curves of the family $\mathbf{X}(t, \alpha)$ corresponding to values of α which differ by a small amount $\Delta\alpha$; find their intersection, and define the envelope as the locus of all such intersections in the limit when $\Delta\alpha \to 0$. Clearly, this definition would be quite inappropriate for the case of the envelope of the osculating circles of a curve, since, as was just seen in the preceding section, none of the circles intersect.[2]

It is not difficult to find a *necessary* condition for the existence of an envelope of the curve family $\mathbf{X}(t, \alpha) = (x_1(t, \alpha), x_2(t, \alpha))$. It is simply that

---

[1] A theorem in differential geometry in the large was used here without proof, i.e., the theorem that a straight line furnishes the regular curve of smallest length between any two points of the plane. More will be said about this type of question later.

[2] This amusing observation was taken from the book of Bierberbach [B.1].

the Jacobian of the functions x_1, x_2 must vanish along a locus that is an envelope in the form of a regular curve:

$$(2.31) \qquad \frac{\partial(x_1, x_2)}{\partial(t, \alpha)} = \begin{vmatrix} \dfrac{\partial x_1}{\partial t} & \dfrac{\partial x_2}{\partial t} \\[2ex] \dfrac{\partial x_1}{\partial \alpha} & \dfrac{\partial x_2}{\partial \alpha} \end{vmatrix} = 0.$$

The reasons why this is a necessary condition for an envelope are as follows. If the Jacobian were different from zero at the point $t = t_0$, $\alpha = \alpha_0$, say, it would be possible, by virtue of the implicit function theorem, to determine t and α as functions of x_1 and x_2 in a neighborhood of this point; i.e., the functions $t = t(x_1, x_2)$ and $\alpha = \alpha(x_1, x_2)$ would result by inverting the equations $x_1 = x_1(t, \alpha)$ and $x_2 = x_2(t, \alpha)$ and they would have a certain number of continuous derivatives.   Consider the equations

$$(2.32) \qquad \begin{aligned} \frac{dx_1}{dt} &= f_1(t, \alpha), \\[2ex] \frac{dx_2}{dt} &= f_2(t, \alpha), \end{aligned}$$

which furnish the components of tangent vectors to the curves of the given family in a neighborhood of the values $t = t_0$, $\alpha = \alpha_0$ [in effect, they are the scalar version of the vector equation $\mathbf{X}' = \partial \mathbf{X}/\partial t = (f_1, f_2)$].   Since all curves of the family are assumed to be regular curves it follows that f_1 and f_2 do not vanish simultaneously.   Since t and α are known as functions of x_1 and x_2, the equations (2.32) can be put in the form

$$(2.33) \qquad \begin{aligned} \frac{dx_1}{dt} &= f_1(t(x_1, x_2), \alpha(x_1, x_2)) = g_1(x_1, x_2), \\[2ex] \frac{dx_2}{dt} &= f_2(t(x_1, x_2), \alpha(x_1, x_2)) = g_2(x_1, x_2), \end{aligned}$$

and g_1 and g_2 are functions that are not simultaneously zero.   The theory of ordinary differential equations[1] guarantees that the system of equations (2.33) under the conditions given here will have a regular curve as *uniquely determined* solution in a neighborhood of the point $x_1 = x_1(t_0, \alpha_0)$, $x_2 = x_2(t_0, \alpha_0)$ once initial values of x_1 and x_2 are prescribed for a given value t_0 of t. Thus no envelope could exist, since any regular curve which is everywhere tangent to the direction field defined by (2.33) is a solution curve of the

---

[1] In Appendix B of this book these theorems are discussed. For the uniqueness theorem to hold—and it is the theorem most needed here—it is sufficient that g_1 and g_2 should have continuous derivatives with respect to x_1 and x_2.

system and therefore belongs of necessity to the curve family by virtue of the uniqueness theorem.   It follows that if the condition (2.31) is not satisfied no envelope can exist.

The condition (2.31) can be expressed in a different form.   The partial derivatives of $\mathbf{X}(t, \alpha)$ are given by

$$(2.34) \qquad \mathbf{X}_t = \frac{\partial \mathbf{X}}{\partial t} = \left(\frac{\partial x_1}{\partial t}, \frac{\partial x_2}{\partial t} \right) \quad \text{and} \quad \mathbf{X}_\alpha = \frac{\partial \mathbf{X}}{\partial \alpha} = \left(\frac{\partial x_1}{\partial \alpha}, \frac{\partial x_2}{\partial \alpha} \right).$$

Clearly, the vanishing of the Jacobian given in (2.31) and the linear dependence of $\mathbf{X}_t$ and $\mathbf{X}_\alpha$ are equivalent conditions.   Consequently, the condition

$$(2.35) \qquad\qquad\qquad \mathbf{X}_t \times \mathbf{X}_\alpha = 0$$

is another way of writing a necessary condition for the existence of an envelope.

As an example, consider the curve family $\mathbf{Y}(s, \alpha)$ defined as the family of straight lines normal to a regular curve $\mathbf{X}(s)$:

$$(2.36) \qquad\qquad\qquad \mathbf{Y}(s, \alpha) = \mathbf{X}(s) + \alpha \mathbf{v}_2(s),$$

and assume also that the curvature κ of $\mathbf{X}(s)$ is not zero.   The condition (2.35), with t replaced by s, becomes in this case

$$(\dot{\mathbf{X}} + \alpha \dot{\mathbf{v}}_2) \times \mathbf{v}_2 = (1 - \alpha\kappa)(\mathbf{v}_1 \times \mathbf{v}_2) = 0,$$

in view of the second of the Frenet equations (2.27).   Since $\mathbf{v}_1 \times \mathbf{v}_2$ does not vanish, it follows that along an envelope $\alpha = 1/\kappa = \rho$, the radius of curvature of $\mathbf{X}(s)$.   Thus the locus singled out by the condition (2.35) is given by $\mathbf{Y}(s, s) \equiv \mathbf{Y}(s) = \mathbf{X}(s) + \rho\mathbf{v}_2(s)$ from (2.36).   This is the locus [cf. (2.28)] treated in the previous section; i.e., it is the curve called the evolute, and it is a regular curve provided that $\dot{\kappa} \neq 0$ holds.

A few words of caution are in order concerning the condition for an envelope.   It was carefully stated that this condition is a *necessary* condition, and, in fact, it is not by any means always a sufficient condition for an envelope.   The set of points which satisfies the condition (2.31), or its equivalent (2.35), may contain many types of loci which are not envelopes.   For example, a locus containing double points or singular points of the curve family may appear.   It would be possible to investigate such matters at this point in some detail, but in this book only envelopes are considered, and the attitude is that any loci furnished by the conditions (2.31) or (2.35) are to be examined in each specific case to see whether or not they are indeed envelopes.

16. The Jordan Theorem as a Problem in Differential Geometry in the Large

The distinction between a problem in differential geometry in the large and a problem in the small for plane curves is that a finite extent of a curve is to be considered in the former case, and in addition some property of the curve is generally involved that cannot be studied without doing more than examining the curve in a neighborhood of each of its points. The distinction between the two kinds of problems is best brought out by giving concrete examples, as will be done in this and subsequent sections.

The most important theorem in the large for plane curves is without doubt the Jordan curve theorem. This theorem is more commonly regarded as a theorem in analysis and topology rather than as a theorem in differential geometry in the large, probably because it refers in the usual version to curves which are merely continuous rather than differentiable and regular. The Jordan curve theorem refers to a continuous closed curve C in the plane without double points (or, what is the same thing, the curve is assumed to be the topological image of a circle). The set of points of the plane covered by C is also called C, and the theorem states that *the plane is the sum of three point sets* $C + C_i + C_e$, *with* C_i *and* C_e *each open connected sets with no common points, but such that all of the boundary points of* C_i *and* C_e *are the points of* C. In addition, *the set* C_i *is a bounded set called the interior of* C, *while* C_e *is an unbounded set which contains, among others, all points of the plane outside of a sufficiently large circle, and is hence called the exterior of* C.

This theorem contains most of the usual ingredients of a theorem of differential geometry in the large:

1. It makes a hypothesis in the small (i.e., continuity), which holds at all points.

2. It makes one or more topological assumptions in the large (closure and freedom from double points).

3. Conclusions in the large are drawn (in this case of a topological nature).

It would be out of place in this book to prove the Jordan theorem as formulated above. Instead, a proof will be given here for the far more restricted case[1] in which C is a regular curve in the sense of differential geometry, and that makes it possible to avoid many of the complexities encountered in the general case. However, some elementary notions from point-set

---

[1] Actually, this case, or the slightly more general case in which C is continuous and piecewise regular, is the case which commonly arises in such important applications as those in the theory of integration in the plane, in the theory of functions of a complex variable, and in the mechanics of continuous media.

topology in the plane are needed.   The book of Patterson [P.1] gives a brief and readable account of these matters.   Perhaps the most important is the Heine-Borel theorem; it states that out of any covering of a closed bounded set by open sets a finite number of the open sets can be selected that will also cover it.   Another important concept is that of the *distance* ρ between two sets.   It is defined as the g.l.b. of the lengths of all straight segments joining all pairs of points of the sets.   In particular, a nonzero distance ρ exists between any pair of disjoint closed bounded sets, and the value of ρ is taken on for at least one pair of points of the sets.   A third concept needed in the later discussion is the concept of a connected *component* C of an open set S.   Such a component is singled out in a unique way by any one of its points p as the subset of points of the set S that can be connected with p by a continuous arc in S.

One of the principal tools to be used here in discussing the Jordan theorem is the *winding number* of an arc of a continuous curve with respect to a point not on the curve—an interesting notion in its own right.   This quantity is defined as follows.   The arc C is assumed to be defined by the vector $\mathbf{X}(t) = (x_1(t), x_2(t))$, $t_0 \leq t \leq t_1$, in terms of the coordinates of its points as continuous functions of t.   A vector $\mathbf{Z}(t)$ is then drawn from an initial point (x_1^0, x_2^0) not on C to the point fixed by $\mathbf{X}(t)$, and the angle $\theta(t)$ between $\mathbf{Z}(t)$ and the positive x_1-axis is defined so that it is continuous over the whole arc C.   (This can be done by making use of the Heine-Borel theorem in the manner that was described in Section 9 in the similar case of the inclination angle of the tangent to a regular curve.)   *The winding number τ of C is then defined as the difference $\tau = \theta(t_1) - \theta(t_0)$*; in effect, τ measures the total change of angle of $\mathbf{Z}(t)$ as it traverses the curve.   The angle $\theta(t)$ would satisfy, for example, the relation

$$\sin \theta(t) = \frac{x_2(t) - x_2^0}{\sqrt{(x_1(t) - x_1^0)^2 + (x_2(t) - x_2^0)^2}},$$

or the similar relation for $\cos \theta(t)$; it is then clear that the winding number τ is a continuous function of the initial point (x_1^0, x_2^0), as well as of the curve parameter t, as long as this point avoids the curve C.   However, it is an important fact, to be proved later, that the winding number τ has a jump discontinuity of amount $\pm 2\pi$ when the point (x_1^0, x_2^0) crosses over the regular Jordan curve.

The winding number of a closed regular Jordan curve C, and its jump upon crossing C, are to be used in conjunction with the construction of a special local coordinate system to prove the Jordan theorem.   This construction makes essential use of the regularity of C, and it is this property of C that makes a relatively simple proof of the theorem possible, compared with the proof when continuity alone is prescribed.   The construction in question consists of an embedding of the curve C in a curvilinear coordinate system in the plane that is valid in a neighborhood of C, which means that C is to be embedded in a ring-shaped region that is mapped in a one-to-one and differentiable fashion on a region topologically equivalent to the region between a pair of concentric circles in a plane. Such a ring will be referred to as a coordinate ring.

The construction of a coordinate ring containing C is carried out as follows. The first step is the construction of a coordinate system in a neighborhood of any point p of C; the method of doing that is sketched out and assigned as Problem 15 at the end of this chapter. The desired local coordinate system is obtained by erecting straight-line normals to C at all points of a neighborhood of p for which the arc length s lies in a certain interval $-s_0 < s < s_0$, $s_0 > 0$, with p corresponding to $s = 0$, and measuring off distances σ along the normals, $-\sigma_0 < \sigma < \sigma_0$, $\sigma_0 > 0$ (for σ_0 sufficiently small) on both sides of C. (Here, exceptionally, the rule that lengths along curves are taken to be positive is violated.) The curvilinear coordinate curves near p are thus the curves $s = $ const., $\sigma = $ const. (they form, in fact, two orthogonal families), with $\sigma = 0$ corresponding to C, and the neighborhood of p covered by them is in one-to-one correspondence with the points of a rectangle in a σ,s-plane that is given by $-\sigma_0 < \sigma < \sigma_0$, $-s_0 < s < s_0$. For later purposes it is important to note that the distance ρ (in the sense of point-set topology) from any point p lying in a local coordinate system to the arc of C embedded in it is σ, since that is the length of a straight segment drawn from p and meeting C at right angles.

The existence of local coordinate systems of the kind just described is next used to embed the whole of C in a similar coordinate system. Since C is a continuous closed curve the point set it covers—which is also called C— is closed and bounded, and consequently the Heine-Borel theorem is applicable to it. This theorem makes it possible to select a finite number of the open intervals on C with respect to which local coordinates can be introduced, and such that C is covered by them. Let $2s_0$ be the length, along C, of the shortest of these segments, and note that any segment of this, or shorter, length placed anywhere on C could be embedded in a legitimate local coordinate system. In addition, it is clear that a positive number σ_0 can be chosen so that a local coordinate system exists about an arbitrary point p of C valid for $-s_0 < s < s_0$, $-\sigma_0 < \sigma < \sigma_0$ (with p corresponding to $s = 0$), and with σ_0 independent of p: again, it suffices to choose σ_0 as the smallest of all such values in the finite covering. Thus C has been embedded in the large in a coordinate ring such that a finite number of local "rectangular" patches cover it, and each of the patches can be put in one-to-one continuous correspondence with an open rectangle in an s,σ-plane by identifying points with curvilinear coordinates (s, σ) near C with the points having s and σ as Cartesian coordinates.   However, this mapping would not in general be one-to-one in the large.   In fact, if C had double points such a one-to-one mapping could not exist, and even if C were without double points the coordinate ring might cross over itself, as indicated in Fig. 2.11; however, since C is assumed

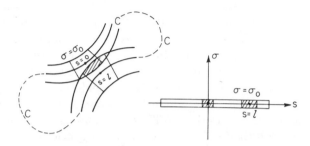

Fig. 2.11   Construction of a coordinate ring.

to be free of double points, it seems clear that such an occurrence can be avoided by choosing the coordinate ring narrow enough.   It will now in fact be shown that σ_0 *can be chosen small enough so that the coordinate ring is mapped in a one-to-one way on the rectangle* $0 \leq s \leq L$, $-\sigma_0 < \sigma < \sigma_0$, *once the end points at* $s = 0$ *and* $s = L$ *with the same σ-coordinates are identified.* The number L is of course the length of C.

The last statement is proved as follows: the curve C is subdivided into k *closed* segments $\bar{S}_n$, $n = 1, 2, \ldots, k$, of equal length $\delta = L/k$, such that δ is less than $2s_0/3$.   The effect of this is, in view of the above stipulation on s_0, that any three consecutive segments $\bar{S}_{n-1}, \bar{S}_n, \bar{S}_{n+1}$ are embedded in a valid local coordinate system—that is, the coordinate system in which $\bar{S}_{n-1} + \bar{S}_n + \bar{S}_{n+1}$ is embedded maps the points covered by it in a one-to-one way on the appropriate rectangle of the s,σ-plane.   Consider any segment $\bar{S}_n$, and the two adjacent *open* segments S_{n-1}, S_{n+1} obtained by deleting the end points of $\bar{S}_{n-1}$ and $\bar{S}_{n+1}$.   The set of curve points obtained by deleting S_{n-1} and S_{n+1} from C is a closed set consisting of $\bar{S}_n$ plus a set $\bar{S}_c$ obtained from C by deleting the open segment $S_{n-1} + \bar{S}_n + S_n$ (see Fig. 2.12).   Since C is

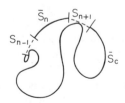

Fig. 2.12   The arcs $\bar{S}_n$ and $\bar{S}_c$.

assumed to be free of double points—and it is here that this assumption is used in an essential way—it follows that $\bar{S}_n$ and $\bar{S}_c$ have no points in common. These two disjoint closed bounded sets (see Fig. 2.12) are therefore at a non-zero distance ρ_n from each other. In the coordinate rectangle containing $\bar{S}_n$—assumed to be centered at $s = 0$—choose a closed subset $\bar{R}_n$ of all points with curvilinear coordinates (s, σ) satisfying $-\delta/2 \leq s \leq \delta/2$, $-\sigma_n \leq \sigma \leq \sigma_n$, with σ_n chosen smaller than σ_0 and also smaller than $\rho_n/3$. Similarly, define a closed bounded set $\bar{R}_c$ containing $\bar{S}_c$ by restricting σ in the same way. It is known, from the nature of the local curvilinear coordinate system, that the distance to $\bar{S}_n$ from all points of $\bar{R}_n$ is at most $\rho_n/3$ and similarly for points of $\bar{R}_c$ in relation to $\bar{S}_c$. It is then clear, by an application of the triangle inequality, that $\bar{R}_n$ and $\bar{R}_c$ have no points in common (in fact, they are at a distance $\geq \rho_n/3$ from each other). Thus by choosing the width of the coordinate ring as less than $\rho_n/3$ the strips $\bar{R}_n$ and $\bar{R}_c$ enclosing $\bar{S}_n$ and $\bar{S}_c$ would not overlap at any point. However, it is not a priori excluded that the band $\bar{R}_c$ might overlap itself or the strips $\bar{R}_{n-1}$ or $\bar{R}_{n+1}$ adjacent to $\bar{R}_n$. On the other hand, it is clear that the strips $\bar{R}_{n-1}$ and $\bar{R}_{n+1}$ do not overlap each other or $\bar{R}_n$ since their union is mapped on the s,σ-plane in a one-to-one way, as was stated above. It follows, therefore, that $\bar{R}_n$ has no points in common with all of the rest of the band containing C, except for the straight segments at its junctions with $\bar{R}_{n-1}$ and $\bar{R}_{n+1}$. This same process can be repeated for each of the k segments $\bar{S}_n$, each time with a certain value σ_n for σ that determines the width of the coordinate band. The smallest of these values σ^* is chosen, finally, and it is then seen that the entire band of width $2\sigma^*$ maps in a one-to-one way on the rectangle in the s,σ-plane, since the points of an arbitrarily chosen coordinate patch have no points in common with any others, except the adjoining pair at the common boundaries.

Consider next any point of the coordinate band that is not on C and for which σ has a positive value, say. It is to be shown that *all points of the band for which the σ-coordinate has a fixed sign*—or, as we shall also say, all points of the band *on the same side of C*—can be connected by a continuous arc which does not cut C. In fact, if p_1 and p_2 correspond to (σ_1, s_1), (σ_2, s_2) with σ_1 and σ_2 of the same sign, it is clear that the two image points can be joined in the σ,s-plane by a straight-line segment that does not cross the line $\sigma = 0$, and such a straight segment has a continuous image in the band containing C which does not touch C, since the mapping of the band into the σ,s-plane is one-to-one. Thus C is seen to have two "sides," in the sense that points near it with $\sigma > 0$ are separated from those near it with $\sigma < 0$ by the curve C.

Now that a coordinate ring or band covering C and having the properties just described has been constructed, it is possible to derive two properties of the winding number τ that lead to the final steps in the proof of the Jordan theorem:

1. *The winding number τ has a constant value for all points (s, σ) of the band that lie on the same side of C*, i.e., for which either $\sigma > 0$, or $\sigma < 0$, holds. First it is clear that τ has the value $2\pi n$ with respect to C for any point (x_1^0, x_2^0) not on it, with n an integer (positive, negative, or zero), simply because C is a closed curve. Since τ is a continuous function of the position coordinates (x_1^0, x_2^0) of a point as long as the point avoids C, and since any pair of points on the same side of C in the band can be joined by a continuous curve that does not touch C, it follows that τ is a constant with respect to all such points since it takes on values that consist solely of integers times 2π.

2. *The values of τ for points in the band on opposite sides of C differ by $\pm 2\pi$*. This is shown with the aid of Fig. 2.13. A normal to C at a point,

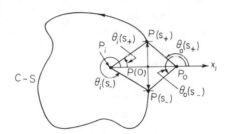

Fig. 2.13   Jump of τ.

which corresponds to $s = 0$, is drawn, and points p_o, p_i on the normal and in the band on opposite sides of C are taken. The x_1-axis is chosen along the normal. Vectors are drawn from p_o to points $p(s_+)$, $p(s_-)$ on C, near to p, and on opposite sides of it. (It is clear that τ for any closed curve has the same value no matter what point p is chosen as starting and end point, since the winding number for a sum of arcs is the sum of their winding numbers.) The angles $\theta_0(s_+)$ and $\theta_0(s_-)$ that these vectors make with the x_1-axis are indicated. The winding number of the arc C–S, obtained by deleting from C the open segment of it between $p(s_-)$ and $p(s_+)$, is clearly $\theta_0(s_-) - \theta_0(s_+)$ when $p(s_+)$ and $p(s_-)$ are taken near to each other, since $\theta_0(s)$ is a continuous function of s. On the other hand, the winding number for C is the sum of the winding numbers of C–S and S. If $p(s_-)$ and $p(s_+)$ approach p the winding number of the arc S between them tends to zero, evidently. Hence that of C–S, which is given by $\theta_0(s_-) - \theta_0(s_+)$, evidently becomes arbitrarily small, and since its value is $2\pi n$, with n an integer, it follows that $\tau = 0$ with respect to p_o. The same considerations applied to the point p_i, and then to the difference $\theta_i(s_-) - \theta_i(s_+)$, are easily seen to lead to the value 2π for $\tau(p_i)$. This proves the statement 2. That the jump in the value of τ upon crossing C in the direction from p_o to p_i is $+2\pi$ comes about because the specific orientation chosen for C was as shown in Fig. 2.13.   With the reverse orientation the sign of the jump would, of course, also be reversed.

A third property of τ, which is not dependent on any assumption of regularity for C, is readily established:

3. *The winding number τ for points in the plane sufficiently far from C has the value 0.*   The reason for this is that since C is bounded the angles $\theta(s)$ measured from a point far away from it differ very little from a constant, the difference $\theta(L) - \theta(0)$ would be very small, and hence necessarily would have the value zero.

The proof of the Jordan theorem is now completed by showing that *the open set D that is the complement of C in the plane, has exactly two components D_i, D_o, each of which has the points of C as its boundary points.*   First of all it is shown that D has at least *two* components.   This follows from (1) and (2) above, since the points on opposite sides of C in the coordinate ring must lie in two different components because the winding numbers with respect to these points are constant on each side, but differ for points on opposite sides: hence two points in different bands (i.e., in bands on opposite sides of C) cannot be connected, but any points in the same one can be connected and hence lie in some uniquely determined component.   Thus at least two components, to be denoted by D_i and D_o, are shown to exist.   It is now to be shown, finally, that D cannot have more than two components.   This is shown by proving that all boundary points of any component D_k of D must lie in C.   If this were not the case, a boundary point $p \subset D_k$ (any component D_k evidently must have boundary points, since D_k would otherwise be the entire plane), but not in C, would be contained in a small circle centered at p that would have no points in common with C, since C is a closed bounded set. But then p itself would clearly lie in the interior of some component of D and not on its boundary.   Thus any boundary point of D_k lies on C, hence points of D_k would exist in the coordinate band containing C and consequently D_k could not be distinct from both of the two components already known to exist.

This completes the proof of the Jordan theorem.   It is not difficult to see that the proof could be extended readily to the more general case of piecewise regular Jordan curves, i.e., curves with a finite number of "vertex" points.

The sets D_o and D_i are distinguished from each other by choosing D_o as that set for which the winding number τ has the value zero, and hence, from (3) above, this set includes all points of the plane outside of a circle large enough to contain the Jordan curve C in its interior.   The set D_o is therefore called the *exterior* of C, and the other component D_i of the complement of C is called the interior.

For many purposes it is desirable to study further properties of C in its relation to the sets D_i and D_o into which it separates the plane.

17.   Additional Properties of Jordan Curves

In many parts of calculus and geometry, and in mechanics and physics, a number of additional properties of regular Jordan curves are much used (but not treated as a rule in topology, and often rather too sketchily elsewhere).   One of the most important of these properties is the relation between a given orientation of the Jordan curve C and the location of the interior and exterior domains relative to the curve.   To investigate this relation consider first a point of the plane far away from C and hence in the exterior domain.   As was shown above, the winding number of C is zero with respect to these points.   The points of the interior domain therefore have as winding number either $+2\pi$ or -2π.   It can be shown that *an orientation of C can always be chosen in such a way that* the winding number with respect to interior points is $+2\pi$, in which case C is said to have a *positive orientation*; in that case, in addition, it is correct and meaningful to say that *the interior domain lies to the left of C.*   This statement means that the normal vector $\mathbf{v}_2$, when chosen relative to the oriented tangent vector $\mathbf{v}_1$ so that $\mathbf{v}_1$ and $\mathbf{v}_2$ have the same orientation as the coordinate axes, cuts into the *interior* of C in the manner shown in Fig. 2.14.   This figure should be compared with Fig. 2.13, in which the curve also was given a positive orientation; the discussion accompanying Fig. 2.13 showed that the orientation of the interior domain was $+2\pi$.

A Jordan curve, even if regular, can be complex to a degree that makes it difficult to say with ease whether a given point situated in the maze defined by it is in its interior or not (cf. Fig. 2.14, which is a relatively simple case), or whether an arrow placed on the curve at a given point determines a

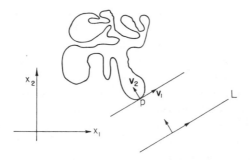

Fig. 2.14   Oriented Jordan curve with interior to left.

positive or negative orientation of the curve. Of course, a decision can be reached in principle by determining the winding number (or also the so-called intersection number described in Courant-Robbins[C.13], p. 267), after having found a point for which it is different from zero. Figure 2.14 indicates another way to fix the orientation definitely. It consists in taking any line L that does not touch the curve and moving it parallel to itself until it touches C at a point p for the first time: that is a meaningful procedure since C as a point set is closed and bounded. The straight line at p is a so-called *support line* of C, i.e., a line such that the whole of C lies in one of the two closed half-planes bounded by the line; the line is also a tangent line to C at p, since C is a regular curve (cf. Problem 4 at the end of this chapter). The normal vector v_2 at p is chosen so that it points into the half-plane containing C, v_1 is then fixed so that v_1 and v_2 are oriented the same as the coordinate axes, and that in turn fixes a positive orientation of C. Because of the existence of a coordinate band inside of C, and the invariance of the winding number for all points in the band, it is seen (again upon recalling the facts brought out with respect to Fig. 2.13) that the orientation induced at p leads to a coherent orientation throughout so that the interior domain always lies to the left. Problem 16 at the end of the chapter proposes a different way of distinguishing the inside and outside of a Jordan curve C that requires only measurements in a neighborhood of C.

A very important property of the interior domain of a Jordan curve is that it, unlike the exterior domain, is *simply connected*. This means that any continuous closed curve in the interior domain can be shrunk down on a point of it in a continuous manner and without leaving the domain. A simple proof of this fact, even for the special case of a regular Jordan curve, seems not to exist in the literature.

A brief and relatively simple, though rather sophisticated and mysterious proof from the intuitive geometric point of view, can be extracted from the book of Ahlfors [A.1] on Complex Analysis. Ahlfors *defines* (p. 112) simple connectivity of an open domain D in the plane as the equivalent of requiring its complement D_c to be connected. That property, however, was established above for the exterior domain. From this, with the aid of the concept of the winding number, it is shown that if $f(z)$ is an analytic function $\neq 0$ of the complex variable z defined in D, then a single-valued branch of $\log f(z)$ can also be defined in D. With these tools, and with the aid of the notion of a normal family of analytic functions, it is shown by Ahlfors (Theorem 10, p. 172) on the basis of a certain maximum problem, that an analytic function $f(z)$ exists such that $w = f(z)$ maps the domain D in a one-to-one way on the interior of a unit circle in the w-plane, and as z tends to C, $|f(z)| \to 1$. This is, of course, the famous Riemann mapping theorem. It proves far more than is necessary to show that S is simply connected. It would be sufficient, for example, to show that $D + C$ could be mapped on the closed-unit circular disk in a one-to-one and *continuous* fashion, i.e., topologically; the simply connected character of D is then rather easy to establish (it is left as an exercise to do so).

It will be shown here in an elementary way that the interior D of a piecewise regular Jordan curve C is *simply connected*. This means that any continuous closed curve $K(t)$, $0 \leq t \leq 1$, $K(0) = K(1)$, in D can be contracted continuously into a point p of D without leaving it; this in turn is understood to mean that the curve $K(t)$ can be embedded in a one-parameter family $K(\sigma, t)$, $0 \leq t \leq 1$, $0 \leq \sigma \leq 1$, lying in D, continuous in σ and t, and such that $K(0, t) = K(t)$, $K(1, t) \equiv p$. Since the set of points in D covered by $K(\sigma, t)$ is closed, it follows that this set is at a finite distance from the Jordan curve C. It follows that a simple closed polygon lying in D can be inscribed in the regular curve C such that the curves $K(\sigma, t)$ lie in its interior: the polygon can, in fact, be placed in the coordinate strip discussed above. It suffices therefore to prove the simple connectivity of the interior of a simple closed *polygon* in order to obtain the desired result.

The proof of simple connectivity for polygons is to be proved by an induction on the number of sides. Since any *convex* polygon is simply connected because a closed curve in it could be shrunk down radially and continuously on an arbitrary interior point of it, it follows, in particular, that every triangle is simply connected. Thus the first step in the proof by induction is valid. Suppose that the theorem were known to be valid for a polygon P_n with n sides; it is to be shown that it holds for a polygon P_{n+1} with $n + 1$ sides. That is done as follows:

1. Take a line L in the plane that does not cut P_{n+1}, and displace it until it touches P_{n+1} for the first time. Such a line necessarily contains at least one vertex point of P_{n+1}. By rotating L, if necessary, about one such vertex a position L' of it can be found such that the line meets P_{n+1} for the first time in two or more distinct vertices that are not end points of the same side of P_{n+1}. The polygon P_{n+1} would of course lie entirely in one of the two closed half-planes bounded by L'. If such a situation did not occur after at most $n + 1$ rotations of the line, it follows that P_{n+1} would be convex, since every side of P_{n+1} would lie in a supporting line of P_{n+1}—and there would be nothing to be proved in that case.

2. Two vertices p_j, p_k of P_{n+1} on L' are selected such that a connected portion of P_{n+1} between the two points has no other vertex point on L' (see Fig. 2.15). A new simple closed polygon P'' is defined by joining the segment $p_k p_j$ to the open polygon with vertices $p_j p_{j+1} \cdots p_{k-1} p_k$. It is clear that P' is a polygon with at least one side less than P_{n+1}, since the polygon $p_j p_{j-1} \cdots p_{k+1} p_k$ has at least two sides. Thus P'' is simply connected by virtue of the

induction hypothesis. The polygon P', which is obtained by adding P_{n+1} to P'', is also simply connected for the same reason.

3. Consider any continuous closed curve $C(t)$ in P_{n+1}. Since it lies also in P', it can be shrunk continuously to a point in P', and hence to a point p in the interior of P_{n+1} (since it is readily seen that the point p can be chosen as any point inside P'). The set of curves $C(\sigma, t)$ in which $C(t)$ is embedded

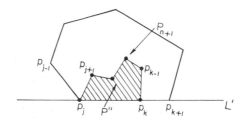

Fig. 2.15   The polygons P_{n+1}, P'', $P' = P_{n+1} + P''$.

can be replaced without loss of generality by a set of polygons $P(\sigma, t)$ with a finite number of sides (in fact, though not necessary for the following, by polygons all having the same number of sides). This can be done by invoking the Heine-Borel theorem, along the following lines. The square $0 \le t \le 1$, $0 \le \sigma \le 1$ in the σ,t-plane can be subdivided into small squares by lines parallel to the axes in such a way that the t-values on all lines $\sigma =$ const. correspond to points (σ, t) in P' that form vertices of a closed polygon in P' for each fixed σ; and also such that any pair of points in any one of the small squares correspond to points inside P' having the same property. These vertices can then be deformed continuously (by a piecewise linear construction) to yield the desired family of polygons $P(\sigma, t)$ with $P(0, t) = P(t)$, $P(1, t) = p$.

4. If it happened that all of the polygons $P(\sigma, t)$ stayed in P_{n+1}, there would be nothing to prove. If some of them do leave it, this occurs because they cross over the open polygon $p_j p_{j+1} \cdots p_{k-1} p_k$—to be called here P_{jk} (see Fig. 2.15)—since they must in any case remain inside P'. Consider any polygon $P(\sigma, t)$ that has points in P_{n+1} as well as points in P'', and the lowest value of t for which $P(\sigma, t)$ cuts through P_{kj}, at a point p_1, on going from P_{n+1} into P''. Follow $P(\sigma, t)$ with increasing t until it crosses P_{kj} again at point p_2, for the first time upon passing from P'' into P_{n+1}. The part of $P(\sigma, t)$ in P'' between p_1 and p_2 is replaced by the portion of P_{jk} between p_1 and p_2 to form a new polygon $P_1(\sigma, t)$ that lies in P_{n+1} except for a part on P_{jk} in its boundary. Next follow $P(\sigma, t)$ from point p_2 into P_{n+1} until it again crosses P_{jk} for the first time and enters P'' at point p_3; continue further until it again crosses back into P_{n+1} at p_4 on P_{jk}. The part of $P(\sigma, t)$ in P''

between p_3 and p_4 is again replaced by the part of P_{jk} between these two points to form a modified polygon $P_2(\sigma, t)$. This process is repeated until $P_m(\sigma, t)$ enters P'' for the first time by crossing P_{jk} at the point p_1 with which the construction started. Evidently the new family of polygons thus constructed would depend continuously on both σ and t. They all lie either inside P_{n+1} or in part on the segment P_{jk} in its boundary; eventually, as $\sigma \to 1$ they shrink down on p without leaving P_{n+1}, since the original set did that in P'. Thus P_{n+1} is shown to be simply connected.

18. The Total Curvature of a Regular Jordan Curve

A theorem in the large that is as intuitively obvious as the Jordan theorem, but like it is also not very easy to prove, is the theorem that the total curvature of a *positively oriented regular Jordan curve is* 2π. If the Jordan theorem is used together with the theorem that the interior domain is simply connected, a proof can be given rather easily. A proof has been given by H. Hopf [H.13] that does not make use of these tools and assumes only the existence of a continuous first derivative, but it is too long to be given here.

The theorem is proved by using a fact proved in the preceding section, i.e., that the interior domain bounded by a regular Jordan curve C is simply connected. The starting point is the existence—proved in Section 16—of a curvilinear coordinate system in a band in the interior of the curve C, supposed given a positive orientation. The coordinate curves are straight-line normals to C together with a family of parallel curves obtained as the locus of points at a fixed distance σ from C measured along the normals toward the inside of C; the parameters are thus s and σ, as in the preceding sections. Consider the field $\mathbf{v}(s)$ of normal vectors to C of small fixed length δ that are drawn from C toward its interior: their end points evidently lie on a curve C_δ of the parameter strip. C_δ is also a regular positively oriented Jordan curve. The total angle $\Delta\theta$ through which these vectors $\mathbf{v}(s)$ turn upon making a circuit around C evidently is the same as the angle turned through by the field of tangent vectors of C. It will be shown that $\Delta\theta$, which is of course the total curvature of C, is equal to the winding number τ_C of C with respect to a point in the interior of C and hence has the value 2π—and thus the theorem would be proved. The method of proof uses the simple connectivity of the interior of the curve C_δ; thus C_δ is supposed embedded in a continuous family $K(\sigma, t)$, $0 \le \sigma, t \le 1$, such that $C_\delta = K(0, t)$, and $K(1, t)$ is a point p in the interior of C_δ. An x_1-axis is taken through p; the angle $\theta(s)$ between it and the vectors $\mathbf{v}(s)$ can then be defined as a continuous function of s. The initial points of the field $\mathbf{v}(s)$ on C are held fixed, but the end points of the vectors are allowed to vary with the deformation defined by $K(\sigma, t)$ by requiring them to be fixed by the same t-value for all σ that they had on C_δ, i.e., for $\sigma = 0$. In doing so the angle $\theta(\sigma, t)$ that the field vectors make with the x_1-axis would vary continuously with σ and t, and therefore also $\varDelta\theta(\sigma, t)$ (it is to be noted that no field vector is ever the zero vector, since the end points of these vectors lie in the interior of the curve C); but the value of $\varDelta\theta$ for each value of σ is always an integer times 2π, and hence is a constant. On the other hand, when $\sigma = 1$ is attained $\varDelta\theta$ is clearly the winding number of C with respect to p, since the field vectors then all pass through the point p, and therefore $\varDelta\theta = 2\pi$, as was shown in Section 16. Thus the theorem is proved.

19. Simple Closed Curves with $\kappa \neq 0$ as Boundaries of Convex Point Sets

The following theorem is to be proved in this section: *a regular closed curve without double points* (that is, a Jordan curve), *for which the curvature κ does not vanish, is the boundary of a convex point set.* Such a curve is often called an oval.

The key point in the proof of this theorem is a theorem which refers solely to convex point sets, and not to an object for which differentiability assumptions are needed. By a convex point set S is meant a set with the property that if p_1 and p_2 are any points of S then the straight-line segment $\overline{p_1 p_2}$ joining them also belongs to S. Sets satisfying this simple condition have a host of interesting properties (see, for example, the books of Bonnesen-Fenchel [B.7] and Yaglom [Y.1]). For example, at every boundary point of such a set there is at least one *support line*, i.e., a line through the point with the property that the entire set lies in one or the other of the two (closed) half-planes bounded by the line. The converse is also true, at least for sets with inner points (i.e., points such that a whole neighborhood of each of them lies in the set): a set with inner points having the property that every boundary point of the set contains a support line of the set is a convex set. This last theorem is capable of a generalization which is needed for present purposes. It is, in fact, not necessary to postulate that a support line for the *entire* set exists at each boundary point, but rather that the support property holds only locally. The following theorem is to be proved:

A bounded set S in the plane, which consists of a connected open set S_1 plus its boundary B is given. At each point $p \subset B$ it is assumed that a straight line exists which is a support line for all points of S in a sufficiently small neighborhood of p. Then the set S is convex.

The following proof is due to Erhard Schmidt (see the book of Bieberbach [B.1], p. 20). (It might be added that the proof is in the essentials identical for sets with inner points in an n-dimensional Euclidean space.) Consider any two points $p,q \subset S_i$. Since S_i is connected it follows that p and q can be connected by a polygon with straight sides which lies in S_i. The vertices of the polygon are numbered $p, p_1, p_2, \ldots, p_r, q$ in the order in which they meet in traversing the polygon from p toward q. The object of the following proof is to show that the vertices can be removed one at a time so that eventually the polygon, while remaining inside S_i, becomes the straight segment joining p and q. To this end, consider the segment $\overline{pp_2}$; if it lies in S_i, then p_1 could be removed. If this process could be repeated up to q, there would be nothing left to prove. Suppose, then, that $\overline{pp_k} \subset S_i$ but that $\overline{pp_{k+1}} \not\subset S_i$. It will be shown that this is in contradiction with the existence of a local support plane at every boundary point; thus the vertex p_k could after all be removed, and this process evidently would lead to the elimination of all vertices. It is therefore to be shown that p_k can be removed, as follows: The points p, p_k, and p_{k+1} can be assumed without loss of generality to form a nondegenerate triangle: otherwise $\overline{pp_{k+1}}$ would belong to S_i and there would be nothing to prove. The proof continues indirectly by assuming that the segment $\overline{pp_{k+1}}$ does not lie in S_i and hence contains at least one boundary point of S_i. Consider the set of segments drawn from p to the points $r \subset \overline{p_k p_{k+1}}$ (cf. Fig. 2.16), and characterized by the angle θ. For θ small

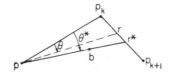

Fig. 2.16   *Erhard Schmidt's construction.*

enough $\overline{pr} \subset S_i$ since $\overline{pp_k} \subset S_i$ and S_i is an open set. Since the boundary points B of S form a closed bounded set, there exists a smallest angle $\theta = \theta^*$ such that the segment $\overline{pr^*}$, $r^* \neq p_k$, contains a point of B since the segment $\overline{pp_{k+1}}$ certainly has that property. On going out from p toward r^* along the segment $\overline{pr^*}$, a first boundary point $b \subset B$ would be encountered on it— again because the set B of boundary points is closed. The situation can now be described as follows: all points of the triangle pp_kr^*, including its sides, are points of S_i, except for b and possibly other points on the segment $\overline{br^*}$. But now it is clear that there can be no support line in the small at the boundary point b: the straight line in the direction fixed by p and b would not serve because there are inner points on $\overline{pb}$ arbitrarily near to b, and a line through b in any other direction penetrates the interior of the triangle, which is composed entirely of inner points. Thus p_k can be removed, and the theorem is proved.

The theorem concerning simple regular closed curves with $\kappa \neq 0$ can now be proved. The curve can be oriented so that $\kappa > 0$ holds. By the Jordan curve theorem proved above the curve is the full boundary of a bounded and connected open set. Furthermore, if the curve is oriented positively, the normal vector $\mathbf{v}_2$ points into the *interior* of the region bounded by the curve, since the curve points themselves lie on that side of the tangent, at least locally (cf. Section 17). The tangent to the curve is thus a local support line at all boundary points of the set enclosed by the curve. The above theorem of E. Schmidt concerning convex point sets then completes the proof of the theorem.

Another way in which the last theorem could be formulated is that a simple closed regular curve which is locally convex is an oval, i.e., it is convex in the large. The assumption that the curve is without double points is essential: in Fig. 2.17 a closed curve with $\kappa > 0$ is shown that is not an oval. However, the analogous theorem for closed *surfaces* in *three-dimensional* space is valid without an assumption of this type. For that case it has been shown by Hadamard [H.2] that a closed regular surface which is locally convex (i.e., is such that all surface points lie on one side of the tangent plane in a neighborhood of the point of tangency) is an ovaloid, i.e., the boundary of a bounded convex set in three-dimensional space. In other words, the closure of the surface and the property of local convexity lead automatically to the nonexistence of double points or double lines of the surface. This theorem, or rather, a generalization of it, is proved in Chapter 8.

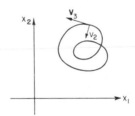

Fig. 2.17   A locally convex closed curve.

20. Four-Vertex Theorem

There is another theorem in the large about ovals in the plane that seems to have fascinated quite a number of people, since many different proofs of it have been given. (The first proof is credited to Mukhopadhyaya by Blaschke [B.5], where further references to the literature on this theorem are given.) The theorem is called the four-vertex theorem. It refers to closed convex curves, i.e., to regular, simple closed curves that are, in addition, such that the curvature κ does not vanish.   From the above discussion it is known that the curve bounds a convex set.   The four-vertex theorem states that such a curve, if it has continuous third derivatives, *contains at least four points at which the curvature is stationary*, or, in other words, four points at which $\dot\kappa = 0$.   An ellipse is a curve of this kind with exactly four such points, i.e., its vertices, so that the number four in the theorem cannot be increased.

The proof begins by remarking that $\kappa = \ddot\theta$, with $\theta(s)$ the angle between the tangent vector and the x_1-axis, must vanish at two points at least on the curve since $\kappa = \dot\theta(s)$ is periodic in s (since the curve is a closed curve) and thus takes on its maximum and minimum.   At these points it is clear that $\dot\kappa$ changes its sign.   In addition, there must be an even number of changes of sign of $\dot\kappa$, since a periodic function changes its sign an even number of times in the period interval.   Thus the theorem can be proved by showing that $\dot\kappa$ changes sign at more than two points.   This in turn is done indirectly by assuming that $\dot\kappa$ changes sign at two points P and Q only, and then showing that a contradiction results.   It is to be observed that under these circumstances $\dot\kappa = \ddot\theta$ has different but fixed signs on the two different arcs between P and Q.   Choose P as the origin of the x_1,x_2-plane and take the x_1-axis through Q, as indicated in Fig. 2.18.   The x_1-axis divides the oval into two parts—one in the upper, and the other in the lower half-plane.   Here the convexity property of the oval is used.   Next consider the line integral $\oint x_2(s)\,\ddot\theta(s)\,ds$ taken once around the curve, assuming now that $\mathbf{X}(s) = (x_1(s), x_2(s))$ has a continuous third derivative so that $\ddot\theta$ is continuous, and that the closed curve is traversed once when s ranges over the interval $s_0 \le s \le s_1$, with the end points both corresponding to the origin.   This integral vanishes, for the following reasons.   Observe first that

$$(2.37) \qquad \int_{s_0}^{s_1} x_2 \ddot\theta \, ds = x_2 \dot\theta \Big|_{s_0}^{s_1} - \int_{s_0}^{s_1} \dot x_2 \dot\theta \, ds.$$

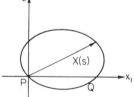

Fig. 2.18   Four vertex theorem.

The integrated term has the value zero since $x_2 = 0$ at s_0 and s_1.   The integral on the right-hand side also has the value zero; this follows because $\dot x_2 = \sin\theta$ and therefore

$$-\int_{s_0}^{s_1} \dot{x}_2\theta \, ds = \int_{s_0}^{s_1} \frac{d}{ds} (\cos \theta) \, ds = (\cos \theta) \Big|_{s_0}^{s_1} = 0,$$

since the curve is a closed curve. Turn next to Fig. 2.18 and recall that $\dot{\kappa} = \ddot{\theta}$ is assumed to have fixed but different signs in the upper and lower half-planes so that the integrand $x_2\ddot{\theta}$ in (2.37) is readily seen to have the same sign over the whole curve. The integral cannot therefore have the value zero and this contradiction establishes the theorem.

Problems

1. Show that the tangent line given by (2.5) is a regular curve with r as parameter if $\mathbf{X}(t)$ is a regular curve.

2. Characterize all regular plane curves for which the curvature is constant.

3. Show that the geometric interpretation of the significance of the sign of κ given in this book is in accord with the interpretation commonly given when the curve is defined by $y = y(x)$, i.e., that κ is positive if the curve is convex toward the x-axis.

4. Show that any straight line through a point p of a regular curve that is not a tangent line of it at p is such that the curve points pass from one side to the other of the line as the curve parameter increases when the curve points cross p; in effect, any line not a tangent line *cuts* the curve at its point of contact. Prove that the osculating circle of a regular curve with a continuous third derivative cuts the curve at any point of osculation where the curvature is not stationary, i.e., where $\dot{\kappa} \neq 0$.

5. Show that the osculating circle at a point of a curve is the circle that fits the curve best at that point in the sense that it is the only circle that has the same first and second derivatives at the point on the curve.

6. Show that the length of a curve expressed in polar coordinates $r = r(\theta)$ is given by

$$L = \int_{\theta_0}^{\theta_1} \sqrt{r^2 + r'^2} \, d\theta.$$

Show also that

$$\kappa = \frac{2r'^2 - rr'' + r^2}{(r^2 + r'^2)^{3/2}}.$$

7. Show that the curve $\mathbf{X}(t) = (a \cos t, b \sin t)$ is an ellipse. Find the coordinates of its centers of curvature.

8. Discuss the evolute of the ellipse of the preceding problem. In particular, show that this curve has cusps as singularities at the points corresponding to the vertices of the ellipse where $\dot{\kappa} = 0$.

9. A regular curve is defined by the vector $\mathbf{X}(s)$ with s as arc length. On each tangent line of the curve a point $\mathbf{Y}(s)$ is fixed by laying off on it a distance $l + s$, $l = $ const., from the point of tangency in the direction opposite to $\dot{\mathbf{X}}(s)$. Show that $\mathbf{Y}(s)$ is a regular curve if the curvature κ of $\mathbf{X}(s)$ is not zero, and that in this case $\mathbf{X}(s)$ is the evolute of $\mathbf{Y}(s)$. How many such involutes $\mathbf{Y}(s)$ are there to a given curve $\mathbf{X}(s)$?

10. A family of plane curves depending on a parameter α is given implicitly by the formula $F(x, y, \alpha) = 0$. Show that the necessary condition (2.35) for the existence of an envelope requires in this case that $F_\alpha(x, y, \alpha) = 0$ in addition to $F(x, y, \alpha) = 0$.

11. Find the envelope of the trajectories of a material particle moving under the action of gravity which is shot from the origin into a vacuum with the same initial speed V_0, but at different angles to the horizontal.

12. Given the family of cycloids $\mathbf{X}(t, \alpha) = (t + \alpha - \sin t, 1 - \cos t)$, examine the loci determined by (2.35) and observe that they include a locus of singular points as well as an envelope.

13. Find the evolute of a cycloid.

14. If the function $\phi(x, y)$ has continuous second derivatives in a neighborhood of the point (x_0, y_0), where $\phi(x_0, y_0) = 0$, and if $\phi_x^2 + \phi_y^2 \neq 0$ at the same point, show that $\phi(x, y) = 0$ defines a locus which is a regular curve in a neighborhood of (x_0, y_0).

15. The vector $\mathbf{v}_2(s)$ is the unit normal to the regular curve $\mathbf{X}(s) = (x_1(s), x_2(s))$, with s as arc length, that forms with the tangent vector $\mathbf{v}_1(s)$ a positively oriented pair of vectors. Consider the curve family

$$\mathbf{Y}(s, \sigma) = \mathbf{X}(s) + \sigma \mathbf{v}_2(s), \qquad -\epsilon \leq \sigma \leq \epsilon.$$

(a) Show that the curves $s = $ const. and $\sigma = $ const. are regular curves if σ is small enough. (b) Show that all of the "parallel" curves $\sigma = $ const. have $\mathbf{v}_2(s)$ as normal, i.e., the curves $s = $ const., $\sigma = $ const. are orthogonal. (c) Show that the curves $s = $ const., $\sigma = $ const. may be chosen as a curvilinear coordinate system by showing that the Jacobian

$$\frac{\partial(y_1, y_2)}{\partial(s, \sigma)} \neq 0;$$

in this $y_1(s, \sigma)$ and $y_2(s, \sigma)$ are the components of $\mathbf{Y}(s, \sigma)$. (d) Show that the number $|\sigma|$ is the shortest distance from a point with curvilinear coordinates (s, σ) to the curve $\mathbf{X}(s)$.

16. A regular Jordan curve C is given in a plane but the only points of the plane accessible to an observer on C are assumed to lie in a narrow band containing it. An orientation of C relative to a coordinate system in the plane is supposed not to be a priori available to the observer. However, he

is supposed to know that he lives in the plane and is familiar with its geometry.   How could the interior of C be identified by measurements in the band alone?   *Hint:* Consider the curves parallel to C in the band containing it, as defined in the preceding problem, measure their lengths, and use the theorem that the total curvature of C is $\pm 2\pi$.   Do the same, using area measurements.

17. By using the existence and uniqueness theorems for ordinary differential equations show that the Frenet equations (2.27) lead to (a) a regular curve $\mathbf{X}(s)$ with $\mathbf{v}_1 = \dot{\mathbf{X}}$, $\kappa \mathbf{v}_2 = \ddot{\mathbf{X}}$ if $\mathbf{v}_1$ and $\mathbf{v}_2$ are chosen initially as orthogonal unit vectors, (b) a unique curve if a point on the curve and the tangent vector there are prescribed—provided that $\kappa(s)$ is a given continuous function of s.

18. A regular curve C is given with curvature that does not vanish.   It is free of double points and extends to ∞ in the plane when traversed in either direction.   Prove the following (not necessarily in the order given):

(a) C has infinite length.

(b) The total curvature of C is at most equal to π.

(c) C has a representation in the form $y = f(x)$, once the coordinate axes are properly chosen, with $f(x)$ single-valued.

(d) C is the full boundary of an unbounded convex set.

(e) Unless the total curvature of C is exactly π, both Cartesian coordinates of the points of C are unbounded, however the coordinate system is chosen.

19. If a closed regular curve has a total curvature κ_T that exceeds 2π, then $\kappa_T \geq 4\pi$ holds, and the curve has at least one double point.   Give examples of closed regular curves with $\kappa_T = 2\pi n$, $n = 0, 1, 2, \ldots$.   Show by examples that $|\kappa_T|$ is not determined uniquely by giving the number of its double points.

CHAPTER III

Space Curves

1. Regular Curves

Many of the basic notions concerning space curves are motivated by the same considerations that were advanced for plane curves; consequently, they are introduced here rapidly and with little comment.

A regular curve is the locus defined by a vector

$$(3.1) \qquad \mathbf{X}(t) = (x_1(t), x_2(t), x_3(t)), \qquad \alpha \leq t \leq \beta,$$

such that the functions $x_i(t)$ have continuous second or third derivatives (depending on individual circumstances) and such that the derivative $\mathbf{X}'(t)$ is not zero:

$$(3.2) \qquad \mathbf{X}'(t) = (x_1', x_2', x_3') \neq 0.$$

The latter condition ensures that the mapping of the t-interval into the x_1, x_2, x_3-space is locally one-to-one; the proof of this fact can be carried out in the same fashion as it was for plane curves.

A new parameter $\tau = \phi(t)$ replacing t can always be introduced without impairing the regularity of the curve $\mathbf{Y}(\tau) = \mathbf{X}(t(\tau))$, provided that ϕ has continuous second or third derivatives and $\phi'(t)$ is not zero.

As with plane curves, *the vector $\mathbf{X}'(t)$ is by definition the tangent vector* of the curve, and it is placed at the point of tangency. The tangent vector is invariant under transformations of coordinates, but not under parameter transformations. However, the notion of the *tangent line*, i.e., the line at $\mathbf{X}(t_0)$ in the direction determined by $\mathbf{X}'(t_0)$ is invariant under parameter transformations.

The choice of a parameter results in a direction of travel along the curve such that when t increases the tangent vector $\mathbf{X}'(t_0)$ points into that one of the two half-spaces determined by a plane perpendicular to $\mathbf{X}'(t_0)$ at t_0 in which the curve points lie for $t > t_0$ and t near enough to t_0. The proof is essentially the same as that given for plane curves.

An interesting special case is that in which the tangent vector is constant:

$$(3.3) \qquad \mathbf{X}'(t) = \mathbf{a}.$$

From this $\mathbf{X}(t) = t\mathbf{a} + \mathbf{b}$, with $\mathbf{b}$ a constant vector, and the curve is therefore a straight line because its Cartesian coordinates $x_i(t)$ are linear in t.

2. Length of a Curve

The length $s(t)$ of the arc of the curve extending from an initial point t_0 to a variable point t is defined by

$$(3.4) \qquad s(t) = \int_{t_0}^{t} \sqrt{\mathbf{X}'(\sigma) \cdot \mathbf{X}'(\sigma)} \, d\sigma = \int_{t_0}^{t} \sqrt{x_1'^2 + x_2'^2 + x_3'^2} \, d\sigma.$$

Just as with plane curves, the arc length s may always be chosen as parameter, since $ds/dt \neq 0$: regularity requires that $\mathbf{X}'(t) \neq 0$ hold. In case the arc length is chosen as parameter it is easily seen from (3.4) that $|\dot{\mathbf{X}}(s)| = 1$. Conversely, if the relation $|\mathbf{X}'(t)| = 1$ holds for all t, then $t = s + \text{const}$. Here, as with plane curves, $s(t)$ is invariant with respect to transformations of the coordinates and also with respect to parameter transformations. Once more it is to be noted that the dot is used to indicate differentiation with respect to the arc length as parameter.

3. Curvature of Space Curves

The total curvature κ_T of a space curve cannot be defined, as was done for plane curves, by introducing the angle of the tangent vector with a fixed direction, because the total curvature $\kappa_T(C) = \kappa_T(C_1 + C_2)$ of a curve C composed of two arcs C_1 and C_2 joined together would not in general equal the sum $\kappa_T(C_1) + \kappa_T(C_2)$ of the total curvatures of the two arcs separately. However, the procedure of introducing the circular image of a plane curve can be applied in an analogous way to space curves, but it is now a *spherical image* of the space curve that is used. To this end the tangent vectors are transported parallel to themselves to the center of a unit sphere. The length of the curve traced out by $\dot{\mathbf{X}}(s)$ on the sphere as s varies over an arc of $\mathbf{X}(s)$ is defined as the total curvature of the arc—and, indeed, this is a rather reasonable measure of the total change in direction of the tangent.[1] Since the

---

[1] It is to be noted that a regular curve may have a spherical image that is not a regular curve, because $\ddot{\mathbf{X}}(s)$ may not be different from zero; nevertheless, the formula for the length of it remains reasonable since $\ddot{\mathbf{X}}(s)$ is assumed to be continuous.

length of any curve is defined by (3.4), it follows that the length of the curve given by $\dot{\mathbf{X}}(s)$ is found from the formula

$$\kappa_T = \int_{s_0}^{s_1} \sqrt{\ddot{\mathbf{X}} \cdot \ddot{\mathbf{X}}} \, ds,$$

and this in turn leads to the following definition for the curvature $\kappa(s)$ at a point

$$(3.5) \qquad\qquad \kappa(s) = \sqrt{\ddot{\mathbf{X}} \cdot \ddot{\mathbf{X}}} = |\ddot{\mathbf{X}}|.$$

In the plane it was possible to give a geometrical significance to a sign of the curvature, but for space curves a similar procedure is not feasible for reasons that will appear later. Consequently the positive square root is taken in (3.5) and thus $\kappa \geq 0$ holds.

The curvature is of course zero if, and only if, $\ddot{\mathbf{X}}$ is the zero vector. An interesting special case is that in which $\kappa = 0$ for all s. In this case $\ddot{\mathbf{X}} \equiv 0$ holds; it follows by integration that $\dot{\mathbf{X}}(s)$ is a constant vector and hence, as was seen above, $\mathbf{X}(s)$ represents a straight line. Thus $\kappa \equiv 0$ characterizes the straight lines.

4. Principal Normal and Osculating Plane

It is convenient here, as it was for plane curves, to introduce a special set of linearly independent orthogonal unit vectors at each point of the curve, to be denoted by $\mathbf{v}_1(s)$, $\mathbf{v}_2(s)$, $\mathbf{v}_3(s)$ with s as the arc-length parameter. The first of these vectors is identified with the tangent vector:

$$(3.6) \qquad\qquad \mathbf{v}_1(s) = \dot{\mathbf{X}}(s),$$

which ensures it to be a unit vector; it will be called *the tangent vector* in what follows. In the theory of plane curves a normal vector $\mathbf{v}_2$ was singled out by requiring it to form with the tangent vector $\mathbf{v}_1$ a positively oriented system. That, evidently, cannot be done in three-space. Instead, a special normal vector, called the *principal normal* $\mathbf{v}_2(s)$, is singled out among all vectors orthogonal to the tangent vector, in the following way. Since $\mathbf{v}_1 \cdot \dot{\mathbf{v}}_1 = 0$ from $\mathbf{v}_1 \cdot \mathbf{v}_1 = 1$, it follows that $\dot{v}_1 = \ddot{\mathbf{X}}$ is a vector orthogonal to the tangent vector. In case $\ddot{\mathbf{X}}$ is different from zero, the unit vector directed along it is defined as the principal normal and is denoted by $\mathbf{v}_2(s)$. However, as (3.5) shows, this process fails if the curvature κ has the value zero. In what follows in this chapter it is usually assumed that $\kappa > 0$ holds; in that case the principal normal vector—often simply called the normal—is therefore defined by the formula [1]

(3.7) $$\mathbf{v}_2 = \kappa^{-1}\dot{v}_1 = \kappa^{-1}\ddot{\mathbf{X}}.$$

The plane determined by $\mathbf{v}_1$ and $\mathbf{v}_2$ is called the *osculating plane*. One of the properties of the osculating plane is that it is the position in the limit of planes containing the tangent $\mathbf{v}_1(s_0) = \dot{\mathbf{X}}(s_0)$ to the curve and the points $\mathbf{X}(s)$ as $s \to s_0$. The proof is as follows (cf. Fig. 3.1). The vector $\mathbf{Y}(s) = \mathbf{X}(s) -$

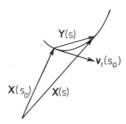

Fig. 3.1   Limit of planes containing the tangent.

$\mathbf{X}(s_0)$ is introduced. Upon applying the "mean value theorem" discussed at the end of Chapter I to this vector [cf. equations (1.30) and (1.31)], the following relation results:

$$\mathbf{Y}(s) = (s - s_0)\,\dot{\mathbf{X}}(s_0) + \tfrac{1}{2}(s - s_0)^2\ddot{\mathbf{X}}^*.$$

This leads to the relation

$$\frac{\mathbf{Y}(s) - (s - s_0)\,\dot{\mathbf{X}}(s_0)}{\tfrac{1}{2}(s - s_0)^2} = \ddot{\mathbf{X}}^*.$$

The left side of the latter equation is a vector in the planes under consideration, since $\mathbf{v}_1 = \dot{\mathbf{X}}$. But, as $s \to s_0$, $\ddot{\mathbf{X}}^* \to \ddot{\mathbf{X}}(s_0)$ and since $\ddot{\mathbf{X}}(s_0) = \kappa\mathbf{v}_2$ the limiting position of the plane is that determined by the vectors $\mathbf{v}_1$ and $\mathbf{v}_2$. This is the reason for giving the name osculating plane to this particular plane: of all planes containing the tangent to the curve it fits the curve best.

---

[1] The need and later usefulness of this formula, and its consequences, justify in the theory of space curves what might seem at first sight to be an unnecessary restriction, which was not required in the theory of plane curves. Although it is true that $\kappa \equiv 0$ leads to the straight line—a perfectly reasonable regular space curve—it is nevertheless also true that if κ is zero without being identically zero the theory of space curves becomes complex, and in general would require special considerations involving the need for derivatives of high order of the vector $\mathbf{X}(s)$ defining it. More on this point will be said later.

5.   Binormal Vector

The relation

(3.8) $$\mathbf{v}_3 = \mathbf{v}_1 \times \mathbf{v}_2$$

defines a third unit vector $\mathbf{v}_3$ called the *binormal*, which is orthogonal to both $\mathbf{v}_1$ and $\mathbf{v}_2$.   It is clear that $\mathbf{v}_1$, $\mathbf{v}_2$, $\mathbf{v}_3$ form of necessity a right-handed system; this system forms what is called the moving trihedral of the curve.   It is perhaps worthwhile to point out that the vector $\mathbf{v}_3$ could not be defined in this way without assuming the curvature κ to be different from zero.

6.   Torsion τ of a Space Curve

The *torsion* τ of a space curve is defined with respect to the binormal $\mathbf{v}_3$ in a way analogous to that in which κ was defined in relation to the tangent $\mathbf{v}_1$.   That is, τ will be defined so that it furnishes a measure of the rate at which the osculating plane turns, while κ serves as a measure of the rate at which a normal plane to the curve turns.   The vector $\mathbf{v}_3$ is therefore transported to the center of a unit sphere and the length of the curve traced out by its end point as s varies along the given curve is taken as a measure of the total torsion of a given arc.   The total torsion τ_T is therefore defined by the formula

$$\tau_T = \int_{s_0}^{s_1} \sqrt{\dot{\mathbf{v}}_3 \cdot \dot{\mathbf{v}}_3} \, ds,$$

and the *numerical* value of the torsion at a point s is then defined quite reasonably by the relation

(3.9) $$|\tau| = \sqrt{\dot{\mathbf{v}}_3^2} = |\dot{\mathbf{v}}_3|.$$

The following considerations lead, finally, to the complete definition of the torsion $\tau(s)$.   From $\mathbf{v}_3 \cdot \mathbf{v}_3 = 1$ follows $\mathbf{v}_3 \cdot \dot{\mathbf{v}}_3 = 0$.   From $\mathbf{v}_1 \cdot \mathbf{v}_3 = 0$ follows $\dot{\mathbf{v}} \cdot \mathbf{v}_3 + \mathbf{v}_1 \cdot \dot{\mathbf{v}}_3 = 0$, or, since $\dot{\mathbf{v}}_1 = \kappa\mathbf{v}_2$, $\kappa\mathbf{v}_2 \cdot \mathbf{v}_3 + \mathbf{v}_1 \cdot \dot{\mathbf{v}}_3 = \mathbf{v}_1 \cdot \dot{\mathbf{v}}_3 = 0$ in view of $\mathbf{v}_2 \cdot \mathbf{v}_3 = 0$, and therefore $\mathbf{v}_1 \cdot \dot{\mathbf{v}}_3 = 0$.   Hence $\dot{\mathbf{v}}_3$ is perpendicular to both $\mathbf{v}_1$ and $\mathbf{v}_3$, and thus $\dot{\mathbf{v}}_3 = \pm \tau\mathbf{v}_2$ follows from (3.9).   The torsion τ is now defined by choosing the negative sign in the last relation:

(3.10) $$\dot{\mathbf{v}}_3 = -\tau\mathbf{v}_2.$$

The reason for this particular choice of sign will appear a little later.   It will also be seen that τ may be of either sign, and of course the sign will have geometrical significance.

An indication concerning the geometric significance of the torsion τ is given by the fact that a *curve is a plane curve* if $\tau \equiv 0$ holds. This is proved as follows. If $\tau \equiv 0$ holds, (3.10) shows that $\dot{\mathbf{v}}_3 = 0$ and consequently $\mathbf{v}_3 = \mathbf{a}$ ($\mathbf{a}$ a constant vector). It follows that $\mathbf{v}_1 \cdot \mathbf{a} = 0$ and from this $d/ds(\mathbf{X}(s) \cdot \mathbf{a}) = \dot{\mathbf{X}} \cdot \mathbf{a} = \mathbf{v}_1 \cdot \mathbf{a} = 0$, so that $\mathbf{X} \cdot \mathbf{a} = b$, with b a scalar constant. This holds along the entire curve. But this means that the curve $\mathbf{X}(s)$ lies in the plane $\mathbf{Y} \cdot \mathbf{a} = b = \text{const.}$; here $\mathbf{Y}$ is a vector to any point in the plane. (That this is indeed the equation of a plane is clear from the fact that it is linear in the components y_1, y_2, y_3 of $\mathbf{Y}$.)

7. The Frenet Equations for Space Curves

Since the vectors $\mathbf{v}_1, \mathbf{v}_2, \mathbf{v}_3$ are mutually orthogonal and thus linearly independent, any other vector can be expressed as a linear combination of these three vectors. In particular, this is true for the derivatives $\dot{\mathbf{v}}_1, \dot{\mathbf{v}}_2, \dot{\mathbf{v}}_3$. The relations $\dot{\mathbf{v}}_1 = \kappa \mathbf{v}_2$, and $\dot{\mathbf{v}}_3 = -\tau \mathbf{v}_2$ have already been obtained [cf. (3.7) and (3.10)]. The equation for $\dot{\mathbf{v}}_2$ is obtained from

$$(3.11) \qquad\qquad \dot{\mathbf{v}}_2 = \alpha \mathbf{v}_1 + \beta \mathbf{v}_2 + \gamma \mathbf{v}_3,$$

where α, β, γ are scalars to be determined. Since $\mathbf{v}_2 \cdot \mathbf{v}_2 = 1$, $\mathbf{v}_2 \cdot \dot{\mathbf{v}}_2 = 0$ and $\beta = 0$ follows by scalar multiplication of (3.11) by $\mathbf{v}_2$. From $\mathbf{v}_2 \cdot \mathbf{v}_1 = 0$ the relation $\mathbf{v}_2 \cdot \dot{\mathbf{v}}_1 + \mathbf{v}_1 \cdot \dot{\mathbf{v}}_2 = 0$ follows; and hence scalar multiplication of (3.11) by $\mathbf{v}_1$ leads to $\alpha = \mathbf{v}_1 \cdot \dot{\mathbf{v}}_2 = -\kappa$, since $\dot{\mathbf{v}}_1 = \kappa \mathbf{v}_2$. Finally from $\mathbf{v}_3 \cdot \mathbf{v}_2 = 0$ the relation $\dot{\mathbf{v}}_3 \cdot \mathbf{v}_2 + \mathbf{v}_3 \cdot \dot{\mathbf{v}}_2 = -\tau + \gamma = 0$ follows, from which $\tau = \gamma$. Hence $\dot{\mathbf{v}}_2 = -\kappa \mathbf{v}_1 + \tau \mathbf{v}_3$. Collecting the three formulas for $\dot{\mathbf{v}}_1, \dot{\mathbf{v}}_2, \dot{\mathbf{v}}_3$, we obtain the following equations, called the *Frenet equations*:

$$
\begin{aligned}
\dot{\mathbf{v}}_1 &= \phantom{-\kappa \mathbf{v}_1} \quad 0 \quad + \kappa \mathbf{v}_2 \quad 0, \\
(3.12) \qquad \dot{\mathbf{v}}_2 &= -\kappa \mathbf{v}_1 \quad \quad 0 \quad + \tau \mathbf{v}_3, \\
\dot{\mathbf{v}}_3 &= \phantom{-\kappa \mathbf{v}_1} \quad 0 \quad - \tau \mathbf{v}_2 \quad 0.
\end{aligned}
$$

These equations will be seen to form the basis for the entire theory of space curves (at least for those with $\kappa \neq 0$).

8. Rigid Body Motions and the Rotation Vector

The equations of Frenet are susceptible of a very interesting geometrical or, perhaps better, kinematical interpretation, which is due to Darboux. To explain this interpretation it is necessary to digress long enough to discuss the kinematics of rigid body motions in general, without reference for the moment to the geometry of space curves. In the present case the rigid body under consideration is thought of as containing a moving trihedral of orthogonal unit vectors $\mathbf{v}_1$, $\mathbf{v}_2$, $\mathbf{v}_3$ fixed in it. In fact, since all points of any rigid body can be located by giving their coordinates relative to Cartesian coordinate axes in the directions of the trihedral, it is clear that anything that can be learned about the motion of rigid bodies can be obtained from a discussion of the motion of a trihedral fixed in it. It is assumed that the trihedral moves so that the origin O' of the trihedral is displaced in time t in a way specified by the vector $\mathbf{X}(t)$ and that it also rotates in space as it moves. It is assumed

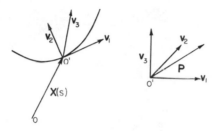

Fig. 3.2   Motion of a rigid body.

also that the origin O' moves at unit speed along the curve $\mathbf{X}(t)$, or, in other words, that the arc length s can be identified with the time t since $ds/dt = 1$. If $\mathbf{P}$ is a vector from O' to any point supposed fixed in the rigid body (cf. Fig. 3.2), then the position vector $\mathbf{Q}$ of that point with respect to a fixed system of coordinates with origin at O is given by

$$\mathbf{Q} = \mathbf{X} + \mathbf{P}.$$

The velocity of the point determined by $\mathbf{Q}$ is, as always in kinematics, defined as $\dot{\mathbf{Q}}$, hence is given by

$$\dot{\mathbf{Q}} = \dot{\mathbf{X}} + \dot{\mathbf{P}}.$$

This in turn can be interpreted to mean that the velocity of the point Q can be regarded as the sum of the velocity $\dot{\mathbf{X}}$ of the point O' and the velocity $\dot{\mathbf{P}}$ which the point Q would have if O' were fixed and the rigid body simply rotated about it. One could also say that the velocity of all points of any rigid body can be regarded as the vector sum of (a) a uniform translatory velocity in the direction of the tangent to the space curve traced out by O' and (b) a velocity that can be studied by assuming O' to be a point fixed in space about which the body rotates.

A study of the velocity $\dot{\mathbf{P}}$ of any point in the rigid body on the assumption that the point O' is fixed will now be made. To begin with, it is rather easy to see that any displacement whatever of the rigid body can in that case be achieved by a rotation about a uniquely determined line through the fixed point O'. In fact, the position of all points of the rigid body will clearly be fixed once any two points of it not on the same line through O' are given: that is a basic fact of Euclidean geometry. Consider two such points P_1 and P_2 at unit distance from O' (cf. Fig. 3.3) and thus in the unit sphere with center

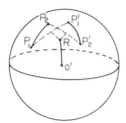

Fig. 3.3   *Motion of a rigid body about a fixed point.*

at O'. A displacement about O' will be characterized once the new positions of P_1 and P_2 are given, say at P_1', P_2'. The great circle arcs on the sphere which are orthogonal to the arcs $\overparen{P_1P_1'}$ and $\overparen{P_2P_2'}$ at their midpoints are drawn to their intersection at R. It is then easily seen that the displacement from one position of the rigid body to the other can be achieved by a rotation about the line $O'R$ through a certain angle.

If the motion depends on the time in a differentiable way it is intuitively clear that the velocity of all points of the body would result from a rotation with a certain angular speed about an axis that would have the limit position of $O'R$ when the positions of $\overparen{P_1P_2}$ and $\overparen{P_1'P_2'}$ are taken for time increments Δt that approach zero. In fact, it will be shown that a uniquely determined vector $\boldsymbol{\omega}$ exists such that the velocity $\dot{\mathbf{P}}$ of any point in the body is given by

$$(3.13) \qquad\qquad \dot{\mathbf{P}} = \boldsymbol{\omega} \times \mathbf{P},$$

with $\boldsymbol{\omega} = \boldsymbol{\omega}(t)$, but independent of the point $\mathbf{P}$. Before proving this statement it is relevant to observe that (3.13) does indeed give the velocity of a rigid body correctly when it rotates with an angular speed $|\boldsymbol{\omega}|$ about a *fixed line* containing $\boldsymbol{\omega}$: from Fig. 3.4 it is seen that $|\dot{\mathbf{P}}|$ should be given by $r|\boldsymbol{\omega}|$ in this case, and this in turn is equal to $|\mathbf{P}|\,|\boldsymbol{\omega}|\sin\theta$. The direction of $\dot{\mathbf{P}}$ is also correctly given by (3.13) if the direction of $\boldsymbol{\omega}$ is chosen in relation to the sense of the rotation by using the right-hand rule.

The existence of the vector $\boldsymbol{\omega}$ can be shown, using a method that seems to be due to Levi-Civita [L.1], as follows. Consider any three mutually orthogonal unit vectors $\mathbf{u}_1$, $\mathbf{u}_2$, $\mathbf{u}_3$ at O' that are considered to be fixed in the rigid body.   Evidently, the motion of the rigid body with O' as a fixed point will be known if the positions of these vectors are given as functions of the time since then the position of an arbitrary point in the body defined by the vector

Fig. 3.4   Rotation about a fixed axis.

$\mathbf{P} = (x_1, x_2, x_3)$ in terms of the coordinates x_1, x_2, x_3 of its end point with respect to the moving coordinate system is given by

$$P = x_1\,\mathbf{u}_1(t) + x_2\,\mathbf{u}_2(t) + x_3\,\mathbf{u}_3(t).$$

It is clear that the vector $d\mathbf{u}_1/dt$ can be written in the form

$$\frac{d\mathbf{u}_1}{dt} = \left(\mathbf{u}_1 \cdot \frac{d\mathbf{u}_1}{dt}\right)\mathbf{u}_1 + \left(\mathbf{u}_2 \cdot \frac{d\mathbf{u}_1}{dt}\right)\mathbf{u}_2 + \left(\mathbf{u}_3 \cdot \frac{d\mathbf{u}_1}{dt}\right)\mathbf{u}_3.$$

Since $\mathbf{u}_1 \cdot \mathbf{u}_1 = 1$, it follows that $\mathbf{u}_1 \cdot d\mathbf{u}_1/dt = 0$, and, since $\mathbf{u}_3 \cdot \mathbf{u}_1 = 0$, it is clear that $\mathbf{u}_3 \cdot d\mathbf{u}_1/dt$ can be replaced by its equal $-\mathbf{u}_1 \cdot d\mathbf{u}_3/dt$.   The last equation therefore can be put in the form

$$\frac{d\mathbf{u}_1}{dt} = \left(\mathbf{u}_2 \cdot \frac{d\mathbf{u}_1}{dt}\right)\mathbf{u}_2 - \left(\mathbf{u}_1 \cdot \frac{d\mathbf{u}_3}{dt}\right)\mathbf{u}_3.$$

Since $\mathbf{u}_3 = \mathbf{u}_1 \times \mathbf{u}_2$, the equation can be put into still another form, as follows:

$$\frac{d\mathbf{u}_1}{dt} = \left[\left(\frac{d\mathbf{u}_3}{dt} \cdot \mathbf{u}_1\right)\mathbf{u}_2 + \left(\frac{d\mathbf{u}_1}{dt} \cdot \mathbf{u}_2\right)\mathbf{u}_3\right] \times \mathbf{u}_1,$$

to which can be added the term $(d\mathbf{u}_2/dt \cdot \mathbf{u}_3)\mathbf{u}_1 \times \mathbf{u}_1$, since this last term is zero.   It is now seen that

$$\frac{d\mathbf{u}_1}{dt} = \boldsymbol{\omega} \times \mathbf{u}_1,$$

once the vector $\boldsymbol{\omega}$ has been defined by the relation

$$\boldsymbol{\omega} = \left(\frac{d\mathbf{u}_2}{dt} \cdot \mathbf{u}_3\right)\mathbf{u}_1 + \left(\frac{d\mathbf{u}_3}{dt} \cdot \mathbf{u}_1\right)\mathbf{u}_2 + \left(\frac{d\mathbf{u}_1}{dt} \cdot \mathbf{u}_2\right)\mathbf{u}_3.$$

Because of the symmetrical character of the right-hand side of this equation (it is unaltered by cyclic interchange of the subscripts) it is at once seen that

$$\frac{d\mathbf{u}_2}{dt} = \boldsymbol{\omega} \times \mathbf{u}_2, \qquad \frac{d\mathbf{u}_3}{dt} = \boldsymbol{\omega} \times \mathbf{u}_3$$

when $\boldsymbol{\omega}$ has the above definition. Thus the velocity $d\mathbf{P}/dt$ of any point with the position vector $\mathbf{P} = (x_1, x_2, x_3)$ relative to the moving axes is given by

$$\frac{d\mathbf{P}}{dt} = x_1 \frac{d\mathbf{u}_1}{dt} + x_2 \frac{d\mathbf{u}_2}{dt} + x_3 \frac{d\mathbf{u}_3}{dt} = \boldsymbol{\omega} \times (x_1 \mathbf{u}_1 + x_2 \mathbf{u}_2 + x_3 \mathbf{u}_3)$$

$$= \boldsymbol{\omega} \times \mathbf{P},$$

and the formula (3.13) is therefore shown to hold. It is left as an exercise to show that the vector $\boldsymbol{\omega}$ is uniquely determined. It ought perhaps to be mentioned that finite rotations cannot be defined as vectors since successive finite rotations cannot in general be replaced by a vector sum. It is also not true that *accelerations* of the points of a rigid body are the result of an angular acceleration about a certain line: only the velocities are obtainable in this way.

9. The Darboux Vector

The results of the previous section are now put in relation to the Frenet equations (3.12). To do so the angular velocity vector $\boldsymbol{\omega}$ of the rigid body that moves along a curve while being attached to the trihedral formed by the vectors $\mathbf{v}_1, \mathbf{v}_2, \mathbf{v}_3$, is expressed in terms of its components with respect to those vectors:

(3.14) $\boldsymbol{\omega} = \omega_1 \mathbf{v}_1 + \omega_2 \mathbf{v}_2 + \omega_3 \mathbf{v}_3.$

From (3.13) the following formulas for the velocities $\dot{\mathbf{v}}_1, \dot{\mathbf{v}}_2, \dot{\mathbf{v}}_3$ are obtained:

$$\dot{\mathbf{v}}_1 = \boldsymbol{\omega} \times \mathbf{v}_1 = \qquad 0 \qquad \omega_3 \mathbf{v}_2 - \omega_2 \mathbf{v}_3,$$

(3.15) $$\dot{\mathbf{v}}_2 = \boldsymbol{\omega} \times \mathbf{v}_2 = -\omega_3 \mathbf{v}_1 \qquad 0 \ + \omega_1 \mathbf{v}_3,$$

$$\dot{\mathbf{v}}_3 = \boldsymbol{\omega} \times \mathbf{v}_3 = \ \ \omega_2 \mathbf{v}_1 - \omega_1 \mathbf{v}_2 \qquad 0.$$

Upon comparison with (3.12) it is seen that the curvature κ and the torsion τ can be interpreted as the components of the angular velocity vector of the trihedral as it moves along the curve; in fact, upon setting

(3.16) $\boldsymbol{\omega} = \tau \mathbf{v}_1 + \kappa \mathbf{v}_3$

the formulas (3.15) are the same as (3.12), provided that the arc length s is interpreted as the time. The vector $\boldsymbol{\omega}$ is sometimes called the Darboux vector in honor of the inventor of this strongly intuitive interpretation of the meaning of the Frenet equations: evidently the formula (3.16) says that the torsion measures the rate at which the osculating plane turns about the tangent to the curve, while the curvature measures the rate at which the normal plane turns about the binormal vector. It is observed also that the convention for the sign of the torsion τ, which was fixed by equation (3.10), leads to a rotation vector $\boldsymbol{\omega}$ that has a *positive* component along the tangent vector if τ is positive.

10. Formulas for κ and τ

The curvature κ has already been given by formula (3.5) in terms of the vector $\mathbf{X}(s)$ that determines the space curve. A formula for the torsion τ will now be derived. From the equation $\dot{\mathbf{v}}_3 = -\tau\mathbf{v}_2$ follows the relation $\tau = -\mathbf{v}_2 \cdot \dot{\mathbf{v}}_3$, since $\mathbf{v}_2$ is a unit vector. From $\mathbf{v}_1 = \dot{\mathbf{X}}$, $\mathbf{v}_2 = \kappa^{-1}\dot{\mathbf{v}}_1$ [cf. (3.7) or the Frenet equations (3.12)], and $\mathbf{v}_3 = \mathbf{v}_1 \times \mathbf{v}_2$ it is found that $\mathbf{v}_3 = \kappa^{-1}(\dot{\mathbf{X}} \times \ddot{\mathbf{X}})$. By differentiation of this formula the following formula results:

$$\dot{\mathbf{v}}_3 = -\dot{\kappa}\kappa^{-2}(\dot{\mathbf{X}} \times \ddot{\mathbf{X}}) + \kappa^{-1}(\dot{\mathbf{X}} \times \dddot{\mathbf{X}}),$$

from which, finally, the formula for the torsion is obtained:

$$(3.17) \qquad\qquad \tau = \kappa^{-2}(\dot{\mathbf{X}} \cdot \ddot{\mathbf{X}} \times \dddot{\mathbf{X}}),$$

upon using $\tau = -\mathbf{v}_2 \cdot \dot{\mathbf{v}}_3 = -\kappa^{-1}\dddot{\mathbf{X}}$. It is to be noted that the assumption $\kappa \neq 0$, which is necessary in deriving the Frenet equations, was used at this point too. In order that τ should be continuous it is clearly necessary to suppose that the curve $\mathbf{X}(t)$ has a continuous third derivative. If a parameter other than the arc length is used to define a space curve the formulas for κ and τ become

$$(3.18) \qquad \begin{aligned} \kappa^2 &= |\mathbf{X}' \times \mathbf{X}''|^2 \cdot |\mathbf{X}'|^{-6} \\ \tau &= (\mathbf{X}' \cdot \mathbf{X}'' \times \mathbf{X}''') \cdot |\mathbf{X}' \times \mathbf{X}''|^{-2} \end{aligned}$$

These formulas are obtained easily by using the fact that $\dot{\mathbf{X}} = \mathbf{X}' \cdot |\mathbf{X}'|^{-1}$.

It is to be noted that the torsion changes its sign if the orientation of the curve is reversed, and also if the orientation of the coordinate system is reversed.

11. The sign of τ

A kinematical interpretation of the meaning of the sign of the torsion has been given. It is of interest to derive still another geometrical interpretation

for this sign.   Taylor's theorem with a remainder of third order is applied to
$\mathbf{X}(s)$:

$$\mathbf{X}(s) = \mathbf{X}(s_0) + (s - s_0)\,\dot{\mathbf{X}}(s_0) + \frac{(s - s_0)^2}{2!}\,\ddot{\mathbf{X}}(s_0) + \frac{(s - s_0)^3}{3!}\,\dddot{\mathbf{X}}*$$

$$= \mathbf{Y}(s) + \frac{(s - s_0)^3}{3!}\,\dddot{\mathbf{X}}*.$$

Since, as usual, $\kappa > 0$ is assumed, $\ddot{\mathbf{X}}(s_0) \neq 0$, and the end point of the vector
$\mathbf{Y}(s)$ lies in the osculating plane at s_0.   The vector $\mathbf{D}(s)$ given by

$$\mathbf{D}(s) = \mathbf{X}(s) - \mathbf{Y}(s) = \frac{(s - s_0)^3}{6}\,\dddot{\mathbf{X}}*$$

is therefore a vector in the direction from the osculating plane to the curve
points.   If $\mathbf{X}(s)$ has a continuous third derivative it is known that $\dddot{\mathbf{X}}* \to \dddot{\mathbf{X}}(s_0)$
when $s \to s_0$, and since $\tau \neq 0$ is tacitly assumed it follows from (3.17) that
$\dddot{\mathbf{X}}(s_0)$ is not the zero vector.   Thus if s is near enough to s_0 and $s - s_0 > 0$
holds, $\dddot{\mathbf{X}}(s_0)$ points into that one of the two half-spaces determined by the
osculating plane in which the curve points near to the point $\mathbf{X}(s_0)$ lie; if
$s - s_0 < 0$ holds into the opposite half space.   If τ is positive formula (3.17)
shows that $\dot{\mathbf{X}}, \ddot{\mathbf{X}}, \dddot{\mathbf{X}}$ form a right-handed system, from which it follows that $\dddot{\mathbf{X}}$
points into the same half-space as the binormal $\mathbf{v}_3$.   The following conclusion
can therefore be drawn: if $\tau > 0$ holds the curve cuts through the osculating
plane in the direction of the binormal; if $\tau < 0$ holds, in the opposite direction (in both cases with increase of s).   If $\tau = 0$ at s_0, no conclusion can be
drawn without investigating derivatives of $\mathbf{X}$ of order higher than third order.

12.   Canonical Representation of a Curve

It is assumed once more that $\kappa > 0$ holds and that the arc length s is the
parameter for a curve $\mathbf{X}(s)$.   It is always possible to choose coordinate axes
and the point from which s is measured in such a way that $\mathbf{X}(0) = (0, 0, 0)$,
$\dot{\mathbf{X}}(0) = (1, 0, 0), \ddot{\mathbf{X}}(0) = (0, \kappa, 0), \dddot{\mathbf{X}}(0) = (\dddot{x}_1(0), \dddot{x}_2(0), \dddot{x}_3(0))$.   From $\tau\kappa^2 =$
$(\dot{\mathbf{X}} \cdot \ddot{\mathbf{X}} \times \dddot{\mathbf{X}})$ [cf. (3.17)] it follows that $\tau\kappa^2 = \kappa\dddot{x}_3(0)$, or $\dddot{x}_3(0) = \kappa\tau$.   The
development of $\mathbf{X}(s)$ in the neighborhood of $s = 0$ is, quite generally, given by

$$\mathbf{X}(s) = \mathbf{X}(0) + s\dot{\mathbf{X}}(0) + \frac{s^2}{2!}\,\ddot{\mathbf{X}}(0) + \frac{s^3}{3!}\,\dddot{\mathbf{X}}(0) + \cdots.$$

In terms of the components of $\mathbf{X}(s)$, as chosen above, this becomes

(3.19)
$$x_1(s) = x + \cdots,$$

$$x_2(s) = \frac{\kappa}{2}\,s^2 + \cdots,$$

$$x_3(s) = \frac{\kappa\tau}{6}\,s^3 + \cdots.$$

From this so-called canonical representatioon of the curve it is easy to determine the nature of the projections of the curve on the x_i,x_j-planes, as indicated in Fig. 3.5, where $\tau > 0$ is taken.   Once more it is seen that a space

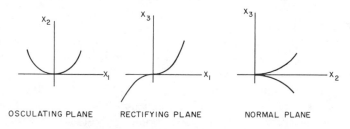

OSCULATING PLANE   RECTIFYING PLANE   NORMAL PLANE

Fig. 3.5   Projections of a curve on various planes.

curve cuts through the osculating plane in the direction of the binormal if τ is positive.   It is of interest to point out that the present discussion would collapse into nothing if κ were zero at $s = 0$, thus illustrating anew the utility of the assumption that $\kappa > 0$ holds.

It is also of interest at this point to give a new characterization of the osculating plane: it is the *only* plane containing the tangent which is cut by the curve.   This is readily proved as follows.   Assume that the x_1,x_2-plane contains the point s_0 and the tangent vector at s_0, and that it is cut by the curve at s_0.   For the x_3 component of the curve in this case the following conditions hold: $x_3(s_0) = \dot{x}_3(s_0) = 0$; hence the development of $x_3(s)$ near $s = s_0$ has the form

$$x_3(s) = \tfrac{1}{2}(s - s_0)^2 \, \ddot{x}_3(s_0) + \cdots.$$

If $\ddot{x}_3(s_0)$ were not zero, the x_1,x_2-plane could not be cut by the curve near $s = s_0$, since $(s - s_0)^2$ is always positive.   Hence $\ddot{x}_3(s_0) = 0$ must hold if the curve cuts the plane, $\ddot{\mathbf{X}}(s_0)$ therefore lies in the x_1,x_2-plane, and this plane is therefore the osculating plane.

The plane determined by the tangent and binormal, like the other two planes determined by the trihedral, has a name; it is called the rectifying plane (see Fig. 3.5).   Only later on can the reason for this name be explained, since an explanation of it requires a discussion concerning developable surfaces and geodetic lines on such surfaces.

13.   Existence and Uniqueness of a Space Curve for Given $\kappa(s)$, $\tau(s)$

The Frenet equations are a system of three linear first-order vector differential equations for the vectors $\mathbf{v}_1$, $\mathbf{v}_2$, $\mathbf{v}_3$ or of nine scalar equations for the nine coordinates of the end points of these vectors. The latter equations may be written in the form

$$\frac{dy_i}{ds} = \sum_{j=1}^{9} \mu_{ij}(s)\, y_j(s), \qquad i = 1, 2, \ldots, 9,$$

where the functions $\mu_{ij}(s)$ are, within sign, either the curvature $\kappa(s)$, or the torsion $\tau(s)$, or zero. If the functions $\kappa(s) > 0$ and $\tau(s)$ are defined over an interval $\alpha \le s \le \beta$, and are such that $\tau(s)$ is continuous and $\kappa(s)$ has a continuous first derivative, then these equations, on the basis of the fundamental existence and uniqueness theorem for systems of ordinary differential equations, possess exactly one solution with continuous first derivatives for which the initial values of the functions y_i are given. It will be shown further that one and only one regular space curve $\mathbf{X}(s)$ (within rigid motions) exists such that $\mathbf{X}(s)$ has a continuous third derivative, with the given functions $\kappa(s)$ and $\tau(s)$ as curvature and torsion in terms of s as arc length.

It is necessary to prove first that if the vectors $\mathbf{v}_i$ form a right-handed orthogonal system of unit vectors initially they will always form such a system. For this purpose the Frenet equations (3.12) are written in the form

$$\dot{\mathbf{v}}_i = \sum_{j=1}^{3} a_{ij}\mathbf{v}_j, \qquad i = 1, 2, 3.$$

The six scalar products $\mathbf{v}_i \cdot \mathbf{v}_k$ are considered, and differential equations are derived for them with the aid of the Frenet equations, as follows:

$$\frac{d}{ds}\left(\mathbf{v}_i \cdot \mathbf{v}_k\right) = \dot{\mathbf{v}}_i \cdot \mathbf{v}_k + \mathbf{v}_i \cdot \dot{\mathbf{v}}_k = \sum_j a_{ij}\mathbf{v}_k \cdot \mathbf{v}_j + \sum_j a_{kj}\mathbf{v}_i \cdot \mathbf{v}_j.$$

This is a set of six homogeneous linear differential equations for the six scalar products with coefficients that are continuous functions of s. Furthermore, these equations can easily be seen to have the solution

$$\mathbf{v}_i \cdot \mathbf{v}_k = \delta_{ik} = \begin{cases} 1, & i = k, \\ 0, & i \ne k, \end{cases}$$

because of the fact that $a_{ik} + a_{ki} = 0$, since the coefficients on the right-hand sides of the Frenet equations have a skew-symmetric matrix. The uniqueness theorem for the solution of this system of differential equations then ensures that these orthogonality relations for the vectors $\mathbf{v}_i$ hold for all s if they hold for the initial value of s. In addition, if the system $\mathbf{v}_i$ is prescribed to have the orientation of the coordinate axes initially, it will preserve the same orientation throughout since $\mathbf{v}_1 \cdot \mathbf{v}_2 \times \mathbf{v}_3 = \pm 1$ in any case, and this determinant is a continuous function of s.

Once the integral of the Frenet equations has been obtained for given

initial vectors $\mathbf{v}_1(0)$, $\mathbf{v}_2(0)$, and $\mathbf{v}_3(0)$ it is possible to set $\dot{\mathbf{X}}(s) = \mathbf{v}_1(s)$ and determine a vector $\mathbf{X}(s)$ uniquely by integration once an initial value $\mathbf{X}(0)$ has been
prescribed. In other words, $\mathbf{X}(s)$ is uniquely determined through the Frenet
equations, once $\mathbf{X}(0)$ and the initial values $\mathbf{v}_i(0)$ have been chosen, by the
equation

$$\mathbf{X}(s) = \int_0^s \mathbf{v}_1(\sigma)\, d\tau + \mathbf{X}(0).$$

It must be verified, finally, that $\mathbf{X}(s)$ defines a regular curve with the given
functions $\kappa(s)$ and $\tau(s)$ as its curvature and torsion as functions of the arc
length. First of all since $\dot{\mathbf{X}} = \mathbf{v}_1(s)$, and since $\mathbf{v}_1(s)$ has been shown to be a
unit vector, it follows that s is indeed the arc length of the regular curve
given by $\mathbf{X}(s)$. That $\mathbf{X}(s)$ has a continuous third derivative follows from
$\ddot{\mathbf{X}}(s) = \dot{\mathbf{v}}_1(s)$, since the vector $\mathbf{v}_1(s)$ has a continuous derivative, and this equation has still another continuous derivative, since $\dot{\mathbf{v}}_1(s) = \kappa(s)\,\mathbf{v}_1(s)$ and $\kappa(s)$
was assumed to have a continuous derivative. By going back to the develop-
ments that led to the Frenet equations, it is readily seen that $\kappa(s)$ and $\tau(s)$ are
indeed the curvature and torsion of the curve $\mathbf{X}(s)$.

It has therefore been proved that if the functions $\tau(s)$ and $\kappa(s) > 0$ are
given there exists one and only one regular curve for which s is the arc length
and κ and τ are the curvature and torsion, once an initial point and an initial
orientation of the trihedral $\mathbf{v}_1$, $\mathbf{v}_2$, $\mathbf{v}_3$ have been prescribed. With this the
theory of space curves has been brought to a close in a certain sense, since
three geometric invariants, i.e., the arc length s, the curvature κ, and the torsion τ have been found, from which a space curve can be constructed uniquely
within rigid motions. If $\kappa(s) = f(s)$, $\tau(s) = \phi(s)$ are given, these equations
are sometimes said to constitute the natural or intrinsic equations of the
curve, since they characterize it and contain only invariants.

14. What about $\kappa = 0$?

These important conclusions from the Frenet equations were based on
the assumption that the curvature does not vanish. On the other hand the
Frenet equations, simply as linear differential equations, make sense whether
κ vanishes or not: the existence of uniquely determined solutions of them is
assured simply if κ and τ are continuous functions. However, that does not
mean that a uniquely determined regular curve would of necessity result if
$\kappa = 0$ is permitted.

In the case of plane curves it was seen that it would not have been at all
restrictive to use the Frenet equations for establishing the existence and
"uniqueness" of the curve once its curvature is given, since it was not neces-

sary to suppose that the curvature always differed from zero. For space curves the situation is different. For example, consider the two curves shown in Fig. 3.6, which lie in the x_1,x_2-plane but are regarded as space curves. They consist of straight-line segments to which are attached curves

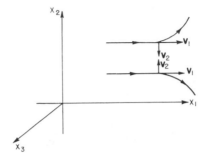

Fig. 3.6   The case $\kappa = 0$ for space curves.

with $\kappa \neq 0$ that are mirror images of each other. Clearly such regular curves exist with as many continuous derivatives as might be desired. As plane curves they are essentially different, since their curvatures differ in sign on the curved parts. As space curves, however, the Frenet equations in three dimensions could be satisfied by choosing $\tau = 0$, $\kappa(s)$ to be always positive or zero, and taking for $\mathbf{X}(s)$ either of the two curves shown in the figure—since $\mathbf{v}_2$ could be taken along the straight segment in either of the positions shown. Thus the restriction $\kappa \neq 0$ is natural for space curves having only three continuous derivatives if they are to be uniquely determined by their curvature and torsion.

15.   Another Way to Define Space Curves

It is perhaps worthwhile to enlarge a little more upon the point raised in the previous section by comparing and contrasting regular curves, as they are treated in this chapter, with the kind of space curves that occur in kinematics and mechanics. It has already been seen that the Frenet equations (3.12) for the vectors $\mathbf{v}_i$ composing the moving trihedral of the curve are a special case of the differential equations (3.15) for the most general motion of such a trihedral in case the angular velocity vector $\boldsymbol{\omega}(t)$ is prescribed. In fact, the proof given above for the existence and "uniqueness" of the trihedral when κ and τ are prescribed holds word for word when the components ω_1, ω_2, ω_3 of $\boldsymbol{\omega}$ are prescribed as arbitrary continuous functions of t. However, to continue by setting $\mathbf{X}'(t) = \mathbf{v}_1(t)$ and integrating once more with the object of obtaining a regular curve with $\kappa > 0$ and a uniquely defined torsion τ would not succeed, as the example in the preceding section shows, although the uniquely defined locus determined by $\mathbf{X}(t)$ might be in many respects a quite reasonable curve.

The regular curves with $\kappa > 0$ are without question quite reasonable and interesting curves to study in differential geometry, but it should be pointed out that such a limitation is not reasonable in some other fields. In mechanics, for example, the natural way in which space curves arise is in connection with the motion in the time t of a particle of mass m under the action of a given force $\mathbf{F}$; the basic law that determines the motion of the particle is Newton's second law:

$$(3.20) \qquad m\mathbf{X}''(t) = \mathbf{F}(\mathbf{X}, \mathbf{X}', t).$$

Here $\mathbf{X}' = d\mathbf{X}/dt$ and $\mathbf{X}'' = d^2\mathbf{X}/dt^2$ are by definition the velocity and acceleration of the particle. The force $\mathbf{F}$ is in general a function—usually a nonlinear function—of the position vector $\mathbf{X}$ and of the velocity $\mathbf{X}'$, as well as of the time t.[1] The general existence and uniqueness theorem of Appendix B for ordinary differential equations then yields a uniquely determined vector $\mathbf{X}(t)$ once the initial position $\mathbf{X}(0)$ and the initial velocity $\mathbf{X}'(0)$ at the time $t = 0$ are given. The velocity $\mathbf{v}(t)$ is by definition $\mathbf{X}'(t)$, and hence in case $\mathbf{X}'(t) \neq 0$ this vector defines what is called in this chapter the tangent vector to the curve; but in mechanics there is no reason at all to exclude the case of zero velocity, and thus to permit $\mathbf{X}'(t) = 0$ as a reasonable occurrence. The acceleration vector $\mathbf{a} = \mathbf{X}''(t)$ can be written as follows, in case it is possible to introduce the arc length s of the locus $\mathbf{X}(t)$ as the parameter:

$$\mathbf{a} = \mathbf{X}'' = \ddot{\mathbf{X}}\left(\frac{ds}{dt}\right)^2 + \dot{\mathbf{X}}\frac{d^2s}{dt^2},$$

with the dots once more referring to differentiation with respect to s. This formula results immediately from $\mathbf{v} = \dot{\mathbf{X}}(ds/dt)$. Since $|\mathbf{v}| = |ds/dt|$ because $\dot{\mathbf{X}}$ is a unit vector this quantity is quite naturally called the *speed* of the particle in its path. On the assumption that the curve $\mathbf{X}(t)$ is a *regular* curve with $\kappa > 0$ the above formula for the acceleration $\mathbf{a}$, from the general theory of space curves developed above, can be put in the form

$$\mathbf{a} = \frac{|\mathbf{v}|^2}{\rho}\mathbf{v}_2 + \frac{d^2s}{dt^2}\mathbf{v}_1$$

[1] Equation (3.20) implies that the force $\mathbf{F}$ is attached to the same point as the position vector $\mathbf{X}$, since that is the case for $\mathbf{X}''$. However, the force $\mathbf{F}$ evidently is applied physically to a particle wherever it happens to be, while the position vector is, in principle, attached to the origin of a fixed coordinate system. To be consistent with the attitude of this book, the acceleration vector $\ddot{\mathbf{X}}$ should be translated to the position of the particle. Such an interpretation of (3.20) constitutes just one of the many complications involved in a logically correct formulation of Newtonian mechanics.

in terms of the tangent vector $\mathbf{v}_1$, the principal normal $\mathbf{v}_2$, and the radius of curvature $\rho = 1/\kappa$. Thus the acceleration vector lies in the osculating plane and has as components the "centrifugal" acceleration $|\mathbf{v}|^2/\rho$ in the direction of the principal normal and the tangential acceleration d^2s/dt^2 along the tangent to the curve, but with a sense fixed of course by the sign of d^2s/dt^2. However, this situation in which the trajectory is a regular curve is the exception rather than the rule in mechanics. For example, consider the case in which the force $\mathbf{F}$ is always zero, and the initial velocity $\mathbf{X}'(0)$ is also zero. The uniquely determined solution of (3.20) is thus $\mathbf{X}(t) = \mathbf{X}(0)$, and the locus $\mathbf{X}(t)$ is a point, not a curve: the particle is at rest in an equilibrium position. Or, to take another example, it is entirely possible that a given curve could be covered any number of times in the course of a motion of a particle: for that, it would suffice to reverse the direction of the force and dispose of its magnitude in an appropriate way. This last is also an example of the fact that a given set of points in space may well carry more than one "curve"— depending on special circumstances. However, experience has shown that restrictions in differential geometry of the type made for regular curves are reasonable since they delimit areas of geometry that have great variety and interest.

16. Some Special Curves

In some of the treatises on differential geometry a great deal of space is taken for discussion of special curves and classes of curves (cf., for example, Eisenhart [E.6]). Here only a few special cases are considered, starting with the case of the circular helix. This curve is given by (cf. Fig. 3.7)

$$(3.21) \qquad \mathbf{X}(s) = (r \cos bs, \, r \sin bs, \, as),$$

with r, a, b constants $\neq 0$. It is assumed for $s = 0$ that the angle $\phi = bs = 0$. The projection of the curve on the x_1,x_2-plane is readily seen to be a circle of radius r; the x_3-coordinate increases linearly with s. The vector $\dot{\mathbf{X}}$ is given by

$$\dot{\mathbf{X}}(s) = (-br \sin bs, \, br \cos bs, \, a),$$

so that $a^2 + b^2r^2 = \dot{\mathbf{X}} \cdot \dot{\mathbf{X}}$ must equal 1 in order that s should represent the arc length of the curve. Further differentiations yield

$$\ddot{\mathbf{X}}(s) = (-b^2r \cos bs, \, -b^2r \sin bs, \, 0)$$

$$\dddot{\mathbf{X}}(s) = (b^3r \sin bs, \, -b^3r \cos bs, \, 0).$$

From these it is seen that $\kappa = |\ddot{\mathbf{X}}(s)| = b^2r$ and $\tau = \kappa^{-2}(\dot{\mathbf{X}} \cdot \ddot{\mathbf{X}} \times \dddot{\mathbf{X}}) = ab$, i.e., both are constant. In view of the fundamental uniqueness theorem derived above in connection with the integration of the Frenet equations, prescribing that κ and τ are constant results always in a circular helix. If $a > 0$ is chosen, the sign of τ is determined by the sign of b. If $\tau > 0$, $\phi = bs$ increases with s and the right-handed screw shown in Fig. 3.7 results, thus illustrating once more the geometric interpretation for the sign of τ.

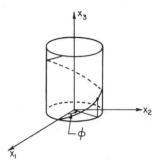

Fig. 3.7   A circular helix.

Since the circular helix is determined by prescribing constant values for κ and τ, it would have been possible to obtain the representation (3.21) by integrating the Frenet equations explicitly in that special case. Instead such a procedure is carried out here in a more general case, i.e., that in which the ratio of τ and κ is a constant c so that $\tau(s) = c\kappa(s)$, with s assumed to be the arc length of the curve. Such curves might be called curves of *constant grade*, for the following reasons. To begin with, the Darboux vector given by (3.16), which fixes the rate of rotation of the trihedral of the curve, evidently has a fixed direction relative to the trihedral for all values of s in the present case. This means, in terms of the kinematics of rigid-body rotations, that the angular velocity vector of a moving trihedral rigidly attached to such a body is fixed in direction in the body; but in that case it can easily be proved that *this vector is also fixed in space*. To this end consider formula (3.13), which furnishes the rate of change $\dot{\mathbf{P}}$ relative to fixed axes of any point $\mathbf{P}$ fixed in a rotating rigid body. Consider a point $\mathbf{Q}$ on the line containing $\boldsymbol{\omega}$ at unit distance, say, from the origin; this point is also fixed in the body. Formula (3.13) applied to point $\mathbf{Q}$ then states that $\dot{\mathbf{Q}}$ is the zero vector since $\mathbf{Q}$ and $\boldsymbol{\omega}$ are in the same line and hence $\boldsymbol{\omega} \times \mathbf{Q} = 0$. In other words, $\mathbf{Q}$ is fixed in space and hence also the direction of $\boldsymbol{\omega}$ is fixed in space. It is thus proved that if the angular velocity vector of a rigid body is fixed in the body it is also fixed in space. (The converse is obviously also correct.) For the space curve with $\tau = c\kappa$ it follows that its tangent vector, which by (3.16) makes a fixed angle with $\boldsymbol{\omega}$, therefore also makes a constant angle with a fixed line in space. If the fixed line were considered to be vertical, this means that the curve goes either uphill or downhill at a constant angle of inclination, and hence the curve can be called quite reasonably a curve of constant grade.

In addition, it is clear that the curve lies on a cylinder whose generators are vertical—it is, in fact, the cylindrical surface generated by moving a vertical line along the curve. In case $\kappa = $ const., $\tau = $ const. holds also, and hence the curve would be a circular helix, as was seen above on the basis of an explicit representation.

An explicit representation can be found for all curves of constant grade. For this purpose the Frenet equations, which in the present case have the form

$$\dot{\mathbf{v}}_1 = \kappa \mathbf{v}_2,$$

$$\dot{\mathbf{v}}_2 = -\kappa \mathbf{v}_1 + c\kappa \mathbf{v}_3,$$

$$\dot{\mathbf{v}}_3 = -c\kappa \mathbf{v}_2,$$

will be integrated explicitly. Since $\kappa(s) \neq 0$ a new regular parameter t can be introduced by setting $t = \int_{s_0}^{s} \kappa(\sigma)\, d\sigma$, since $dt/ds = \kappa \neq 0$. In terms of t as parameter the differential equations are readily seen to take the form

$$\mathbf{v}_1' = \mathbf{v}_2,$$

$$\mathbf{v}_2' = -\mathbf{v}_1 + c\mathbf{v}_3,$$

$$\mathbf{v}_3' = -c\mathbf{v}_2.$$

These equations can be integrated explicitly with ease since they are linear, homogeneous, and have constant coefficients. If the first equation is differentiated twice, the second once, $\mathbf{v}_2$ and $\mathbf{v}_3$ can be eliminated from the three equations to obtain the following equation for $\mathbf{v}_1'$ alone:

$$\mathbf{v}_1''' + \lambda^2 \mathbf{v}_1' = 0, \qquad \lambda^2 = 1 + c^2.$$

For $\mathbf{v}_1'$, therefore, the following general solution is obtained:

(3.22) $$\mathbf{v}_1' = \mathbf{a}\cos\lambda t + \mathbf{b}\sin\lambda t,$$

with $\mathbf{a}$ and $\mathbf{b}$ constant, but arbitrary, vectors. Since

$$\mathbf{v}_1' = \kappa^{-1}(s)\dot{\mathbf{v}}_1 = \kappa^{-2}(s)\,\ddot{\mathbf{X}}(s),$$

it is possible to obtain $\mathbf{X}(s)$ by an integration making use of (3.22). The details are left to be carried out as one of the problems at the end of this chapter.

PROBLEMS

1. For the circular helix of Fig. 3.7 prove the following:

(a) The principal normal is parallel to the x_1,x_2-plane and is directed through the x_3-axis,

(b) The tangent and binormal make constant angles with the generators of the cylinder which contains the curve.

2. Prove: If the points of two curves are in one-to-one correspondence so that the tangents at corresponding points are parallel, the principal normals are also parallel at corresponding points.

3. If all the tangents to a curve go through a point, the curve is a straight line.

4. If all the osculating planes of a curve go through a point, the curve is a plane curve.

5. If $\Delta\theta$ is the angle between the directions of the tangent vectors $\dot{X}(s)$ and $\dot{X}(s + \Delta s)$, show that $\kappa = \lim \Delta\theta/\Delta s$.

6. If $\Delta\theta$ is the angle between the osculating planes of a space curve at s and $s + \Delta s$, show that $|\tau| = \lim \Delta\theta/\Delta s$.

7. Starting from (3.22) obtain an explicit representation for all curves $X(s)$ of constant grade. Find the relations which must hold between the integration constants in order that $\kappa(s)$ and $\tau(s)$ should represent the curvature and torsion in terms of the arc length of $X(s)$.

8. The center of curvature of a space curve is defined as the point at distance $\rho = 1/\kappa$ from the curve along the normal v_2. Show that if $\kappa = $ const. the locus of centers of curvature is orthogonal to the osculating plane of the curve at the corresponding point and is also a curve of constant curvature.

9. Prove that a tangent to the locus of centers of curvature of a space curve is orthogonal to the corresponding tangent to the space curve, but that it does not in general fall along the principal normal of the curve.

10. Prove directly from the Frenet equations that the tangents to a space curve along which $\kappa(s) = c\tau(s)$, with c a constant, make a constant angle with a fixed line in space.

11. Prove that the angular velocity vector $\boldsymbol{\omega}$ in (3.13) is uniquely determined.

12. Prove that a point of a rigid body can always be found such that the velocity of the point falls along the same line as the angular velocity vector, i.e., a "screw motion" of a rigid body furnishes the most general momentary velocity field possible for its points.

CHAPTER IV

The Basic Elements of Surface Theory

1. Regular Surfaces in Euclidean Space

Familiar examples of surfaces are the plane, the surface of a sphere, the cylindrical surface. In this chapter, however, such surfaces are considered only incidentally, since they are examples of surfaces *in the large* while this chapter is devoted to surfaces defined only in some neighborhood of one of their points, or *in the small*, as it is said. As will be seen in a later chapter, the concept of a surface in the large is introduced by building it up as a set of overlapping patches, each of which is a surface in the small. But quite aside from this use of the concept of a surface defined locally it turns out, as with curves, that geometrical studies of surfaces in the small are in themselves very interesting and rewarding.

A *regular surface* is defined as follows:

1. A surface is a locus of points in Euclidean three-space defined by the end points of a vector $\mathbf{X} = \mathbf{X}(u, v)$, depending on two real parameters u, v:

$$(4.1) \qquad \mathbf{X}(u, v) = (x_1(u, v), x_2(u, v), x_3(u, v)),$$

with $x_i(u, v)$ the components of the vector. These real functions are assumed to be defined over an open connected domain of a Cartesian u,v-plane and to have continuous fourth partial derivatives there (although the existence of continuous second derivatives would suffice for many of the considerations of this chapter).

2. The vectors $\mathbf{X}_u = \mathbf{X}_1 = \partial\mathbf{X}/\partial u$, $\mathbf{X}_v = \mathbf{X}_2 = \partial\mathbf{X}/\partial v$, to be called *coordinate vectors*, are assumed to be linearly independent:

$$(4.2) \qquad \frac{\partial\mathbf{X}}{\partial u} \times \frac{\partial\mathbf{X}}{\partial v} = \mathbf{X}_1 \times \mathbf{X}_2 \neq 0.$$

This condition can also be formulated conveniently in terms of the matrix

$$(4.2)_1 \qquad M = \begin{bmatrix} \dfrac{\partial x_1}{\partial u} & \dfrac{\partial x_2}{\partial u} & \dfrac{\partial x_3}{\partial u} \\[2mm] \dfrac{\partial x_1}{\partial v} & \dfrac{\partial x_2}{\partial v} & \dfrac{\partial x_3}{\partial v} \end{bmatrix}, \qquad \text{rank } M = 2.$$

In fact [cf. (1.21)], the condition (4.2) is the same as the requirement that this matrix should be of rank two, i.e., that at least one of the three second-order determinants (all having the form of Jacobians) that can be formed from it should be different from zero. If this condition is satisfied, it is easy to show that it is always possible to introduce at least one pair of the coordinates x_i as regular surface parameters in a certain neighborhood of a point. To show this, suppose that the Jacobian $\partial(x_1, x_2)/\partial(u, v)$ is different from zero at a point (u_0, v_0). It follows from the theorem on implicit functions (for which see any standard book on analysis) that u and v can be expressed in a neighborhood of $x_1(u_0, v_0)$, $x_2(u_0, v_0)$) as single-valued and differentiable functions of x_1 and x_2; or, as it can also be put, this condition guarantees that a one-to-one and differentiable mapping of a neighborhood of the point (u_0, v_0) on a neighborhood of its image point in the x_1,x_2-plane is defined. Thus the surface could be given, in a neighborhood of a point, in the form

$$\mathbf{X} = (x_1, x_2, f(x_1, x_2)).$$

The coordinate vectors are given in this case by the relations

$$\mathbf{X}_1 = \left(1, 0, \frac{\partial f}{\partial x_1}\right),$$

$$\mathbf{X}_2 = \left(0, 1, \frac{\partial f}{\partial x_2}\right),$$

and these vectors are evidently linearly independent. Thus condition (4.2) is satisfied, and this in turn means that a surface given in what is perhaps the most familiar representation, i.e., $x_3 = f(x_1, x_2)$, yields a regular surface as long as the function $f(x_1, x_2)$ has a few continuous derivatives. This of course means that the vertical coordinate x_3 is given as a function of the other two; thus the surface points are in one-to-one correspondence with a domain of the x_1,x_2-plane through an orthogonal projection on the plane.

One of the main reasons for the requirement (4.2) or (4.2)$_1$ is that it ensures the two-dimensional character of the surface in the sense that, if it is fulfilled the mapping from the parameter plane into the three-dimensional space furnished by $\mathbf{X}(u, v)$ is locally one-to-one and hence topological in the small. In fact, this has already been made clear since, as was just seen, this condition ensures that the surface has always a one-to-one orthogonal projection on at least one of the coordinate planes.

2. Change of Parameters

As with curves, the surface parameters are to a considerable degree arbitrary; in fact, parameters u, v can be replaced by new parameters through the equations

(4.3) $$\qquad\qquad \bar{u} = \phi_1(u, v), \qquad \bar{v} = \phi_2(u, v),$$

provided that the Jacobian $\partial(\bar{u}, \bar{v})/(u, v)$ is different from zero. The theory of implicit functions states that the equations (4.3) can be solved in the small in that case for u and v to yield $u = \psi_1(\bar{u}, \bar{v})$, $v = \psi_2(\bar{u}, \bar{v})$, say, and that u and v will possess continuous partial derivatives if that is true of the functions ϕ_i. By the transformation $\mathbf{X}(u, v)$ goes into $\mathbf{X}(\psi_1, \psi_2) \equiv \bar{\mathbf{X}}(\bar{u}, \bar{v})$, since by definition the locus determined by $\mathbf{X}$ should be unchanged. It remains to be seen that $\bar{u}$ and $\bar{v}$ can be used as regular parameters. Suppose that $\partial(x_1, x_2)/\partial(u, v)$ differs from zero—an assumption (cf. the preceding section) that can be made without loss of generality—and hence (x_1, x_2) can be used as regular parameters. It is known that

$$\frac{\partial(x_1, x_2)}{\partial(u, v)} = \frac{\partial(x_1, x_2)}{\partial(\bar{u}, \bar{v})} \cdot \frac{\partial(\bar{u}, \bar{v})}{\partial(u, v)},$$

and since $\partial(\bar{u}, \bar{v})/\partial(u, v) \neq 0$ is assumed it follows that $\partial(x_1, x_2)/\partial(\bar{u}, \bar{v})$ does not vanish. Thus $\bar{u}$ and $\bar{v}$ can be used as regular parameters.

Obviously the parameter transformation $\bar{u} = u + c_1$, $\bar{v} = v + c_2$, c_1 and c_2 constants—a translation of the origin in the u,v-plane—is a legitimate parameter transformation. It is therefore always possible to investigate the geometry in the neighborhood of any point of a surface on the assumption that the point corresponds to $u = 0$, $v = 0$, and it is frequently convenient to do that.

3.   Curvilinear Coordinate Curves on a Surface

It is customary to speak of the parameters u, v as determining a curvilinear coordinate system on the surface (cf. Fig. 4.1). Suppose, for example,

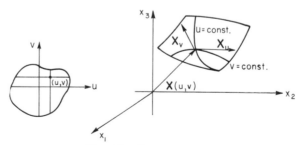

Fig. 4.1   Parameter curves.

that $u = c = \text{const}$. In that case $\mathbf{X}(c, v)$ represents a curve that lies on the surface. This curve is a regular space curve in the sense used in the previous chapter if the surface $\mathbf{X}(u, v)$ is regular, since $\mathbf{X}_v$ cannot vanish without violating the condition (4.2). The curves $u = \text{const.}$, $v = \text{const.}$—the parallels to the coordinate axes in the u,v-plane—thus correspond to a net of regular space curves covering a region on the surface. These curves are often called the u-curves and the v-curves, with the property that no u-curve has the same tangent as a v-curve at an intersection, since $\mathbf{X}_u \times \mathbf{X}_v \neq 0$ is one of the basic regularity conditions.

4. Tangent Plane and Normal Vector

Consider $u = u(t)$, $v = v(t)$ as functions of t in a t-interval, with continuous derivatives of a certain order, and such that $u'(t)$, $v'(t)$ are not simultaneously zero. This regular curve in the parameter plane has as an image on the regular surface S a curve $\mathbf{X}(u(t), v(t)) \equiv \mathbf{X}(t)$ that is also regular, since

$$\mathbf{X}'(t) = \mathbf{X}_u u' + \mathbf{X}_v v',$$

and the right-hand side cannot be the zero vector because $\mathbf{X}_u$ and $\mathbf{X}_v$ are tangents to the parameter curves and are linearly independent. Hence $\mathbf{X}'(t)$ lies in the plane determined by $\mathbf{X}_u$ and $\mathbf{X}_v$. *This plane is defined as the tangent plane.* An important theorem can now be stated, which is obvious from these remarks, i.e., that *the tangents to all regular curves at a point of the surface lie in the tangent plane.*

It has some point to notice that the tangent vectors of a surface at a point are defined by the pair of values u', v' that figure in the above equation for $\mathbf{X}'(t)$: these numbers define a resolution of the tangent vector into components along the coordinate vectors $\mathbf{X}_u$, $\mathbf{X}_v$. It is also clear that u' and v' could have any real values whatever, since $u = c_1 t$, $v = c_2 t$ are regular curves in the parameter plane for any constant values c_1, c_2, provided only that $c_1^2 + c_2^2 \neq 0$ holds.

The normal vector to the surface is defined as *the unit vector* $\mathbf{X}_3$ *that is perpendicular to the tangent plane.* It is therefore given by the equation

$$(4.4) \qquad \mathbf{X}_3 = \frac{\mathbf{X}_u \times \mathbf{X}_v}{\sqrt{(\mathbf{X}_u \times \mathbf{X}_v)^2}} = \frac{\mathbf{X}_1 \times \mathbf{X}_2}{\sqrt{(\mathbf{X}_1 \times \mathbf{X}_2)^2}}.$$

It is also of interest to express $\mathbf{X}_3$ in the following form, using the components of $\mathbf{X}_1$ and $\mathbf{X}_2$:

$$(4.4)_1 \qquad \sqrt{(\mathbf{X}_1 \times \mathbf{X}_2)^2}\, \mathbf{X}_3 = \left[\frac{\partial(x_2, x_3)}{\partial(u, v)}, \frac{\partial(x_3, x_1)}{\partial(u, v)}, \frac{\partial(x_1, x_2)}{\partial(u, v)} \right],$$

which states that appropriate Jacobians obtained from the matrix [cf. (4.2)]

$$\begin{bmatrix} \dfrac{\partial x_1}{\partial u} & \dfrac{\partial x_2}{\partial u} & \dfrac{\partial x_3}{\partial u} \\[2mm] \dfrac{\partial x_1}{\partial v} & \dfrac{\partial x_2}{\partial v} & \dfrac{\partial x_3}{\partial v} \end{bmatrix}$$

are direction numbers for the normal vector.

5. Length of Curves and First Fundamental Form

Consider a regular curve $\mathbf{X}(t) = \mathbf{X}(u(t), v(t))$ on the surface and define its length $s(t)$ by the integral

$$(4.5) \qquad\qquad s(t) = \int_{t_0}^{t} \sqrt{(\mathbf{X}'(\sigma))^2}\, d\sigma$$

in conformity with what was done generally for space curves in the previous chapter. Since $\mathbf{X}'(t) = \mathbf{X}_u\, du/dt + \mathbf{X}_v\, dv/dt$, it follows that

$$(4.6) \qquad \left(\frac{ds}{dt}\right)^2 = \mathbf{X}' \cdot \mathbf{X}' = E\left(\frac{du}{dt}\right)^2 + 2F\frac{du}{dt}\frac{dv}{dt} + G\left(\frac{dv}{dt}\right)^2,$$

with

$$(4.7) \qquad E = \mathbf{X}_u^2 = \mathbf{X}_u \cdot \mathbf{X}_u, \qquad F = \mathbf{X}_u \cdot \mathbf{X}_v, \qquad G = \mathbf{X}_v^2 = \mathbf{X}_v \cdot \mathbf{X}_v.$$

The quadratic form on the right-hand side of (4.6) is called the *first fundamental form*; it is sometimes denoted by the Roman numeral I. This quadratic form is evidently positive definite, i.e., it vanishes only when u' and v' vanish simultaneously. For this to hold it is necessary that the inequalities $E > 0$, $G > 0$, and $EG - F^2 > 0$ should be satisfied. The third inequality follows readily from the identity

$$Eu'^2 + 2Fu'v' + Gv'^2 = \frac{1}{E}[(Eu' + Fv')^2 + (EG - F^2)v'^2].$$

Another proof that $EG - F^2 > 0$ must be satisfied can be obtained from the identity of Lagrange:

$$(\mathbf{X}_a \times \mathbf{X}_b) \cdot (\mathbf{X}_c \times \mathbf{X}_d) = (\mathbf{X}_a \cdot \mathbf{X}_c)(\mathbf{X}_b \cdot \mathbf{X}_d) - (\mathbf{X}_b \cdot \mathbf{X}_c)(\mathbf{X}_a \cdot \mathbf{X}_d).$$

By setting $\mathbf{X}_a = \mathbf{X}_c = \mathbf{X}_1$, $\mathbf{X}_b = \mathbf{X}_d = \mathbf{X}_2$ this identity leads to

$$\mathbf{X}_1 \times \mathbf{X}_2 \cdot \mathbf{X}_1 \times \mathbf{X}_2 = (\mathbf{X}_1 \cdot \mathbf{X}_1)(\mathbf{X}_2 \cdot \mathbf{X}_2) - (\mathbf{X}_1 \cdot \mathbf{X}_2)^2 = EG - F^2,$$

or

$$(4.8) \qquad\qquad (\mathbf{X}_1 \times \mathbf{X}_2)^2 = EG - F^2.$$

This last expression for the discriminant of the first fundamental form is often useful.

6. Invariance of the First Fundamental Form

It is clear that the functions E, F, and G are invariants as regards changes of coordinates. However, these quantities are evidently not invariants with respect to transformations in the parameters (u, v), while on the

other hand the first fundamental form clearly ought to be invariant with re-
pect to both parameter and coordinate transformations in view of its geometric significance.   A type of question is touched upon here that will come
up for discussion again later on in a systematic way in introducing the notion
of tensors in Riemannian manifolds.   Here it is shown that the line element
defined by (4.6) is indeed invariant with respect to parameter transforma-
tions given by $\bar{u} = \bar{u}(u, v)$, $\bar{v} = \bar{v}(u, v)$ but which it is convenient now to denote by $\bar{u}_i = \bar{u}_i(u_1, u_2)$, $i = 1, 2$.   At the same time it is convenient to investigate the invariance properties of other entities already introduced.

The coordinate vectors $\mathbf{X}_i$ transform as follows:

$$(4.9) \qquad \overline{\mathbf{X}}_i = \sum_{l=1,2} \mathbf{X}_l \frac{\partial u_l}{\partial \bar{u}_i}, \qquad i = 1, 2,$$

hence are seen *not* to be invariants under parameter transformations; on the
other hand, the tangential plane is such an invariant since it is determined in
the new parameters by the two vectors $\overline{\mathbf{X}}_1$, $\overline{\mathbf{X}}_2$ and these are linearly independent vectors which are themselves linear combinations of $\mathbf{X}_1$ and $\mathbf{X}_2$, as
(4.9) shows.   It follows that the normal vector $\mathbf{X}_3$, defined by (4.4), is an invariant under parameter transformations only under the restriction that the
Jacobian $\partial(\bar{u}, \bar{v})/\partial(u, v)$ should remain positive, since otherwise the orientation of the coordinate vectors in the tangent plane would be changed—as a
calculation using (4.9) shows—and (4.4) then shows that $\mathbf{X}_3$ would be reversed
in direction.

Consider now the line element defined by (4.6) and (4.7).   It is convenient to write it in the form

$$(4.6)' \qquad \left(\frac{ds}{dt}\right)^2 = \mathbf{X}' \cdot \mathbf{X}' = \sum_{i,k} g_{ik}(u_1, u_2) u_i' u_k',$$

$$(4.7)' \qquad g_{ik} = g_{ki} = \mathbf{X}_i \cdot \mathbf{X}_k,$$

and note that all sums are taken over 1, 2.   The quantities g_{ik} thus introduced constitute a standard notation for the coefficients of the first fundamental form.   From (4.9) the transformed coefficients $\bar{g}_{ik}$ are given by the
formulas

$$\bar{g}_{ik} = \overline{\mathbf{X}}_i \cdot \overline{\mathbf{X}}_k = \left(\sum_l \mathbf{X}_l \frac{\partial u_l}{\partial \bar{u}_i}\right) \cdot \left(\sum_m \mathbf{X}_m \frac{\partial u_m}{\partial \bar{u}_k}\right)$$

$$= \sum_{l,m} \mathbf{X}_l \cdot \mathbf{X}_m \frac{\partial u_l}{\partial \bar{u}_i} \frac{\partial u_m}{\partial \bar{u}_k} = \sum_{l,m} g_{lm} \frac{\partial u_l}{\partial \bar{u}_i} \frac{\partial u_m}{\partial \bar{u}_k}.$$

Thus it is clear that the coefficients g_{ik} of the first fundamental form are not invariants in general under parameter transformations. For the line element $(d\bar{s}/dt)^2$ in the parameters $\bar{u}_i$ the following formula results:

$$\left(\frac{d\bar{s}}{dt}\right)^2 = \bar{\mathbf{X}}' \cdot \bar{\mathbf{X}}' = \sum_{i,k} \bar{g}_{ik} \bar{u}'_i \bar{u}'_k$$

$$= \sum_{i,k,l,m,j,n} g_{lm} \frac{\partial u_l}{\partial \bar{u}_i} \frac{\partial u_m}{\partial \bar{u}_k} \frac{\partial \bar{u}_l}{\partial u_j} u'_j \frac{\partial \bar{u}_k}{\partial u_n} u'_n$$

upon using the formula just derived for the quantities $\bar{g}_{ik}$ and noting that $\bar{u}'_i = \sum_j (\partial \bar{u}_i / \partial u_j) u'_j$. (It is understood, of course, that all sums are for indices 1, 2 and all are to be carried out independently.) Since, however, the transformations from the parameters u_i to the parameters $\bar{u}_i$ are inverse to each other, it follows that

$$\sum_i \frac{\partial u_l}{\partial \bar{u}_i} \frac{\partial \bar{u}_i}{\partial u_j} = \delta_{jl} = \begin{cases} 0, & j \neq l, \\ 1, & j = l. \end{cases}$$

This comes about because the left-hand side represents $\partial u_l / \partial u_j$, since $u_l = u_l(\bar{u}_1, \bar{u}_2) = u_l(\bar{u}_1(u_1, u_2), \bar{u}_2(u_1, u_2)) \equiv u_l$. As a consequence the line element in the new variables takes the form

$$\left(\frac{d\bar{s}}{dt}\right)^2 = \sum_{l,m,j,n} g_{lm} \delta_{jl} \delta_{mn} u'_j u'_n$$

$$= \sum_{j,n} g_{jn} u'_j u'_n,$$

and the latter is the same as $(4.6)'$ since the names given to the summation indices obviously have no effect on the value of the sum. Thus the line element is invariant under parameter transformations although its individual coefficients are not.

In Riemannian geometry the reversed position is taken: in order to ensure that the analogous first fundamental form remains invariant its coefficients are *required* to transform by the same formula as above. Here, the invariance comes out of itself because of the nature of the Euclidean geometry of the space in which the surface is embedded.

7. Angle Measurement on Surfaces

It is an important fact that the angle between two curves at a point on the surface can be calculated in terms of the coefficients E, F, G of the first fundamental form. Consider the two curves $u = u_1(t)$, $v = v_1(t)$ and $u =$

$u_2(\tau)$, $v = v_2(\tau)$ with $u_1(0) = u_2(0)$, $v_1(0) = v_2(0)$.   Their tangent vectors are given by

$$\frac{d\mathbf{X}}{dt} = \mathbf{X}_u \frac{du_1}{dt} + \mathbf{X}_v \frac{dv_1}{dt} = \mathbf{X}_t,$$

$$\frac{d\mathbf{X}}{d\tau} = \mathbf{X}_u \frac{du_2}{d\tau} + \mathbf{X}_v \frac{dv_2}{d\tau} = \mathbf{X}_\tau,$$

and hence

$$(4.10) \quad \cos \phi = \frac{\mathbf{X}_t \cdot \mathbf{X}_\tau}{\sqrt{(\mathbf{X}_t)^2 (\mathbf{X}_\tau)^2}}$$

$$= \frac{E \dfrac{du_1}{dt}\dfrac{du_2}{d\tau} + F\left(\dfrac{du_1}{dt}\dfrac{dv_2}{d\tau} + \dfrac{du_2}{d\tau}\dfrac{dv_1}{dt}\right) + G \dfrac{dv_1}{dt}\dfrac{dv_2}{d\tau}}{\sqrt{E\left(\dfrac{du_1}{dt}\right)^2 + 2F\dfrac{du_1}{dt}\dfrac{dv_1}{dt} + G\left(\dfrac{dv_1}{dt}\right)^2} \cdot \sqrt{E\left(\dfrac{du_2}{d\tau}\right)^2 + 2F\dfrac{du_2}{d\tau}\dfrac{dv_2}{d\tau} + G\left(\dfrac{dv_2}{d\tau}\right)^2}}$$

Evidently, the coefficients E, F, G suffice to determine the angle once the curves are given.   If the two curves are, in particular, the parameter curves $u = $ const., $v = $ const. themselves, it follows that $du_1/dt = 0$, $dv_1/dt = 1$; $du_2/d\tau = 1$, $dv_2/d\tau = 0$; and hence

$$(4.11) \qquad\qquad\qquad \cos \omega = \frac{F}{\sqrt{EG}}.$$

From this formula it is clear that the *curvilinear coordinates are orthogonal if and only if $F = 0$.*

An orientation of the parameter plane is achieved by ordering the coordinates $u \to v$, and this can be carried over to the tangent planes of the surface through the vectors $\mathbf{X}_u$ and $\mathbf{X}_v$, thus achieving by definition an orientation of the surface.   The positive sense for measuring angles is then, of course, defined as that from $\mathbf{X}_u$ toward $\mathbf{X}_v$.   The sense of the normal vector $\mathbf{X}_3$, as given by (4.4), is then fixed by this convention.   If any two tangent vectors $\mathbf{X}_1'$, $\mathbf{X}_2'$ of the surface are given, the sine of the angle ϕ through which $\mathbf{X}_1'$ must be turned in order to bring it into coincidence with $\mathbf{X}_2'$ is given numerically by the formula

$$(4.12) \qquad\qquad\qquad |\mathbf{X}_1'|\, |\mathbf{X}_2'|\, |\sin \phi| = |\mathbf{X}_1' \times \mathbf{X}_2'|,$$

while the sign of ϕ is determined by noting whether the vector product $\mathbf{X}_1' \times \mathbf{X}_2'$ has the same sense as $\mathbf{X}_3$ or not.   In particular, the sine of the angle ω between the tangents of the coordinate curves is given by

$$(4.13) \qquad\qquad\qquad \sin \omega = \frac{\sqrt{EG - F^2}}{\sqrt{EG}},$$

as is seen from (4.7) and (4.8).

8. Area of a Surface

The *area* of a surface is defined by making use once more of the functions E, F, G. By definition, the area of a curved surface is given by the integral [see (4.8)]:

$$(4.14) \qquad A = \iint_R \sqrt{EG - F^2} \, du \, dv = \iint_R \sqrt{(\mathbf{X}_1 \times \mathbf{X}_2)^2} \, du \, dv$$

over the region R of the u, v-plane that is mapped into the surface. It could be shown that this is the only possible definition for the area of a curved surface that satisfies the following requirements: (a) it is given by an integral of the form $\iint f \, du \, dv$ in which f depends only upon u, v, $\mathbf{X}$, $\mathbf{X}_1$, $\mathbf{X}_2$; (b) it is invariant with respect to orthogonal transformations of the coordinate axes and also to parameter transformations that preserve the orientation of the surface; (c) it furnishes the value 1 for the area of a square of side length l in the plane. These are, obviously, rather reasonable conditions to impose. The notion of area defined by (4.14) satisfies these conditions. Condition (a) is obviously satisfied. That the expression for A is invariant under orthogonal transformation is also clear. That it is invariant under orientation-preserving parameter transformations can be seen as follows. The functions g_{ik} at a given point transform so that $\sum g_{ik} u_i' u_k' = \sum \bar{g}_{lm} \bar{u}_l' \bar{u}_m'$, i.e., they behave like the coefficients of a quadratic form under a linear transformation of the variables (in this case the variables are u_1', u_2'). It is well known that in such a case

$$g_{11} g_{22} - g_{12}^2 = (\bar{g}_{11} \bar{g}_{22} - \bar{g}_{12}^2) \left(\frac{\partial(\bar{u}_1 \cdot \bar{u}_2)}{\partial(u_1, u_2)} \right)^2.$$

But this in conjunction with the general rule for the transformation of multiple integrals shows that the expression for A has the required invariance property with respect to parameter transformations. As for condition (c), take a plane surface given by

$$\mathbf{X}(u_1, u_2) = \mathbf{X}(0) + u_1 \mathbf{v}_1 + u_2 \mathbf{v}_2$$

with $\mathbf{v}_1$, $\mathbf{v}_2$ any constant orthogonal unit vectors; it is readily verified that a regular surface is defined with the use of the parameters u_i, and it is a plane, since the locus is a linear subspace of the three-space. The functions g_{ik} are in this case $g_{11} = g_{22} = 1$, $g_{12} = 0$ and hence condition (c) is seen to hold.

That the conditions (a), (b), (c) determine the form of the integrand defining A is also not very difficult to prove, but the proof is omitted here. Instead, the following intuitive discussion indicates why (4.14) is a reasonable definition for the area of a surface. Consider the "parallelogram" on the surface enclosed by the parameter curves u, v, $u + du$, $v + dv$ and let ds_u, ds_v

represent the lengths of the sides indicated in Fig. 4.2. From (4.6) the formulas $ds_u = \sqrt{E}\, du$, $ds_v = \sqrt{G}\, dv$ result, and from (4.13) the sine of ω is determined. Hence the "element of area" is given rather reasonably by the formula

$$(4.15) \qquad\qquad dA = \sqrt{EG - F^2}\, du\, dv.$$

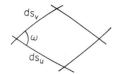

Fig. 4.2   *Element of area.*

9. A Few Examples

The first example that illustrates some of the concepts introduced above is the *plane*. It is defined here as the locus

$$\mathbf{X}(u, v) = (u, v, 0),$$

so that the coordinate vectors are

$$\mathbf{X}_u = (1, 0, 0), \qquad \mathbf{X}_v = (0, 1, 0),$$

i.e., they are orthogonal unit vectors in the direction of the u- and v-axes. The line element is given by

$$ds^2 = du^2 + dv^2,$$

as it should be.

A second example is the *sphere* given in spherical coordinates $u = \theta$, $v = \phi$ (cf. Fig. 4.3). Thus the parameter curves are the meridians and the parallels of latitude. This surface is clearly given by the vector

$$\mathbf{X} = (r \sin \theta \cos \phi,\ r \sin \theta \sin \phi,\ r \cos \theta).$$

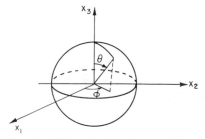

Fig. 4.3   *The sphere in polar coordinates.*

Differentiation yields

$$\mathbf{X}_\theta = \mathbf{X}_1 = (r \cos \theta \cos \phi, \, r \cos \theta \sin \phi, \, -r \sin \theta),$$

$$\mathbf{X}_\phi = \mathbf{X}_2 = (-r \sin \theta \sin \phi, \, r \sin \theta \cos \phi, \, 0).$$

The element of length is found to be

$$ds^2 = r^2 \, d\theta^2 + r^2 \sin^2 \theta \, d\phi^2,$$

so that

$$E = r^2, \qquad F = 0, \qquad G = r^2 \sin^2 \theta,$$

thus verifying the fact, obvious from the geometry of the sphere, that the parameter curves are orthogonal. Also $\sqrt{EG - F^2} = r^2 \sin \theta$ in the present case, which leads to the familiar area element of the sphere. Finally, the unit normal vector is given by

$$\mathbf{X}_3 = \frac{\mathbf{X}_1 \times \mathbf{X}_2}{\sqrt{EG - F^2}}$$

$$= (r^2 \sin \theta)^{-1} (r^2 \sin^2 \theta \cos \phi, \, r^2 \sin^2 \theta \sin \phi, \, r^2 \sin \theta \cos \theta)$$

$$= \left(\frac{1}{r}\right)\mathbf{X},$$

as it obviously should be in this case. It is to be noted that the poles ($\theta = 0, \pi$) are points for which $\mathbf{X}_1 \times \mathbf{X}_2 = 0$. This is due, evidently, not to a singularity of the locus, but to a singularity in the coordinate system.

Another interesting simple example (cf. Fig. 4.4) is the circular cylinder. It is given by

$$\mathbf{X}(u, v) = (r \cos u, \, r \sin u, \, v).$$

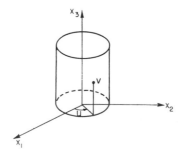

Fig. 4.4   The cylinder in cylindrical coordinates.

The coordinate vectors are

$$\mathbf{X}_1 = (-r \sin u, r \cos u, 0),$$

$$\mathbf{X}_2 = (0, 0, 1).$$

The parameter net on the surface consists of the generators of the cylinder and the circles cut out of it by planes orthogonal to them. The coefficients of the first fundamental form are $E = r^2$, $F = 0$, $G = 1$, and hence $ds^2 = r^2 du^2 + dv^2$. If new parameters are introduced by setting $\bar{u} = ru$, $\bar{v} = v$ (note that the Jacobian does not vanish for this transformation so that it yields new parameters in terms of which the surface is regular), the result is $ds^2 = d\bar{u}^2 + d\bar{v}^2$ for the line element in the new parameters. This has the following geo- metrical significance: the points of the surface and those of a $\bar{u},\bar{v}$-plane are in one-to-one correspondence in such a way that lengths of corresponding curves are the same. The cylinder is therefore said to be *isometric* to the plane.

The problem of deciding whether surfaces are isometric, i.e., whether their points can be put into a one-to-one correspondence such that lengths of corresponding curves are the same, and its general implications for geometry is a problem that will be formulated and discussed more generally later on. However, it should already be pointed out here that such isometric mappings cannot be constructed for arbitrary pairs of surfaces. For example, the sphere cannot be mapped isometrically on the plane for the following reasons. A triangle formed of great circle arcs on the sphere, if mapped isometrically on the plane, should go into a triangle with straight sides since great circle arcs on the sphere are arcs of shortest length on it (this fact is assumed here without proof) and hence would of necessity map into arcs of shortest length in the plane. But angles are also preserved in isometric mappings, since angles are measured solely with the aid of the first fundamental form. But the sum of the interior angles of all triangles in the plane is π, but on the sphere it is well known that this sum is greater than π; consequently, the sphere cannot be mapped isometrically on the plane.

10. Second Fundamental Form of a Surface

The next step in the development of the theory of surfaces is taken in studying the deviation of the surface from its tangent plane in the neighbor- hood of the point of tangency, and this in turn leads to various developments touching on the curvature of the surface. It is convenient to introduce the function $\rho(u, v) = [\mathbf{X}(u, v) - \mathbf{X}(0, 0)] \cdot \mathbf{X}_3(0, 0)$, where $\mathbf{X}_3$ is the unit normal vector. Thus $\rho(u, v)$ is the perpendicular distance (with an appropriate sign) from the tangent plane to the point of the surface S fixed by $\mathbf{X}(u, v)$ (cf. Fig. 4.5). Assuming $\mathbf{X}(u, v)$ to have continuous third derivatives, it can be developed as follows:

$$\mathbf{X}(u, v) = \mathbf{X}(0, 0) + \mathbf{X}_1 u + \mathbf{X}_2 v + \tfrac{1}{2}(\mathbf{X}_{11} u^2 + 2\mathbf{X}_{12} uv + \mathbf{X}_{22} v^2) + \cdots;$$

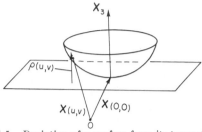

Fig. 4.5   *Deviation of a surface from its tangent plane.*

here the vector functions $\mathbf{X}_i = \partial\mathbf{X}/\partial u_i$ and $\mathbf{X}_{ij} = \partial^2\mathbf{X}/\partial u_i \partial u_j$ are evaluated at $(0, 0)$ and the dots refer to terms that are cubic at least in u and v. Hence $\rho(u, v)$ is given by

$$\rho(u, v) = \tfrac{1}{2}(\mathbf{X}_{11}\cdot\mathbf{X}_3 u^2 + 2\mathbf{X}_{12}\cdot\mathbf{X}_3 uv + \mathbf{X}_{22}\cdot\mathbf{X}_3 v^2) + \cdots,$$

since $\mathbf{X}_i\cdot\mathbf{X}_3 = 0$. The quantities $L = \mathbf{X}_{11}\cdot\mathbf{X}_3 = L_{11}$, $M = \mathbf{X}_{12}\cdot\mathbf{X}_3 = L_{12}$, $N = \mathbf{X}_{22}\cdot\mathbf{X}_3 = L_{22}$ are introduced, and the *second fundamental form* II is defined as the quadratic form

(4.16) $\mathrm{II} = Lu^2 + 2Muv + Nv^2.$

For small values of u and v the function $2\rho(u, v)$ is approximated by the quadratic form II with errors of third or higher order in u and v; a study of the form II will therefore give information about the shape of the surface S near the point of tangency, as will be seen in the next and later sections.

The coefficients $L_{ik} = \mathbf{X}_{ik}\cdot\mathbf{X}_3 = \mathbf{X}_{ik}\cdot(\mathbf{X}_1 \times \mathbf{X}_2)/\sqrt{EG - F^2}$ of the second fundamental form II are invariant under coordinate transformations that preserve the orientation of the axes, but they change sign if the orientation is reversed. Like the coefficients g_{ik}, they are not invariant under parameter transformations. However, the second fundamental form itself is an invariant under parameter transformations with a positive Jacobian, as will be seen later on.

11.   Osculating Paraboloid

It is of interest to consider the following surface in a Euclidean u,v,ρ-space:

(4.17) $\rho = \tfrac{1}{2}(Lu^2 + 2Muv + Nv^2).$

This quadratic surface in the u,v,ρ-space is, for obvious reasons, called the osculating paraboloid of the surface S. The nature of this paraboloid determines, in certain qualitative respects at least, the nature of the surface in the neighborhood of the point of tangency. It is appropriate to distinguish four cases, depending on the sign of the determinant $LN - M^2$ of the second fundamental form II:

(a) *Elliptic case:* $LN - M^2 > 0$. The osculating paraboloid is an elliptic paraboloid; the form II is definite. Hence the surface [in a neighborhood of $\mathbf{X}$ (0, 0) lies wholly on one side of the tangent plane, touching it only at $\mathbf{X}$ (0, 0), by virtue of the discussion of the previous section. If II is positive definite, the surface lies on the side toward which $\mathbf{X}_3$ points; if II is negative definite, on the other side, as is clear from the way in which $\rho(u, v)$ was defined.

(b) *Hyperbolic case:* $LN - M^2 < 0$. The quadric is a hyperbolic paraboloid; the form II is indefinite. The two distinct real straight lines given by

$$L\left(\frac{u}{v}\right)^2 + 2M\left(\frac{u}{v}\right) + N = 0$$

divide the tangent plane into four regions in which the osculating paraboloid lies alternately above and below that plane. The surface with II as second fundamental form has the same qualitative appearance.

(c) *Parabolic case:* $LN - M^2 = 0$, but not all of the coefficients L, M, N separately vanish. The quadric is a parabolic cylinder, the form II is semidefinite (it is a perfect square), and the cylinder lies entirely on one side of the tangent plane. The surface S behaves similarly.

(d) *Planar case:* $L = M = N = 0$. The osculating paraboloid degenerates to a plane. The shape of S near such a point is not defined by the process used here in the sense that nothing can be said with its aid concerning the sign of $\rho(u, v)$.

It is of interest to prove that *a surface having only planar points is a plane.* To do so it is noted that the following identity holds in general:

$$\mathbf{X}_{ij}\cdot\mathbf{X}_3 + \mathbf{X}_i\cdot\mathbf{X}_{3j} = 0;$$

this follows from $\mathbf{X}_i\cdot\mathbf{X}_3 = 0$ by differentiation. Since $L_{ij} = \mathbf{X}_{ij}\cdot\mathbf{X}_3$, the above equation yields the useful relations

(4.18) $L_{ij} = \mathbf{X}_{ij}\cdot\mathbf{X}_3 = -\mathbf{X}_i\cdot\mathbf{X}_{3j},$

which also hold quite generally. In the present special case $L_{ij} = 0$ holds everywhere by assumption, and hence $\mathbf{X}_i\cdot\mathbf{X}_{3j} = 0$ for all values (u, v) and for all combinations of the values 1, 2 of i and j. Since $\mathbf{X}_3\cdot\mathbf{X}_3 = 1$, and hence

$\mathbf{X}_3 \cdot \mathbf{X}_{3j} = 0$, it follows that $\mathbf{X}_{3j}$ is a vector in the tangent plane, and from $\mathbf{X}_i \cdot \mathbf{X}_{3j} = 0$ it follows that $\mathbf{X}_{3j} = 0, j = 1, 2$, since the vectors $\mathbf{X}_i$ are linearly independent.   Thus $\mathbf{X}_3 = \mathbf{a}$, with $\mathbf{a}$ a constant vector, in this case.   It follows that $\mathbf{X}' \cdot \mathbf{a} = (\mathbf{X} \cdot \mathbf{a})' = 0$, when the differentiation is taken with respect to any curve on the surface.   Consequently, $\mathbf{X} \cdot \mathbf{a} = b = \text{const}$; and this is the equation of a plane.

12.   Curvature of Curves on a Surface

The above considerations serve only in a limited degree to describe the deviation of the surface from its tangent plane.   In order to obtain more precise information, the curvature of individual curves of the surface through a particular point is studied next.   Let $\mathbf{X}(s) = \mathbf{X}(u(s), v(s))$ be a curve on the surface for which s is the arc length; such curves will be regarded here as regular space curves in the sense of the preceding chapter, and hence their curvature, if not zero, will always be positive.   By differentiation the following formulas are found:

$$\dot{\mathbf{X}} = \mathbf{X}_1 \dot{u} + \mathbf{X}_2 \dot{v},$$

$$\ddot{\mathbf{X}} = \mathbf{X}_1 \ddot{u} + \mathbf{X}_2 \ddot{v} + (\mathbf{X}_{11} \dot{u}^2 + 2\mathbf{X}_{12} \dot{u}\dot{v} + \mathbf{X}_{22} \dot{v}^2),$$

and, by taking the scalar product of both sides of the latter equation with $\mathbf{X}_3$ the result is, in view of the definitions of the quantities $L, M,$ and N:

$$\mathbf{X}_3 \cdot \ddot{\mathbf{X}} = L\dot{u}^2 + 2M\dot{u}\dot{v} + N\dot{v}^2,$$

since $\mathbf{X}_3 \cdot \mathbf{X}_1 = \mathbf{X}_3 \cdot \mathbf{X}_2 = 0$.   But $\sqrt{\ddot{\mathbf{X}} \cdot \ddot{\mathbf{X}}} = \kappa$ [cf. (3.5)], so that $\mathbf{X}_3 \cdot \ddot{\mathbf{X}} = \kappa \mathbf{X}_3 \cdot \mathbf{v}_2 = \kappa \cos \theta$, in which κ is the curvature of the curve $\mathbf{X}(s)$ and θ is the angle between the unit normal $\mathbf{X}_3$ and the principal normal $\mathbf{v}_2$ of the curve [cf. (3.7)].   Consequently, the above equation takes the form

$$\kappa \cos \theta = L\dot{u}^2 + 2M\dot{u}\dot{v} + N\dot{v}^2$$

$$= (Lu'^2 + 2Mu'v' + Nv'^2)\left(\frac{dt}{ds}\right)^2$$

upon introduction of a new parameter t for the curve in place of s.   But [see (4.6)]

$$\left(\frac{ds}{dt}\right)^2 = Eu'^2 + 2Fu'v' + Gv'^2 = \mathrm{I},$$

from the definition of the first fundamental form (which is denoted by the symbol I), with the result, finally,

$$(4.19) \qquad \kappa \cos \theta = \frac{Lu'^2 + 2Mu'v' + Nv'^2}{Eu'^2 + 2Fu'v' + Gv'^2} = \frac{\mathrm{II}}{\mathrm{I}}.$$

This equation leads to the following theorem:

All regular curves on the surface that have the same tangent and principal normal vectors (hence also the same osculating plane) at a point P of the surface have at P the same curvature, provided only that cos θ ≠ 0, in other words, provided that the principal normal of the curve is not a tangent vector of the surface.

The proof is simple. Two curves for which the tangents fall along the same line have the same value for the ratio u'/v' and hence the right side of (4.19) is the same for both. If the two curves have, in addition, the same principal normal, cos θ is the same for both. Hence κ is the same for both, and the theorem is proved. This discussion implies, since κ is always positive, that θ is an acute or an obtuse angle depending upon whether the form II is positive or negative.

Thus complete information on the possible values of κ for regular curves through a given point can be obtained by narrowing the investigation to *plane* curves cut out of the surface at the point. Once κ has been determined for a plane curve cut out by a plane containing the *normal* of the surface (a so-called normal section), the values of κ for all other sections by planes containing the same tangent can be calculated at once from (4.19). It suffices therefore for present purposes to consider the curvature of the normal sections alone, implying that θ = 0 or θ = π in (4.19). It is convenient to set

$$(4.20) \qquad\qquad k = \frac{\text{II}}{\text{I}}$$

so that $|k|$ is the curvature of the normal plane sections (regarded as space curves) determined by given direction numbers u' and v' in the tangent plane. From the above discussion it is known that the normal section lies on the side of the tangent plane toward which $\mathbf{X}_3$ points if k is positive, on the other side if k is negative, since the first fundamental form I is always positive, and hence k has the sign of the second fundamental form II. The quantity k is thus not the curvature as defined earlier for space curves, since it changes sign with the form II. It is also not the curvature as defined for plane curves, since, for example, k does not change sign when the direction of travel along the curve is reversed: I and II are homogeneous quadratic forms that do not change sign when both u' and v' change sign. However, $|k|$ is always the numerical value of the curvature, as was remarked above.

It is now clear on geometric grounds that the second fundamental form must be invariant with respect to parameter transformations that have a positive Jacobian J: II = kI has, in any case, a numerical value that is invariant, and if $J > 0$ holds, II has always the same sign when the parameters are changed.

For the purpose of studying the quantity k in greater detail the following special parameters (which are often useful in other cases as well) are introduced. The tangent plane of the regular surface is taken as the parameter plane with the origin at the point of tangency, and with $u = x_1$, $v = x_2$. [It is seen from $(4.4)_1$ that the surface is regular when (x_1, x_2) are chosen as parameters since $\partial(x_1, x_2)/\partial(u, v) \neq 0$ holds in this case; thus the surface is defined by the vector $\mathbf{X}(x_1, x_2) = (x_1, x_2, f(x_1, x_2))$.] Having done so, it follows at once that the vectors $\mathbf{X}_1$ and $\mathbf{X}_2$ are given by $\mathbf{X}_1 = (1, 0, 0)$, $\mathbf{X}_2 = (0, 1, 0)$, since $\partial f/\partial x_1$ and $\partial f/\partial x_2$ vanish at the origin, and hence that $E = 1$, $F = 0$, $G = 1$ at the point of tangency. Consider a curve $\mathbf{X}(s) = \mathbf{X}(u(s), v(s))$ on the surface through the point of tangency with s as arc length. Then $\dot{\mathbf{X}}(s) = \dot{u}\mathbf{X}_1 + \dot{v}\mathbf{X}_2$ with $\dot{u}^2 + \dot{v}^2 = 1$ and consequently $\dot{u} = \cos\phi$, $\dot{v} = \sin\phi$, with ϕ the angle in polar coordinates that $\dot{\mathbf{X}}$ makes with the vector $\mathbf{X}_1$. Consequently for this special choice of parameters the formula (4.20) leads to

$$k = \frac{L\dot{u}^2 + 2M\dot{u}\dot{v} + N\dot{v}^2}{E\dot{u}^2 + 2F\dot{u}\dot{v} + G\dot{v}^2}$$

$$= L\cos^2\phi + 2M\sin\phi\cos\phi + N\sin^2\phi.$$

It is customary to set $|k| = 1/r^2$ and obtain

(4.21) $\pm 1 = L(r\cos\phi)^2 + 2M(r\cos\phi)(r\sin\phi) + N(r\sin\phi)^2$

$$= Lx_1^2 + 2Mx_1x_2 + Nx_2^2$$

upon setting $x_1 = r\cos\phi$, $x_2 = r\sin\phi$. The equation (4.21) gives loci (they are conic sections) in the x_1,x_2-plane such that the length of any line segment from the center of the conic to a point on it is the reciprocal square root of the curvature of a normal section containing the segment (cf. Fig. 4.6). The sign of the left-hand side of (4.21) is to be taken in accordance with that of the right-hand side. The conic section (or a pair of conjugate hyperbolas if II is indefinite) given by (4.21) is called the *Dupin-indicatrix*. [It is evident

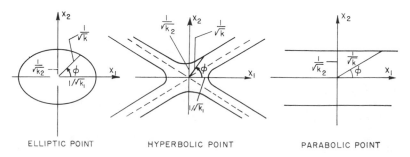

Fig. 4.6   The Dupin indicatrix.

also by comparison of (4.17) with (4.21) that the indicatrix is similar to the curves cut from the osculating paraboloid by planes parallel to the tangent plane and thus furnishes an approximation to the shape of such plane sections of the surface itself.]

13. Principal Directions and Principal Curvatures

Consider once more the four possible cases discussed in Section 11 above, but this time with reference to the indicatrix:

(a) *Elliptic case* ($LN - M^2 > 0$). The indicatrix is an ellipse.

(b) *Hyperbolic case* ($LN - M^2 < 0$). The indicatrix consists of a pair of conjugate hyperbolas, the common asymptotes being directions for which $k = 0$ (i.e., for which r is infinite).

(c) *Parabolic case* ($LN - M^2 = 0$, $L^2 + N^2 + M^2 \neq 0$). The right side of (4.21) is a perfect square and the indicatrix is a pair of parallel lines in the direction of which $k = 0$.

(d) *Planar case* ($L = M = N = 0$). The indicatrix does not exist.

Hence, excluding the planar case and the special elliptic case in which the indicatrix becomes a circle, there exists exactly one pair of orthogonal directions for which the values of k take on maximum and minimum values k_1, k_2; they are normal curvatures—which may have either sign—for normal sections in the directions of the principal axes of the indicatrix. These directions are defined as the *principal directions* on the surface, while the corresponding normal curvatures are defined as the *principal curvatures*. In the two exceptional cases all directions are regarded as principal directions, and obviously the normal curvatures are then the same for all directions.

It is of interest to obtain the principal curvatures by a direct calculation based on equation (4.20):

$$(4.22) \qquad k(u', v') = \frac{\mathrm{II}}{\mathrm{I}} = \frac{Lu'^2 + 2Mu'v' + Nv'^2}{Eu'^2 + 2Fu'v' + Gv'^2}.$$

It was seen in the previous section that special coordinates may be chosen at any one point in such a way that $u'^2 + v'^2 = 1$, and, since the denominator I in (4.22) never vanishes except for $u' = v' = 0$, it follows that a maximum and a minimum of k exist since k is continuous over the closed bounded domain $u'^2 + v'^2 = 1$. For the extrema of k when u' and v' vary (the coefficients g_{ik} and L_{ik} of the fundamental forms are constants in the present discussion) the equations

$$(L - kE)u' + (M - kF)v' = 0,$$
$$(M - kF)u' + (N - kG)v' = 0,$$

obtained by setting $\partial k/\partial u' = 0$ and $\partial k/\partial v' = 0$, must hold. The quantity k must be determined so that these homogeneous linear equations are satisfied for values of u' and v' that are not both zero. For the extreme values, then, the determinant of the coefficients must vanish:

$$(4.23) \qquad \begin{vmatrix} L - kE & M - kF \\ M - kF & N - kG \end{vmatrix} = 0,$$

and this "secular equation" is a quadratic equation (since $EG - F^2 > 0$) that possesses two real roots denoted by k_1 and k_2. The values k_1 and k_2 when inserted in the linear homogeneous equations yield the principal directions, which consist of exactly two orthogonal directions if $k_1 \neq k_2$, or of all possible directions if $k_1 = k_2$. In the latter case the point is a *planar point* if $k_1 = k_2 = 0$ [as is readily seen from (4.22)] and is called an *umbilical point* if $k_1 = k_2 \neq 0$. An umbilical point is characterized by the conditions $L/E = M/F = N/G$, as can be seen from (4.22) and the fact that k is independent of u' and v'.

In the above discussion known facts about the conic sections were used to justify the conclusions, but it is perhaps not amiss to remark here that the above process of determining the principal directions and curvatures is a special case of a purely algebraic process capable of generalization to higher dimensions and of application to a wide variety of geometrical and physical problems. Involved in these problems is a pair of quadratic forms $Q_a = \sum_{i,k=1}^{n} a_{ik}x_ix_k$ and $Q_b = \sum_{i,k=1}^{n} b_{ik}x_ix_k$, with coefficient matrices $A = (a_{ik})$, $B = (b_{ik})$, which may be assumed to be symmetrical without loss of generality. In the present case $A = (L_{ik})$, $B = (g_{ik})$ with u' and v' playing the role of x_1, x_2. The problem is to find a new orthogonal coordinate system with respect to which Q_a and Q_b are simultaneously transformed into quadratic forms free of all cross-product terms, or, as it can also be put, the matrices of both of the transformed forms are diagonal matrices. In linear algebra it is shown that this can always be achieved if one of the quadratic forms, Q_b say, is positive definite. Furthermore, the new coordinate axes are furnished by vectors $\mathbf{X}$ which satisfy the linear homogeneous equations $A\mathbf{X} = \lambda B\mathbf{X}$, with λ a constant, a so-called eigenvalue, to be determined. It can be shown that there are n such values of λ, which are all real if the coefficients a_{ik} and b_{ik} are real (and Q_b is a definite form), and that the corresponding vectors $\mathbf{X}_i$ are mutually orthogonal. The eigenvalues λ_i are the roots of the secular equation, or determinantal equation, $|a_{ik} - \lambda b_{ik}| = 0$.

14. Mean Curvature *H* and Gaussian Curvature *K*

Upon expanding the determinant (4.23) the result is

$$(4.24) \qquad\qquad k^2 - 2Hk + K = 0,$$

with the coefficients H and K defined as follows:

$$(4.25) \qquad H = \frac{1}{2}\frac{EN - 2FM + GL}{EG - F^2} = \frac{1}{2}(k_1 + k_2),$$

$$(4.26) \qquad K = \frac{LN - M^2}{EG - F^2} = k_1 k_2.$$

The important quantities H and K defined by (4.25) and (4.26) are called the *mean* curvature and the *Gaussian* curvature, respectively. Each one of them has in certain respects an analogy with the notion of curvature of a plane curve, as will be seen later on.

Since k_1 and k_2 are invariants with respect to parameter transformations that preserve the orientation (i.e., have a positive Jacobian) it is clear that H and K also have this property. In addition K is invariant without the restriction with regard to the sign of the Jacobian since k_1 and k_2 both change sign with it. It will be seen later that K has still another highly important and interesting invariance property.

15.   Another Definition of the Gaussian Curvature K

Of the two notions of curvature introduced above, the Gaussian curvature plays by far the more important role in surface theory. First of all it is to be noted that the sign of K fixes in a qualitative way the shape of a surface near a given point, in view of (4.26), as was just discussed above: if K is positive, the point is elliptic, if K is negative hyperbolic, and if K is zero either parabolic or planar. It is shown next that the Gaussian curvature K can be defined by using the spherical image of the normal to the surface in a manner exactly analogous to the procedures used to define the curvature and torsion of space curves. The unit normal vectors $\mathbf{X}_3(u, v)$ of a region R of a surface S given by $\mathbf{X}(u, v)$ are translated to a point 0; the locus of the end points of these vectors is a point set Ω on the unit sphere called *the spherical image* of the portion R of S (see Fig. 4.7). The total curvature K_T of R is by definition

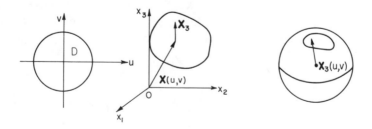

Fig. 4.7   The spherical image of a surface.

the area covered by Ω; this quantity (cf. (4.14) with $\mathbf{X}(u, v)$ replaced by $\mathbf{X}_3(u, v)$) is therefore defined by

$$K_T = \iint\limits_{D} \sqrt{(\mathbf{X}_{31} \times \mathbf{X}_{32})^2} \, du \, dv.$$

The domain D is, of course, the image of R in the parameter plane. Gauss proved that K_T *is the integral over R of the Gaussian curvature K with respect to the area of R*, i.e., that the formula

$$K_T = \iint\limits_{R} K \, dA = \iint\limits_{D} K \sqrt{(\mathbf{X}_1 \times \mathbf{X}_2)^2} \, du \, dv$$

holds. This fact is proved by establishing the much more restrictive formula

(4.27) $$\mathbf{X}_{31} \times \mathbf{X}_{32} = K(\mathbf{X}_1 \times \mathbf{X}_2),$$

which is valid for all regular surfaces. Once this is done it is clear that *the ratio of the elements of area of the spherical image and of the corresponding element of area of the surface is the Gaussian curvature.* The formula (4.27) also shows that the spherical image $\mathbf{X}_3(u, v)$ is itself a regular surface if K is different from zero, and also that the orientation fixed on a surface by the coordinate vectors $\mathbf{X}_1$ and $\mathbf{X}_2$ is the same, or the reverse, of that of its spherical image according to whether K is positive or negative.

To prove that (4.27) holds it is noted first that $\mathbf{X}_3 \cdot \mathbf{X}_{31} = \mathbf{X}_3 \cdot \mathbf{X}_{32} = 0$, since $\mathbf{X}_3 \cdot \mathbf{X}_3 = 1$, and therefore the vectors $\mathbf{X}_{3i}$ lie in the tangent plane. Hence $\mathbf{X}_{31}$ and $\mathbf{X}_{32}$ can be expressed, with the aid of certain scalar coefficients a_i^j, in the form

$$\mathbf{X}_{31} = a_1^1 \mathbf{X}_1 + a_1^2 \mathbf{X}_2,$$

$$\mathbf{X}_{32} = a_2^1 \mathbf{X}_1 + a_2^2 \mathbf{X}_2,$$

and consequently

$$\mathbf{X}_{31} \times \mathbf{X}_{32} = \begin{vmatrix} a_1^1 & a_1^2 \\ a_2^1 & a_2^2 \end{vmatrix} (\mathbf{X}_1 \times \mathbf{X}_2),$$

as an easy calculation shows. It is convenient to write $\mathbf{X}_{3i} = \sum_j a_i^j \mathbf{X}_j$ and recall that $g_{ij} = \mathbf{X}_i \cdot \mathbf{X}_j$ and $L_{ij} = \mathbf{X}_3 \cdot \mathbf{X}_{ij} = -\mathbf{X}_{3i} \cdot \mathbf{X}_j$ [cf. (4.18)]. As a result, the following relations hold: $\mathbf{X}_k \cdot \mathbf{X}_{3i} = -L_{ik} = \sum_j a_i^j g_{jk}, i, k, j = 1, 2$. It follows immediately from the rule for the product of two determinants that

$$|L_{ik}| = |a_i^j| \cdot |g_{ik}|,$$

or that

$$|a_i^j| = \frac{|L_{ik}|}{|g_{ik}|} = \frac{LN - M^2}{EG - F^2} = K,$$

which proves the correctness of (4.27) in view of (4.26).

16. Lines of Curvature

A large and interesting part of surface theory deals with families of curves on a surface that have special geometrical properties. The discussion above concerning the manner of variation of the normal curvatures at a point of a surface leads rather naturally to the introduction of two special families of curves that are called *lines of curvature* and *asymptotic lines*. These are, as the names indicate, curves on the surface that are at all of their points *either tangent to a direction of principal curvature* or *tangent to a direction of an asymptote of the Dupin indicatrix*, i.e., tangent to a direction for which the normal curvature is zero.

Consider first the case of the lines of curvature. It was seen above that there are exactly two principal directions at any point, which are in addition orthogonal, provided that the point is neither an umbilical point nor a planar point: in the latter cases all directions are principal directions. Thus, excluding these exceptional cases, a pair of uniquely determined orthogonal principal directions at every point is defined at each point of a surface. These directions are given by two different pairs of values $(u', v') \neq (0, 0)$ that satisfy the equations preceding (4.23) when k is given one or the other of the principal values k_1 and k_2. This in turn requires the vanishing of the following determinant:

$$\begin{vmatrix} Lu' + Mv' & Eu' + Fv' \\ Mu' + Nv' & Fu' + Gv' \end{vmatrix} = 0.$$

For later purposes, it is of advantage to put this condition in the following equivalent form:

$$(4.28) \qquad \begin{vmatrix} v'^2 & -u'v' & u'^2 \\ E & F & G \\ L & M & N \end{vmatrix} = 0.$$

Since u' and v' do not vanish simultaneously, this equation can be interpreted as a quadratic equation in one or the other of the ratios $v'/u' = dv/du$ or $u'/v' = du/dv$, say the former, in which case (4.28) can be replaced in the small by two *first-order differential equations* of the form

$$(4.29) \qquad \frac{du}{dv} = f_1(u, v), \qquad \frac{du}{dv} = f_2(u, v), \qquad f_1 \neq f_2.$$

What was done here in arriving at the differential equations (4.29) is a special case of an important general process of constant occurrence in differential geometry; and thus such processes warrant some discussion. The differential equations (4.29) are obviously *defined in the parameter plane*, where they define at each point the slope of a line.   They were obtained by considering direction fields on a surface that was represented by the vector $\mathbf{X}(u, v)$.   If the surface is assumed to have continuous third derivatives it is clear from (4.28) that the functions $f_1(u, v)$ and $f_2(u, v)$ have continuous first derivatives with respect to u and v, since the coefficients E, F, G, L, M, N have that property.   It follows from the theory of ordinary differential equations (cf. Appendix B) that a family of solution curves of the equations (4.29) exists in the small, each curve being uniquely determined when an initial point (u_0, v_0) is chosen.   The curve family is therefore given by $u = u(v; u_0, v_0)$ such that $u = u_0$ for $v = v_0$.   These solution curves are regular curves in the u,v-plane simply because they are given in the form $u = u(v; u_0, v_0)$ and they have continuous first derivatives in all three arguments.   Each such curve has as image curve on the surface the curve $\mathbf{Y}(v) \equiv \mathbf{X}(u(v; u_0, v_0), v)$ with a tangent vector $\mathbf{Y}' = d\mathbf{Y}/dv = \mathbf{X}_u(du/dv) + \mathbf{X}_v$, and this vector in the present case falls in a principal direction.   Thus first-order differential equations of the type (4.29) defined in the parameter plane give rise to vector fields on the surface, and hence also to fields of undirected line elements.   In this sense equations such as (4.29) are regarded as differential equations defining direction fields on the surface.   When the field is differentiable the regular solution curves of it in the parameter plane correspond to regular curves on the surface.   If such a curve is given by $u = u(t)$, $v = v(t)$ in the parametric form the corresponding curve $\mathbf{X}(t) = \mathbf{X}(u(t), v(t))$ has as tangent vector $\mathbf{X}'(t) = \mathbf{X}_u u' + \mathbf{X}_v v'$, and either $u'/v' = du/dv$ or $v'/u' = dv/du$ are defined as differentiable functions of u and v.   In the case of the lines of curvature, for example, it should be noted, however, that while these two families are orthogonal on the surface, it is not true in general that their images in the parameter plane are orthogonal in the Euclidean metric of that plane.

Thus the theory of ordinary differential equations guarantees the existence of one and only one pair of lines of curvature through a given point. Ruled out in this discussion are planar points, for which $L = M = N = 0$, and umbilical points, for which $L/E = M/F = N/G$.   Such points are called *singular points of the direction field*, since the field has no definite direction at such points, as is seen immediately from (4.28).   The existence and uniqueness theorems for the differential equations still apply, but a solution of the differential equations containing such a point does not in general represent a regular curve in the sense of differential geometry.   (The questions touched upon here are discussed at greater length in Chapter VIII with respect to the interrelation between vector fields on a surface and in the parameter plane, and also with respect to singularities of vector fields.)

The lines of curvature have various interesting geometrical properties. To begin with, they can be characterized as the only curves along which the derivative $\mathbf{X}_3'$ of the normal vector falls along the tangent $\mathbf{X}'$ of the curve or, as it could also be put, they are the only curves with tangents parallel to the tangents of their own spherical images. The proof of this, in turn, leads to an interesting and useful formula due to Rodrigues, obtained as follows. Set $u = u_1$, $v = u_2$ and introduce again the notation $g_{ij} = \mathbf{X}_i \cdot \mathbf{X}_j$, $L_{ij} = \mathbf{X}_3 \cdot \mathbf{X}_{ij} = -\mathbf{X}_{3i} \cdot \mathbf{X}_j$ [cf. (4.18)] for the coefficients of the two fundamental forms. The formula (4.22) for the curvature k of the normal sections then takes the form

$$ k = \frac{\sum_{i,j} L_{ij} u_i' u_j'}{\sum_{i,j} g_{ij} u_i' u_j'}, \qquad i,j = 1,2. $$

The derivatives of k with respect to u' and v' are

$$ \frac{1}{2} \frac{\partial k}{\partial u_j'} = \frac{\sum_i (L_{ij} - k g_{ij}) u_i'}{\sum_{i,j} g_{ij} u_i' u_j'} = -\frac{\sum_i (\mathbf{X}_{3i} + k\mathbf{X}_i) u_i' \cdot \mathbf{X}_j}{\sum_{i,j} g_{ij} u_i' u_j'}, $$

when the formulas for the coefficients L_{ij} and g_{ij} are used. But the following formula is valid:

$$ \sum_i (\mathbf{X}_{3i} + k\mathbf{X}_i) u_i' = \mathbf{X}_3' + k\mathbf{X}' $$

since, for example, $\mathbf{X}' = \mathbf{X}_1 u_1' + \mathbf{X}_2 u_2' = \sum_i \mathbf{X}_i u_i'$. If the pair of values u_1', u_2' determines a principal direction, the equations $\partial k/\partial u_j' = 0$ must hold, so that $(\mathbf{X}_3' + k\mathbf{X}') \cdot \mathbf{X}_j = 0$ for $j = 1,2$. But since the vectors $\mathbf{X}_j$ in the tangent plane are linearly independent and the vector $\mathbf{X}_3' + k\mathbf{X}'$ is in their plane, it follows that

(4.30) $\mathbf{X}_3' + k\mathbf{X}' = 0$

holds in a principal direction. Of course, k must be given the value that belongs to that direction. This is the *formula of Rodrigues*. To complete the discussion it is shown that if (4.30) holds for a certain direction it must be a principal direction; thus the condition (4.30) would be shown to be characteristic for such directions. To do so it is convenient to choose parameters in such a way that $\mathbf{X}_1$ and $\mathbf{X}_2$ are themselves orthogonal and in principal directions. [For example, this could be arranged in the way it was done in deriving (4.21).] Thus the equations $\mathbf{X}_{31} + k_1\mathbf{X}_1 = 0$, $\mathbf{X}_{32} + k_2\mathbf{X}_2 = 0$ would hold. For *any* direction fixed by (u_1', u_2') the following equations then result:

(4.31)
$$ \mathbf{X}' = \mathbf{X}_1 u_1' + \mathbf{X}_2 u_2', $$
$$ \mathbf{X}_3' = \mathbf{X}_{31} u_1' + \mathbf{X}_{32} u_2' = -k_1 u_1' \mathbf{X}_1 - k_2 u_2' \mathbf{X}_2. $$

Hence $\mathbf{X}'$ and $\mathbf{X}_3'$ are parallel if and only if

$$\begin{vmatrix} u_1' & u_2' \\ k_1 u_1' & k_2 u_2' \end{vmatrix} = (k_2 - k_1)u_1'u_2' = 0.$$

If $k_1 = k_2$, i.e., at planar and umbilical points, all directions are principal directions; if $k_1 \neq k_2$, then either $u_1' = 0$ or $u_2' = 0$, i.e., the direction is a principal direction since it falls along either $\mathbf{X}_1$ or $\mathbf{X}_2$. Thus the condition (4.30) has been shown to be characteristic for the principal directions.

17. Third Fundamental Form

It is possible to derive another interesting formula with the aid of the formulas (4.31)—again assuming the coordinate vectors $\mathbf{X}_1$ and $\mathbf{X}_2$ to be in orthogonal principal directions. By combining these formulas the following relations result with respect to differentiation along an arbitrary curve:

$$\mathbf{X}_3' + k_1 \mathbf{X}' = a\mathbf{X}_2,$$
$$\mathbf{X}_3' + k_2 \mathbf{X}' = b\mathbf{X}_1,$$

with a and b certain scalars, the values of which play no role in what follows. Scalar multiplication of these equations with one another yields the equation

$$(4.32) \qquad \mathbf{X}_3' \cdot \mathbf{X}_3' + (k_1 + k_2)\mathbf{X}_3' \cdot \mathbf{X}' + k_1 k_2 \mathbf{X}' \cdot \mathbf{X}' = 0,$$

since $\mathbf{X}_1 \cdot \mathbf{X}_2 = 0$. The *third fundamental form III* is defined, rather naturally, by the formula

$$(4.33) \qquad\qquad III = \mathbf{X}_3' \cdot \mathbf{X}_3' = (\mathbf{X}_3')^2.$$

The form III evidently furnishes the line element of the spherical image of the surface. In a moment it will be shown that $\mathbf{X}_3' \cdot \mathbf{X}' = -II$, and since $I = \mathbf{X}' \cdot \mathbf{X}'$, (4.32) yields the following identity that holds for the three fundamental forms:

$$(4.34) \qquad\qquad III - 2HII + KI = 0,$$

since $H = \frac{1}{2}(k_1 + k_2)$ and $K = k_1 \cdot k_2$ [cf. (4.25) and (4.26)]. Thus the three forms are not independent. It is still to be shown that $II = -\mathbf{X}_3' \cdot \mathbf{X}'$, but this is done easily by the following calculation:

$$II = \sum_{i,j} \mathbf{X}_{ij} \cdot \mathbf{X}_3 u_i' u_j' = - \sum_i \mathbf{X}_i u_i' \cdot \sum_j \mathbf{X}_{3j} u_j' = -\mathbf{X}' \cdot \mathbf{X}_3'.$$

It is of interest to observe that (4.34) is a relation that is invariant with respect to parameter transformations since the forms III, I, and the Gaussian curvature K have that property, and although the form II changes its sign if the Jacobian of the transformation is negative, so also does H: thus while the *derivation* of the formula made use of special parameters the end result is seen to be invariant with respect to all parameter transformations.

18. Characterization of the Sphere as a Locus of Umbilical Points

Earlier it was shown that the only regular surfaces with continuous second derivatives all of whose points are planar points are the planes. With the aid of the formula of Rodrigues it will now be shown that the only regular surfaces with continuous third derivatives, *all of whose points are umbilical points are the spheres.* In this case every regular curve on the surface is a line of curvature (cf. Section 16) and therefore $\mathbf{X}_3' + k\mathbf{X}' = 0$ along all curves. At each point k is the same for all directions, but it is not clear a priori that it is constant in the parameters u, v. This is shown, however, as follows. Along the parameter curves themselves the following equations hold [cf. (4.30)]:

$$\mathbf{X}_{31} + k\mathbf{X}_1 = 0,$$
$$\mathbf{X}_{32} + k\mathbf{X}_2 = 0.$$

These are identities in the parameters u and v and hence, by differentiation with respect to v and u, two further identities result:

$$\mathbf{X}_{312} + k_v\mathbf{X}_1 + k\mathbf{X}_{12} = 0,$$
$$\mathbf{X}_{321} + k_u\mathbf{X}_2 + k\mathbf{X}_{21} = 0.$$

By assumption $\mathbf{X}$ has continuous third derivatives and therefore $\mathbf{X}_{12} = \mathbf{X}_{21}$ and $\mathbf{X}_{312} = \mathbf{X}_{321}$. Thus a subtraction of one of the two equations from the other yields the following identity:

$$k_v\mathbf{X}_1 - k_u\mathbf{X}_2 = 0.$$

From this it follows that $k_u = k_v = 0$ for all u, v since $\mathbf{X}_1$ and $\mathbf{X}_2$ are linearly independent. This proves that k is constant over the surface. Once this is known it follows from the validity of $\mathbf{X}_3' + k\mathbf{X}' = 0$ along all curves on the surface, that the equation $\mathbf{X}_3 + k\mathbf{X} = \mathbf{a}$, where $\mathbf{a}$ is a constant vector, is valid at all points on the surface. Hence

$$\mathbf{X} = \frac{\mathbf{a} - \mathbf{X}_3}{k},$$

and $\mathbf{X}$ is seen to determine a sphere of radius $1/k$ and center $\mathbf{a}/k$ since it is a locus of points at the fixed distance $1/k$ from the point located by the vector $\mathbf{a}/k$.

19. Asymptotic Lines

The asymptotic lines are defined as curves that are everywhere tangent to an asymptote of the Dupin indicatrix, i.e., to directions for which the normal curvature k is zero, and are thus in directions such that $\mathrm{II} = 0$ [cf. (4.20)]. The differential equation for these curves is therefore

$$(4.35) \qquad Lu'^2 + 2Mu'v' + Nv'^2 = 0.$$

Only in hyperbolic and parabolic points are the direction fields furnished by (4.35) real: in the elliptic case the discriminant $LN - M^2$ is negative and (4.35) then holds for no real values of u' and v' except $u' = v' = 0$. In the hyperbolic case there are two distinct sets of solution curves; in the parabolic case only one.

It was seen [cf. (4.19) and the subsequent discussion] that

$$\kappa \cos \theta = L\dot{u}^2 + 2M\dot{u}\dot{v} + N\dot{v}^2,$$

where κ is the curvature of any curve on the surface with a tangent defined by $\dot{u}$ and $\dot{v}$ (the arc length is parameter, and θ is the angle between $\mathbf{X}_3$ and $\mathbf{v}_2$, the principal normal of the curve). If the curvature κ of an asymptotic line is not zero, it follows that $\cos \theta = \mathbf{X}_3 \cdot \mathbf{v}_2 = 0$, or in other words the osculating plane of the curve coincides with the tangent plane of the surface. In case $\kappa = 0$ all along a curve on a surface, it is a straight line, and hence all straight lines on a surface are asymptotic lines.

20. Torsion of Asymptotic Lines

It is of interest to calculate the torsion of an asymptotic line that has nonvanishing curvature. For the moving trihedral of the curve (with the arc length as parameter) the following formula holds [cf. (3.7)]:

$$\mathbf{v}_3 = \mathbf{v}_1 \times \mathbf{v}_2 = \frac{1}{\kappa}(\dot{\mathbf{X}} \times \ddot{\mathbf{X}}).$$

One of the Frenet equations [cf. (3.12)] states that

$$\dot{\mathbf{v}}_3 = -\tau\mathbf{v}_2.$$

For an asymptotic line with $\kappa \neq 0$ the binormal $\mathbf{v}_3$ is defined and $\mathbf{v}_3 = \mathbf{X}_3$, as was proved in the preceding section. But [cf. (4.34)] the following identity relating the three fundamental forms holds:

$$\mathrm{III} + 2H\mathrm{II} + K\mathrm{I} = 0,$$

and since $\text{III} = \dot{\mathbf{X}}_3^2 = (\dot{\mathbf{v}}_3)^2 = \tau^2$, $\text{II} = 0$ (in an asymptotic direction), and $\text{I} = \dot{\mathbf{X}}^2 = 1$ it follows that

$$\tau^2 + K = 0,$$

or

(4.36) $$|\tau| = \sqrt{-K},$$

a formula attributed to Beltrami and Enneper. If $K \neq 0$, it could be shown that the two distinct asymptotic curves, which exist in that case, have opposite signs for τ.

21. Introduction of Special Parameter Curves

It is often very useful for particular purposes to introduce special sets of curves on a surface as parameter curves. Such curves are almost invariably the solution curves of certain differential equations, e.g., those of the lines of curvature or of the asymptotic lines. In these cases, the fundamental existence theorem for ordinary differential equations (cf. Appendix B) yields [if $\mathbf{X}(u, v)$ has continuous third derivatives]:

(a) *Lines of curvature:* Two orthogonal sets of regular curves, except at planar and umbilical points, which are singularities of the direction field given by the differential equation.

(b) *Asymptotic lines:* Two distinct sets of intersecting curves in hyperbolic points, one set in parabolic points, no (real) curves in elliptic points. Planar points are the only singularities.

In differential geometry much use is made of the following theorem:

Any two families of regular intersecting curves (i.e., two families such that no curve of one set is tangent to a curve of the other) *may be chosen as parameter curves, in case each such family arises from the integration of a differentiable field of directions on the surface.*

As a preliminary to proving this theorem it is advisable to review parts of the discussion of Section 16 above concerning first-order differential equations defined in the parameter plane, the solutions of which determine regular curves on the surface that are everywhere tangent to a given differentiable direction field on the surface. On the basis of that discussion it is clear that the proof of the above theorem can be carried out in the parameter plane. Thus two first-order ordinary differential equations are given in the u,v-plane, and they can be assumed without loss of generality to have the form $du/dv = f_1(u, v)$, $du/dv = f_2(u, v)$, with f_1 not equal to f_2 at any point of a neighborhood of $u = 0$, $v = 0$. It is convenient to choose the u and v axes as the bisectors of the angle between the directions of the two fields at the origin, so that both families, since they are continuous, cut both axes under nonzero angles in a neighborhood of the origin. Consider the curve family defined by $du/dv = f_1(u, v)$. A uniquely determined curve of the family results from integration of the differential equation when $u = u_0$ is prescribed for $v = 0$, i.e. when an initial point on the u-axis is chosen; in addition it is known (and not difficult to prove) that these curves fill a neighborhood of the origin, again because of the existence of a unique solution curve through any point (u, v) in a certain such neighborhood. The solution curves are thus regular plane curves given in the form $u = g(v, u_0)$ with $u_0 = g(0, u_0)$. Since the solutions depend in a differentiable way on the initial conditions and since $\partial g/\partial u_0 = 1$ for $v = 0$, it follows from the implicit function theorem that the equation $u = g(v, u_0)$ can be solved for u_0 to obtain $u_0 = p(u, v)$ and that p has continuous first derivatives with respect to u and v. The function $p(u, v)$ is uniquely defined in a certain neighborhood of the origin, since the solution curves cover such a neighborhood simply. Finally it is clear that

$$\frac{\partial p}{\partial u}\bigg|_{v=0} = 1 \quad \text{and} \quad \frac{\partial p}{\partial v}\bigg|_{v=0} = 0,$$

since $p(u, 0) \equiv u$ holds. In the same fashion the other set of field curves can be integrated and represented in the form $q(u, v) = \text{const.}$ with $q(0, v) = v$, so that $q_u = 0$ and $q_v = 1$ at the origin. The Jacobian

$$\begin{vmatrix} p_u & p_v \\ q_u & q_v \end{vmatrix}$$

therefore has the value one at the origin, which proves that $p = \text{const.}$ and $q = \text{const.}$ can be introduced as new parameter curves.

In particular, it now follows that the lines of curvature and the asymptotic lines can be introduced as parameter curves in a neighborhood of any point where these pairs of curves exist and are distinct, provided that the surface given by $\mathbf{X}(u, v)$ has continuous derivatives of third order, since that suffices to ensure that the theorems on the existence, uniqueness, and differentiable dependence on initial conditions of the appropriate differential equations hold. If $\mathbf{X}(u, v)$ has higher derivatives the new parameter transformation will also have higher derivatives—however, it is seen that the geometric quantities specified in the new parameters may have fewer derivatives with respect to them than with respect to the original parameters. This is one of the reasons why many writers in differential geometry prefer to work with surfaces of class C^∞, i.e., surfaces with derivatives of all orders.

The theorems about differential equations used in this and the following sections are formulated in Appendix B.

22.   Asymptotic Lines and Lines of Curvature as Parameter Curves

If the (u, v) curves themselves are the asymptotic lines, i.e., if $u' = 0$, $v' = 0$ yield these curves, it follows [cf. (4.35)] that $L = N = 0$; conversely, if $L = N = 0$, the parameter curves are the asymptotic lines. Thus $L = N = 0$ characterizes the curves.

If the lines of curvature are the parameter curves, it follows from (4.28) that $FN - GM = 0$ and $EM - FL = 0$. But $F = 0$ since these curves are orthogonal, and since E and G are always positive, it follows that $M = 0$. Thus a characterization for *the lines of curvature as parameter curves is that* $F = M = 0$. From this fact a useful conclusion can be drawn from the formula (4.20) for the normal curvatures: clearly, the principal curvatures k_1, k_2 are given by

$$(4.37) \qquad\qquad k_1 = \frac{L}{E}, \qquad k_2 = \frac{N}{G},$$

when the lines of curvature are used as parameter curves, and conversely if (4.37) holds the parameter curves are in principal directions.

23.   Embedding a Given Arc in a System of Parameter Curves

Later on more than one situation will be encountered in which it will be convenient to introduce a special set of parameter curves in the neighborhood of an arc of a regular curve on the surface; it is therefore useful to deal with this matter once for all at this point. Suppose that the curve C on the surface S is the image of the regular curve $C_0: u = u(s)$, $v = v(s)$ in the parameter plane; the curve C_0 is also given by the vector $\mathbf{x}(s) = (u(s), v(s))$ in that plane. What is desired is that the image of C of C_0 given by $\mathbf{X}(u(s), v(s))$ on the surface S should be embedded in one of a set of parameter curves on S in the following way. On C the parameter s should represent the arc length, and a set of regular curves orthogonal to C and lying in $\mathbf{X}(u, v)$ is then to be defined so that it covers a neighborhood of C (with an obvious reservation at its end points); on the curves orthogonal to C the arc length σ is to be taken as parameter with $\sigma = 0$ on C. The family of parameter curves to which C itself belongs is then the set of curves determined by $\sigma = $ const. In effect, it is to be shown that the surface can be represented in terms of (s, σ) as regular parameters, i.e., $\mathbf{X} = \mathbf{X}(s, \sigma)$ for $s_0 \le s \le s_1$, and for $|\sigma|$ sufficiently small, such that the curves $s = $ const. are orthogonal to C. To prove that this is always possible, consider the directions at the points $\mathbf{x}(s)$ of the curve C_0 in the parameter plane that correspond to the directions orthogonal to the image C of C_0 on S,[1] and denote unit vectors in these directions by $\mathbf{y}(s)$; clearly $\dot{\mathbf{x}}(s)$ and $\mathbf{y}(s)$ are linearly independent vectors. A regular representation of the plane near C_0 can then be given in the form $\mathbf{x}(s, \sigma_1) = \mathbf{x}(s) + \sigma_1 \mathbf{y}(s)$. In fact, $\mathbf{x}_s \times \mathbf{x}_{\sigma_1} = (\dot{\mathbf{x}} + \sigma_1 \dot{\mathbf{y}}) \times \mathbf{y}$, and since $\dot{\mathbf{x}} \times \mathbf{y} \neq 0$ it follows that $\mathbf{x}_s \times \mathbf{x}_{\sigma_1} \neq 0$ for σ_1 sufficiently small; this in turn is seen to mean that the Jacobian $\partial(u, v)/\partial(s, \sigma_1)$ does not vanish. Thus (s, σ_1) are legitimate parameters in a neighborhood of C_0 in the u,v-plane, and hence also in a neighborhood of C on S; thus the surface is given by $\mathbf{X}(s, \sigma_1)$ for $s_0 < s < s_1$ and $-\epsilon < \sigma_1 < \epsilon$. However, σ_1 is not in general the arc length along the curves $s = $ const. on S. To achieve that a new parameter σ is introduced in place of σ_1 by setting $\sigma_1 = f(\sigma, s)$, with f to be determined so that σ represents the arc lengths on the images of the curves $\mathbf{x}(s, \sigma_1)$, $s = $ const., on S. For this it suffices to choose $f_\sigma = 1/|\mathbf{X}_{\sigma_1}|$ since from $\mathbf{X}_\sigma = \mathbf{X}_{\sigma_1} f_\sigma$ it follows that $\mathbf{X}_\sigma$ would be a unit vector. But the equation $f_\sigma = |\mathbf{X}_{\sigma_1}(s, f(\sigma, s))|^{-1}$ is an ordinary differential equation (containing s as a parameter) for the function $f(\sigma, s)$ to which the existence and uniqueness theorems apply, since $\mathbf{X}_{\sigma_1} \neq 0$; in addition, $f(0, s) = 0$ is prescribed as initial condition for all values of s because of the fact that $\mathbf{X}(s, 0)$ corresponds to C, i.e., to $\sigma_1 = 0$, and thus $f(\sigma, s)$ is uniquely determined and differentiable in both arguments. Finally, the transformation $s = s$, $\sigma_1 = f(\sigma, s)$ is legitimate, since its Jacobian is f_σ and f_σ is different from zero. Thus the surface can be represented in the form $\mathbf{X}(s, \sigma)$ with parameter curves of the kind desired.

It is perhaps worth remarking that an *orthogonal* curvilinear coordinate system can always be introduced over any portion of a surface—even in the large—that is given in terms of regular parameters. It suffices, in fact, to take one of the two given families of parameter curves and join to them their orthogonal trajectories as a second family of parameter curves, since the orthogonal field of directions is differentiable. Thus it is always possible to assume that the coefficient F of the first fundamental form is everywhere zero if that is convenient. (It is perhaps worth remarking that orthogonal coordinate systems do not in general exist for surfaces of dimension greater than two.)

24. Analogues of Polar Coordinates on a Surface

It is possible to introduce in the small coordinate curves on a surface that are analogous to those of polar coordinates in the plane, and it is some-times quite useful to do so. All that is necessary is to introduce a family of

---

[1] If the tangent vector to C is given by $\dot{\mathbf{X}} = X_u \dot{u} + X_v \dot{v}$, then an orthogonal vector is obtained [cf. (4.10)] as $X' = X_u u' + X_v v'$ by choosing any pair of values $(u', v') \neq (0, 0)$ such that $Euu' + F(uv' + vu') + Gvv' = 0$. (It is an easy exercise to show that such values u', v' always exist.) The direction numbers (u', v') fix directions in the parameter plane that correspond to orthogonal directions to C on S.

regular curves issuing from a point P of the surface with each curve of the family fixed by the angle of inclination θ of its tangent at P with the coordinate vector $\mathbf{X}_1$ at P and a second family as the loci of points at a fixed distance from P along the curves of the first family. It is assumed that $s \geq 0$ represents the arc length of each curve issuing from P, with $s = 0$ at P. Thus the curve family is given by $u = u(s, \theta)$, $v = v(s, \theta)$. It is assumed that the functions u and v are defined and have continuous second derivatives with respect to s and θ for s sufficiently small and for $0 \leq \theta \leq 2\pi$. Under these circumstances it is to be shown that a neighborhood of P on S is simply covered by these curves, i.e., that every point other than P in the neighborhood lies on one and only one of the curves issuing from P, and that s and θ are parameters in terms of which the surface is regular except at $s = 0$, i.e., at P itself. To accomplish this new variables x, y are introduced by the transformation T given by

$$x = s \cos \theta,$$

$$y = s \sin \theta,$$

in which s and θ are regarded as polar coordinates in an x,y-plane so that T and its inverse T^{-1} are nonsingular except for $s = 0$. It follows that the functions $u(s(x, y), \theta(x, y)) \equiv u(x, y)$ and $v \equiv v(x, y)$ have continuous derivatives with respect to x and y except possibly at point P itself. It will be shown, however, that the assumptions made above ensure that the first derivatives of u and v with respect to x and y are continuous at P, and that the Jacobian $\partial(u, v)/\partial(x, y)$ does not vanish at P. In other words, the situation is qualitatively the same as it is for polar coordinates in the plane.

To prove these statements it is convenient to begin by choosing parameters u and v (as was done in Section 12 above) such that, at point P, $u = v = 0$ and $E = G = 1$, $F = 0$. In other words, the coordinate vectors $\mathbf{X}_1$ and $\mathbf{X}_2$ are orthogonal unit vectors at P. It is readily seen under these circumstances that

(4.38) $$\left.\frac{\partial u}{\partial s}\right|_{0, \theta} = \cos \theta, \quad \left.\frac{\partial v}{\partial s}\right|_{0, \theta} = \sin \theta,$$

since s represents the arc length on the regular curves $u = u(s, \theta)$, $v = v(s, \theta)$ for a given value of θ. For $s \neq 0$ the following equations hold, upon using the inverse T^{-1} of T:

$$u_x = u_s \cos \theta - \frac{u_\theta \sin \theta}{s},$$

$$u_y = u_s \sin \theta + u_\theta \frac{\cos \theta}{s}.$$

These equations may be rewritten in the following slightly different form:

(4.39)

$$u_x - 1 = (u_s - \cos \theta) \cos \theta - \left(\frac{u_\theta}{s} + \sin \theta\right) \sin \theta,$$

$$u_y = (u_s - \cos \theta) \sin \theta + \left(\frac{u_\theta}{s} + \sin \theta\right) \cos \theta.$$

From equations (4.38) above it is seen that

$$\lim_{s \to 0} (u_s - \cos \theta) = 0$$

uniformly in θ since u_s is assumed to be continuous for sufficiently small values of s and for $0 \le \theta \le 2\pi$. Also, since $u_\theta(0, \theta) = 0$ for all θ and $\sin \theta = -u_{s\theta}(0, \theta)$ from (4.38), it follows that

$$\frac{u_\theta}{s} + \sin \theta = \frac{u_\theta(s, \theta) - u_\theta(0, \theta)}{s} - u_{s\theta}(0, \theta).$$

From this it is seen that

$$\lim_{s \to 0} \left(\frac{u_\theta}{s} + \sin \theta\right) = u_{\theta s}(0, \theta) - u_{s\theta}(0, \theta) = 0,$$

since the second derivatives of $u(s, \theta)$ are assumed to be continuous; this limit also holds uniformly in θ. It follows therefore from (4.39) that $u_x \to 1$ and $u_y \to 0$ as $(x, y) \to 0$. It is known that the derivatives of u and v with respect to x and y are continuous except possibly at P. It will now be shown that these derivatives exist and are continuous at P. This follows from the following lemma: If $f(\xi)$ is continuous in $0 \le \xi \le \xi_0 (\xi_0 > 0)$ and if $f'(\xi)$ is continuous for $0 < \xi < \xi_0$, and $\lim_{\xi \to 0} f'(\xi) = a$, then $f'(0)$ exists and has the value a. To prove this lemma it need only be observed that

$$\frac{f(\xi) - f(0)}{\xi} = f'(\xi^*), \qquad 0 < \xi^* < \xi,$$

by the mean value theorem [which does not require that $f'(\xi)$ be defined at $\xi = 0$], and the lemma follows by allowing ξ to approach 0. All the conditions of the lemma are fulfilled by the functions u and v, and consequently their first derivatives exist at the origin and have as values either zero or one. They are also continuous at the origin, since the limits are the same for these derivatives on any sequences approaching the origin. Thus the functional determinant $\partial(u, v)/\partial(x, y)$ is defined and continuous at the origin, where its value is one, and therefore a neighborhood of the point P on the surface is mapped in a one-to-one way on a neighborhood of the origin of the x,y-plane.

This construction of a local parameter system resembling polar coordinates was done without reference to a differential equation. However, in using it later on the curves $u(s, \theta)$, $v(s, \theta)$ will be defined as solutions of a pair of second-order ordinary differential equations that are uniquely determined once θ is given at $s = 0$; in addition, the differentiability of these functions will also be a consequence of the fact that they satisfy a differential equation on a surface with properties such that derivatives of an appropriate order exist.

Later on it will be quite important to make use of all of the types of parameter curves discussed above. In fact, all books on the subject do so, but they usually assume the validity of such parameter transformations without proof. It is true that the matter has been disposed of without too much difficulty—although some fairly delicate arguments were needed. Nevertheless it seems reasonable and appropriate to carry out the proofs in this book since such important matters as the proof of the existence and uniqueness of the arcs of shortest length on a surface in a neighborhood of a point, the derivation of a necessary condition for such arcs of shortest length, and the discussion of the properties of spaces of constant Gaussian curvature that lead to models of the non-Euclidean geometries all are dealt with conveniently through the use of one or the other of these types of coordinate systems.

PROBLEMS

1. A surface is given by an equation of the form $z = f(x, y)$:

(a) Determine the first and second fundamental forms.

(b) Find the differential equations for the lines of curvature and the asymptotic lines.

2. Determine the asymptotic lines on the hyperboloid of one sheet. Show that $K < 0$ holds for this surface.

3. Find sufficient conditions for a relative maximum or minimum of a function $f(x, y)$.

4. Denote by $k(\phi)$ the normal curvature of a surface in a direction fixed by the angle ϕ in the tangent plane. Show that the mean curvature H is given by $H = \frac{1}{2}(k(\phi) + k(\phi + \pi/2))$ for all values of ϕ. Also show that $H = (1/2\pi) \int_0^{2\pi} k(\phi) \, d\phi$. These results show that H is quite properly called the mean curvature.

5. Prove that asymptotic lines on which $\kappa \neq 0$ can be characterized as curves that are orthogonal to their spherical images.

6. Prove that a closed regular surface in three-dimensional space cannot have Gaussian curvature that is everywhere negative. Prove that it must have points of positive Gaussian curvature. *Hint:* to prove the first statement bring up a plane from infinity until a first point of contact with the surface occurs. To prove the second statement enclose the surface in a large sphere.

7. Prove that the integral of the Gaussian curvature over any closed convex surface is 4π.

8. Show that the integral of the Gaussian curvature over the whole of any elliptic paraboloid is 2π.

9. Show that the line element of the spherical image of a surface (that is, the third fundamental form) is given by $k_1^2\,du^2 + k_2^2\,dv^2$ when the lines of curvature are chosen as parameters.

10. If the line element is given by $ds^2 = E\,du^2 + 2F\,du\,dv + G\,dv^2$ when the u and v curves are the asymptotic lines, show that the third fundamental form is given by

$$-K\,(E\,du^2 - 2F\,du\,dv + G\,dv^2).$$

CHAPTER V

Some Special Surfaces

This chapter is largely concerned with *surfaces of revolution*—whose definition is obvious—and *developable surfaces*, which are defined as surfaces with Gaussian curvature K that is everywhere zero. These two special classes of surfaces are quite interesting for their own sakes, and they also serve well as concrete illustrations and applications of the general theory developed in the preceding chapter.

1. Surfaces of Revolution

A surface of revolution is defined as the locus of points generated by the rotation of a regular plane curve $x_2 = f(t)$, $x_3 = g(t)$ about the x_3-axis (cf. Fig. 5.1). In terms of the angle ϕ, which locates a meridian plane, the surface is given by the vector function $\mathbf{X}(t, \phi)$ in the form

(5.1) $$\mathbf{X}(t, \phi) = (f(t) \cos \phi, f(t) \sin \phi, g(t)).$$

The parameter curves $t = $ const. are called circles, or parallels, of latitude, and the curves $\phi = $ const. are called meridians.

To investigate whether there might be singular points in this representation $\mathbf{X}_t \times \mathbf{X}_\phi$ is calculated, with the result

$$\mathbf{X}_t \times \mathbf{X}_\phi = f \cdot (-g' \cos \phi, -g' \sin \phi, f').$$

Since $f'^2 + g'^2 \neq 0$, because of the assumed regularity of the meridian curve, it follows that $\mathbf{X}_t$ and $\mathbf{X}_\phi$ are linearly independent if $f \neq 0$, i.e., everywhere except possibly on the axis of rotation. Even on the axis of rotation the surface is regular if the meridians cut it at right angles. In that case the surface can be represented in a neighborhood of the axis of rotation by $x_3 = h(r)$, with $h'(0) = 0$, $r^2 = x_1^2 + x_2^2$, and the function x_3 is differentiable at $r = 0$ with respect to x_1 and x_2 as many times as it is with respect to r, as is easily seen. From $\mathbf{X} = (x_1, x_2, x_3(x_1, x_2))$ the vector $\mathbf{X}_1 \times \mathbf{X}_2 = (0, 0, 1)$ results

109

for $r = 0$, and the surface is therefore regular in the parameters x_1, x_2. This is an example in which one set of surface parameters—t and ϕ, that is—do not furnish a regular representation at a certain point, but the surface is seen to be regular if appropriate new parameters are introduced.

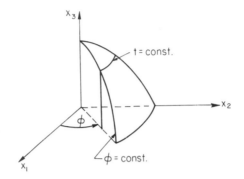

Fig. 5.1   A surface of revolution.

Consider the representation of the surface in the form (5.1). The parameter curves are lines of curvature in this case. This could of course be proved by showing that $F = M = 0$, but it is also easy to verify geometrically, as follows. The normal $\mathbf{X}_3$ to the surface along a meridian has a derivative $\mathbf{X}_3'$ that clearly lies in the direction of $\mathbf{X}'$ along that curve. The normal $\mathbf{X}_3$ passes through a fixed point of the axis of rotation when the point to which it belongs moves along a parallel; from this it is again seen that $\mathbf{X}'$ and $\mathbf{X}_3'$ are parallel along such a curve, since in that case $\mathbf{X} = \mathbf{b} + a\mathbf{X}_3$, with $\mathbf{b}$ a constant vector and a a constant scalar (cf. Fig. 5.2). Thus for both sets of curves $\mathbf{X}_3' + k\mathbf{X}' = 0$, and the formula of Rodrigues (cf. Chapter IV, Section 16) characterizes these directions as principal directions.

It is of interest to consider the location of possible planar and umbilical points, since they are singularities for the fields of principal directions on the surface. If the meridian curve cuts the axis of rotation at right angles it was seen that the surface is regular there, and it is clear that such a point is either a planar point or an umbilical point since all directions through it are evidently principal directions. It is easily proved that a planar point results if the curvature of the meridian curve vanishes at the axis; otherwise, the point is an umbilical point. For the paraboloid $x_3 = x_1^2 + x_2^2$ the origin is an umbilical point, but for $x_3 = (x_1^2 + x_2^2)^2$ the origin is a planar point.

There can be planar points and umbilical points that do not lie on the axis of rotation. To investigate these and other possibilities it is to be recalled that the principal curvatures k_1 and k_2 are the curvatures of normal sections that are either tangent to the parallels or to the meridians, since these curves are tangent to the principal directions.   Suppose, for instance, that k_1 is the curvature of the normal section that is tangent to a parallel. The formula of Rodrigues

$$(5.2) \qquad \qquad \mathbf{X}_3' + k_1 \mathbf{X}' = 0$$

then holds along the parallel.   But k_1 is constant along this curve, since the surface is a surface of revolution, and as a consequence an integration can be performed along the curve to obtain

$$(5.3) \qquad \qquad \mathbf{X}_3 + k_1 \mathbf{X} = \mathbf{a},$$

with $\mathbf{a}$ a constant vector.

It is useful to distinguish two cases, i.e., $k_1 = 0$ and $k_1 \neq 0$.   In the former case $\mathbf{X}_3 = \mathbf{a}$, with $\mathbf{a}$ a constant vector, would hold along the parallel curve, which is possible only if $\mathbf{X}_3$ is parallel to the axis of rotation (cf. Fig. 5.2), and this in turn requires the tangent to the meridian curve at such a point to be orthogonal to the axis of rotation.   Conversely it is to be noted that k_1 *must* be zero if $\mathbf{X}_3$ is parallel to the axis of rotation since $\mathbf{X}_3'$ vanishes along the curve in that case, and since $\mathbf{X}'$ does not vanish it follows from (5.2) that $k_1 = 0$.   Thus such points are certainly always either parabolic or planar points.   They will be planar points if at the same time the curvature of the meridian curve at such a point is zero, since then k_2 as well as k_1 would be zero, and this means that the curvatures of all normal sections are zero. Parabolic points clearly occur at any points where the curvature of a meridian curve vanishes.

To investigate the second case, in which $k_1 \neq 0$ is assumed, it is convenient to write (5.3) in the form

$$\mathbf{X} + \frac{1}{k_1} \mathbf{X}_3 = \mathbf{b},$$

with $\mathbf{b}$ a constant vector.   This formula holds along any parallel; it therefore follows, because of rotational symmetry, that $\mathbf{b}$ locates the point Q on the axis of rotation where the normal vector $\mathbf{X}_3$ cuts that axis, and that $|1/k_1|$ is the distance $\overline{PQ}$ from this point to the surface point P with $\mathbf{X}_3$ as normal (cf. Fig. 5.2).   Thus the principal curvature k_1 is different in numerical value from the curvature $1/r$ of the corresponding line of curvature (which is a circle of radius r) unless the normal $\mathbf{X}_3$ on it is orthogonal to the axis of rotation. (This is a special case of a general fact, i.e., that the principal curvature has only exceptionally the same value as the curvature of the corresponding line of curvature.)

It is of interest to locate the elliptic points $(K > 0)$ and the hyperbolic points $(K < 0)$ on surfaces of revolution.   Points with $K > 0$ are generated

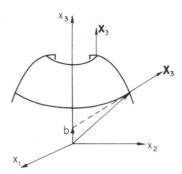

Fig. 5.2   Principal curvatures of surfaces of revolution.

by segments of meridian curves that are *concave* toward the axis of rotation, while for $K < 0$ the opposite holds (cf. Fig. 5.3).   The most direct way to verify this is to observe that the tangent plane has its point of tangency as the only contact with the surface points in a neighborhood of that point when $K > 0$ holds but cuts the surface when $K < 0$ holds.   In addition, as long as $\mathbf{X}_3$ is not parallel to the axis of rotation (and that is meant to be implied in the requirement calling for convexity or concavity) the principal curvature k_1 clearly does not vanish, and k_2 also does not vanish since the meridian segment is supposed to have nonzero curvature.   It is a worthwhile exercise

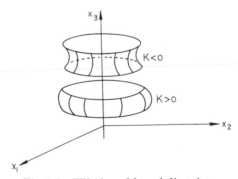

Fig. 5.3   Elliptic and hyperbolic points.

also to visualize what happens with respect to the spherical image of the surface in the two cases, and, in particular, to observe that the tangent vector $\mathbf{X}_3'$ of the image of a parallel curve has the same direction as the tangent to that curve in either case, but the vector $\mathbf{X}_3'$ is the same in direction as the tangent to the meridian curve if $K > 0$ holds, but is opposite in direction if $K < 0$ holds.   Thus the orientation of the spherical image is the same as that of the surface for $K > 0$, opposite to it for $K < 0$.   The situation quite generally is illustrated by Fig. 5.4—without restriction particularly to surfaces of revolution.

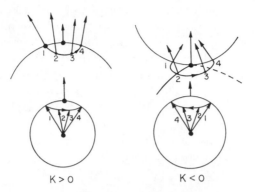

$$K > 0 \qquad\qquad K < 0$$

Fig. 5.4   Orientation of the spherical image.

The case of the torus, i.e., the surface generated by rotating a circle about a line in its plane that does not cut it (cf. Fig. 5.5), is of particular interest. It contains two parallels made up of *parabolic points*, i.e., the curves generated by the high and low points of the circle, since $k_2 \neq 0$ at these points. The formula (4.30) of Beltrami and Enneper applies to these curves on the torus, since they are asymptotic lines with $k \neq 0$; thus the torsion of these curves is zero—and they are indeed plane curves. Consider next the spherical

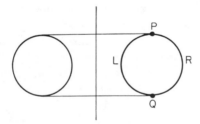

Fig. 5.5   The torus.

image of the torus. Clearly, the unit sphere is covered twice in this mapping by means of the normal $\mathbf{X}_3$ since the images of the surface points due to rotation of the arc PRQ alone cover the whole sphere, and the same is true for points on the arc PLQ. The north and south poles of the sphere correspond in this mapping to the uppermost and lowermost circles of latitude of the torus—and this is not surprising, since the Gaussian curvature K is zero there and the spherical image has been shown to be a one-to-one mapping of the surface only in the neighborhood of points for which $K \neq 0$ holds. The total curvature $K_T = \iint K \, da$ taken over the whole torus can now be seen to be zero, since the two coverings of the sphere are such that K is positive on the one which corresponds to the convex part of the torus, and negative on the other, and K_T is on the other hand the area of the spherical image [cf. (4.27)] taken with the appropriate sign. The parabolic points of the torus, which map into the north and south poles of the spherical image, thus correspond to branch points of the spherical image mapping through which it is necessary to pass in order to go continuously from one sheet of the spherical image to the other. Later on it will be seen that any simple closed surface of the same genus as the torus, i.e., any regular surface that is the topological image of a torus, also has $K_T = 0$—in fact, the number K_T is a *topological* invariant of the simple closed surfaces.

2.   Developable Surfaces in the Small Made Up of Parabolic Points

Developable surfaces are defined here as the *regular surfaces for which the Gaussian curvature K is everywhere zero.* A special case was already treated in the preceding chapter, i.e., the case in which K is not only zero, but in which all coefficients of the second fundamental form are everywhere zero; in this case all points are planar points and the surface itself was then found to be a plane.

Before treating the general case it is useful and interesting to consider another special case, i.e., the case in which a point of the developable surface is a parabolic point, and to examine the surface in a neighborhood of such a point that is small enough so that all points of it are parabolic points. Let, then, p_0 be a parabolic point on the surface; it follows that one of the two principal curvatures at p_0 is zero, the other not, and hence by continuity that $k_1(p) \neq 0$, say, while $k_2(p) = 0$ for all points p in a neighborhood of p_0. In that case the lines of curvature can be introduced as parameter curves (see the remarks about this toward the end of Chapter IV), since these curve families result from differentiable direction fields that have no singularities. At the same time the family of lines of curvature along which $k_2 = 0$ is also a family of asymptotic lines. It is assumed that $u = $ const. on the lines of curvature corresponding to $k_2 = 0$ (which are therefore also asymptotic lines), that $v = $ const. on those corresponding to $k_1 \neq 0$, and that p_0 corresponds to $u = 0$, $v = 0$ (cf. Fig. 5.6). In addition, u and v are chosen in such a way that they represent the arc length on the pair of curves $u = 0$, $v = 0$ through p_0[1]: this is always possible because a transformation of the form

---

[1] On the other curves of the two families u and v usually do not represent the arc length.

$u = u(\bar{u})$, $v = v(\bar{v})$ is legitimate if $du/d\bar{u}$ and $dv/d\bar{v}$ are both different from zero, since the Jacobian $\partial(u, v)/\partial(\bar{u}, \bar{v})$ is given by $(du/d\bar{u})(dv/d\bar{v})$ in such a case. In effect, a nonessential change of scale in u and v is thus introduced without otherwise changing the curvilinear coordinate curves on the surface.

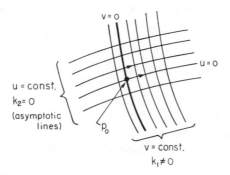

Fig. 5.6   Parameter curves near a parabolic point.

The formula of Rodrigues yields the following equations:

(5.4)
$$\mathbf{X}_{3u} + k_1\mathbf{X}_u = 0 \qquad \text{on } v = \text{const.},$$
$$\mathbf{X}_{3v} = 0 \qquad\qquad \text{on } u = \text{const.}$$

Thus $\mathbf{X}_3$, the normal vector, *is constant along the asymptotic lines* $u = \text{const.}$, since v alone varies on these lines. Some further consequences of these formulas are needed. First of all, $M = F = 0$, since the parameter curves are the lines of curvature (cf. Section 22, of Chapter IV), and also $N = 0$ since $k_2 = 0$ [cf. (4.37)]. The equations 5.4 are identities since they hold at all points of a neighborhood U of p_0 because the curves $u = \text{const.}$, $v = \text{const.}$ cover such a neighborhood (see Section 21 of Chapter IV); consequently $\mathbf{X}_3$ can be eliminated from these identities by differentiation to obtain

(5.5)
$$(k_1\mathbf{X}_u)_v = k_{1v}\mathbf{X}_u + k_1\mathbf{X}_{uv} = 0,$$

since $\mathbf{X}_{3vu} = \mathbf{X}_{3uv} = 0$. Upon taking the scalar product with $\mathbf{X}_v$ the following relation holds:

(5.6)
$$k_1\mathbf{X}_{uv} \cdot \mathbf{X}_v = 0,$$

and it is valid throughout the neighborhood U of p_0. Since $\mathbf{X}_v \cdot \mathbf{X}_v = 1$ for $u = 0$, because v was chosen as the arc length on this curve, it follows that

(5.7)
$$\mathbf{X}^2_v = 1$$

throughout U because $(\mathbf{X}^2_v)_u$ vanishes identically from (5.6) since k_1 does not vanish. Thus v represents the arc length on all of the asymptotic lines $u = \text{const.}$, and in addition it is clear that $\mathbf{X}_v \cdot \mathbf{X}_{vv} = 0$ holds identically, i.e., that

X_{vv} is orthogonal to X_v throughout U. From $F = X_u \cdot X_v = 0$ in U, $X_{vv} \cdot X_u + X_v \cdot X_{uv} = 0$ follows, and therefore $X_u \cdot X_{vv} = 0$, since $X_v \cdot X_{uv} = 0$ holds from (5.6). Finally, $N = X_{vv} \cdot X_3 = 0$ holds in U, since $k_2 = 0$. Thus X_{vv} has been proved to be orthogonal to X_u, X_v, and X_3, and therefore

$$(5.8) \qquad\qquad\qquad X_{vv} = 0$$

holds throughout the neighborhood U of p_0. It follows at once that *the asymptotic lines*, along which v varies, are *straight lines*. Thus it has now been proved that *the surface in a neighborhood of the point p_0 is covered by a family of straight lines along which the normal vector is constant*, or, as it could also be put, *the tangent planes to the surface are in contact with it along a one-parameter family of straight lines that cover it*. Such straight lines are referred to in what follows as *generators*.

The surface could therefore also be regarded as the envelope of a one-parameter family of planes, since it is tangent to a set of planes along its straight asymptotic lines while not itself belonging to any member of the family of planes. Later on in this chapter the general problem of treating envelopes of one-parameter families of surfaces—including some cases leading to developables—will be discussed briefly.

The above analysis of the situation in a neighborhood of a parabolic point p_0 shows that the surface can be given the following representation in terms of the parameters u, v employed above:

$$(5.9) \qquad\qquad\qquad X(u, v) = Y(u) + vZ(u).$$

Here $Y(u)$ represents the curve $v = 0$, with u as arc length on it. The curve $Y(u)$ is the line of curvature through p_0 on which $k_1 \neq 0$, while $Z(u)$ represents a family of unit vectors orthogonal to this curve (and thus in the direction of the asymptotic lines) so that v represents the arc length along them. It is known that the normal vector X_3 is constant along the asymptotic lines $u = $ const.; hence the vector $(\dot{Y} + v\dot{Z}) \times Z$, which determines the direction of X_3, must therefore have a fixed direction independent of v. Since $\dot{Y} + v\dot{Z} = X_u$ is in the tangent plane, which is spanned by $\dot{Y}$ and Z, it follows that $\dot{Z}$ is necessarily a linear combination of $\dot{Y}$ and Z. Since $\dot{Z}$ and Z are orthogonal, it follows that $\dot{Z}$ and $\dot{Y}$ fall on the same line since $\dot{Y}$ and Z are also orthogonal. A necessary condition, in order that the vector $X(u, v)$ given by (5.9) should represent a developable—in addition to those already stipulated with respect to the vectors $Y(u)$ and $Z(u)$—is therefore given by the relation

$$(5.10) \qquad\qquad\qquad \dot{Z}(u) = \phi(u) \, \dot{Y}(u),$$

with $\phi(u)$ an appropriately defined scalar function of u.

It is readily seen that any surface given by (5.9) with $|\dot{\mathbf{Y}}| = |\dot{\mathbf{Z}}| = 1$ and $\dot{\mathbf{Y}} \cdot \mathbf{Z} = 0$ is a regular surface for v sufficiently small. It is, in fact, a so-called ruled surface, since the curves $u = $ const. are straight lines that lie in the surface. However, such a surface is not in general a developable surface since the condition (5.10) need not necessarily hold; for example, some of the quadric surfaces, such as the hyperboloid of one sheet, are ruled surfaces but not developables—on them the Gaussian curvature is negative (cf. Problem 2 of the preceding chapter). However, if (5.10) holds, it is easily seen upon reversing the argument above that the tangent plane is a fixed plane all along the lines $u = $ const., and it will be seen in a moment that $K = 0$ in this case: in other words, (5.10) is a sufficient condition ensuring that (5.9) represents a developable. That K does vanish under these circumstances follows almost immediately from the known relation between the three fundamental forms [cf. (4.34)]:

$$\text{III} - 2H\text{II} + K\text{I} = 0.$$

In the present case $\text{II} = 0$, and it is assumed that $\mathbf{X}_3' = 0$ along straight-line generators; therefore $K = 0$ follows, since the first fundamental form I is never zero. (Quite generally, $K < 0$ holds along any straight line in a surface since $\text{II} = 0$, $\text{I} > 0$, $\text{III} \geq 0$ imply that K cannot be positive.) Conversely *any straight line in a developable is also a generator of it*, since $\text{III} = 0$ holds along the line, and that implies that $\mathbf{X}_3' = 0$ also holds along the line.

Before continuing to discuss the character of developables in general, thus allowing the possibility that the surface may contain both planar and parabolic points, it is interesting to consider a few special cases that arise for developables consisting entirely of parabolic points. These special cases result upon making various restrictive assumptions regarding the function $\phi(u)$ in (5.10), as follows:

Case 1. $\phi \equiv 0$. In this case $\mathbf{Z}$ is a constant vector, and since $\dot{\mathbf{Y}} \cdot \mathbf{Z} = 0$ it follows that $\mathbf{Y}$ represents a plane curve with $\mathbf{Z}$ perpendicular to the plane of the curve. Thus the developable is a cylinder.

Before introducing other special cases it is noted that

$$\mathbf{X}_u \times \mathbf{X}_v = (\dot{\mathbf{Y}} + v\dot{\mathbf{Z}}) \times \mathbf{Z} = (1 + v\phi(u))\dot{\mathbf{Y}} \times \mathbf{Z},$$

so that singularities of the surface occur only where $v = -1/\phi(u)$.

Case 2. $\phi \equiv c = $ const. $\neq 0$. Upon integrating (5.10), $\mathbf{Z}(u) = c\mathbf{Y}(u) + \mathbf{a}$ results, with $\mathbf{a}$ a constant vector. Hence $\mathbf{X}(u, -1/c) = -(1/c)\mathbf{a}$, as an easy calculation shows. Thus all of the straight generators $u = $ const. pass through the point $-(1/c)\mathbf{a}$ and the surface is a cone with this point as a singular point. The curve $\mathbf{Y}(u)$ lies on a sphere centered at the vertex of the cone.

Case 3. $\phi'(u) \neq 0$. In this case it will be shown that the points $\mathbf{X}(u, -1/\phi(u))$, where singular points are to be expected, fill out a regular space curve whose tangents are the generators of the developable. This locus of points is a curve given by the following vector $\mathbf{t}(u)$:

$$\mathbf{t}(u) = \mathbf{Y}(u) - \phi^{-1}(u)\,\mathbf{Z}(u).$$

From this the tangent vectors are given by

$$\mathbf{t}'(u) = \dot{\mathbf{Y}}(u) + \frac{\dot{\phi}}{\phi^2}\,\mathbf{Z}(u) - \phi^{-1}\,\dot{\mathbf{Z}}(u) = \frac{\dot{\phi}}{\phi^2}\,\mathbf{Z}(u),$$

in view of (5.10); thus $\mathbf{t}'(u)$ is not the zero vector. The locus given by the vector $\mathbf{t}(u)$ therefore represents a regular space curve, and its tangents fall along the generators of the developable. A partial converse of this is also true: *the tangents of a regular space curve on which the curvature κ does not vanish form a developable that is regular everywhere except on the curve.* The required locus is evidently given by the equation

(5.11) $$\mathbf{X}(s, t) = \mathbf{Y}(s) + t\dot{\mathbf{Y}}(s),$$

when $\mathbf{Y}(s)$ defines the space curve. From this it follows that

$$\mathbf{X}_s = \dot{\mathbf{Y}} + t\ddot{\mathbf{Y}}, \ \mathbf{X}_t = \dot{\mathbf{Y}} \quad \text{so that} \quad \mathbf{X}_s \times \mathbf{X}_t = -t\dot{\mathbf{Y}} \times \ddot{\mathbf{Y}}.$$

Thus the normal vector of the surface is fixed in direction along the straight lines $s = \text{const.}$, and the surface is regular except for $t = 0$ since $\ddot{\mathbf{Y}} \neq 0$ in view of the asumption that $\kappa \neq 0$.

Cases in which $\phi'(u)$ is sometimes zero, but not identically zero, are evidently not covered by the treatment here.

3.   Edge of Regression of a Developable

It is of interest to investigate the nature of the developable near the space curve $\mathbf{Y}(s)$. In particular, the surface is shown to be singular there. For this purpose it is convenient to make use of the canonical representation of a space curve given in Section 12 of Chapter III:

$$y_1(s) = s \cdots,$$

$$y_2(s) = \frac{\kappa}{2}\,s^2 + \cdots,$$

$$y_3(s) = \frac{\kappa\tau}{6}\,s^3 + \cdots,$$

in which y_1, y_2, y_3 are coordinates taken along the tangent, principal normal, and binormal of the space curve $\mathbf{Y}(s)$, and the representation is valid near the point $s = 0$. From (5.11), with the same choice of coordinate axes it follows that

$$x_1(s, t) = (s + \cdots) + t(1 + \cdots),$$

$$x_2(s, t) = \left(\frac{\kappa}{2} s^2 + \cdots\right) + t(\kappa s + \cdots),$$

$$x_3(s, t) = \left(\frac{\kappa\tau}{6} s^3 + \cdots\right) + t\left(\frac{\kappa\tau}{2} s^2 + \cdots\right),$$

is a representation of the surface $\mathbf{X}(s, t) = (x_1, x_2, x_3)$ valid near $t = 0, s = 0$. Consider the curve cut out of the surface by the plane $x_1 = 0$, which is of course normal to the curve $\mathbf{Y}(s)$ at $s = 0$. From the expression above for $x_1(s, t)$ it is clear that $x_1(s, t) = 0$ requires that $t = -s + \cdots$ for s and t small, and hence that the curve of intersection with the plane $x_1 = 0$ is given by

$$x_2 = -\frac{\kappa}{2} s^2 + \cdots,$$

$$x_3 = -\frac{\kappa\tau}{3} s^3 + \cdots.$$

This curve has a cusp with branches tangent to the x_2-axis, i.e., to the principal normal of the curve defined by $\mathbf{Y}(s)$. The curve $\mathbf{Y}(s)$ is called the *edge of regression* of the developable; along it, two different sheets of the developable are tangent to each other, each one of them formed by the half-tangents issuing from $\mathbf{Y}(s)$ in each of the two possible senses (cf. Fig. 5.7). Thus the developable has an actual singular locus along the edge of regression, and not a singularity that could be removed by a change of parameters, unless $\mathbf{Y}(s)$ represents a plane curve.

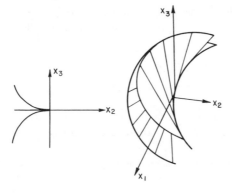

Fig. 5.7   Edge of regression of a developable.

4. Why the Name Developable?

It will be shown in a later chapter that all surfaces with $K \equiv 0$ are right-fully called developable surfaces since they can be, as it is the custom to say, "developed on the plane." By this is meant that a one-to-one mapping of the surface on the plane can be achieved in such a way that all curves that correspond in the mapping have equal lengths. (Such a mapping is called an *isometric* mapping.) This fact will be proved generally later on in Chapter VIII, but it is interesting to show at this point that at least the special types of developables considered so far really do have the property that makes it reasonable to give them such a name. Thus it is now to be shown that any surface consisting of parabolic points can, in the small at least, be developed on the plane. In fact, the mapping will be defined in such a way that the straight-line generators of these special developables—the lines $u = $ const. in the representation (5.9) that is generally valid for the present cases—are the isometric images of straight lines in a plane that, in turn, are orthogonal to a certain curve $\mathbf{y}(u)$ in the plane that is the isometric image of the curve $\mathbf{Y}(u)$. Thus it is as if the developable were to be rolled out on the plane with $\mathbf{Y}(u)$ falling on $\mathbf{y}(u)$, with the correspondence defined by the coincidence of the points of the surface and those of the plane.[1]

As a first step in this construction consider a regular curve $\mathbf{y}(s)$ in the u,v-plane, with s as arc length, and the points of that plane defined by the following vector function $\mathbf{x}(u, v)$:

$$(5.12) \qquad \mathbf{y}(s) + t\mathbf{z}(s) = (u(s, t), v(s, t)) \equiv \mathbf{x}(u, v),$$

with $\mathbf{z}(s)$ a unit vector orthogonal to the curve, so that $\mathbf{z} \cdot \mathbf{z} = 1$ and $\mathbf{z} \cdot \dot{\mathbf{y}} = 0$. It was shown at the end of the preceding chapter that such a band about the curve $\mathbf{y}(s)$ in the u,v-plane is mapped on a rectangle $s_0 < s < s_1, -\epsilon < t < \epsilon$, $\epsilon > 0$, in an s,t-plane in a one-to-one way. It is in any case easily checked that the Jacobian $\partial(u, v)/\partial(s, t)$ is different from zero for ϵ small enough.

The curve $\mathbf{y}(s)$ is now to be chosen so that the developable, assumed given in the form (5.9), i.e., by

$$\mathbf{X}(u, v) = \mathbf{Y}(u) + v\mathbf{Z}(u),$$

with $\mathbf{Z} \cdot \mathbf{Z} = 1$ and $\mathbf{Z} \cdot \dot{\mathbf{Y}} = 0$, will correspond isometrically to the band containing $\mathbf{y}(s)$ when $u = s$, $v = t$ defines the correspondence of surface points

---

[1] This process could be treated kinematically by regarding the developable as a rigid body constrained to roll without slipping on the plane in such a way that the curves $\mathbf{Y}(u)$ and $\mathbf{y}(u)$ come into coincidence. At each moment the generators would then fall along the angular velocity vector of the rolling developable.

and points in the u,v-plane. The coefficients E, F, G, and e, f, g of the first fundamental forms of $\mathbf{X}(u, v)$ and $\mathbf{x}(s, t)$ are

$$E = [1 + v\phi(u)]^2, \qquad e = (\dot{\mathbf{y}} + t\dot{\mathbf{z}})^2,$$
$$F = 0, \qquad\qquad f = 0,$$
$$G = 1, \qquad\qquad g = 1,$$

as can be seen readily upon using (5.9), (5.10), (5.12), and the relations $|\dot{\mathbf{Y}}| = |\mathbf{Z}| = 1$, $\dot{\mathbf{Y}} \cdot \mathbf{Z} = 0$, $|\mathbf{z}| = 1$, and $\dot{\mathbf{y}} \cdot \mathbf{z} = 0$. Evidently, if $\mathbf{y}(s)$ were to be chosen so that $\dot{\mathbf{z}} = \phi(s)\dot{\mathbf{y}}$, it would follow that $e = [1 + t\phi(s)]^2$, since $|\dot{\mathbf{y}}| = 1$, and this is equal to E for $u = s$, $v = t$; in this way an isometric mapping would be achieved. To define a plane curve $\mathbf{y}(s)$ with the required property, its derivative $\dot{\mathbf{y}}(s)$ is defined by

$$\dot{\mathbf{y}}(s) = \mathbf{u}_1 \cos \psi + \mathbf{u}_2 \sin \psi,$$

with $\mathbf{u}_1$ and $\mathbf{u}_2$ unit vectors along the u and v axes and $\psi(s)$ a function defined by $\psi(s) = -\int_{s_0}^{s} \phi(\sigma)\, d\sigma$, so that $\dot{\psi}(s) = -\phi(s)$. A vector $\mathbf{y}(s)$ determined by

$$\mathbf{y}(s) = \mathbf{u}_1 \int_{s_0}^{s} \cos \psi(\sigma)\, d\sigma + \mathbf{u}_2 \int_{s_0}^{s} \sin \psi(\sigma)\, d\sigma,$$

represents a regular curve in the plane since $\dot{\mathbf{y}}(s) = 1$. Since $\mathbf{z} \cdot \dot{\mathbf{y}} = 0$ and $|\mathbf{z}| = 1$ are to hold the function $\mathbf{z}(s)$ is given by

$$\mathbf{z}(s) = -\mathbf{u}_1 \sin \psi(s) + \mathbf{u}_2 \cos \psi(s);$$

and hence

$$\dot{\mathbf{z}}(s) = (-\mathbf{u}_1 \cos \psi - \mathbf{u}_2 \sin \psi)\dot{\psi}$$
$$= \phi(s)\, \dot{\mathbf{y}}(s),$$

since $\dot{\psi}(s)$ was chosen equal to $-\phi(u)$. Thus the portion of the plane represented in the form (5.12) is in isometric correspondence with the developable given by (5.9) since $E \equiv e$ holds. It might be noted that the curvature of the plane curve $\mathbf{y}(s)$ is $\phi(s)$. (Problem 5 at the end of the chapter is of interest in this connection.)

5. Developable Surfaces in the Large [1]

The discussion in Section 2 above is stated in nearly all books on differential geometry to prove that the developables consist exclusively of the

---

[1] This discussion follows rather closely the author's paper [S.10] (see also the paper of Massey [M.1]), which was based on a paper by Hartman and Nirenberg [H.4]. In the latter paper a discussion of other relevant literature is given, including references to papers by Chern and Lashof [C.6], and Pogorelov.

plane, the *cylinder,* the *cone,* or the *tangent surface of a space curve.* This is not true since these surfaces resulted upon assuming that they consist either entirely of planar points or entirely of parabolic points, and there is no reason to exclude surfaces with $K \equiv 0$ on which both types of points occur. For instance, in such cases it is not true in the large that a straight line in the surface along which the tangent plane is fixed can be continued indefinitely. Figure 5.8 illustrates a counterexample that consists of a triangle to which

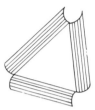

Fig. 5.8   A developable with planar and parabolic points.

cylinders have been attached along each of the sides; clearly, this could be done in such a way that the surface would be regular and have as many continuous derivatives as desired.   A generator starting from an interior point of the triangle evidently would not lie on the surface after crossing a side of the triangle.   It will be seen, however, that the general situation with regard to the distribution of planar points and parabolic points is typified in a certain sense by the type of surface shown in Fig. 5.8.

Since this section belongs, in the main, to differential geometry in the large it might well have been placed in Chapter VIII, where the appropriate basic ideas—abstract two-dimensional manifolds, their use as parameter domains for the purpose of defining surfaces in the large in three-dimensional space, etc.—are introduced.   But the present section can get along with very simple ideas of this sort, and it is suitable to treat developables in the large in this chapter, particularly since the material presented is quite elementary in character.   The surfaces to be treated here are simply defined as surfaces in three-space given by a vector function $\mathbf{X}(u, v)$ defined over an open connected domain of the u, v-plane such that each point of that domain has a neighborhood within which $\mathbf{X}(u, v)$ is defined as a regular surface with Gaussian curvature zero.   The theorems obtained here are thus somewhat less general than would be the case if the formulations of Chapter VIII were to be employed.

The material of this section is based on a theorem due to Hartman and Nirenberg [H.4].   The statement of the theorem involves a classification of points of four different kinds (not all of which are mutually exclusive) on a developable surface, as follows: (a) parabolic points, (b) planar points, (c) essentially parabolic points, and (d) flat points.   Only (c) and (d) require definitions.   An *essentially parabolic point* is defined as *a limit point of parabolic points*; it might be either a parabolic or a planar point, but all parabolic points are evidently also essentially parabolic points.   *A flat point is a planar point that has a neighborhood consisting entirely of planar points*; such a neighborhood therefore lies in a plane (cf. Section 11 of Chapter IV).   Any point that is not an essentially parabolic point is a flat point, since it has a neighborhood free of parabolic points.   The set of all flat points of S is an open set, the set of all essentially parabolic points is a closed set, and, as was just noted, the union of these sets consists of all points of S.   Finally, it is convenient to make use of the term *generator* for every open connected segment of a straight line that lies in the surface; as was shown above, the tangent plane at any point of such a line is also the tangent plane at all of its points, simply because $K = 0$ holds on S.

The theorem of Hartman and Nirenberg is:

1. *Every essentially parabolic point p_0 on a developable surface S contains a uniquely determined generator $l = l(p_0)$ that either extends to infinity or has one or both of its end points on the boundary of S.*

2. *All points on $l(p_0)$ are essentially parabolic.*

3. *The points of $l(p_0)$ are either all parabolic points or all planar points.*

It is of interest to point out the obvious fact that the uniqueness statement of the above theorem does not hold for generators through flat points.

The theorem is proved first for the special case of a *parabolic* point p_0. It was shown, in fact, that (5.9) in this case gives a representation locally of a developable in which a generator of the developable is a uniquely determined straight-line segment $l = l(p_0)$ through p_0.   It is to be shown that $l(p_0)$ can be extended in both directions as a generator on S until it either goes to ∞ or reaches the boundary of S.   To this end the formula $k_1 = L/E$ [cf. (4.37)], and the representation (5.9) with p_0 given by $u = 0$, $v = 0$, are used.   The coefficient L of the second fundamental form of S is given by $L(0, v) = \mathbf{X}_3 \cdot \mathbf{X}_{uu} = \mathbf{X}_3 \cdot (\ddot{\mathbf{Y}} + v\ddot{\mathbf{Z}})$, and from (5.10) the relations $\mathbf{X}_3 \cdot \ddot{\mathbf{Z}} = \mathbf{X}_3 \cdot (\phi \dot{\mathbf{Y}})^{\cdot} = \phi \mathbf{X}_3 \cdot \ddot{\mathbf{Y}}$ hold since $\mathbf{X}_3 \cdot \dot{\mathbf{Y}} = 0$.   Here the dots refer to differentiations with respect to u.   Consequently $L(0, v)$ is given by

$$L(0, v) = \mathbf{X}_3 \cdot \ddot{\mathbf{Y}}(1 + v\phi(0))$$

$$= L(0, 0)(1 + v\phi(0)), \qquad L(0, 0) \neq 0,$$

since $\mathbf{X}_3$ is independent of v along a curve $u = $ const. on S. Since E is given by $(1 + v\phi(u))^2$ (cf. the preceding section), the principal curvature $k_1(0, v)$ is given by

$$(5.13) \qquad\qquad k_1(0, v) = \frac{L(0, v)}{E(0, v)} = \frac{L(0, 0)}{1 + v\phi(0)}.$$

Consider any connected segment of $l(p_0)$. Clearly $k_1(0, v) \neq 0$, no matter how large the distance v along $l(p_0)$ from p_0 to another point p is; consequently $l(p_0)$ cannot terminate at an inner point p of S, since such a point would be a parabolic point, and hence $l(p_0)$ could be extended beyond p. Thus $l(p_0)$ either goes to ∞ or terminates in a boundary point of S.

A proof of the theorem is completed by showing that it holds when the essentially parabolic point p_0 is a planar point. This point is the limit of a sequence p_n of parabolic points. Through each point p_n there exists, as was just seen, a uniquely determined generator $l_n = l(p_n)$ extending on S to infinity or to its boundary and with the normal $\mathbf{X}_3$ of S fixed in direction along it. The lines l_n have at least one limit direction—determined, say, by taking a line from p_0 to any accumulation point of the intersections of the lines l_n with a sphere centered at p_0. The line through p_0 in this direction contains a generator $l(p_0)$ of S. This follows—by choosing a converging subsequence from the set l_n—from the fact that limit points of points of S belong either to S or to its boundary. Since the generator $l(p_0)$, the existence of which is now established, is the limiting position of the generators $l(p_n)$ of a converging sequence, all of whose points are parabolic points, it follows that $l(p_0)$ consists of essentially parabolic points; at the same time it is clear that $l(p_0)$ either extends to infinity or to the boundary of S, since that holds for the generators l_n. Furthermore, all points on $l(p_0)$ are planar points; otherwise, as was seen above, $l(p_0)$ would consist entirely of parabolic points, and that is not possible since p_0 was assumed to be a planar point. To complete the proof of the theorem it is necessary to show that the generator $l(p_0)$ is *uniquely* determined. The generator $l(p_0)$ was the limit of a special sequence $l_n = l(p_n)$ of generators on parabolic points. Suppose that a different generator $l'(p_0)$ could exist through p_0. Since the sequence p_n converges to p_0 and since the lines l_n all lie on S, it follows that $l'(p_0)$ would of necessity intersect the lines $l(p_n)$ for n sufficiently large—as can be seen most readily perhaps by considering the images of these lines near the image of p_0 in the parameter plane. This in turn implies that $l'(p_0)$ would contain parabolic points—in contradiction with the fact that p_0 is a planar point. This completes the proof of the theorem.

At first sight it might be thought that the distribution of parabolic and planar points on developable surfaces could be rather arbitrary, but the theorem just proved shows that that is not true. For example, isolated planar points cannot occur since such a point would be an essentially parabolic point and therefore would contain a generator—but that would not be possible since such a generator contains only planar points. The fact is that the theorem permits rather far-reaching conclusions to be drawn regarding the manner in which the various types of points occur on developables. In the paper of Hartman and Nirenberg [H.4] these possibilities are analyzed. Here the theorem is used to prove another theorem, which singles out the cylinders as those developables that have no boundary points on generators. As was noted earlier, this is the situation when the developable is assumed to be made up entirely of parabolic points, since the cones and the tangent surfaces to space curves have singularities at finite distances as measured along generators; however, it is not desired here to eliminate the occurrence of planar points.

Although the main object of this section is to discuss certain developables in the large, it has not been necessary up to this point to invoke any very essential hypotheses in the large. It now becomes necessary to do so, particularly since the developables are open, not closed, surfaces (in Chapter 4 it was noted that all regular closed surfaces must have points of *positive* Gaussian curvature), and some condition is usually needed in such cases to rule out the possibility that the surface is not just a piece of some still larger surface. The assumption usually made for an open surface is that the surface has a complete metric in the sense of Hopf and Rinow [H.16]: a concept to be discussed at length in Chapter 8. One of a number of equivalent definitions of this concept is that every divergent regular curve on the surface —that is, the image on the surface of a half-open interval of the parameter plane—should be infinitely long. Hartman and Nirenberg [H.4] show that a *complete* developable (in the sense of Hopf and Rinow) is a cylinder, but the proof requires more sophisticated tools than will be used here. (Hartman and Nirenberg also generalize the theorem for n-dimensional developables in $(n + 1)$-dimensional space.) Pogorelov gave the first proof in the two-dimensional case, but for surfaces without continuous second derivatives; consequently this proof also requires a more sophisticated discussion than is necessary here.

The "completeness" assumption in the large made here is that *no connected portion of a generator in S ends in a boundary point of S.* In view of the theorem of Hartman and Nirenberg, this assumption in the large leads to the conclusion that *all generators of S through essentially parabolic points* are infinitely long in both directions. For generators through flat points, however, that need not be true: a generator starting in the interior of the triangle of Fig. 5.8 would end on reaching a side of the triangle. Thus the "completeness" assumption made here does not rule out the possibility that a generator might terminate at an *interior point* of S. (In the books of Hicks [H.8] and O'Neill [O.1], it is assumed in treating this same problem that the surface is closed in the space E^3; this is, however, a more restrictive condition than the condition of completeness of Hopf and Rinow since examples of complete surfaces are easily given that are not closed in E^3.)

The following theorem is to be proved:

THEOREM. *A "complete" and connected developable surface in three-dimensional space is a cylinder, with infinite straight lines as generators.*

It is perhaps worth noting that this theorem does include surfaces that are not complete in the sense of Hopf and Rinow, e.g., the circular cylinder erected over an open connected segment of a circle. The only cylinders complete in the sense of Hopf and Rinow are those erected over a plane *closed* curve or over an open plane curve that is infinite in length in both directions.

The proof of the theorem takes the following course. It will be shown that (1) through every point on the surface there is a uniquely determined generator that extends to infinity in both directions, and that (2) every such generator has a neighborhood on S made up entirely of generators parallel to each other. Once these two things are proved, it is readily seen in standard fashion, with the aid of the Heine-Borel theorem, that S is a cylinder. This follows since any pair p, q of points of S can, by an assumption of the theorem, be connected by a continuous curve C in S. By selecting a finite number of open neighborhoods covering C it is seen that the uniquely determined generators through p and q are parallel, since the generators, being uniquely determined, coincide where two neighborhoods overlap.

Since the first statement has already been proved in general for essentially parabolic points, it remains to prove it for *flat points* p of S; it is to be recalled that these are planar points that also have a neighborhood consisting entirely of like points, and therefore such a neighborhood lies in a plane. In the set F of all flat points of S—clearly an open set—consider a *component* F_p containing some point p of F, i.e., the set of all points of F that can be connected to the point p. The component F_p lies in a plane E. It is to be proved that this set is *an infinite plane strip with boundaries that are parallel generators of S made up of essentially parabolic points.* Thus for flat points *both* of the two statements above would be proved. The plane strip might be the entire plane E—but then S itself would clearly be a plane and the theorem would be proved. Otherwise, on going out from p along some ray in the plane E of F_p a first boundary point $\bar{p}$ of F_p would be encountered on that ray, since the set of boundary points of F_p is a closed set. Such a point belongs to S because of the "completeness" hypothesis of the theorem, and it is a planar point as well as an essentially parabolic point. Through it therefore a unique generator G of S exists that is infinitely long in both directions, because of the theorem of Hartman and Nirenberg and the "completeness" of S. The line G necessarily lies in the plane E containing F_p, since E is a tangent plane to S all along the segment $p\bar{p}$. It is to be shown next that all points of G are boundary points of F_p, as follows. From p draw a straight line in E until it intersects G in any point $\bar{p}_1$, say (cf. Fig. 5.9). If $\bar{p}_2$ rather than $\bar{p}_1$ were the first boundary point of F_p encountered on $p\bar{p}_1$ in going out from p, then an infinite generator G_1 of S would exist through $\bar{p}_2$ in E which would of necessity be parallel to G since $\bar{p}_2$ would be an essentially parabolic point, hence G_1 would contain only essentially parabolic points, and G and G_1 cannot intersect since both are uniquely determined by any one of their

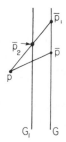

Fig. 5.9   Component containing a flat point.

points.   Arbitrarily near to G_1 there are infinite generators of S consisting of parabolic points, since G_1 consists of essentially parabolic points.   It follows that the segment $p\bar{p}$ would be cut by some of these generators (in fact, near the intersection of $p\bar{p}$ with G_1); but this is in contradiction with the fact that all points on the segment $p\bar{p}$ are planar points.   Thus G consists entirely of boundary points of F_p.   Also, since this discussion makes it clear that the segment $p\bar{p}_1$ belongs, except for $\bar{p}_1$, in F_p, it follows that the entire plane strip bounded by G and a straight line through p parallel to G belongs to F_p. If there were no other boundary points of F_p than those on G, it is clear that F_p would be a half-plane.   If there were a boundary point of F_p not on G, then that point would contain an infinite generator G_3 on S in the boundary of F_p and in E, as was just seen, and this generator would be parallel to G, since G_3 and G are in the same plane and cannot intersect.   As before, all points in E joining p with G_3 would belong to F_p.   Hence F_p would be an infinite strip with parallel straight lines as boundaries.   Thus, if p is a flat point on S (and S is not the entire plane), it contains a unique infinite generator on S—i.e., one parallel to G—and it is in addition embedded in an open set of parallel infinite generators.   Thus for generators through flat points both of the statements (1) and (2) are established.

Since statement (1) holds for generators through essentially parabolic points, by virtue of the "completeness" hypothesis, and since statements (1) and (2) have been shown to hold for generators through flat points, the proof of the theorem will be completed once it is shown that statement (2) holds for generators through essentially parabolic points[1]; i.e., that they also have neighborhoods made up of parallel generators.

An essential step in the proof of this statement is the remark that the set of uniquely defined generators covering S, now known to exist, depends continuously at an essentially parabolic point p on the points of S, i.e., if a sequence p_n converges to p then the generators $l(p_n)$ converge to the generator $l(p)$. This was proved in essence above while proving the uniqueness statement of the Hartman-Nirenberg theorem for essentially parabolic points. Here again it follows that the end points of unit vectors along the generators $l(p_n)$, when translated to a fixed point, would have at least one accumulation point, and the corresponding direction would in turn determine a generator of S; but since the generators are uniquely determined at essentially parabolic points by any point on them it follows that the original sequence $l(p_n)$ converges at such points to $l(p)$. At flat points the generators, being parallel near it, are obviously continuous. Since all points of S are either flat points or essentially parabolic points it follows that the generators are continuous at all points.

This continuity property of generators is now used to show, finally, that a generator through an essentially parabolic point p_0 is embedded in a family of generators parallel to the generator $l(p_0)$. This is done as follows. Since the tangent plane to S at p_0 contains the generator $l(p_0)$, it follows that a neighborhood of $l(p_0)$ on S has a one-to-one orthogonal projection on that plane. Such a neighborhood can be represented in Cartesian coordinates in the form $z = z(x, y)$, with the origin at p_0, the x-axis along $l(p_0)$, and the z-axis along the normal to S. Consider a normal section $C(s)$ of S at p_0 and take any point p of $l(p_0)$, and a plane P orthogonal to it at p. The generators $l(s)$ for all points of $C(s)$ near to p_0 cut the plane P, since the generators $l(s)$, being continuous in s, make angles with the plane of $C(s)$ that differ little from $\pi/2$. The locus generated by the intersection of the generators $l(s)$ with P is a continuous curve $N(s)$ near p, since these straight lines depend continuously on s, and the locus $N(s)$ is at the same time in an orthogonal plane section of S at p; it follows that the locus is this section N of S and hence it is a regular plane curve on S. Since the normal vector is constant along each generator, it follows that $N(s)$ has a one-to-one orthogonal projection on the tangent plane of S at p since this coincides with the tangent plane at p_0. Thus a fixed set of generators covering a neighborhood of p_0 that has a one-to-one

---

[1] The remainder of the proof would be much easier if the representation (5.9) in terms of the lines of curvature were available near essentially parabolic points that are planar points, since it could then be shown that $k_1(0, v)$, as given by (5.13), would be independent of v; in that case the generators would be readily seen to be parallel. However, Hartman and Wintner [H.5] show that such a coordinate system is not generally available at such points.

projection on the tangent plane containing $l(p_0)$ can be extended so that it covers a neighborhood of an arbitrary point p on the generator $l(p_0)$. These generators necessarily project into straight lines in the x,y-plane that extend to ∞ in both directions in that plane. Furthermore, the projections of the generators on the x,y-plane must be parallel lines since an intersection of two of them would mean that their images on the surface intersect; which is not possible since the generators on S are uniquely determined; it thus follows that the partial derivatives z_x and z_y of $z(x, y)$ are constant along each generator, since the normal to S is fixed in direction along each of these lines. It follows that z_x and z_y are functions of y alone throughout a neighborhood of $l(p_0)$, i.e., $z_x = a(y)$, $z_y = b(y)$, since the x-axis was chosen parallel to the generators. From this it follows that $z_{xy} = z_{yx} = 0$, so that z_x is independent of y and thus has the value zero everywhere near $l(p_0)$ since that is its value for $y = 0$, i.e., on $l(p_0)$. Consequently z is everywhere independent of x and hence $z = \beta(y)$ holds in a neighborhood of $l(p_0)$, thus all generators on S in a neighborhood of $l(p_0)$ are parallel lines, and S is cylindrical there.

This concludes the proof of the theorem that a "complete" developable is a cylinder.

6.   Developables as Envelopes of Planes

A family of regular surfaces depending on a single parameter s is assumed to be defined by $\mathbf{X}(u, v; s)$. Such a family of surfaces may have an *envelope*, which is defined to be a regular surface that is not a member of the family, but which is covered by a family of regular curves, called *characteristics*, along each of which it is tangent to a member of the family. It is not proposed to treat here the problem of determining envelopes in a complete way, but to derive a rather obvious necessary condition for their existence, and then to use the condition to study the special case of envelopes of planes.

A triply infinite set of regular surfaces $\mathbf{X}(u, v, w)$, such that $\mathbf{X}(u, v, c)$, for example, yields regular surfaces in the parameters (u, v) for all fixed values of c (and similarly for the other pairs of parameters), determines a curvilinear coordinate system in three-dimensional space if $\mathbf{X}_u \cdot \mathbf{X}_v \times \mathbf{X}_w \neq 0$, since this scalar triple product is at the same time the Jacobian $J = \partial(x_1, x_2, x_3)/\partial(u, v, w)$. It follows therefore that a family $\mathbf{X}(u, v; s)$ of surfaces could not possibly have an envelope unless

$$(5.14) \qquad \mathbf{X}_s \cdot \mathbf{X}_u \times X_v = 0,$$

since otherwise u, v, and s could be utilized as coordinates in space. More specifically, the family of coordinate surfaces $u = $ const., for example, is topologically equivalent to a family of parallel planes in a neighborhood of any point where J does not vanish, and the three sets of coordinate surfaces cut each other at nonzero angles. Thus, if an envelope exists (5.14) must hold on it. Of course, as in the case of envelopes of plane curves (which were discussed in Chapter 2), the loci determined by the condition (5.14) might well contain many things other than envelopes; evidently, the loci include, for example, any points where $\mathbf{X}_i$, or $\mathbf{X}_i \times \mathbf{X}_j$, vanish, i.e., points that may be singular points of surfaces of the family.

A developable that is made up of essentially parabolic points is clearly the envelope of its tangent planes, and the characteristics are clearly also the generators of the developable. Consider, as a concrete example in the other direction, the one-parameter family of planes consisting of the "rectifying planes" of a regular space curve; that is, the family of planes determined by the tangent $\mathbf{v}_1$ and the binormal $\mathbf{v}_3$ of the curve (see Chapter 3). In this case the family of surfaces is defined as follows:

$$(5.15) \qquad \mathbf{X}(u, v; s) = \mathbf{X}(s) + u\mathbf{v}_1(s) + v\mathbf{v}_3(s),$$

with $\mathbf{v}_1$ and $\mathbf{v}_3$ the tangent and binormal of the curve $\mathbf{X}(s)$, and with s, the parameter singling out each member of the family of planes, assumed to represent the arc length of the curve. The condition (5.14) is in this case

$$(\dot{\mathbf{X}} + u\dot{\mathbf{v}}_1 + v\dot{\mathbf{v}}_3) \cdot \mathbf{v}_1 \times \mathbf{v}_3 = 0.$$

Since $\dot{\mathbf{X}} = \mathbf{v}_1$, $\dot{\mathbf{v}}_1 = \kappa\mathbf{v}_2$, and $\dot{\mathbf{v}}_3 = -\tau\mathbf{v}_2$ (from the Frenet equations derived in Chapter 3), it follows that the condition for an envelope is $(u\kappa - v\tau)\mathbf{v}_2 \cdot \mathbf{v}_1 \times \mathbf{v}_3 = 0$, and this requires that

$$(5.16) \qquad\qquad\qquad u\kappa - v\tau = 0.$$

The points of the family (5.15) where this condition holds therefore are given by a vector $\mathbf{Y}(s, v)$ as follows:

$$(5.17) \qquad \mathbf{Y}(s, v) = \mathbf{X}(s) + \frac{v}{\kappa}(\tau\mathbf{v}_1 + \kappa\mathbf{v}_3).$$

For this surface it is easily found that

$$\mathbf{X}_s \times \mathbf{X}_v = -\mathbf{v}_2\left[1 + v\left(\frac{\tau}{\kappa}\right)\right],$$

and it is regular if $\kappa \neq 0$ and v is sufficiently small. In addition, it is clear from the last formula that the normal vector of the envelope is constant along the lines $s = $ const., where it has the direction of the binormal $\mathbf{v}_2$ of $\mathbf{X}(s)$, and thus from (5.17) it is constant along the straight line in the direction of the vector $\tau\mathbf{v}_1 + \kappa\mathbf{v}_3$. Thus there is an envelope in this case that is a developable whose generators fall along the Darboux vector of the curve $\mathbf{X}(s)$. The envelope $\mathbf{Y}(s, v)$ is called the rectifying developable of the curve $\mathbf{X}(s)$ because an isometric mapping of the developable on the plane would lead to a straight line as the image of the curve $\mathbf{X}(s)$. Why this must be so will be seen easily later on, after studying the curves called *geodesics*, which are curves satisfying a necessary condition for curves of shortest length joining pairs of points on a surface. It will be seen that $\mathbf{X}(s)$ is such a geodesic on $\mathbf{Y}(s, v)$, and hence when $\mathbf{Y}(s, v)$ is mapped isometrically on the plane the curve $\mathbf{X}(s)$ goes into a shortest arc in the plane, i.e., into a straight line.

It is of interest to discuss the families of planes determined by the normal planes and the osculating planes of a space curve. These cases are reserved as problems, but it is of interest to remark that in the latter case the characteristics turn out to be the tangents to the space curve.

PROBLEMS

1. Using the representation for surfaces of revolution given by (5.1), derive the following formulas:

$$g_{11} = (f')^2 + (g')^2, \qquad g_{12} = 0, \qquad g_{22} = f^2,$$

$$L_{11} = \frac{f'g'' - f''g'}{\sqrt{(f')^2 + (g')^2}}, \qquad L_{12} = 0, \qquad L_{22} = \frac{fg'}{\sqrt{(f')^2 + (g')^2}},$$

$$K = \frac{g'(f'g'' - f''g')}{f((f')^2 + (g')^2)^2}, \qquad H = \frac{g'(f')^2 + (g')^2 + f(f'g'' - f''g')}{f((f')^2 + (g')^2)^{3/2}},$$

$$k_1 = \frac{g'}{f\sqrt{(f')^2 + (g')^2}}, \qquad k_2 = \frac{f'g'' - f''g'}{((f')^2 + (g')^2)^{3/2}}.$$

2. Prove that the only regular surfaces of revolution for which $H = 0$ are generated by catenary curves.

3. Discuss the envelopes of the normal and osculating planes of a space curve.

4. Prove that the lines of curvature on a surface are characterized by the statement: the normals to the surface along such a curve form a developable surface.

5. A developable is formed as the locus of straight lines along unit vectors $\mathbf{Z}(u)$ normal to a space curve $\mathbf{Y}(u)$, where u is the length along the curve. If $\alpha(u)$ is the angle between $\mathbf{Z}$ and the principal normal $\mathbf{v}_2$ of $\mathbf{Y}$, show that $\dot{\alpha}(u) = -\tau$, where τ is the torsion of $\mathbf{Y}$. From this prove that there are an infinite number of developables with generators perpendicular to a given space curve, any two of which cut at a constant angle. This result means that the generators along $\mathbf{Y}(u)$ turn about the tangent to the curve $\mathbf{Y}(u)$ in the opposite sense to that of the osculating plane but at the same rate. Since τ is a measure of the deviation of $\mathbf{Y}(u)$ from the osculating plane, it is as

though the generators of the developable attempt to stay as much as possible in a plane while constrained to be orthogonal to $\mathbf{Y}(u)$.

6. Prove the following:

(a) If two surfaces intersect each other along a curve under equal angles, and the curve of intersection is a line of curvature on one of the surfaces, it is a line of curvature on the other also.

(b) If two surfaces intersect along a curve that is a line of curvature on both, the surfaces cut at a constant angle along the curve.

(c) A plane or sphere intersects a surface at a constant angle along a curve if and only if the curve is a line of curvature of the surface.

7. The surface $\mathbf{X}(s, t) = \mathbf{X}(s) + t\mathbf{Z}(s)$, with $\dot{\mathbf{X}}\cdot\dot{\mathbf{X}} = 1$, $\mathbf{Z}\cdot\mathbf{Z} = 1$, but $\mathbf{Z}\cdot\dot{\mathbf{X}}$ not necessarily zero, is a regular surface (called a ruled surface) if $\dot{\mathbf{X}} \times \mathbf{Z} \neq 0$, i.e., if $\mathbf{Z}$ is not tangent to the space curve $\mathbf{X}(s)$. Show that the condition under which $\mathbf{X}(s, t)$ is a developable is $\dot{\mathbf{X}}\cdot\mathbf{Z}x\dot{\mathbf{Z}} = 0$. (*Hint*: Make use of the expansion formula for the vector product of two vector products.)

8. From the result of Problem 7 show that a Möbius strip free of singularities cannot be constructed as a developable if the curve $\mathbf{X}(s)$ is a closed *plane* curve and $\mathbf{Z}(s)$ is nowhere tangent to the curve. (*Hint*: Take the curve in the x_1,x_2-plane and find a differential equation for the x_3-component of $\mathbf{Z}$.) The student is advised to take a thin strip of paper, twist it and join the ends to form a Möbius strip. A straight line drawn parallel to the edges of the strip will be seen to go into a twisted space curve, and attempts to force the curve into a plane result in folds in the paper, i.e., in singularities in the developable.

CHAPTER VI

The Partial Differential Equations of Surface Theory

1. Introduction

The principal purpose of this chapter is to derive partial differential equations for surfaces that play a role somewhat analogous to the role of the Frenet equations for space curves. The analogy is, however, far from perfect, as will be seen, though the procedures for arriving at the differential equations are similar in the two cases.

At each point of a regular surface three linearly independent vectors have been defined: the tangent vectors $\mathbf{X}_1$ and $\mathbf{X}_2$ of the parameter curves, and the normal vector $\mathbf{X}_3$. Since every vector can be expressed as a linear combination of these three vectors, it follows in particular that the derivatives $\mathbf{X}_{ij} = \partial^2\mathbf{X}/\partial u_i\,\partial u_j$ and $\mathbf{X}_{3j} = \partial\mathbf{X}_3/\partial u_j$ of these three vectors can be so expressed. This was, of course, the sort of procedure used to obtain the Frenet equations. As a preliminary step in making the calculations it is useful to introduce the matrix G of the coefficients g_{ik} of the first fundamental form:

$$(6.1) \qquad G = \begin{pmatrix} g_{11} & g_{12} \\ g_{21} & g_{22} \end{pmatrix},$$

and its inverse G^{-1}:

$$(6.2) \qquad G^{-1} = \begin{pmatrix} g^{11} & g^{12} \\ g^{21} & g^{22} \end{pmatrix} = \frac{1}{|G|}\begin{pmatrix} g_{22} & -g_{12} \\ -g_{21} & g_{11} \end{pmatrix}.$$

Here $|G|$ denotes the determinant of G, which never vanishes since it is assumed that the first fundamental form is positive definite. Since G is symmetric, i.e., $g_{12} = g_{21}$, it follows that G^{-1} is also symmetric. Since G and G^{-1} are inverses, the product GG^{-1} is the identity matrix and hence

$$(6.3) \qquad a_i^k = \sum_{j=1}^{2} g_{ij}g^{jk} = \delta_i^k = \begin{cases} 1, & i = k, \\ 0, & i \neq k \end{cases}$$

133

by the rule for forming the product of two matrices. It is possible and reasonable to be rid of the summation sign in (6.3) and similar expressions by making the stipulation that summation over a repeated index such as j in (6.3) is always to be understood, and in this chapter always over the numbers 1 and 2. This convention, as well as the systematic introduction of super-scripts as well as subscripts, are earmarks of the tensor calculus. For the time being no use is made of the tensor calculus as such, though some of its notations and formal procedures are taken over.[1]

2. The Gauss Equations

The first group of partial differential equations to be derived is due to Gauss; these equations express the second derivatives X_{ij} of $X(u^1, u^2)$ in terms of the vectors X_i, $i = 1,2$, and X_3. These vectors are written, with undermentioned coefficients, as the following equations:

(6.4) $$X_{ik} = \Gamma_{ik}^l X_l + a_{ik} X_3, \qquad i,k = 1,2,$$

for the second derivatives of $X(u^1, u^2)$. The coefficients Γ_{ik}^l, which of necessity depend on three indices, are called the Christoffel symbols of the second kind. (Christoffel used the notation $\{_l^{i\,k}\}$ for them; the notation adopted here is due to Cartan.) It is to be noted again with respect to the first term that a summation on l is to be taken over 1, 2. The coefficients a_{ik} are easily determined: scalar multiplication with X_3 yields $X_{ik} \cdot X_3 = a_{ik}$, since $X_l \cdot X_3 = 0$; it follows that $a_{ik} = L_{ik}$, with L_{ik} the coefficients of the second fundamental form. It remains to calculate the values for the Christoffel symbols. To this end the scalar product of both sides of (6.4) with X_m is taken, with the result

(6.5) $$X_{ik} \cdot X_m = \Gamma_{ik}^l g_{lm},$$

since $X_l \cdot X_m = g_{lm}$ and $X_l \cdot X_3 = 0$. The equations (6.5) form a system of linear nonhomogeneous equations for the coefficients Γ_{ik}^l, which can be solved explicitly with the aid of the quantities g^{ik} introduced above. In fact, multiplication of both sides of (6.5) with g^{lm} after replacing the summation

---

[1] Actually, some deviations from the usual notations of tensor calculus have occurred here and there in earlier chapters. For example, the surface parameters would be denoted in the tensor calculus by u^1, u^2 instead of by u_1, u_2. In fact, the superscripts indicate what is called contravariant, the subscripts covariant, character of a given entity, but this distinction is not at present important; these matters are the concern of Chapter 9. A rederivation of the material of this chapter is given in Chapter 10, in terms of the formalism of E. Cartan, which uses invariant differential forms, their wedge products, and their exterior derivatives to obtain the basic equations in a very concise invariant form.

index l on the right by r (always a legitimate procedure), leads to the equations

(6.6)
$$\mathbf{X}_{ik} \cdot \mathbf{X}_m g^{lm} = \Gamma^r_{ik} g_{rm} g^{lm}$$
$$= \Gamma^l_{ik} \delta^l_r = \Gamma^l_{ik};$$

upon using (6.3).

3.   The Christoffel Symbols Evaluated

The Christoffel symbols Γ^l_{ik} defined by (6.6) depend only upon the coefficients g_{ik} of the first fundamental form and their first derivatives. The appropriate formulas exhibiting this dependence result from calculation of the quantities $\mathbf{X}_{ik} \cdot \mathbf{X}_m$ as follows.   Starting from $g_{im} = \mathbf{X}_i \cdot \mathbf{X}_m$ and differentiating with respect to u^k results in the equations

$$\frac{\partial g_{im}}{\partial u^k} = \mathbf{X}_{ik} \cdot \mathbf{X}_m + \mathbf{X}_i \cdot \mathbf{X}_{mk},$$

$$\frac{\partial g_{mk}}{\partial u^i} = \mathbf{X}_{mi} \cdot \mathbf{X}_k + \mathbf{X}_m \cdot \mathbf{X}_{ki},$$

$$\frac{\partial g_{ki}}{\partial u^m} = \mathbf{X}_{km} \cdot \mathbf{X}_i + \mathbf{X}_k \cdot \mathbf{X}_{im}.$$

Addition of the first two and subtraction of the last of these equations leads to the relation

(6.7)
$$\mathbf{X}_{ik} \cdot \mathbf{X}_m = \frac{1}{2}\left(\frac{\partial g_{im}}{\partial u^k} + \frac{\partial g_{mk}}{\partial u^i} - \frac{\partial g_{ki}}{\partial u^m}\right) = \Gamma_{ikm},$$

in which the new quantity Γ_{ikm} defined in it is called the Christoffel three-index symbol of the first kind (denoted by $[^i_m{}^k]$ by Christoffel).   From (6.6) the following equation serves to connect the two different kinds of Christoffel symbols:

(6.8)
$$\Gamma^l_{ik} = g^{lm}\Gamma_{ikm}.$$

In this way the coefficients Γ^l_{ik} in (6.4) are defined in terms of the coefficients g_{ik} of the first fundamental form and their first derivatives.   Thus (6.4) leads, finally, to the partial differential equations introduced by Gauss:

(6.9)
$$\mathbf{X}_{ik} = \Gamma^l_{ik}\mathbf{X}_l + L_{ik}\mathbf{X}_3,$$

with Γ^l_{ik} defined through (6.8), (6.7), and (6.2).

It is very important to emphasize that the quantities Γ^l_{ik} and Γ_{ikl} depend only upon the coefficients of the first fundamental form.   It is also useful to observe that the following symmetry relations hold in view of (6.7) and $\mathbf{X}_{ik} = \mathbf{X}_{ki}$:

$$(6.10) \qquad \begin{aligned} \Gamma_{ikm} &= \Gamma_{kim}, \\ \Gamma_{ik}^m &= \Gamma_{ki}^m. \end{aligned}$$

4. The Weingarten Equations

Differential equations involving the derivatives of the normal vector $\mathbf{X}_3$ are next derived.   Since $\mathbf{X}_3$ is chosen as a unit vector normal to the surface it follows that its derivatives $\mathbf{X}_{3i}$ lie in the tangent plane, and consequently they may be written in the form

$$(6.11) \qquad \mathbf{X}_{3i} = -L_i^k \mathbf{X}_k, \qquad i = 1,2,$$

with undetermined scalar coefficients $-L_i^k$.   To fix the value of the coefficients a scalar multiplication of (6.11) with $\mathbf{X}_j$ is performed, with the result

$$(6.12) \qquad \mathbf{X}_{3i} \cdot \mathbf{X}_j = -L_i^k g_{kj}$$

since $\mathbf{X}_j \cdot \mathbf{X}_k = g_{kj}$.   But since $\mathbf{X}_3 \cdot \mathbf{X}_j = 0$ it follows that

$$(6.13) \qquad \mathbf{X}_{3i} \cdot \mathbf{X}_j + \mathbf{X}_3 \cdot \mathbf{X}_{ij} = \mathbf{X}_{3i} \cdot \mathbf{X}_j + L_{ij} = 0,$$

because $L_{ij} = \mathbf{X}_3 \cdot \mathbf{X}_{ij}$, and hence from (6.12) the following formula results:

$$(6.14) \qquad L_{ij} = L_i^k g_{kj} = L_i^r g_{rj}.$$

This set of linear equations for the quantities L_i^r can be solved through multiplication with g^{kj}, which thus implies a summation on j, followed by use of (6.3); the result is

$$(6.15) \qquad L_{ij}g^{kj} = L_i^r g_{rj} g^{kj} = L_i^r \delta_r^k = L_i^k,$$

and the coefficients L_i^k are thus defined in terms of the coefficients of the two fundamental forms.   The partial differential equations (6.11) now have known coefficients:

$$(6.16) \qquad \mathbf{X}_{3i} = -L_{ij}g^{kj}\mathbf{X}_k, \qquad i = 1,2.$$

These differential equations are called the Weingarten equations.

5. Some Observations About the Partial Differential Equations

The equations 6.9 and 6.16 are usually referred to as the fundamental partial differential equations of surface theory.   They express the first partial derivatives of the trihedral X_1, X_2, X_3 of surface vectors, defined at all points of the surface, in terms of these vectors. To this extent they are analogous to the Frenet equations for space curves. However, there are quite important differences. In the case of the Frenet equations, their coefficients κ and τ are invariants under all orientation-preserving parameter transformations, as well as under transformations of coordinates in three-space. In the present case the coefficients of the differential equations depend on the functions $g_{ik} = X_i \cdot X_k$ and $L_{ik} = X_3 \cdot X_{ik}$ and these are not invariant under parameter transformations, as is seen immediately, and the same is true of the quantities Γ_{ik}^l and L_i^k.

In the case of space curves it was proved in Chapter 3 that regular curves exist as integrals of the Frenet equations when the curvature κ (assumed to be positive) and the torsion τ are any arbitrarily given functions of the distance s along the curve (aside from smoothness requirements). In addition, all curves obtained in this way for given functions κ and τ are congruent—that is, they differ at most by rigid motions in three-space. For surfaces it is natural to ask, by analogy, the following fundamental questions:

1. Will a regular surface exist if the coefficients g_{ik} and L_{ik} of the two fundamental forms are arbitrarily prescribed functions of the surface parameters (aside from differentiability conditions on these functions)?

2. In case the surface exists will it be uniquely determined within motions?

The answer to the first question is No, in general. The basic reason for this is that the system of partial differential equations (6.9) and (6.16) is what is called an *overdetermined system* since there are more scalar equations than there are scalar functions to be determined—in fact there are fifteen differential equations for nine scalar functions. Thus it should occasion no surprise that the differential equations of surface theory need not, as was the case with the Frenet equations, have a solution for arbitrarily given coefficients $g_{ik}(u^1, u^2)$ and $L_{ik}(u^1, u^2)$ of the fundamental forms. For that to be true certain identities, called compatibility conditions, that connect the functions g_{ik} and L_{ik} must hold. The situation is exactly analogous to that involved in determining a single function $u(x, y)$ when its partial derivatives u_x and u_y are both prescribed:

$$u_x = f(x, y), \qquad u_y = g(x, y).$$

In calculus it is shown that a solution $u(x, y)$ with continuous second derivatives will exist in a neighborhood of a point (x_0, y_0) if, and only if, the compatibility condition $f_y = g_x$ is satisfied at all points of the neighborhood and that this solution is uniquely determined in a simply connected domain if the value of u is prescribed at the point (x_0, y_0). The necessity of the condition is clear since it follows from $u_{xy} = u_{yx}$.

One of the principal objects of this chapter is to derive the compatibility conditions associated with the partial differential equations (6.9) and (6.16). There are three such compatibility conditions; they are rather complicated nonlinear identities involving the functions g_{ik}, L_{ik}, and first and second derivatives of them. These identities result from the assumption that the vector $\mathbf{X}(u, v)$ defining the surface has continuous third derivatives. It is then a basically important fact of differential geometry that *if the three compatibility equations are satisfied for the functions g_{ik} and L_{ik} there exist functions $\mathbf{X}_1$, $\mathbf{X}_2$, $\mathbf{X}_3$ that satisfy the partial differential equations.* Afterward, because of the special form of these differential equations, it will be shown in Section 12 that a regular surface exists with $\mathbf{X}_1$, $\mathbf{X}_2$, $\mathbf{X}_3$ as coordinate and normal vectors, and with the given functions g_{ik} and L_{ik} as coefficients of its fundamental forms.

The second question raised above has a quite satisfactory answer, i.e., that if a regular surface exists and satisfies the partial differential equations it is determined uniquely within rigid motions. This question, which is easily disposed of, is dealt with in the next section, to be followed by sections dealing with the other questions raised here.

6.   Uniqueness of a Surface for Given g_{ik} and L_{ik}

It is to be shown that the surface $\mathbf{X}(u, v)$, if it exists at all with given g_{ik} and L_{ik}, is uniquely determined within rigid motions. Suppose that two such surfaces $\mathbf{X}(u, v)$ and $\mathbf{Y}(u, v)$ exist. By a rigid motion of $\mathbf{Y}(u, v)$ they can be brought together at the same point $\mathbf{X}(u_0, v_0)$ and with the same normal vector. In addition, since the angle between the coordinate vectors of both surfaces is determined from the functions g_{ik} alone, it follows that a rotation of $\mathbf{Y}(u, v)$ about the normal can be achieved so that $\mathbf{X}_i(u_0, v_0) = \mathbf{Y}_i(u_0, v_0)$, since these vectors have the same length—again because the lengths of them are fixed by the functions g_{ik}. Consider any regular curve C: $u = u(t)$, $v = v(t)$ through (u_0, v_0) in the parameter plane, and its images on $\mathbf{X}(u, v)$ and $\mathbf{Y}(u, v)$. The image of the curve C on $\mathbf{X}(u, v)$ satisfies the equations

$$\mathbf{X}' = \mathbf{X}_1 u' + \mathbf{X}_2 v',$$

$$\mathbf{X}'_1 = \mathbf{X}_{11} u' + \mathbf{X}_{12} v',$$

$$\mathbf{X}'_2 = \mathbf{X}_{21} u' + \mathbf{X}_{22} v',$$

in which the prime means differentiation with respect to t. In the second and third of these equations the second derivatives $\mathbf{X}_{ik}$ are replaced by the first derivatives $\mathbf{X}_1$ and $\mathbf{X}_2$ through the use of (6.9)—after replacing $\mathbf{X}_3$ by

$\mathbf{X}_1 \times \mathbf{X}_2/|\mathbf{X}_1 \times \mathbf{X}_2|$. The result is a set of three ordinary vector differential equations of first order:

$$\mathbf{X}' = R(\mathbf{X}_1, \mathbf{X}_2, t),$$
$$\mathbf{X}'_1 = S(\mathbf{X}_1, \mathbf{X}_2, t),$$
$$\mathbf{X}'_2 = T(\mathbf{X}_1, \mathbf{X}_2, t),$$

for the three vector functions $\mathbf{X}(t)$, $\mathbf{X}_1(t)$, $\mathbf{X}_2(t)$. The function R is determined by the curve C in the parameter plane alone, while S and T depend upon C and the functions g_{ik} and L_{ik}. Consequently, the same differential equations hold for $\mathbf{Y}(t)$, $\mathbf{Y}_1(t)$, $\mathbf{Y}_2(t)$, and since these vectors are assumed to be the same as $\mathbf{X}$, $\mathbf{X}_1$, and $\mathbf{X}_2$ at the point (u_0, v_0) they are the same for all t, by the uniqueness theorem for ordinary differential equations. Thus, since the images of all curves in the parameter plane coincide on the two surfaces it follows that the two surfaces are congruent, and the uniqueness statement is proved.

7. The *theorema egregium* of Gauss

The *theorema egregium* of Gauss is without much question the most important single theorem in differential geometry. It states that the Gaussian curvature K, originally defined in terms of the coefficients L_{ik} of the second fundamental form, really *depends only upon the coefficients g_{ik} of the first fundamental form and certain of their first and second derivatives*. This result has a long series of consequences. For the present it is noted that such a relation constitutes at least one of the abovementioned compatibility conditions for the functions g_{ik} and L_{ik} that are necessary if a surface $\mathbf{X}(u, v)$ with continuous third derivatives is to exist with these functions as coefficients of its fundamental forms: evidently, if $L_{11}L_{22} - L_{12}^2 = F(g_{ik})$ is an identity for all surfaces $\mathbf{X}(u, v)$, it is not possible to prescribe the functions L_{ik} in a completely arbitrary way once the functions g_{ik} have been given.

A first direct derivation—due to R. Baltzer—of the theorem is given here (several others will be given later), starting with the definition given in Chapter 4:

$$K = \frac{LN - M^2}{EG - F^2}.$$

From $L_{ik} = \mathbf{X}_{ik} \cdot \mathbf{X}_3$, $\mathbf{X}_3 = (\mathbf{X}_1 \times \mathbf{X}_2)/\sqrt{EG - F^2}$, this relation is written in the form

$$K(EG - F^2)^2 = (\mathbf{X}_{11} \cdot \mathbf{X}_1 \times \mathbf{X}_2)(\mathbf{X}_{22} \cdot \mathbf{X}_1 \times \mathbf{X}_2) - (\mathbf{X}_{12} \cdot \mathbf{X}_1 \times \mathbf{X}_2)^2.$$

On the right-hand side the product of two determinants and the square of a determinant occur. These can be expanded, by the usual rule for the product of determinants, to obtain the equation

$$K(EG - F^2)^2 =$$

$$\begin{vmatrix} \mathbf{X}_{11} \cdot \mathbf{X}_{22} & \mathbf{X}_{11} \cdot \mathbf{X}_1 & \mathbf{X}_{11} \cdot \mathbf{X}_2 \\ \mathbf{X}_1 \cdot \mathbf{X}_{22} & E & F \\ \mathbf{X}_2 \cdot \mathbf{X}_{22} & F & G \end{vmatrix} - \begin{vmatrix} \mathbf{X}_{12} \cdot \mathbf{X}_{12} & \mathbf{X}_{12} \cdot \mathbf{X}_1 & \mathbf{X}_{12} \cdot \mathbf{X}_2 \\ \mathbf{X}_{12} \cdot \mathbf{X}_1 & E & F \\ \mathbf{X}_{12} \cdot \mathbf{X}_2 & F & G \end{vmatrix}$$

$$= (\mathbf{X}_{11} \cdot \mathbf{X}_{22} - \mathbf{X}_{12} \cdot \mathbf{X}_{12}) \begin{vmatrix} E & F \\ F & G \end{vmatrix} + \begin{vmatrix} 0 & \mathbf{X}_{11} \cdot \mathbf{X}_1 & \mathbf{X}_{11} \cdot \mathbf{X}_2 \\ \mathbf{X}_1 \cdot \mathbf{X}_{22} & E & F \\ \mathbf{X}_2 \cdot \mathbf{X}_{22} & F & G \end{vmatrix}$$

$$- \begin{vmatrix} 0 & \mathbf{X}_{12} \cdot \mathbf{X}_1 & \mathbf{X}_{12} \cdot \mathbf{X}_2 \\ \mathbf{X}_{12} \cdot \mathbf{X}_1 & E & F \\ \mathbf{X}_{12} \cdot \mathbf{X}_2 & F & G \end{vmatrix}$$

(As an aid in checking this calculation, it may be helpful to observe that the elements of the product of two determinants can be regarded as the scalar products of vectors whose components are appropriate rows and columns of the two factors.) From $\mathbf{X}_1 \cdot \mathbf{X}_1 = E$, $\mathbf{X}_1 \cdot \mathbf{X}_2 = F$, $\mathbf{X}_2 \cdot \mathbf{X}_2 = G$ the following identities result by differentiation:

$$\mathbf{X}_{11} \cdot \mathbf{X}_1 = \tfrac{1}{2} E_1,$$

$$\mathbf{X}_{12} \cdot \mathbf{X}_1 = \tfrac{1}{2} E_2,$$

$$\mathbf{X}_{22} \cdot \mathbf{X}_2 = \tfrac{1}{2} G_2,$$

$$\mathbf{X}_{12} \cdot \mathbf{X}_2 = \tfrac{1}{2} G_1,$$

$$\mathbf{X}_{11} \cdot \mathbf{X}_2 = F_1 - \tfrac{1}{2} E_2,$$

$$\mathbf{X}_{22} \cdot \mathbf{X}_1 = F_2 - \tfrac{1}{2} G_1.$$

Differentiation of the fourth of these equations with respect to u and of the fifth with respect to v, followed by a subtraction, yields the identity

$$\mathbf{X}_{11} \cdot \mathbf{X}_{22} - \mathbf{X}_{12} \cdot \mathbf{X}_{12} = -\tfrac{1}{2} G_{11} + F_{12} - \tfrac{1}{2} E_{22}.$$

(At this step the existence of third derivatives of $\mathbf{X}(u, v)$ was needed.) In this way all terms on the right-hand side of the above equation have been expressed in terms of the functions E, F, G, and certain of their first and second derivatives, and this proves the theorem of Gauss.

For later purposes it is useful to write down this identity for the special case of an orthogonal curvilinear coordinate system.   In this case $F = 0$, and a formal calculation shows that K is given by the following equation:

$$(6.17) \qquad K = - \frac{1}{2\sqrt{EG}} \left[\frac{\partial}{\partial v} \left(\frac{E_v}{\sqrt{EG}} \right) + \frac{\partial}{\partial u} \left(\frac{G_u}{\sqrt{EG}} \right) \right].$$

A few comments about (6.17) are in order.   It is a convenient and compact way to express the curvature K, and the right hand side of the equation has the virtue of being a so-called *divergence expression* since it is in such a form that an integration of it over a domain of a surface can be transformed into a line integral over the boundary of the domain.   This fact will be used in Chapter 8 in the course of deriving an important formula, the Gauss-Bonnet formula.   It should be noted, however, that the formula (6.17) does not express the curvature K in an invariant form with respect to parameter transformations, although K is an invariant.   In fact, it is not possible to express K in an invariant form that is also a divergence expression: this fact comes out in connection with the Gauss-Bonnet formula.   An invariant form for K is given in Section 9 of this chapter.

8.   How Gauss May Have Hit upon His Theorem

Gauss [G.1] proved his theorem by a long calculation that is not easy to motivate.   It has been conjectured that he himself fell upon the theorem by considering it in terms of his definition (as given in Chapter 4) of K in terms of the spherical image, and that he made it plausible by considering not a regular curved surface but rather a surface in the form of a polyhedron made

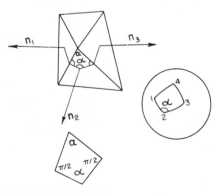

Fig. 6.1   Spherical image of a polyhedron.

up of plane triangles joined together at their vertices to make a convex polyhedral angle, as in Fig. 6.1. One of the most important consequences of the theorem of Gauss is that the Gaussian curvature K is invariant under any deformations of the surface that preserve the lengths of all curves on it: this follows immediately from the fact that K depends only on the first fundamental form, i.e., only on the functions g_{ik}. Or, put in another way, the curvature K is not changed by bending a surface in space as long as it is not stretched in the process: a fact that is not at all obvious intuitively. But for the case of the polyhedral surface it can be seen that a length preserving deformation, which consists in keeping the individual faces rigid but permits them to rotate about the edges where they are joined, preserves what one might reasonably define as its Gaussian curvature, although the surface can change its shape in space.

The Gaussian curvature of such a surface is defined here as the area of the map on the unit sphere of the outward normals on *all support planes through the vertex of the polyhedral surface. These are planes that contain the surface in one of the half-spaces defined by them.* In the case of a convex vertex it can be seen that these points fill out a polygon on the unit sphere whose sides are great circle arcs joining points which are the images of the normals on the faces of the vertex, and in the same order as the faces occur in making a circuit about the vertex. In fact, the sides of the polygon forming the spherical image of a vertex of the polyhedron are the images of normals on support planes that contain a fixed one of the edges of the vertex. It is now claimed that the area of the spherical image defined in this way is not changed by any deformation of the polyhedron that leaves the faces rigid. The reason for this can be seen on the basis of Fig. 6.1, which shows a vertex with four faces—one with three faces only would be rigid. (This example was taken from the book by Hilbert and Cohn Vossen [H.10]). Normals n_1, n_2, n_3 are shown. On the spherical image the side 1–2, for example, is on a great circle which is in a plane parallel to that fixed by n_1 and n_2, or also, in a plane orthogonal to the edge common to the faces 1 and 2. In similar fashion the side 2–3 of the spherical image is in a plane orthogonal to the edge common to the faces 2 and 3. Since the angle between the edges of a given face at the vertex is fixed, it follows that the angles of the spherical polygon are also fixed since they are the supplements of the angles between the edges; for example, $\alpha = \pi - a$, as indicated in the figure. It is, however, a well-known fact of spherical trigonometry (which we will prove later in Chapter 8 to be a special case of a very general theorem) that the area of a spherical polygon is fixed by the sum of its interior angles, and this gives the result that was wanted. Presumably Gauss imagined a curved surface as a limit of approximating polyhedra, which is a rather dangerous procedure in some connections, but the analogy does lead in this case to a correct theorem.

9. Compatibility Conditions in General

The theorem of Gauss gives only one of the compatibility conditions associated with the partial differential equations of surface theory. In this and the next two sections all of them are obtained by using a systematic procedure that determines all possible identities of this kind. The differential equations are [cf. (6.9) and (6.16)]

$$(6.18) \qquad \mathbf{X}_{ik} = \Gamma^l_{ik}\mathbf{X}_l + L_{ik}\mathbf{X}_3, \qquad i, k = 1, 2,$$

$$(6.19) \qquad \mathbf{X}_{3i} = -L^k_i\mathbf{X}_k, \qquad\qquad i = 1, 2.$$

The Christoffel symbols Γ^l_{ik} are defined in terms of the functions g_{ik} and g^{ik} by means of equations (6.7) and (6.8), and the coefficients L^k_i are defined by (6.15). Again it is to be observed that the summation convention applies, i.e. that a summation on such indices as l and k over 1 and 2 is intended above because these indices are repeated in a term.

It has already been stated that the compatibility conditions for the overdetermined system given by (6.18) and (6.19) result from the assumption that the vector function $\mathbf{X}(u^1, u^2)$, which should represent a regular surface, has continuous third derivatives. Thus $\partial \mathbf{X}_{ik}/\partial u^j = \partial \mathbf{X}_{ij}/\partial u^k$ must hold or, in shorter notation, $\mathbf{X}_{ikj} = \mathbf{X}_{ijk}$ is a consequence of the assumed continuity of third derivatives. Compatibility conditions are derived from (6.18) by differentiating the equation with respect to u_j, followed by a replacement of second derivatives through the use of (6.18) and (6.19). For $\mathbf{X}_{ikj}$ the following equation results:

$$\mathbf{X}_{ikj} = \frac{\partial \Gamma^l_{ik}}{\partial u^j}\mathbf{X}_l + \Gamma^l_{ik}\mathbf{X}_{lj} + L_{ik,j}\mathbf{X}_3 + L_{ik}\mathbf{X}_{3j}$$

$$= \frac{\partial \Gamma^l_{ik}}{\partial u^j}\mathbf{X}_l + \Gamma^l_{ik}\Gamma^n_{lj}\mathbf{X}_n + \Gamma^l_{ik}L_{lj}\mathbf{X}_3 + L_{ik,j}\mathbf{X}_3 - L_{ik}L^n_j\mathbf{X}_n$$

$$= \mathbf{X}_n[\frac{\partial \Gamma^n_{ik}}{\partial u^j} + \Gamma^l_{ik}\Gamma^n_{lj} - L_{ik}L^n_j] + \mathbf{X}_3[L_{ik,j} + \Gamma^l_{ik}L_{lj}],$$

after replacing the second derivatives $\mathbf{X}_{ij}$ by using (6.18) and (6.19). Interchange of j and k and subsequent subtraction yields $\mathbf{X}_{ikj} - \mathbf{X}_{ijk} = 0$, and in view of the linear independence of the vectors $\mathbf{X}_n$ and $\mathbf{X}_3$, the following identities are obtained:

$$(6.20) \qquad \frac{\partial \Gamma^n_{ik}}{\partial u^j} - \frac{\partial \Gamma^n_{ij}}{\partial u^k} + \Gamma^l_{ik}\Gamma^n_{lj} - \Gamma^l_{ij}\Gamma^n_{lk} = L_{ik}L^n_j - L_{ij}L^n_k,$$

$$(6.21) \qquad \frac{\partial L_{ik}}{\partial u^j} - \frac{\partial L_{ij}}{\partial u^k} + \Gamma^l_{ik}L_{lj} - \Gamma^l_{ij}L_{lk} = 0,$$

in which $i, j, k, n = 1, 2$, and summations on the index l are to be taken.

10. Codazzi-Mainardi Equations

Consider first equations (6.21). They are satisfied identically if $j = k$.
It is sufficient in order to obtain all of the independent identities, to take
$j = 1, k = 2$, since an interchange of j and k multiplies the left side of (6.21)
by -1. There are therefore only two essentially different identities in the
group (6.21); they are obtained by setting $i = 1, i = 2$ successively:

(6.22)
$$\frac{\partial L_{12}}{\partial u^1} - \frac{\partial L_{11}}{\partial u^2} + \Gamma_{12}^l L_{l1} - \Gamma_{11}^l L_{l2} = 0,$$

$$\frac{\partial L_{22}}{\partial u^1} - \frac{\partial L_{21}}{\partial u^2} + \Gamma_{22}^l L_{l1} - \Gamma_{21}^l L_{l2} = 0.$$

Equations (6.22) are called the Codazzi-Mainardi equations.

11. The Gauss *theorema egregium* Again

Consider now equations (6.20). It is convenient to introduce a quantity
r^n_{ijk} as follows:

(6.23)
$$r^n_{.ijk} = \frac{\partial \Gamma_{ik}^n}{\partial u^j} - \frac{\partial \Gamma_{ij}^n}{\partial u^k} + \Gamma_{ik}^l \Gamma_{lj}^n - \Gamma_{ij}^l \Gamma_{lk}^n.$$

Equations (6.20) then become

(6.24)
$$r^n_{.ijk} = L_{ik} L_j^n - L_{ij} L_k^n.$$

A new set of quantities is defined as follows:

(6.25)
$$r_{mijk} = g_{nm} r^n_{.ijk},$$

and (6.24) is replaced by

(6.26)
$$r_{mijk} = L_{ik} L_{mj} - L_{ij} L_{mk},$$

This is correct since $g_{nm} L_j^n = L_{mj}$ [cf. (6.14)]. In (6.26) is contained the
special case

(6.27)
$$r_{2121} = L_{11} L_{22} - L_{12}^2.$$

This equation is another way of expressing the Gauss *theorema egregium*,
since the quantities r_{ijkl} contain only the functions g_{ik} and certain of their
first and second derivatives, and since $K = (g_{11} g_{22} - g_{12}^2)^{-1} r_{2121}$. Although it
is not obvious from this discussion, it is a fact that this last expression for K
[unlike (6.17)] is invariant under parameter transformations. It would be
possible, though tedious, to prove this by a direct calculation. Later on the

quantities r^n_{ijk} and r_{mijk} will appear in Chapter 9 as components of the Riemannian curvature tensor for a two-dimensional Riemannian manifold; as such, they then appear as invariants.

It is to be shown next that the many equations given by (6.20) either vanish identically or reduce essentially to (6.27). From relations of the type

$$g_{nm}\Gamma^n_{ik} = (g_{nm}\Gamma^n_{ik})_j - \Gamma^n_{ik}\frac{\partial g_{nm}}{\partial u^j},$$

the definition of r_{mijk}, and the relation $g_{nm}\Gamma^n_{ij} = \Gamma_{ljm}$ [obtained from (6.8) by a familiar process] the following identities result:

$$r_{mijk} = \Gamma_{ikm} - \Gamma_{ijm} + \Gamma^l_{ik}\Gamma_{ljm} + \Gamma^l_{ij}\Gamma_{lkm}$$

$$- \Gamma^n_{ik}\frac{\partial g_{nm}}{\partial u^j} + \Gamma^n_{ij}\frac{\partial g_{nm}}{\partial u^k}.$$

This reduces, upon using (6.7) and the following relations which come from it:

(6.28)
$$\frac{\partial g_{nm}}{\partial u^j} = \Gamma_{njm} + \Gamma_{mjn},$$

$$\frac{\partial g_{nm}}{\partial u^k} = \Gamma_{nkm} + \Gamma_{mkn}$$

to the equations

$$r_{mijk} = \frac{1}{2}\left(\frac{\partial^2 g_{nm}}{\partial u^i \, \partial u^j} + \frac{\partial^2 g_{ij}}{\partial u^k \, \partial u^m} - \frac{\partial^2 g_{ik}}{\partial u^j \, \partial u^m} - \frac{\partial^2 g_{jm}}{\partial u^i \, \partial u^k}\right)$$

$$+ g_{nl}(\Gamma_{ijl}\Gamma_{mkn} - \Gamma_{ikl}\Gamma_{mjn}).$$

From these last equations the following symmetry relations are seen to be valid:

(6.29)
$$r_{mijk} = r_{jkmi},$$

$$r_{mijk} = -r_{imjk},$$

$$r_{mijk} = -r_{mikj}.$$

Hence it follows that

$$r_{11jk} = r_{22jk} = r_{mi22} = r_{mi11} = 0,$$

and only

$$r_{1212} = -r_{2112} = -r_{1221} = r_{2121}$$

can be different from zero. Since the right-hand side of (6.26) also has the symmetries expressed in (6.29), it follows that only four of these equations are not identically satisfied. But these four follow from any one of them such as (6.27).

Finally, the compatibility equations for (6.19) are to be found from $\partial L_1^k X_k / \partial u^2 = \partial L_2^k X_k / \partial u^1$. It turns out without difficulty that these are again the Codazzi-Mainardi equations (6.22). In other words, the full set of compatibility equations for (6.18) and (6.19) consists of the Codazzi-Mainardi equations and the analytical expression of Gauss's *theorema egregium*.

12. Existence of a Surface with Given g_{ik} and L_{ik}

In Section 5 it was stated that a regular surface is uniquely determined within rigid motions if the coefficients g_{ik} (with $g_{11}g_{22} - g_{12}^2 > 0$) and L_{ik} satisfy the compatibility conditions (6.22) and (6.26), that is, the equation expressing Gauss's theorem, and the Codazzi-Mainardi equations. (The functions g_{ik} should have continuous second derivatives and L_{ik} continuous first derivatives.)

The essential point in proving the existence statement is that the over-determined system of partial differential equations (6.18) and (6.19) for the three vector functions $\mathbf{X}_1$, $\mathbf{X}_2$, $\mathbf{X}_3$ have under the given circumstances a uniquely determined solution with continuous third derivatives once initial values for these vectors have been prescribed at a point (u_0, v_0). This follows from the general theory of such systems, as derived in Appendix B.

It still remains to be shown, however, that a surface defined by a vector $\mathbf{X}(u, v)$ with continuous third derivatives exists such that $\partial \mathbf{X} / \partial u = \mathbf{X}_1$, $\partial \mathbf{X} / \partial v = \mathbf{X}_2$, and with $\mathbf{X}_3$ as normal vector, when $\mathbf{X}_1$, $\mathbf{X}_2$, $\mathbf{X}_3$ are solutions of the partial differential equations. These solutions are assumed to result by choosing as initial conditions at (u_0, v_0) vectors $\mathbf{X}_1(u_0, v_0)$, $\mathbf{X}_2(u_0, v_0)$, $\mathbf{X}_3(u_0, v_0)$ such that

$$(6.30) \qquad \mathbf{X}_i \cdot \mathbf{X}_k |_{u_0, v_0} = g_{ik}(u_0, v_0), \qquad i,k = 1,2,$$

$$(6.31) \qquad \mathbf{X}_3(u_0, v_0) = \frac{\mathbf{X}_1 \times \mathbf{X}_2}{|\mathbf{X}_1 \times \mathbf{X}_2|_{u_0, v_0}}.$$

The initial normal vector can be defined in this way, since (6.30) guarantees that $\mathbf{X}_1$, $\mathbf{X}_2$ are linearly independent because $g_{11}g_{22} - g_{12}^2 > 0$ is assumed to hold.

From (6.18) it is clear that $\mathbf{X}_{ik} = \mathbf{X}_{ki}$. Thus by setting $\mathbf{X}_u = \mathbf{X}_1$, $\mathbf{X}_v = \mathbf{X}_2$ this overdetermined system for the function $\mathbf{X}(u, v)$ is such that the compatibility conditions are satisfied. It follows that a uniquely determined vector $\mathbf{X}(u, v)$ results when $\mathbf{X}(u_0, v_0) = \mathbf{X}_0$ is prescribed; in fact, $\mathbf{X}(u, v)$ is given by

$$(6.32) \qquad \mathbf{X}(u, v) = \int_{u_0}^{u} \mathbf{X}_1(\xi, v_0)\, d\xi + \int_{v_0}^{v} \mathbf{X}_2(u_0, \eta)\, d\eta + \mathbf{X}_0.$$

Of course, $\mathbf{X}_1$ and $\mathbf{X}_2$ are the solutions of the Gauss and Weingarten equations that are uniquely determined by the initial conditions (6.30) and (6.31).

Finally, it is to be shown that the surface given by (6.32) has the vectors $\mathbf{X}_1$, $\mathbf{X}_2$, $\mathbf{X}_3$ as coordinate and normal vectors, and the given functions g_{ik} and L_{ik} as coefficients of its fundamental forms. To this end consider the quantities $(\mathbf{X}_i \cdot \mathbf{X}_k)$, $(\mathbf{X}_3 \cdot \mathbf{X}_k)$, i, $k = 1,2$, and $(\mathbf{X}_3 \cdot \mathbf{X}_3)$. By differentiation the following partial differential equations for these scalar functions result:

$$(6.33) \quad (\mathbf{X}_i \cdot \mathbf{X}_k)_j = \mathbf{X}_{ij} \cdot \mathbf{X}_k + \mathbf{X}_i \cdot \mathbf{X}_{kj}$$

$$= \Gamma_{ij}^m \mathbf{X}_m \cdot \mathbf{X}_k + \Gamma_{kj}^m \mathbf{X}_m \cdot \mathbf{X}_i + L_{ij} \mathbf{X}_3 \cdot \mathbf{X}_k + L_{kj} \mathbf{X}_3 \cdot \mathbf{X}_i,$$

$$(6.34) \quad (\mathbf{X}_3 \cdot \mathbf{X}_k)_j = \mathbf{X}_{3j} \cdot \mathbf{X}_k + \mathbf{X}_3 \cdot \mathbf{X}_{kj}$$

$$= - L_j^m \mathbf{X}_m \cdot \mathbf{X}_k + \Gamma_{kj}^m \mathbf{X}_m \cdot \mathbf{X}_3 + L_{kj} \mathbf{X}_3 \cdot \mathbf{X}_3,$$

$$(6.35) \quad (\mathbf{X}_3 \cdot \mathbf{X}_3)_j = 2\mathbf{X}_3 \cdot \mathbf{X}_{3j} = -2L_j^k \mathbf{X}_k \cdot \mathbf{X}_3.$$

In this calculation equations (6.18) and (6.19) were used. These three sets of equations again form an overdetermined system for the scalar products as unknown functions, and the compatibility conditions are obviously satisfied since the vector functions $\mathbf{X}_1$, $\mathbf{X}_2$, $\mathbf{X}_3$ exist and have derivatives. Hence a solution for the scalar products is uniquely determined if their values at (u_0, v_0) satisfy (6.30) and $\mathbf{X}_3 \cdot \mathbf{X}_k = 0$, $\mathbf{X}_3 \cdot \mathbf{X}_3 = 1$. But it can be easily seen that *the scalar products satisfy these conditions for all values of u and v*, as follows. From (6.28) and (6.8) it follows that

$$\frac{\partial g_{ik}}{\partial u_j} = (\mathbf{X}_i \cdot \mathbf{X}_k)_j = \Gamma_{ijk} + \Gamma_{kji}$$

$$= \Gamma_{ij}^m g_{mk} + \Gamma_{kj}^m g_{mi},$$

and (6.14) can be written in the form

$$L_{kj} = L_j^m g_{mk}.$$

It is then easily seen that (6.33), (6.34), and (6.35) are satisfied identically by setting $\mathbf{X}_i \cdot \mathbf{X}_k = g_{ik}$, $\mathbf{X}_k \cdot \mathbf{X}_3 = 0$, $\mathbf{X}_3 \cdot \mathbf{X}_3 = 1$, and hence these constitute the only solutions of that system. Thus the solutions $\mathbf{X}_1$, $\mathbf{X}_2$, $\mathbf{X}_3$ of (6.18) and (6.19) form the trihedral of surface vectors at all points of the surface $\mathbf{X}(u, v)$ defined by (6.32). In addition, it has been shown that the functions g_{ik} are the coefficients of the first fundamental form of the surface. By taking the scalar product of equation (6.18) with $\mathbf{X}_3$ it is seen at once that $\mathbf{X}_3 \cdot \mathbf{X}_{ik} = L_{ik}$, and therefore the functions L_{ik} are the coefficients of the second fundamental form of the surface. Thus the general theorem has been proved.

13.   An Application of the General Theory to a Problem in the Large

It is evidently important to have available the fundamental differential equations (6.18) and (6.19), and the compatibility conditions, since the result of the previous section is obtained.   It is perhaps also of interest at this point to give a concrete example illustrating how the theory can be applied.

It is to be proved that *the sphere is the only regular closed surface in three-dimensional space that has constant Gaussian curvature K*.   This theorem of course belongs to differential geometry in the large, and perhaps would belong more properly in Chapter 8, where the appropriate concepts are introduced. However, in the present instance the essential argument is concerned with a neighborhood of a point.   Thus it is assumed here that a closed regular surface is defined, the term *closed* meaning that the surface, as a point set, is a closed set.   However, each point of the surface is assumed to have an open neighborhood within which the surface is regular.   It is to be noted first that the curvature K must be positive, since no closed regular surface can exist in three-space with $K < 0$ or $K = 0$ everywhere.

The theorem in question was first proved by Liebmann [L.6] and afterward in a simpler way by Hilbert [H.9].   It was noticed by Chern [C.5] that the proof of Hilbert can be used to prove a more general theorem concerning what are called Weingarten surfaces.   These are surfaces upon which a functional relation between the principal curvatures exists: e.g., the surfaces with $K = k_1 k_2 = $ const., or $H = \frac{1}{2}(k_1 + k_2) = $ const. at all points. The theorem of Chern is as follows:

A closed regular surface of positive Gaussian curvature is given in E^3 for which the principal curvatures k_1, k_2 with $k_1 > k_2$ are so related that $k_2 = f(k_1)$ and f is a decreasing function of k_1.   Such a surface is a sphere.

Two corollaries of this theorem are immediate:

1. If S and S' are isometric and regular and S is a sphere, then S' is also a sphere.   This follows at once from the fact that K is constant for both; hence $k_2 = c/k_1$ with $c > 0$, and Chern's theorem applies.

2. If a closed regular surface has positive Gaussian curvature and constant mean curvature, then it is a sphere.   Here $k_2 = c - k_1$ and again Chern's theorem applies.   This theorem was proved first by Liebmann [L.5]. The closed surfaces with constant mean curvature (but not necessarily taken to be convex at the outset) form an interesting class of surfaces concerning which many recent investigations have been made by A. D. Alexandrov, H. Hopf, and others.   An account of this can be found in the lecture notes by H. Hopf [H.15].

The proof of Chern's theorem results from the following lemma, due to

Hilbert, of a local character: *If $K > 0$ holds on a surface and if k_1 has a maximum and k_2 a minimum at a certain point, then $k_1 = k_2$ at that point.* The proof of the lemma is carried out indirectly by supposing that $k_1 \neq k_2$ at the point in question. In that case it is legitimate to introduce the lines of curvature as local parameters, since the point is not an umbilical point. Gauss's *theorema egregium* and the Codazzi-Mainardi equations then have the following forms [cf. (6.17) and (6.22) with $L_{12} = M = 0$]:

$$K = - \frac{1}{2\sqrt{EG}} \left[\left(\frac{E_v}{\sqrt{EG}} \right)_v + \left(\frac{G_u}{\sqrt{EG}} \right)_u \right],$$

$$L_v = \frac{E_v}{2} \left(\frac{L}{E} + \frac{N}{G} \right) = \frac{E_v}{2} (k_1 + k_2),$$

$$N_u = \frac{G_u}{2} \left(\frac{L}{E} + \frac{N}{G} \right) = \frac{G_u}{2} (k_1 + k_2),$$

since $k_1 = L/E$, $k_2 = N/G$ from (4.37).   From $L = Ek_1$ it follows that

$$L_v = E_v k_1 + E(k_1)_v$$

and the first of the Codazzi-Mainardi equations can be put into the form

$$E_v = - \frac{2E}{k_1 - k_2} (k_1)_v.$$

In similar fashion it follows that

$$G_u = \frac{2G}{k_1 - k_2} (k_2)_u.$$

If these relations are substituted in Gauss's equation, the result can be written in the form

(6.36)
$$-2EGK = - \frac{2E}{k_1 - k_2} (k_1)_{vv} + \frac{2G}{k_1 - k_2} (k_2)_{uu}$$
$$+ f(u, v)(k_1)_v + g(u, v)(k_2)_u.$$

In this equation $f(u, v)$ and $g(u, v)$ are certain bounded functions that need not be specified in detail.   Since $K > 0$ is assumed, the left-hand side of this equation is negative (and not zero).   At the point where k_1 has its assumed maximum and k_2, therefore, its minimum the following conditions hold:

$$(k_1)_v = (k_2)_u = 0, \qquad (k_1)_{vv} \leq 0, \qquad (k_2)_{uu} \geq 0.$$

Since $k_1 - k_2 > 0$ is assumed to hold, it follows that the right-hand side of equation (6.36) is not negative, and this establishes the contradiction that proves the lemma.

The theorem of Chern now results from the lemma in a fairly obvious way: On a closed surface k_1 attains its maximum at a certain point p and k_2 therefore attains its minimum at the same point, since k_2 is a decreasing function of k_1.   (It is here that the hypothesis of a closed surface is used, and in a decisive way.)   By the lemma $k_1(p) = k_2(p)$.   However, for all points q on the surface the following sequence of inequalities is valid: $k_1(p) \geq k_1(q) \geq k_2(q) \geq k_2(p) = k_1(p)$, and hence $k_1(q) = k_2(q)$.   In other words, all points of the surface are umbilical points and the surface is therefore a sphere, as was shown in Chapter IV.

Problems

1. Show that Gauss's *theorema egregium* can be put in the form

$$\sqrt{EG - F^2}\, K = \frac{\partial}{\partial u}\left(\frac{FE_v - EG_u}{2E\sqrt{EG - F^2}}\right) + \frac{\partial}{\partial v}\left(\frac{2EF_u - FE_u - EE_v}{2E\sqrt{EG - F^2}}\right).$$

2. Derive the identity between the three fundamental forms by using the Weingarten equations.

3. Derive the Codazzi-Mainardi equations as the compatibility conditions for the Weingarten equations.

CHAPTER VII

Inner Differential Geometry in the Small from the Extrinsic Point of View

1. Introduction. Motivations for the Basic Concepts

Hitherto properties of curves and surfaces in three-dimensional space have been considered that are invariant under transformations of the co-ordinate axes. However, the *theorema egregium* of Gauss indicates that it might well be of interest to consider properties of surfaces that are invariant under a wider class of transformations, i.e., transformations that preserve the lengths of curves lying on the surfaces, but not necessarily lengths be-tween pairs of points in the space: such transformations are called *isometric* transformations. For a surface in three-dimensional space this means that bending of the surface, and hence a change in its shape, is disregarded as long as it results in no changes in the lengths of curves on the surface. Thus it would not be necessary to think of the surface as embedded in three-dimen-sional space, and the geometry to be studied under these circumstances could be referred to quite properly as "inner" or intrinsic differential geometry since it would suffice to study only what occurs *in* the surface without regard to what that might mean with respect to the form of the surface in three-dimensional space. Such an intrinsic geometry is not at all lacking in content, as will be seen.

There are various ways available to introduce the concepts appropriate for the study of the inner differential geometry of surfaces. Since the form of the surface in three-dimensional space is not basically relevant to such a study, it would be logical and reasonable to avoid an embedding of the surface in three-space altogether. This in effect means that the surface would be studied by confining attention to the parameter plane, in which a line element $ds^2 = g_{ik}du^i\,du^k$ is defined by introducing functions $g_{ik}(u^1, u^2)$ so that this quadratic differential form is positive definite. However, what to do next is by no means obvious since it is clear that any reasonable treatment of the differential geometry of such important things as curves in the surface, for example, requires the introduction of the notion of the tangent vector to

151

such curves, and this in turn raises the general question of how to introduce properly the concept of vectors lying in the curved surface. By "properly" would be meant, quite naturally, that the entities to be identified as vectors should have properties more or less of the sort made familiar in linear algebra and in Euclidean geometry—in particular they should be invariants in some appropriate sense with respect to arbitrary changes in curvilinear coordinates since there would be no reason to single out any special coordinate systems. This in itself poses a problem since vectors (and their generalization, the tensors) are defined as invariants in linear spaces with respect to changes in *Cartesian coordinates*, and a bridge must be found to the more general invariance that is in mind here.

A reasonable way to accomplish these objectives was first pointed out by Riemann in his classic lecture "Über die Hypothesen welche der Geometrie zu grundeliegen" [R.3]. The idea is to study the geometry of an n-dimensional continuum by dealing first with the geometry in the neighborhood of a point with the use of differential relations and differential equations in a fashion analogous to the way continuum mechanics is treated in Euclidean spaces as a means of arriving at mathematical formulations of the physical laws in terms of differential equations. In carrying out this idea in a concrete way Riemann made use of the concepts and techniques of Euclidean geometry by postulating that the geometrical continuum under investigation should be approximately Euclidean in the neighborhood of each of its points. One of the convenient ways to learn how to do that in a concrete way is to begin by studying the continuum when it is embedded in a Euclidean space. This then furnishes clues regarding how the concepts should be introduced with respect to a neighborhood of each point without reference to such an embedding, as will be done later on.

The kind of geometry introduced by Riemann would lose a great deal of its interest if it dealt only with geometry in the neighborhood of each individual point. What is wanted in addition is a means to compare the character of the geometry at *different* points in the continuum. In Euclidean spaces this problem is made simple because of the existence of Cartesian coordinate systems. Thus if $v_1, v_2, \ldots, v_n$ are basis vectors in such a space attached to a point O as origin, it is possible to describe the space with respect to any other point O' as origin by taking as a basis at O' vectors parallel to the vectors v_i. An arbitrary point P of the space is then given by a vector X referred to the basis at O and X' when referred to that at O' and $X = X' + a$, with a as the constant vector (i.e., a vector independent of the point P) that locates O' in the coordinate system based on O. Or, as it is also put, parallel translations of basis vectors are assumed to be freely possible, and two geometrical entities described with reference to each of two such systems can be defined as *congruent* when their corresponding points are fixed by vectors with differences that are constant for all points; afterward, this notion of congruence can be extended by introducing rigid rotations and reflections (that is, orthogonal transformations) in addition to translations.

On a curved surface, or continuum of higher dimension, such a procedure is not generally available for connecting the geometry of the space near one point to the geometry near another since nothing like a Cartesian coordinate system valid over the whole surface exists, and, in addition, no obvious notion of parallelism of a vector with another at a distant point is available either.   However, Levi-Civita [L.2] introduced a notion of parallelism of vectors—often called, in fact, parallelism in the sense of Levi-Civita—that is reasonable and appropriate, and this notion of parallelism was shortly afterward recognized by H. Weyl [W.1] as an essential tool needed to clarify the ideas of Riemannian geometry and to make reasonable various analogies and resemblances with Euclidean geometry as well as the various ways in which they differ.   Of course, Levi-Civita's notion of parallelism coincides with the classical one for a plane surface, i.e. for Euclidean geometry.   Levi-Civita introduced his definition of parallel fields of vectors for surfaces embedded in a Euclidean space.   Technically, this procedure manifests itself through the employment of the normal to the surface and the second fundamental form as essential elements in the discussion.   Afterward it can be verified rather easily that the fields of vectors defined by him were determined solely in terms of the coefficients g_{ik} of the first fundamental form, and hence that such fields really did not depend on the form of the surface in three-space.   Levi-Civita's development of the theory will be followed in the main in the present chapter since it helps very much to motivate the more subtle procedure of Weyl.   In fact, throughout this chapter, as its title indicates, the inner or intrinsic geometry of surfaces will be studied extrinsically, and for two reasons:

1. It is both interesting and important for its own sake to characterize intrinsic properties extrinsically.

2. Such a procedure, as was remarked earlier, can be carried out in a fashion that furnishes clues for the generalization to Riemannian geometry in a manifold of any dimension without the necessity of embedding in a Euclidean space.

A purely intrinsic treatment of Riemannian geometry is desirable for obvious reasons, and also for a reason rooted in physics: it would seem rather strange to treat Einstein's general theory of relativity, which is basically the Riemannian geometry of a certain four-space, by first embedding it in a higher dimensional Euclidean space, since Einstein's object was to investigate the character of the actual space in which we live—and to find that it is not Euclidean.   These matters will be the concern of Chapter IX, but the present chapter is written so that some of the concepts in the more general case can be introduced later without detailed motivation.

One of the most important concepts in the study of the intrinsic geometry of surfaces is that of the shortest arc connecting pairs of points on the surface: after all, such lines would be in some sense the natural analog of the straight lines in Euclidean spaces. This subject has many facets, and it can be treated in several ways. The usual approach to it is by way of the calculus of variations. This leads to a differential equation of second order, the solutions of which furnish any possible curves that minimize the length integral. Such curves are called *geodetic lines*. In this chapter the notion of parallel fields of vectors is chosen as the point of departure for the study of intrinsic geometry, and hence a different way of defining geodetic lines is chosen: they are, in fact, defined as curves on the surface whose tangent vectors form a parallel field in the sense of Levi-Civita. Clearly this also makes these curves the analog of the straight lines in Euclidean spaces, which as seen in earlier chapters can also be considered as those regular curves whose tangents at all points are parallel in the ordinary sense. Afterward this characterization of the geodetic lines is found to coincide with that characterizing them as candidates for arcs of shortest length.

In the previous chapter it has already been seen that the Gaussian curvature K, defined extrinsically as the product of the principal curvatures, is really an isometric invariant and thus it furnishes an intrinsic property of a surface. There is an exceedingly important and interesting relation between the curvature K and the concept of parallel fields of vectors. It turns out that if a parallel field is defined along a closed curve on the surface that the vector at an arbitrarily chosen initial point of the curve will not in general coincide with the vector that results by parallel transport once around the curve, but will make a certain non-zero angle α with it. In case the curve is a simple closed curve bounding a simply connected domain of the surface it will be seen that the angle α is given by the integral of K over the area of the domain enclosed by the curve. If K happened to be zero everywhere then α would also vanish. The converse is also true, i.e., if α vanishes for every closed curve then K is of necessity zero everywhere. Surfaces in three-space with this property are developables (cf. Chapter V), and as the word implies, they can be mapped isometrically on the plane. Thus the intrinsic geometry on a surface with $K \equiv 0$ is exactly the same (in the small, at least) as that of the plane when the geodesics are defined as the straight lines in this geometry. It also means that free motions of "rigid bodies" in such a surface are just as possible as in the plane, i.e. that a given geometrical object anywhere on the surface can be moved about in it freely without changing the lengths of any curves in it. The surfaces with $K \equiv 0$ are not the only surfaces having this latter property: all surfaces with $K \equiv$ constant have it, and these surfaces lead to models of the so-called non-Euclidean geometries. If K is a positive constant this geometry is essentially that of the sphere, while if K is negative it is the Lobachefsky geometry in which an infinity of straight lines can be drawn parallel to a given line through any point not on it. This model of non-Euclidean geometry will be treated in the present chapter.

These remarks perhaps suffice to give an idea of the richness and variety of the subject matter of the intrinsic geometry of surfaces.

2. Approximate Local Parallelism of Vectors in a Surface

The purpose of this section is to introduce a provisional definition of *approximate* local parallelism in a surface. As was said in the previous section this is to be done in the first instance by dealing with the surface in Euclidean three-space. It is assumed, therefore, that the surface is a regular surface defined by a vector $\mathbf{X}(u, v) = (x_1(u, v), x_2(u, v), x_3(u, v))$, just as in Chapter IV.

It is clear that the only reasonable way to find candidates for vectors *in* a surface at a point P is to take them as vectors that lie in the tangent plane at P. If, then, at the point P fixed by parameter values (u_0, v_0) a tangent vector $\mathbf{v}$ is given, the first step to be taken is to define what is meant by a vector approximately parallel to $\mathbf{v}$ at point Q in a neighborhood of P. A plausible way to do that is to take as vector "parallel" to $\mathbf{v}$ at Q the vector $\mathbf{v}^*$ that results by moving $\mathbf{v}$ parallel to itself in three-dimensional space to the point Q and then projecting it orthogonally onto the tangent plane to the surface at Q (cf. Fig. 7.1). The vector $\mathbf{v}^*$ is thus defined in such a way that it is the uniquely determined vector in the tangent plane at Q *that makes the smallest angle α with* $\mathbf{v}$. In effect, $\mathbf{v}^*$ is as nearly parallel to $\mathbf{v}$ as is compatible with its being a tangent vector of the surface.

Attention is restricted to points Q near P in the sense that the angle θ between the tangent plane T^* at Q and the tangent plane T at P (again see

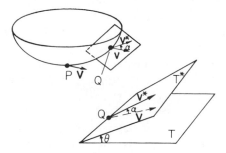

Fig. 7.1   *Approximate local parallelism on a surface.*

Fig. 7.1) should be regarded as a small quantity of first order. It is then readily seen that **v** and **v*** differ in direction by a quantity of first order and in magnitude by a quantity of second order. It has some point to assume from now on that the vectors **v** and **v*** are of unit length, so that the difference **v***–**v** is a normal vector to the surface within second-order terms—as can be readily seen. The angle θ is a continuous function of the surface parameters

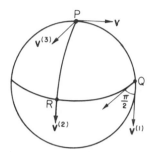

Fig. 7.2   *Parallel displacement on a sphere.*

(u, v) for the point Q, since the normal vector of the surface is assumed to have that property. Thus if a point Q is given by $(u_0 + \Delta u, v_0 + \Delta v)$, the quantities Δu, Δv, and θ are all small of the same order. It is also readily seen that if two different vectors $\mathbf{v}_1$ and $\mathbf{v}_2$ are taken at P their "parallel" vectors $\mathbf{v}_1^*$ and $\mathbf{v}_2^*$ at Q make the same angle with each other as that between $\mathbf{v}_1$ and $\mathbf{v}_2$ within *second-order terms* in θ; thus a set of unit tangent vectors at P corresponds in this process of "parallel transport" to a set at Q congruent to it within errors of second order.

Thus given **v** at P, it is possible to define a unique field of tangent vectors to the surface in a neighborhood of P which would form at least approximately a parallel field. However, this concept of parallelism if examined more closely leads to some curious happenings, which occur if one tries to extend it over a finite region of the surface by a step-wise process of extension from one point to another. As a concrete example of what may happen, it is useful to consider a special case, i.e., that of the unit sphere (cf. Fig. 7.2). The point P is taken at the north pole, and a vector **v** tangent to the great circle arc PQ (of length $\pi/2$) is selected. At all points of the arc PQ it is clear that the approximately parallel field obtained stepwise by the method described above in going from point to point would furnish vectors tangent to the arc PQ at all points along it; thus the vector $\mathbf{v}^{(1)}$ at Q would result. This process is repeated with Q as the starting point and $\mathbf{v}^{(1)}$ as initial vector to obtain a field along the great circle arc QR, again of length $\pi/2$, which is orthogonal to PQ. Obviously, all of these vectors would be orthogonal to the arc QR—in fact, they are all exactly parallel to each other in the usual sense—in particular, such a vector $\mathbf{v}^{(2)}$ is shown at point R. This process is repeated, starting from R and considering the field vectors "parallel" to $\mathbf{v}^{(2)}$ obtained by proceeding stepwise along the great circle arc from R back to P; these vectors are clearly tangent vectors to the arc RP. Hence the vector $\mathbf{v}^{(3)}$ which is "parallel" to $\mathbf{v}^{(2)}$ is orthogonal to $\mathbf{v}$, the vector with which the whole process began. Thus the process of defining a parallel field, when carried out stepwise from point to point, leads to discrepancies that are at first sight rather disturbing: it is seen, in fact, that the vectors at P that are parallel to $\mathbf{v}^{(1)}$ and to $\mathbf{v}^{(2)}$. are turned at $90°$ to each other, though $\mathbf{v}^{(1)}$ and $\mathbf{v}^{(2)}$ are actually exactly parallel in three-space. Or, to put it in a different way, the result of proceeding from P to extend the vector $\mathbf{v}$ by parallelism back to P by proceeding pointwise over the closed curve $PQRP$ has led to a vector $\mathbf{v}^{(3)}$ at P differing from $\mathbf{v}$ at P by $90°$.

Thus the provisional definition of local parallelism, given here, while giving a unique field locally, fails to do that if a stepwise extension from point to point in the large is attempted.

3. Parallel Transport of Vectors Along Curves in the Sense of Levi-Civita

It is not appropriate, as was done above, to introduce a concept of parallelism of vectors in a surface by defining the concept in a whole neighborhood of a point at one blow. Instead—and that is the basic and fruitful idea due to Levi-Civita—the proper and fundamentally useful concept is that of a parallel field of vectors *defined along a specific curve issuing from a given point.* That was what was done, really, in the discussion above based on Fig. 7.2.

This does not mean that all of the discussion of the preceding section is wasted. Rather, it is used as a motivation for the introduction of *a pair of ordinary differential equations, the solution of which defines a unique field of "parallel" vectors along a given regular curve once an initial vector has been chosen arbitrarily.* This is done as follows. A curve C on the surface through a point P as initial point is defined by the equations $u = u(t)$, $v = v(t)$; the functions $u(t)$ and $v(t)$ are assumed to be differentiable over the interval $t_0 \le t \le t_1$, with P corresponding to $t = t_0$. Points near to P on the curve are given by $u(t_0 + \Delta t)$, $v(t_0 + \Delta t)$, and Δt, and the angles $\theta(t_0 + \Delta t)$ employed in the previous section can be regarded as small of the same order for points Q on C near to P. As in that section, a field $\mathbf{v}^*(t)$ "parallel" to the vector $\mathbf{v}(t_0)$ is defined; this field $\mathbf{v}^*(t)$ is evidently a differentiable field with respect to t, since the surface is assumed to be regular. In particular, the derivative

$$(7.1) \qquad \frac{d\mathbf{v}}{dt}\bigg|_{t=t_0} = \lim_{\Delta t \to 0} \frac{\Delta \mathbf{v}}{\Delta t}$$

exists at point P, and since $\Delta v = \mathbf{v}^* - \mathbf{v}$ is by definition always a normal vector to the surface, within second-order terms in Δt, it follows that $d\mathbf{v}/dt$ is a vector normal to the surface.   Levi-Civita's idea is to make use of this fact as the motivating principle for his definition of a parallel field of vectors $\mathbf{v}(t)$ along a curve, which consists simply in requiring that *the derivative $d\mathbf{v}/dt = \dot{\mathbf{v}}$ along the curve should be always normal to the surface.*   In terms of the two linearly independent coordinate vectors $\mathbf{X}_1$ and $\mathbf{X}_2$ (i.e., the tangent vectors $\partial\mathbf{X}/\partial u = \partial\mathbf{X}/\partial u_1$ and $\partial\mathbf{X}/\partial v = \partial\mathbf{X}/\partial u_2$ to the parameter curves on the surface) the condition of Levi-Civita can be expressed as follows:

$$(7.2) \qquad \begin{cases} \dot{\mathbf{v}}\cdot\mathbf{X}_1 = 0, \\ \dot{\mathbf{v}}\cdot\mathbf{X}_2 = 0, \end{cases}$$

since these two conditions are necessary and sufficient conditions ensuring that $\dot{\mathbf{v}} = d\mathbf{v}/dt$ should be orthogonal to the surface.

It is easy to show that the conditions (7.2) lead to a pair of first-order ordinary differential equations determining the vector $\mathbf{v}(t)$ along any pre-scribed regular curve $C: u_1 = u_1(t)$, $u_2 = u_2(t)$, as follows.   The vector $\mathbf{v}(t)$, being a tangent vector of the surface, can be expressed in the form

$$(7.3) \qquad \mathbf{v}(t) = v^1(t)\,\mathbf{X}_1(t) + v^2(t)\,\mathbf{X}_2(t) = v^i(t)\,\mathbf{X}_i(t),$$

with summation over 1 and 2, in terms of its components $(v^1(t)\,v^2(t))$ with respect to the coordinate vectors $\mathbf{X}_i(t)$ of the surface along C.   The two equations (7.2) can therefore be written as follows:

$$\dot{\mathbf{v}}\cdot\mathbf{X}_j = \dot{v}^i g_{ij} + v^i \dot{\mathbf{X}}_i\cdot\mathbf{X}_j$$
$$= \dot{v}^i g_{ij} + v^i \alpha_{ij} = 0, \qquad j = 1,2.$$

In these equations the functions g_{ij} are the coefficients $\mathbf{X}_i\cdot\mathbf{X}_j$ of the first fundamental form of the surface and the functions α_{ij} are given by $\dot{\mathbf{X}}_i\cdot\mathbf{X}_j$. Along a given curve C it is clear that the coefficients g_{ij} and α_{ij} are known functions of the parameter t.   Thus the conditions in (7.2) take the form

$$(7.4) \qquad \dot{v}^i g_{ij} + v^i \alpha_{ij} = 0, \qquad j = 1,2.$$

Evidently, this is a pair of first-order linear ordinary differential equations for the two components $v^1(t)$, $v^2(t)$ of $\mathbf{v}(t)$ which determine [cf. (7.3)] that vector with respect to the surface coordinate vectors $\mathbf{X}_i(t)$.   If the surface has continuous derivatives of second order, and the curve $C(t)$ derivatives of first order, the differential equations in (7.4) have (cf. Appendix B) a uniquely determined solution along the whole curve $C(t)$ once an initial vector $\mathbf{v}(t) = (v^1(0), v^2(0))$ has been chosen at any point of C.   (The theorem in the Appendix is not directly applicable to the system (7.4), but becomes so upon observing that the determinant $|g_{ij}|$ of the coefficients of the quantities $\dot{v}^i$

is not zero since the line element is positive definite; thus the equations can be solved for the quantities $\dot{v}^i$—as will be done explicitly a little later.) The resulting field of vectors $\mathbf{v}(t)$ is called a *parallel field along C in the sense of Levi-Civita*. Thus one step in Riemann's program has been achieved by deriving a differential equation for parallelism along a curve through using a local approximation, based on Euclidean geometry, followed by a passage to a limit.

It is of value to discuss once more the relation between this concept of a parallel field and the concept of approximate parallelism in a neighborhood of a point P introduced in the previous section. To this end consider an initial vector $\mathbf{v}(0) = (v^1(0), v^2(0))$ at P and a one-parameter family of surface curves issuing from P and covering a neighborhood of P. (Such a family could be obtained, for example, by taking the set of all plane sections cut out of the surface by planes containing the surface normal at P.) By integration of (7.4) along all of these curves fields of vectors parallel to $\mathbf{v}(0)$ would be defined in a unique way in a neighborhood of P. It is clear that this field of vectors would differ from the field originally defined by ortho- gonal projection by quantities of second order in the increments Δu and Δv of the surface parameters. However, it does not follow that the process of integration of (7.4) along every one-parameter family of curves through P covering a neighborhood of P would lead to the same field: in fact, that is simply not true in general, as was already seen in the case of a spherical surface. Indeed, if the differential equations (7.4) are integrated along two *different* curves $C_1(t)$, $C_2(t)$ that start at P and end in the same point Q there is no a priori reason why their solutions $v^i(t)$ should be the same at Q for both curves since the coefficients $g_{ij}(t)$ and $\alpha_{ij}(t)$ in (7.4) are in general different along the two curves. In fact, the result of parallel *propagation*, or *trans- port*, of $\mathbf{v}(0)$ along C_1 and C_2, as this process is called, will in general lead to vectors at Q that differ by a finite amount when the parameter t ranges over a finite interval. Only under exceptional circumstances is it true that the pro- cess of parallel propagation along arbitrary curves joining P with the same point Q leads to a uniquely determined vector. In such a case it is said that the field is integrable, and, as was indicated above and will be proved later on, this is a property that holds for all pairs of points on a surface only when the surface is a developable.

Thus it lies in the nature of things that parallel transport of vectors is defined in the first instance along curves and not over two-dimensional regions, and this at first sight seeming defect in Levi-Civita's concept of parallelism really turns out to be a virtue since it has interesting and far- reaching consequences for the inner geometry of curved surfaces.

An analysis of the properties of parallel fields follows. These properties result from discussion of the defining conditions (7.2), or (7.4). First of all it is clear on geometrical grounds from the equations (7.2), which are equivalent to (7.4), that parallel transport of vectors yields the same field independent of any special choice of curvilinear coordinates. This fact will be proved more formally later on, since it is obviously a requirement that must be fulfilled if the concept of parallel transport is to be an intrinsic property. It is equally important to show that the equations (7.4), while formulated here by working with the surface in three-dimensional space, in reality depend only upon the functions g_{ik} (and their first derivatives)—that is to say only upon the line element of the surface—and hence the concept of a parallel field belongs to the intrinsic geometry of the surface.

4.   Properties of Parallel Fields of Vectors Along Curves

First of all, it is to be shown that parallel fields of vectors along a curve C, as fixed by solutions of the differential equations (7.4), are determined by the coefficients g_{ik} of the first fundamental form and, of course, by the curve C; this means that such fields belong to the intrinsic geometry of the surface and are independent of an embedding. To prove the statement, it clearly suffices to show that the coefficients α_{ij} in (7.4) are fixed by the coefficients g_{ik}. Since $\alpha_{ij} = \dot{\mathbf{X}} \cdot \mathbf{X}_j$ and

$$\dot{\mathbf{X}}_i = \mathbf{X}_{ij}\dot{u}_j, \qquad \mathbf{X}_{ij} = \frac{\partial^2 \mathbf{X}}{\partial u_i\, \partial u_j},$$

it suffices to show that the scalar products $\mathbf{X}_{ij} \cdot \mathbf{X}_k$ are expressible in terms of the functions g_{ij}. But this was already done in Chapter VI, where it was shown [cf. (6.7), (6.8), and (6.3)] that

$$\mathbf{X}_{ij} \cdot \mathbf{X}_k = \Gamma^l_{ij} g_{lk},$$

in terms of the Christoffel symbols Γ^l_{ij}, which are given in terms of first derivatives of the coefficients g_{ij} of the first fundamental form. In terms of these symbols the differential equations (7.4) take the form

$$\dot{v}^l g_{lk} + v^i \dot{u}_j \Gamma^l_{ij} g_{lk} = 0, \qquad k = 1,2.$$

Multiplication of these equations by the quantities g^{rk}—the elements of the matrix inverse to that of the functions g_{ik}—followed by a summation on k leads, since $g_{lk} g^{rk} = \delta^r_l$, to the equations

(7.5) $$\dot{v}^r + v^i \dot{u}_j \Gamma^r_{ij} = 0, \qquad r = 1,2.$$

As was remarked earlier, the general existence and uniqueness theorems for systems of ordinary linear differential equations (cf. Appendix B) now ensure that *a unique parallel field is obtained by integration of these equations along a curve C once an initial vector has been chosen.*   The curve C should be a continuous piecewise regular curve; hence its arc length s can be, and commonly is, used as curve parameter.   It is clear geometrically from the defining property of parallel fields formulated in equations (7.2), and which is expressed in an equivalent form by (7.5), that the parallel field depends only upon the surface and the curve C and not upon a particular parameter representation of the surface.   It can be proved by a somewhat tedious formal calculation that the left-hand side of (7.5) is invariant under transformation of coordinates; this fact will emerge in a natural way from the discussion in Chapter IX, where such invariance properties come about because they are formulated in terms of tensors.

All properties of parallel fields can in principle be worked out with reference to the equations (7.5).   It is sometimes more efficient, however, to make use of the conditions (7.2), which served as the starting point for the derivation of (7.5)—and that is also appropriate here since a restriction to the intrinsic point of view is not intended in this chapter.   Some basic properties of parallel fields are:

PROPERTY 1. If $\mathbf{v}(s)$ and $\mathbf{w}(s)$ denote two parallel fields along the same curve C—that is, fields starting with two different initial vectors—*then the scalar product* $\mathbf{v} \cdot \mathbf{w}$ *has the same value all along* C.   To prove this statement consider the relation

$$\frac{d}{ds}\,(\mathbf{v}\cdot\mathbf{w}) = \dot{\mathbf{v}}\cdot\mathbf{w} + \mathbf{v}\cdot\dot{\mathbf{w}}.$$

Since $\mathbf{v}$ and $\mathbf{w}$ are both tangent vectors of the surface, and since $\dot{\mathbf{v}}$ and $\dot{\mathbf{w}}$ are both normal to the surface in view of the basic defining property of parallel fields, it follows that $d/ds\,(\mathbf{v}\cdot\mathbf{w}) = 0$, which proves the statement.

PROPERTY 2.   *The length of the vectors* $\mathbf{v}(s)$ *resulting from parallel propagation of a vector along a curve is constant.*   The correctness of this statement is seen immediately to be a consequence of the first property: it suffices to set $\mathbf{v} = \mathbf{w}$, whence $(d/ds)(\mathbf{v}\cdot\mathbf{v}) = 0$ follows.

PROPERTY 3.   *The angle between the vectors* $\mathbf{v}(s)$ *and* $\mathbf{w}(s)$ *of two sets of vectors both forming parallel fields along C is a constant.*   This is an immediate conclusion from the first two properties.   Another way to state this is to say that any bundle of vectors remains congruent to itself in parallel transport along a curve.

PROPERTY 4.   *In the Euclidean plane vectors that are parallel in the sense of Levi-Civita are parallel in the ordinary sense.*   This of course is seen immediately upon reverting to the motivation of Section 2, where approximate parallelism was defined locally by using a parallel transport in three-space followed by projection onto the tangent plane: here all tangent planes coincide. It is also of interest to verify that the same conclusion follows from (7.5). In the Euclidean plane orthogonal Cartesian coordinates are chosen (as can be done since parallel transport is independent of the choice of a coordinate system), so that the line element has the coefficients $g_{11} = 1$, $g_{12} = 0$, $g_{22} = 1$, and hence all Christoffel symbols have the value zero. Thus (7.5) yields $\dot{v}^1 = \dot{v}^2 = 0$ and hence the components of the field vectors are constant with respect to a fixed Cartesian system. Thus they are parallel in the usual sense.

5. Parallel Transport is Independent of the Path only for Surfaces Having $K \equiv 0$

It has already been seen that parallel transport of a vector **v** starting at an initial point P along two different curves both ending in a point Q will in general lead to two different vectors at Q. In this section it is to be shown that *the only surfaces for which the result of parallel transport is independent of all paths joining all pairs of points are the surfaces with Gaussian curvature K everywhere zero.*

That this property holds for $K \equiv 0$ is known for the special case of the Euclidean plane, for which K does of course vanish everywhere. The proof of the general statement is as follows. An arbitrary unit vector is chosen at any point P, and a field of vectors is defined by parallel transport of the vector to all points on the surface; since this process furnishes by hypothesis a uniquely defined field at all surface points it follows that functions $v^1(u_1, u_2)$, $v^2(u_1, u_2)$ exist such that the differential equations (7.5) hold for all curves on the surface. If the surface has derivatives of sufficiently high order, these functions v^1, v^2, as solutions of ordinary differential equations will also have derivatives of order as high as might be wanted. It follows that $\dot{v}^s$ in (7.5) can be replaced by $(\partial v^s / \partial u_1)\,\dot{u}_1 + (\partial v^s / \partial u_2)\dot{u}_2$ to obtain

$$\left(\frac{\partial v^s}{\partial u_i} + v^j \Gamma_{ji}^s\right)\dot{u}_i = 0, \qquad s = 1,2.$$

This holds by hypothesis along *every* curve, i.e., for arbitrary functions $\dot{u}_i$. It follows that

$$(7.6) \qquad \frac{\partial v^s}{\partial u_i} + v^j \Gamma_{ji}^s = 0, \qquad s = 1, 2, \qquad i = 1, 2,$$

and this overdetermined set of four first-order linear partial differential equations for two functions must satisfy the compatibility conditions that arise because the relations $\partial^2 v^s / \partial u_i \partial u_j = \partial^2 v^s / \partial u_j\,\partial u_i$ must hold. This in turn yields the conditions

$$\frac{\partial}{\partial u_k}\left(v^j \Gamma_{ji}^n\right) - \frac{\partial}{\partial u_i}\left(v^j \Gamma_{jk}^n\right) = 0.$$

Upon carrying out the differentiations and using the partial differential equations (7.6) once more this latter condition can be written in the form

$$v^j\left(\frac{\partial \Gamma_{ji}^n}{\partial u_k} - \frac{\partial \Gamma_{jk}^n}{\partial u_i}\right) - (\Gamma_{sk}^j \Gamma_{ji}^n - \Gamma_{si}^j \Gamma_{jk}^n)v^s = 0$$

or, what is the same thing,

$$v^j\left(\frac{\partial \Gamma_{ji}^n}{\partial u_k} - \frac{\partial \Gamma_{jk}^n}{\partial u_i} + \Gamma_{ji}^s \Gamma_{sk}^n - \Gamma_{jk}^s \Gamma_{si}^n\right) = v^j r^n_{.jki} = 0.$$

But integrals of the partial differential equations are assumed to exist for arbitrarily given initial values of v^1 and v^2, and for any initial points. Consequently the coefficient of v^j in the last equation must vanish at all points of the surface. This coefficient is, however, the quantity $r^n_{.jki}$—the so-called Riemann tensor—which was discussed in the preceding chapter [cf. (6.23)], and its vanishing means that the Gaussian curvature is zero. Thus K is everywhere zero if the parallel transport of vectors is independent of the path. Later on it will be seen more precisely how the values of the Gaussian curvature on the surface affect parallel transport on it, and that the condition $K = 0$ for general independence of the path in parallel transport of vectors, which has just been seen to be a necessary condition, is also a sufficient condition (at least on a simply connected portion of the surface); the key point here will be seen to be that the overdetermined system (7.6) has solutions if the compatibility conditions are satisfied.

6. The Curvature of Curves in a Surface: The Geodetic Curvature

It is clearly of interest to give, if possible, a reasonable definition for the curvature of curves *in* a surface that has analogies with the concept of the curvature of curves in the plane. For this purpose it is not reasonable to make use of the curvature κ of the curve, defined in an earlier chapter, as a twisted curve in space: clearly, this curvature would not in general remain unchanged if the surface containing the curve were to be deformed by bending in the space. The discussion carried out in the first two sections of the present chapter hints that the right way to define this concept would be to project the given curve C orthogonally on to the tangent plane to the surface at a given curve point and take the curvature of this plane curve at the point as the *intrinsic curvature* of C. This notion of curvature, called the *geodetic curvature* κ_g, introduced thus by making use of an embedding in three-space, does indeed prove to be a fruitful concept.

However, it is possible to introduce the idea of geodetic curvature in a different way that makes its intrinsic character clear from the outset—a little later, the two different ways will be shown to give the same quantity. It is possible to accomplish this because the notion of a parallel field of vectors along a curve is available. In the plane the curvature κ of a curve was defined by $\kappa = d\theta/ds$, with $\theta(s)$ the angle formed by the oriented tangent to the curve with any fixed direction—usually taken as the positive x-axis—and with s the arc length of the curve. Evidently, the angle $\theta(s)$ could also be defined as the angle between the oriented tangent to the curve and a field vector of an arbitrary parallel field in the sense of Levi-Civita along it, since such a parallel field consists of vectors that are parallel in the ordinary sense. This concept of curvature in the plane is now generalized to curves on arbitrary surfaces by defining κ_g as the value of $d\theta(s)/ds$, with $\theta(s)$ now of course meaning the angle between the oriented tangent of curve C and the field vector $\mathbf{v}(s)$ of a parallel field defined along the curve. The angle $\theta(s)$ is given a sign: it is taken positive if the sense of rotation of $\mathbf{v}(s)$ into the tangent vector $\dot{\mathbf{X}}(s)$ of C is the same as that of the first coordinate vector $\mathbf{X}_1$ of the surface into the second $\mathbf{X}_2$. Since angles are determined solely by the first fundamental form, it is clear that this definition of the geodetic curvature κ_g makes it belong to the inner differential geometry of the surface.[1]

It is of interest to express the geodetic curvature κ_g in terms of the vector that describes the form of the surface in three-space. For this purpose consider a curve C on the surface given by the vector $\mathbf{X}(s)$ in terms of arc length as parameter, and a parallel field $\mathbf{v}(s)$ of unit vectors along it. The angle $\theta(s)$ satisfies the equation $\cos \theta = \mathbf{v} \cdot \dot{\mathbf{X}}$, and hence $-\dot{\theta} \sin \theta = \dot{\mathbf{v}} \cdot \dot{\mathbf{X}} + \mathbf{v} \cdot \ddot{\mathbf{X}}$. The characteristic property of a parallel field is that $\dot{\mathbf{v}}$ is orthogonal to the tangent plane; and hence $\dot{\mathbf{v}} \cdot \dot{\mathbf{X}} = 0$. It is always possible to choose the field $\mathbf{v}(s)$ at any one point of C so that it is orthogonal to $\dot{\mathbf{X}}$ at that point and so oriented that $\sin \theta = +1$. Hence $\mathbf{v}$ can be written as $\dot{\mathbf{X}} \cdot \mathbf{X}_3$ in terms of the normal vector $\mathbf{X}_3$ of the surface since $\mathbf{v}$ and $\dot{\mathbf{X}}$ are oriented the same as the coordinate vectors $\mathbf{X}_1$ and $\mathbf{X}_2$. Thus it follows that $-\dot{\theta} = \mathbf{v} \cdot \ddot{\mathbf{X}} = \dot{\mathbf{X}} \times \mathbf{X}_3 \cdot \ddot{\mathbf{X}} = -\dot{\mathbf{X}} \cdot \ddot{\mathbf{X}} \times \mathbf{X}_3$ and the following formula for κ_g results:

$$(7.7) \qquad \kappa_g = \frac{d\theta(s)}{ds} = \dot{\mathbf{X}} \cdot \ddot{\mathbf{X}} \times \mathbf{X}_3.$$

Another formula of some interest results upon introducing the trihedral of vectors $\mathbf{v}_1, \mathbf{v}_2, \mathbf{v}_3$ of the space curve C, as it was defined in Chapter III. Since $\dot{\mathbf{X}} = \mathbf{v}_1$, $\ddot{\mathbf{X}} = \kappa \mathbf{v}_2$, it follows that $\kappa_g = \kappa(\mathbf{v}_1 \cdot \mathbf{v}_2 \times \mathbf{X}_3) = \kappa(\mathbf{v}_1 \times$

---

[1] It should be pointed out that the definition given here for κ_g would not be of much value if parallel fields did not have the property 3 described earlier, i.e., that the angle between pairs of field vectors of two different parallel fields along the same curve is a constant.

$v_2 \cdot X_3) = \kappa v_3 \cdot X_3$ in terms of the curvature κ of C and its binormal vector v_3. Thus the following formula is obtained:

$$(7.8) \qquad \kappa_g = \kappa \cos \alpha,$$

with α the angle between v_3 and X_3. From the formula (7.8) it is intuitively clear that κ_g is the curvature of the plane curve that results when C is projected orthogonally onto the tangent plane of the surface. The proof is left as one of the problems at the end of the chapter to be solved as exercises.

7.   First Definition of Geodetic Lines: Lines with $\kappa_g = 0$

The discussion of the preceding sections makes almost inevitable the introduction of a class of curves on surfaces that are analogous to the straight lines in the plane: evidently such curves could reasonably be singled out as *those having vanishing geodetic curvature*; they are called *geodetic lines*, or also simply *geodesics*. Since $\kappa_g = \dot{\theta}(s)$, it follows that the tangent vectors to any such geodetic line form a parallel field: a property they have in common with the tangent vectors of straight lines in the plane.

From formula (7.8) it is seen that if $\kappa_g = 0$ holds along a curve C the orthogonality of the binormal of C and the surface normal is implied, and hence that *the principal normal of C falls along the surface normal*, provided that the curvature of κ of C as a space curve is not zero. Conversely, if the principal normal falls along the surface normal, it is clear that $\kappa_g = 0$. This property of geodesics makes it possible to identify them as such in a variety of interesting special cases. For example, it is clear that all meridian curves on surfaces of revolution are geodesics, and in particular that all great circle arcs on a sphere are geodesics. Helices on a circular cylinder are now seen to be geodetic lines of the cylinder since such helices have the required property. In addition, the formula (7.8) makes it obvious that straight lines on any surface are geodetic lines of the surface; in particular, the generators of a developable are geodetic lines, as are also the straight lines on ruled surfaces.

The above definition of geodetic lines, which makes them analogous to straight lines through the property of conserving their direction at all points, is for some considerations in geometry the basically important definition. For example, that is the case for the models of non-Euclidean geometry obtained in terms of the inner geometry of certain curved surfaces, since the useful concept of a straight line in this connection is that of a line that preserves its direction. For other purposes in differential geometry a generalization of a different characteristic property of the straight line is wanted, i.e., its property of being the *curve of shortest length* joining pairs of points in the plane, as compared with all other continuous and piecewise regular curves joining the same pair of points. Thus the shortest arcs in the plane are the straight lines, that is to say the geodesics, on such surfaces. For general surfaces the matter is more complicated. It remains true, as will be seen shortly, that the only reasonable candidates for shortest arcs on general surfaces are the geodesics, i.e., the curves along which $\kappa_g = 0$. However, it can happen that some pairs of points are not even joinable by such a curve; or even if they are that the curve may not be the shortest arc.

The questions thus raised give rise to much that is of interest; they will be dealt with in subsequent sections of this chapter, but also in the following chapter, where problems of differential geometry in the large are treated. It is perhaps useful at this point to make one observation that foreshadows much of the discussion to follow in the next few sections, i.e., that the condition $\kappa_g = 0$ leads to *a pair of ordinary differential equations characterizing geodetic lines*. That comes about through the fact that these lines are defined as those whose tangent vectors form a parallel field. Thus if such a curve is defined by the relations $u_1 = u_1(s)$, $u_2 = u_2(s)$, it follows that the functions v^1 and v^2 in the differential equations (7.5) for a parallel field can be identified with $\dot{u}_1(s)$, $\dot{u}_2(s)$, siice these furnish the components of its unit tangent vector, and hence the following equations hold:

$$(7.9) \qquad \ddot{u}_r + \dot{u}_i \dot{u}_j \Gamma_{ij}^r = 0, \qquad r = 1, 2.$$

These equations are a pair of second-order ordinary differential equations for the determination of the functions $u_1(s)$, $u_2(s)$. Evidently, they express what must be true on purely geometric grounds, i.e., that the concept of a geodetic line belongs to the intrinsic geometry of a surface, since only the arc length and the Christoffel symbols enter into (7.9). Also, since the differential equations are of second order, it follows (cf. Appendix B) that if $u_1(0)$, $u_2(0)$, $\dot{u}_1(0)$, $\dot{u}_2(0)$ are prescribed arbitrarily—that is to say if an initial point of the curve and an initial direction there are prescribed—then a uniquely determined regular[1] curve satisfying (7.9) exists provided that the functions defining the surface have a few continuous derivatives. Thus a uniquely determined bundle of geodesics emanates from a given initial point in all directions. It is, however, not in general true that a solution curve of the differential equations (7.9) exists when the curve is required *to pass through two given points*, nor if it does that it is unique. The required number of subsidiary conditions to be adjoined to (7.9) in such a case is the same as for the initial value problem, i.e., four, since u_1 and u_2 are then prescribed at two

---

[1] The existence theorem ensures only that the solutions $u_1(s)$, $u_2(s)$ are differentiable; however $\dot{u}_1^2 + \dot{u}_2^2 \neq 0$ holds all along the curve, if $\dot{u}_1^2(0) + \dot{u}_2^2(0) \neq 0$, since it is known that parallel transport of a vector preserves its length.

different points; but the resulting problem is a *boundary value problem* for a nonlinear system of differential equations, and that is basically the reason for the inherent complexities of this problem.

In the sections immediately following, the discussion will center around (7.9), or its equivalent $\kappa_g = 0$, because it will be seen that this condition is at least a *necessary condition* for an arc to be a shortest join—though, as was already indicated, it is not in general a sufficient condition to guarantee it.

8.   Geodetic Lines as Candidates for Shortest Arcs

It might seem odd, but it is inevitable, that the study of curves of shortest length is not begun by giving conditions with respect to a surface which make it possible to prove that every pair of points on it could be joined by a regular curve which is shorter than all others joining the same points.   This is a most important problem of differential geometry in the large that is not dealt with fully in this chapter.   That the problem would not be without difficulties can be readily seen by considering the following simple example, i.e., the u,v-plane itself with the Euclidean metric, but with the origin deleted.   The problem then is to connect the points $(-1, 0)$ and $(+1, 0)$ with a regular curve of shortest length.   It is supposed known—it will be proved shortly—that a straight line in the plane is the shortest regular arc joining any pair of its points.   It is now intuitively clear that there is no shortest regular curve joining the two points $(-1, 0)$ and $(+1, 0)$ since the "hole" at the origin prevents the straight line segment joining these points from being such a connection.   By bypassing the origin with tiny arcs it is evidently possible to find curves of length as near to 2 as might be desired, but there is no regular arc connecting the two points that has this length.   Thus shortest regular arcs will not exist always if the surface has holes—nor if it has boundaries at finite distances from any point; in other words some hypothesis about the surface would be needed that rules out such possibilities before a general proof of the existence of a shortest arc could be given.

The problem of finding shortest arcs belongs to the branch of mathematics called the calculus of variations.   In the present case the relevant variational problem can be formulated as follows.   Two points P and Q are given in the interior of a domain D of the u,v-plane within which a line element ds^2 is defined:

$$(7.10) \qquad ds^2 = E \, du^2 + 2F \, du \, dv + G \, dv^2.$$

An admissible curve is defined as a continuous curve C in D given by $u = u(t)$, $v = v(t)$, $t_0 \leq t \leq t_1$, joining P and Q, and such that $u'(t)$ and $v'(t)$ are piecewise continuous with $u'^2 + v'^2 \neq 0$. Such curves are rectifiable and their lengths are defined by

$$(7.11) \qquad L = \int_{t_0}^{t_1} \sqrt{Eu'^2 + 2Fu'v' + Gv'^2}\, dt,$$

in which the functions E, F, and G are of course to be evaluated on C. The problem is to find an admissible curve C such that $L(C)$ is smaller than it is for any other admissible curve joining the same pair of points. This is clearly not a problem in maxima and minima of the type encountered in elementary calculus, since it is a pair of unknown *functions* that is to be found so that a certain quantity is made a minimum. In the calculus a function of one or more variables is given and a maximum or minimum of it is to be found by giving proper values to those variables.

In this book the calculus of variations cannot be treated at length; in fact, for the present at least, no attempt will be made to discuss the question of the *existence* in general of an arc minimizing the length integral. Instead a general *necessary condition* will be derived in order that a regular arc should minimize the length L. This will lead along a different path to the same *geodetic lines* introduced earlier.

9.   Straight Lines as Shortest Arcs in the Euclidean Plane

Even though the general question of the existence of a shortest arc on curved surfaces is left aside for the time being, it is worthwhile to consider it in the simplest special case, i.e., that in which the surface is the Euclidean plane itself—in fact, the whole of that plane. Also the discussion is restricted to curves in the plane that are given, not in the parametric form, but in the form $y = y(x)$. In this case the integral to be minimized is

$$L = \int_{x_0}^{x_1} \sqrt{1 + y'^2}\, dx = \int_{x_0}^{x_1} f(y')\, dx,$$

in which $y'(x)$ of course refers to the derivative of $y(x)$. Desired is an admissible curve ("admissible" being understood in the sense defined above) $y = y(x)$ with $y_0 = y(x_0)$, $y_1 = y(x_1)$, which makes L smaller than for any other admissible curve satisfying the same end conditions. The discussion begins by finding the necessary condition mentioned above for such a minimizing arc. To this end it is assumed that $y = y(x)$, $x_0 \leq x \leq x_1$, does furnish the minimum. It follows that the set of curves

$$(7.12) \qquad y(x;\, \epsilon) = y(x) + \epsilon\eta(x), \qquad x_0 \leq x \leq x_1,$$

in an admissible curve family passing through (x_0, y_0), (x_1, y_1) for any value of the parameter ϵ if $\eta(x)$ is any given function that is continuous and has a piecewise continuous first derivative, and if

(7.13)
$$\eta(x_0) = \eta(x_1) = 0.$$

For this family of curves, with $\eta(x)$ held fixed, the length L of the curve is a function of ϵ only, and it is given by

(7.14)
$$L(\epsilon) = \int_{x_0}^{x_1} f(y' + \epsilon\eta')\, dx.$$

For the curve $y(x)$ itself the integral has the value $L(0)$, and if $L(0)$ is to be the smallest value for all admissible curves—as was assumed—it must in particular be the smallest value among those furnished by (7.12).   But this presents an ordinary problem in maxima and minima.   Thus $L(0)$ can be a minimum only if $L'(0) = 0$, and this fact leads to the necessary condition that is wanted.   Since $f(y')$ has a continuous derivative in the present case it follows that the integral in (7.14) can be differentiated under the integral sign with respect to ϵ, with the result

(7.15)
$$L'(\epsilon) = \int_{x_0}^{x_1} f_{y'}(y' + \epsilon\eta')\eta'\, dx,$$

and hence that

(7.16)
$$L'(0) = \int_{x_0}^{x_1} f_{y}'\, (y'(x))\, \eta'(x)\, dx = 0$$

holds for an arbitrarily chosen exemplar of the functions $\eta(x)$.

A necessary condition for a minimum now follows from the so-called *fundamental lemma of the calculus of variations*: Suppose that $M(x)$ is a piecewise continuous function defined in the interval $x_0 \leq x \leq x_1$.   *If the integral*

$$\int_{x_0}^{x_1} M(x)\, \eta'(x)\, dx$$

vanishes for every function $\eta(x)$ *that has the properties already given above, then* $M(x)$ *is a constant.*   To prove this it is noted that

$$\int_{x_0}^{x_1} [M(x) - C]\eta'(x)\, dx = 0$$

for any constant C, since $\eta(x) = 0$ at x_0 and x_1.   Choose for $\eta(x)$ the special function

$$\eta(x) = \int_{x_0}^{x} M(\xi)\, d\xi - C(x - x_0)$$

with C a constant given by the equation

$$C(x_1 - x_0) = \int_{x_0}^{x_1} M(x)\,dx;$$

clearly, this particular function $\eta(x)$ vanishes at the end points and has also the desired regularity properties. But for this particular function $\eta(x)$ the equation $\eta'(x) = M(x) - C$ holds except at possible points of discontinuity of $M(x)$. Thus

$$\int_{x_0}^{x_1} [M(x) - C]^2\,dx = 0,$$

and it is clear that $M(x) = C$ since the integrand is piecewise continuous and does not change sign.

With the aid of this lemma it can now be proved that the straight line joining (x_0, y_0) and (x_1, y_1) furnishes the arc of shortest length among all admissible arcs of the form $y = y(x)$. From (7.16) and the lemma the following equation holds along a minimizing arc:

$$f_{y'} = \frac{y'}{\sqrt{1 + y'^2}} = C = \text{const.};$$

therefore the slope y' of the curve is a constant and the curve is thus a straight line. Consider next any other admissible curve

$$Y = y(x) + \eta(x), \qquad \eta(x_0) = \eta(x_1) = 0,$$

obtained by "varying" $y(x)$ by the amount $\eta(x)$, and call its length $L + \delta^2 L$, with L the length corresponding to $y = y(x)$, or $\eta(x) \equiv 0$. The length of the varied curve is given by

$$L + \delta^2 L = \int_{x_0}^{x_1} [f(y' + \eta')]\,dx,$$

and Taylor's theorem with remainder yields the relation

$$\delta^2 L = \int_{x_0}^{x_1} f_{y'}(y')\eta'\,dx + \frac{1}{2} \int_{x_0}^{x_1} f_{y'y'}(y' + \theta\eta')\eta'^2\,dx,$$

with θ some function of x with values between 0 and 1. From the basic lemma it is known that $f_{y'} = C = \text{const.}$; the first integral therefore vanishes since $\eta(x_1) = \eta(x_0) = 0$, and consequently $\delta^2 L$ is given by

$$\delta^2 L = \frac{1}{2} \int_{x_0}^{x_1} f_{y'y'}(y' + \theta\eta')\eta'^2\,dx.$$

This so-called second variation is always positive in the present case for $\eta' \neq 0$ since $f_{y'y'} = (1 + y'^2)^{-3/2}$. Thus the varied curve is always longer than the straight segment unless $\eta'(x)$ is zero everywhere; but if $\eta'(x) \equiv 0$ holds, it follows that $\eta(x)$ is 0 everywhere since it vanishes at x_0 and x_1. Thus a straight line segment is the shortest admissible arc. This completes the existence proof, since every pair of points in the plane can be joined by a straight line.

10.   A General Necessary Condition for a Shortest Arc

This rather simple way of treating the case of the straight line as the shortest connection in the Euclidean plane is in some ways typical in the calculus of variations. However, it is seldom easy to show in other cases that an admissible arc satisfying the necessary condition and passing through a pair of arbitrary points exists, and also usually still less easy to show that the second variation relative to such an arc is positive for all comparison arcs. In this section a *necessary condition* will be obtained for an arc that minimizes the integral given by (7.11) on *any* surface. In doing so, the requirement for admissibility of a curve is that *it should have a continuous second derivative*, although a curve on the surface is rectifiable if it has only piecewise continuous first derivatives. However, it is proved in the calculus of variations that any curve that satisfies a necessary condition of the sort discussed in the special case above will have automatically a continuous second derivative: in effect, the minimizing process tends to select a particularly smooth curve. (In fact, the solution of the special problem in the plane treated above, i.e., the straight line, is actually an analytic curve, although the minimizing curves were sought in the class of functions having only a piecewise continuous first derivative.) Thus there is no loss of generality, really, in making this assumption.

It is supposed therefore that a regular curve given by the vector function $C_0(s)$, with continuous second derivatives joining points P and Q on the surface exists that minimizes the integral (7.11); the parameter s, $s_0 \leq s \leq s_1$ is assumed to represent the arc length. It is also assumed that the surface containing $C_0(s)$ lies in three-dimensional space, even though such an assumption would not be basically necessary. It would be possible to deal directly with the integral in the form given by (7.11), but it is more convenient to replace it by another that results from the choice of a special coordinate system arranged to contain the curve C_0 as a member of one of the families of parameter curves. At each point of C_0 a regular curve on the surface orthogonal to C_0 is taken, with the arc length σ as parameter on each such curve, and with $\sigma = 0$ on C_0. It was shown at the end of Chapter IV that parameters

s and σ can be defined so that they furnish a regular parameter system—if $|\sigma|$ is sufficiently small. Thus the surface, in a neighborhood of $\mathbf{C}_0$, is given by $\mathbf{X} = \mathbf{X}(s, \sigma)$. An admissible set of curves given by the vectors $\mathbf{C}_\tau$ in a neighborhood of $\mathbf{C}_0$ is defined as follows:

$$(7.17) \qquad \mathbf{C}_\tau = \mathbf{X}(s, \tau\Phi(s)); \qquad \Phi(s_0) = \Phi(s_1) = 0,$$

with $\Phi(s)$ a function defined for $s_0 \leq s \leq s_1$ with a continuous second derivative, but otherwise arbitrary; evidently the curves $\mathbf{C}_\tau$ are located by measuring off the distances $\sigma = \tau\Phi(s)$ on the parameter curves $s = $ const., i.e., on the orthogonal trajectories to $\mathbf{C}_0$. It is clear that each curve of the family passes through the points P and Q, and that $\mathbf{C}_0$ corresponds to $\tau = 0$. The curves $\mathbf{C}_\tau(s) = \mathbf{X}(s, \tau\Phi(s))$ have a first derivative with respect to s given by

$$\mathbf{C}_\tau' = \mathbf{X}_s + \mathbf{X}_\sigma \cdot \tau\Phi_s.$$

Since $\mathbf{C}_\tau'$ does not vanish for $\tau = 0$ [since $\mathbf{X}_s(s, 0) \neq 0$ holds], and consequently $\mathbf{C}_\tau' \neq 0$ for τ sufficiently small, say for $|\tau| \leq \tau_0$, which is assumed from now on, it follows that the curves of the family are regular curves. Thus the curves $\mathbf{C}_\tau$ are seen to be admissible curves. The lengths $L(\tau)$ of the curves $\mathbf{C}_\tau = \mathbf{X}(s, \tau\Phi(s))$ are given by

$$(7.18) \qquad L(\tau) = \int_{s_0}^{s_1} \sqrt{[\mathbf{X}'(s, \tau\Phi(s))]^2}\, ds,$$

and a necessary condition for $L(0)$ to furnish the minimum length is that $dL/d\tau|_{\tau=0} = 0$. Upon differentiating under the integral sign, the following formula is obtained:

$$(7.19) \qquad \frac{dL}{d\tau}\bigg|_{\tau=0} = \int_{s_0}^{s_1} \left(\mathbf{X}_s \cdot \frac{\partial \mathbf{X}_s}{\partial \sigma}\right) \frac{d\sigma}{d\tau}\bigg|_{\tau=0} ds = \int_{s_0}^{s_1} \left(\mathbf{X}_s \cdot \mathbf{X}_\tau\right)\bigg|_{\tau=0} ds$$

$$= \int_{s_0}^{s_1} (\mathbf{X}_s \cdot \mathbf{X}_{s\sigma})\big|_{\tau=0}\, \Phi(s)\, ds,$$

since $\mathbf{X}_s(s, 0)$ is a unit vector because s represents the arc length on $\mathbf{C}_0$. The vectors $\mathbf{X}_\sigma(s, 0)$ and $\mathbf{X}_s(s, 0)$ are by construction orthogonal unit vectors, and consequently

$$(7.20) \qquad \mathbf{X}_\sigma(s, 0) = -\mathbf{X}_s(s, 0) \times \mathbf{X}_3$$

in terms of the normal vector $\mathbf{X}_3 = \mathbf{X}_s \times \mathbf{X}_\sigma$ of the surface. The integral in (7.19) is transformed by an integration by parts, which is effected by observing that

$$\mathbf{X}_s \cdot \mathbf{X}_{s\tau} = (\mathbf{X}_s \cdot \mathbf{X}_\tau)_s - \mathbf{X}_{ss} \cdot \mathbf{X}_\tau.$$

This is legitimate since the curves have by assumption continuous second derivatives. Thus the integral, with use of $\mathbf{X}_\tau = \mathbf{X}_\sigma \varPhi(s)$, becomes

$$\left. \frac{dL}{d\tau} \right|_{\tau = 0} = \mathbf{X}_s \cdot \mathbf{X}_\sigma \varPhi(s) \left.\right|_{s_0}^{s_1} - \int_{s_0}^{s_1} \mathbf{X}_{ss} \cdot \mathbf{X}_\sigma \varPhi(s) \, ds,$$

with σ set equal to zero in the derivatives of $\mathbf{X}(s, \sigma)$. But $\varPhi(s) = 0$ at s_0 and s_1. Thus the following condition results, in view of (7.20):

$$\left. \frac{dL}{d\tau} \right|_{\tau = 0} = \int_{s_0}^{s_1} \mathbf{X}_{ss} \cdot (\mathbf{X}_s \times \mathbf{X}_3) \, \varPhi(s) \, ds$$

$$= - \int_{s_0}^{s_1} (\dot{\mathbf{X}} \cdot \ddot{\mathbf{X}} \times \mathbf{X}_3) \, \varPhi(s) \, ds = 0,$$

in which the dot now means differentiation with respect to the arc length s along $\mathbf{C}_0$. The desired necessary condition results from the fact that $\varPhi(s)$ is an arbitrary function that characterizes the variations of $\mathbf{C}_0$. Since $\varPhi(s)$ can be chosen with a large degree of arbitrariness it is possible to show that the above relation holds only if the function in the integrand multiplying it vanishes; thus a necessary condition for a minimum is obtained in the form

(7.21) $$\kappa_g = \dot{\mathbf{X}} \cdot \ddot{\mathbf{X}} \times \mathbf{X}_3 = 0,$$

in view of (7.7). It is therefore seen that the curve $\mathbf{C}_0$, if it is of minimum length, must be a geodetic line: a fact already announced earlier.

The proof that (7.21) follows from the formula preceding it is carried out by a standard argument in the calculus of variations. One assumes the contrary, i.e., that κ_g is not zero at some point $s = s^*$, say. Since it is a continuous function of s it would therefore be different from zero in some interval about s^*. The function $\varPhi(s)$ is now chosen to be zero everywhere except in that interval, where it should be of one sign; that a function with a continuous second derivative satisfying this condition exists is easily seen. But in that case it is clear that $L'(0)$ could not be zero. Hence the assumption that κ_g does not vanish at all points of $\mathbf{C}_0$ leads to a contradiction.

It was shown in the preceding section [cf. equation (7.9)] that equation (7.21) is equivalent to a pair of second-order ordinary differential equations for the geodetic line $\mathbf{C}_0$ given by $u_1 = u_1(s)$, $u_2 = u_2(s)$:

(7.21)$_1$ $$\ddot{u}_r + \dot{u}_i \dot{u}_j \varGamma_{ij}^r = 0, \qquad r = 1, 2.$$

These equations are the so-called Euler equations that result in dealing with the problem of minimizing the length integral given by (7.11). Evidently, it must be possible to obtain them directly from (7.11) by the variational process (it will be carried out in Chapter IX in a much more general context)—in which case there would be no need to think of the surface as embedded in three-space. However, it was already seen that the geodetic curvature is an isometric invariant, so that the intrinsic character of the curves satisfying (7.21) is known, although it is not obvious from the formula for κ_g as given in (7.21).

11. Geodesics in the Small and Geodetic Coordinate Systems

It is now known that a regular curve of shortest length joining two points on the surface must be a geodesic. Thus it is natural to inquire whether a solution of the differential equations (7.21) could not be found that would determine such a curve. This poses a so-called boundary value problem, as was remarked earlier, since a solution curve through two different points is desired. The count of the number of conditions available in the case of the boundary value problem makes its solution seem at least possible since there are four such conditions on the functions $u_1(s)$, $u_2(s)$, i.e., their values at the two end points of the curve, and the order of the system $(7.21)_1$ of differential equations is four, so that its solution depends on four constants. However, as was already remarked, this problem—which is a nonlinear boundary value problem—is not easy to solve, and in fact it need not have a solution in the large. In addition, even if the boundary value problem has a solution, the resulting curve need not necessarily be unique nor need it furnish an arc of shortest length. For example, consider the unit sphere. Any great circle arc is a geodesic, as was noted above, but if an arc of a great circle of length $> \pi$ is taken it is clear that such an arc is not the arc of shortest length joining its end points. If the length is π, there are infinitely many geodesics joining the two points since they are antipodal points. Consider also the circular cylinder, and two points on a generator of it. These points can be joined by infinitely many helices, all of which are geodesics, but only the straight line segment of the generator furnishes the shortest arc. The interesting questions which these examples bring to the fore cannot be dealt with at this point. In fact, a general theory in the domain of the calculus of variations, which is based on solving the boundary value problem posed by even a single Euler variational equation of second order, leads very far afield; an interestingly written introduction to this subject, the development of which owes very much to Jacobi and Weierstrass, can be found in the Carus Monograph of Bliss [B.7]. In Chapter VIII this theory will be developed for the case of geodesics.

The boundary value problem associated with $(7.21)_1$ thus presents difficulties in the large, but in the small the difficulties can be overcome without too much trouble. It will be shown, in fact, that *any pair of points that lie close enough to each other can be joined by a unique geodesic that furnishes the shortest regular arc connecting them.* The proof of this theorem is most readily given by making use of a special coordinate system in itself worthwhile for later purposes.

Before carrying this out, however, it is of interest to consider a related but different question which can also be dealt with through the use of a special coordinate system. The system of coordinate curves wanted here is the type of system used in the discussion that ended with equation (7.21). It was determined by a regular curve C and a set of orthogonal trajectories of it. In the present case the orthogonal trajectories to a given geodesic C are also chosen as geodesics. Since a geodesic is uniquely determined by giving a point and a direction for it at the point, it follows that this procedure does indeed lead to a regular coordinate system in a neighborhood of C (as explained at the end of Chapter IV), since the geodesics, as solutions of differential equations, also satisfy the needed regularity conditions. This system of curvilinear coordinates is called a *geodetic parallel coordinate system*.

Along the geodesics orthogonal to C, v_1 is taken as the arc length, with $v_1 = 0$ on C. Along C the parameter v_2 is assumed to be the arc length. For v_1 sufficiently small v_1 and v_2 furnish a parameter system with C given by $v_1 = 0$, and with geodesics for the curves $v_2 = $ const. It is now to be proved that this is an *orthogonal coordinate system*; i.e., that not only $v_1 = 0$ is orthogonal to the geodesics $v_2 = $ const., but that all curves $v_1 = $ const. cut these geodesics orthogonally. Once this is shown, it becomes clear that the line element in these coordinates has the particularly simple form

$$(7.22) \qquad ds^2 = dv_1^2 + G(v_1, v_2)\, dv_2^2,$$

since v_1 has the meaning of arc length on the curves $v_2 = $ const., and the coordinate curves are presumably orthogonal. The validity of (7.22) is proved as follows. From $\mathbf{X}_1 = \partial\mathbf{X}/\partial v_1$, $\mathbf{X}_2 = \partial\mathbf{X}/\partial v_2$, $g_{12} = \mathbf{X}_1\cdot\mathbf{X}_2$, the relation $\partial g_{12}/\partial v_1 = \mathbf{X}_1\cdot\mathbf{X}_{21} + \mathbf{X}_{11}\cdot\mathbf{X}_2$ holds. From $g_{11} = \mathbf{X}_1\cdot\mathbf{X}_1 = 1$ follows $\mathbf{X}_1\cdot\mathbf{X}_{21} = 0$. The vectors $\mathbf{X}_1(s)$ are tangent vectors to a geodesic and thus form a parallel field; as a consequence $\mathbf{X}_{11}$ is orthogonal to $\mathbf{X}_2$, since $\mathbf{X}_2$ is a vector in the surface, and thus $\mathbf{X}_{11}\cdot\mathbf{X}_2 = 0$. Consequently $\partial g_{12}/\partial v_1 = 0$ on each geodesic and hence that holds at all points of the curvilinear coordinate system. It follows that g_{12} is constant on each geodesic $v_2 = $ const., and hence is everywhere zero since it is zero on C. Thus (7.22) holds.

The converse statement is also true, i.e., *if a coordinate system is such that the line element has the form* (7.22) *then the lines $v_2 = $ const. are geodesics.* The proof of this statement is as follows. Since $g_{12} = \mathbf{X}_1\cdot\mathbf{X}_2 = 0$ everywhere, the relation $\mathbf{X}_1\cdot\mathbf{X}_{21} + \mathbf{X}_{11}\cdot\mathbf{X}_2 = 0$ holds, and from $g_{11} = \mathbf{X}_1\cdot\mathbf{X}_1 \equiv 1$ it follows that $\mathbf{X}_1\cdot\mathbf{X}_{11} = \mathbf{X}_1\cdot\mathbf{X}_{12} = 0$, $\mathbf{X}_2\cdot\mathbf{X}_{11} = 0$, and by (7.2) the field vectors $\mathbf{X}_1$ along any curve $v_2 = $ const. form a parallel field. Consequently the coordinate curves $v_2 = $ const. are geodesics.

An interesting geometric fact follows from (7.22) i.e., that *any pair* $v_1 = c_1$, $v_1 = c_2$ *of the orthogonals to the family of geodesics* $v_2 = $ const. *cuts off equal lengths on the geodesics.* The proof is almost immediate. The two orthogonals to the geodesics are given by $v_1 = c_1$, $v_1 = c_2$, and the length of an arc $v_2 = $ const. between these two curves is given by $L = \int_{c_1}^{c_2} dv_1 = c_2 - c_1$, since (7.22) holds, and this means that the two v_1-curves cut off the same length on all of the geodesics. However, it is not in general true that v_2 has the significance of arc length on the curves of the family $v_1 = $ const.: if this were true it would require $G(v_1, v_2)$ to have the value one everywhere, and the line element would be that of the Euclidean plane.

It is perhaps worthwhile to make still another observation regarding the construction of geodetic parallel coordinate systems. Such a system can always be determined in a neighborhood of a point in which a one-parameter family of geodesics forms a simple covering of the neighborhood since the orthogonal trajectories of such a *field*, as it is called, then form a second set of parameter curves and the line element must have the form (7.22), as was just shown.

In the calculus of variations a set of curves of the sort we considered here, i.e., a family of curves that satisfies the Euler differential equation of a variational problem, and is such that the family covers a domain simply, is called quite generally a *field*. Such fields of curves that satisfy the Euler differential equation are an important ingredient of the Weierstrass theory of variational problems of second order. In the present instance a special case of a general theorem concerning such fields is treated, i.e., the theorem that a solution of the Euler equation that is embedded in the curves of a field furnishes an extremum of the variational problem relative to any other curves which lie in a domain covered by the field and have the same end points. It is to be proved, in fact, that *a geodesic embedded in a field is the shortest regular arc connecting any pair of its points among all arcs lying in the region covered by the field.* The proof is an almost immediate consequence of (7.22), which holds for a field of geodesics and their orthogonal trajectories. Take any two points P, Q on a geodesic in the field and an arc $v_1 = v_1(t)$, $v_2 = v_2(t)$, $t_0 \leq t \leq t_1$, joining them and lying in the field. The length L_1 of the arc is given by

$$L_1 = \int_{t_0}^{t_1} \sqrt{v_1'^2 + Gv_2'^2}\, dt.$$

Since G is positive it follows that $L_1 \geq \int_{t_0}^{t_1} \sqrt{v_1'^2}\, dt$ holds, which in turn means that $L_1 \geq L$ holds, with L the length of the geodesic joining P and Q, and it is clear that $L_1 = L$ only for that geodesic. Thus the theorem is proved.

The above theorem is rather weak in a sense since it does not permit the conclusion that a shortest arc joining a *given* pair of points exists even if the two points are close to each other. This problem is now attacked by introducing still another special coordinate system, which is a generalization of the polar coordinate system in the plane. Toward the end of Chapter IV it was shown that a set of regular curves issuing from a point P in the direction θ (relative to the coordinate vector $\mathbf{X}_1$ at P, say), and with the arc length σ as parameter, forms in a neighborhood of P a regular parameter system except at P, provided that the functions $u(\sigma, \theta)$ and $v(\sigma, \theta)$, which define the curves, have continuous second derivatives with respect to σ and θ. In the present case these curves are selected as geodesics issuing from P in the direction given by θ, and the properties required for $u(\sigma, \theta)$ and $v(\sigma, \theta)$ follow from the theorems on the differentiable dependence of solutions of differential equations with respect to the independent variable and the initial conditions: for these theorems to hold it suffices to require the surface to have continuous derivatives of a proper order. Since σ is the arc length on a family of geodesics that covers a neighborhood of P, the situation resembles closely that of the geodetic parallel coordinates dealt with above; thus it is to be expected that the curves $\sigma = $ const. will turn out to be orthogonal to the geodesics $\theta = $ const. This is in fact true, so that the line element in the geodetic polar coordinate system is given by

$$(7.23) \qquad\qquad ds^2 = d\sigma^2 + G(\sigma, \theta)\, d\theta^2.$$

To prove that (7.23) holds it must be shown that $\mathbf{X}_1 = \mathbf{X}_\sigma$, $\mathbf{X}_2 = \mathbf{X}_\theta$ are orthogonal. Since $\mathbf{X}_1 \cdot \mathbf{X}_1 = 1$, it follows that $\mathbf{X}_1 \cdot \mathbf{X}_{12} = 0$. But $\mathbf{X}_1(\sigma) = \mathbf{v}(\sigma)$ are unit tangent vectors to a geodesic, and so they form a parallel field of unit vectors; hence $\dot{\mathbf{v}} = \mathbf{X}_{11}$ is orthogonal to $\mathbf{X}_2$ or $\mathbf{X}_{11} \cdot \mathbf{X}_2 = 0$. It therefore follows that $(\mathbf{X}_1 \cdot \mathbf{X}_2)_\sigma = \mathbf{X}_{11} \cdot \mathbf{X}_2 + \mathbf{X}_1 \cdot \mathbf{X}_{21} = 0$ and thus that $\mathbf{X}_1 \cdot \mathbf{X}_2$ is a constant along every geodesic; but this vanishes at the point P, since there $\mathbf{X}_2 = \mathbf{X}_\theta = 0$, obviously. Hence $\mathbf{X}_1 \cdot \mathbf{X}_2 = 0$ everywhere and (7.23) therefore defines the line element in these coordinates. This shows also that the orthogonal trajectories to the geodesics issuing from P are closed curves—a fact that is not at all obvious at the outset; it certainly would not be true for an arbitrary family of regular curves issuing from P. The curves $\sigma = $ const. are often referred to as *geodetic circles*.

For later purposes it is of interest to calculate the geodetic curvature κ_g of the geodetic circles $\sigma = $ const., and with (7.23) as first fundamental form. It is, of course, possible to carry that out intrinsically, but it is rather easier to do it with the use of the formula $\kappa_g = (\dot{\mathbf{X}} \cdot \ddot{\mathbf{X}} \times \mathbf{X}_3)$ given in Section 6. In the present case the differentiations of $\mathbf{X}(\sigma, \theta)$ are to be taken with respect to the arc length on a curve $\sigma = $ const.; it follows that

$$\dot{\mathbf{X}} = \mathbf{X}_\theta \cdot \frac{1}{\sqrt{G}},$$

$$\ddot{\mathbf{X}} = \frac{1}{\sqrt{G}}\left(\mathbf{X}_{\theta\theta} - \frac{1}{2}\mathbf{X}_\theta \frac{G_\theta}{G}\right).$$

The Gauss equation [cf. (6.9)] for $\mathbf{X}_{\theta\theta}$ is

$$\mathbf{X}_{\theta\theta} = A\mathbf{X}_\sigma + B\mathbf{X}_\theta + N\mathbf{X}_3$$

in terms of scalar coefficients A, B, N, of which only A needs to be calculated, since κ_g is found, after a little calculation, to have the value A. From the formulas (6.7), (6.8) for the Christoffel symbols and the definition of A given in terms of them, the following formula for κ_g is obtained:

$$\kappa_g = -\frac{1}{2}\frac{G_\sigma}{G}.$$

12. Geodesics as Shortest Arcs in the Small

The tools are now available to show that *a given point P on the surface has always a neighborhood such that any point Q in it can be joined to P by a geodesic the length of which is smaller than that of any other regular arc joining P and Q.*

From the discussion of the preceding section it is known that a positive number σ_0 exists such that the line element can be expressed in the form

$$ds^2 = d\sigma^2 + G(\sigma, \theta)\, d\theta^2 \quad \text{for} \quad 0 \le \sigma \le \sigma_0,$$

with $\sigma = 0$ corresponding to P. The length L of any regular arc $\mathbf{X}(\sigma(t), \theta(t))$ joining P with a point Q in this neighborhood is given by

$$L = \int_{t_0}^{t_1} \sqrt{\sigma'^2 + G\theta'^2}\, dt$$

if the arc lies entirely in the neighborhood. It follows that the following inequality holds:

$$L \ge \int_{t_0}^{t_1} \sqrt{\sigma'^2}\, dt = \sigma_1 < \sigma_0,$$

in which σ_1 is the "geodetic radius" drawn from P to Q; since G is positive the equality sign obviously holds only for the geodesic $\theta = $ const. joining P with Q. Hence the geodesic is the shortest of all regular arcs joining P and Q among those which lie within the neighborhood of P defined by $0 \le \sigma \le$

σ_0. It is also easy to see that this arc is shorter than *all* regular arcs on the surface joining P and Q, since any such curve that does not lie entirely in the closed neighborhood $0 \leq \sigma \leq \sigma_0$ must have a length greater than σ_0: such an arc must, by the Jordan curve theorem, go outside the geodetic circle $\sigma = \sigma_0$, and since lengths of curves are always positive, the correctness of the statement follows at once.

Thus the center of a sufficiently small geodetic circle can be joined by a geodetic arc of shortest length to any point inside the circle—in fact, the geodetic radius furnishes that connection. It is not true, however, that arbitrary pairs of points inside such a circle can be joined by an arc of shortest length that lies inside the circle. A geodetic circle of radius greater than $\pi/2$ on the unit sphere furnishes a counterexample. In Section 8 of the next chapter it will be shown that *sufficiently small* geodetic circles have this property of being geodetically convex.

For the sake of later developments in the next chapter, where shortest arcs in the large are treated, it is important to observe that *a quantitative lower bound for the positive number σ_0 can be found which depends only upon bounds for the coefficients of the differential equation of the geodesics and certain of their derivatives*—and hence only upon *bounds for certain derivatives of the surface itself.* This follows from the general discussion about this type of coordinates (cf. Chapter IV) and from Appendix B. This will be used in Chapter VIII to establish the existence of a uniform positive lower bound in the large for σ_0 in certain cases.

13. Further Developments Relating to Geodetic Coordinate Systems

It is useful for later purposes to obtain some further consequences of the existence in the small of geodetic parallel coordinates, in terms of which the line element has the form given by (7.22). This coordinate system was defined by an arbitrary geodesic C and geodesics orthogonal to it. On the orthogonal geodesics to C the parameter v_1 was defined as the arc length measured from C, while v_2 represented the arc length along C. The curves $v_1 = \text{const.}$ will not be geodesics in general, except for $v_1 = 0$. Thus $G = \mathbf{X}_2 \cdot \mathbf{X}_2 = 1$ on C, while $\mathbf{X}_1 \cdot \mathbf{X}_2 = 0$ everywhere. As a consequence $(\partial/\partial v_2)$ $(\mathbf{X}_1 \cdot \mathbf{X}_2) = \mathbf{X}_1 \cdot \mathbf{X}_{22} + \mathbf{X}_2 \cdot \mathbf{X}_{21} = 0$. On C, $\mathbf{X}_1 \cdot \mathbf{X}_{22} = 0$ since C is a geodesic, and hence the vectors $\mathbf{X}_2$ form a parallel field. It follows that $(\partial/\partial v_1)(\sqrt{G}) = \mathbf{X}_2 \cdot \mathbf{X}_{21}/\sqrt{G} = 0$ since $\mathbf{X}_2 \cdot \mathbf{X}_{21} = 0$. Thus for the function $\sqrt{G}$ on C the following conditions hold:

$$(7.24) \qquad \begin{cases} \sqrt{G}\big|_{v_1 = 0} = 1, \\ \dfrac{\partial \sqrt{G}}{\partial v_1}\bigg|_{v_1 = 0} = 0. \end{cases}$$

These conditions become quite interesting when put into relation with the fact that the function $\sqrt{G}$ is the solution of a second-order partial differential equation, and the conditions (7.24) then furnish initial data along C for that differential equation. The differential equation is in fact the equation that is the statement of Gauss's *theorema egregium*. For any orthogonal coordinate system the Gaussian curvature K is given in general by the equation [cf. (6.17)]

$$K = -\frac{1}{2\sqrt{EG}}\left[\left(\frac{E_v}{\sqrt{EG}}\right)_v + \left(\frac{G_u}{\sqrt{EG}}\right)_u\right].$$

Since $E = 1$ in the present case, it follows that

$$K = -\frac{1}{2\sqrt{G}}\frac{\partial}{\partial v_1}\left(\frac{1}{\sqrt{G}}\frac{\partial G}{\partial v_1}\right)$$

$$= -\frac{1}{\sqrt{G}}\frac{\partial^2\sqrt{G}}{\partial v_1^2}.$$

It has some point to write this equation in the form

$$(7.25) \qquad\qquad \frac{\partial^2}{\partial v_1^2}(\sqrt{G}) + K(\sqrt{G}) = 0.$$

Thus a partial differential equation for the function $\sqrt{G}$ has been found, together with a pair of initial conditions (7.24) satisfied by it along the geodesic C. Since the initial conditions furnish the value of $\sqrt{G}$ and its normal derivative along C, they are reasonable initial conditions for such a partial differential equation as (7.25). In fact, this initial value problem leads directly to interesting and important conclusions in the special case in which the Gaussian curvature is constant, as will be seen.

It is also of interest to consider the behavior of the coefficient $G(\sigma, \theta)$ when the line element is given in terms of geodetic polar coordinates. At the end of Chapter IV, where the proof of the validity of this type of curvilinear coordinate system was given, an auxiliary coordinate system was introduced by setting $x = \sigma \cos \theta$, $y = \sigma \sin \theta$, and this system was shown to be a regular parameter system in a neighborhood of $x = y = 0$. When, as in the present instance, the curves $\theta = $ const. are geodesics, the coordinates (x, y) are called *Riemann normal coordinates*. It is assumed that the line element is given in these coordinates:

$$ds^2 = g_{11}\,dx^2 + 2g_{12}\,dx\,dy + g_{22}\,dy^2.$$

(Although the Riemann normal coordinates are somewhat analogous to Cartesian coordinates in the plane, they need not be orthogonal.) It is known that

$$ds^2 = d\sigma^2 + G\,d\theta^2.$$

Since the differentials $d\sigma$ and $d\theta$ satisfy the equations

$$d\sigma = \frac{x\,dx + y\,dy}{\sigma}, \qquad d\theta = \frac{x\,dy - y\,dx}{\sigma^2},$$

it is easily found that

$$g_{11} = \frac{x^2}{\sigma^2} + G\frac{y^2}{\sigma^4}, \qquad g_{22} = \frac{y^2}{\sigma^2} + G\frac{x^2}{\sigma^4},$$

and these lead to the identity

$$x^2(g_{11} - 1) = y^2(g_{22} - 1),$$

when $\sigma^2 = x^2 + y^2$ is used. From this identity it follows that

$$g_{11} - 1 = \alpha y^2 + \cdots,$$

$$g_{22} - 1 = \alpha x^2 + \cdots,$$

in which the dots denote terms in x and y of third order at least, and α denotes a constant. These and other relations to follow are valid since the functions g_{ik} have derivatives of a sufficiently high order at $\sigma = 0$ (that is, for $x = y = 0$) if the original surface has derivatives of an appropriate order—which was assumed to be the case. Since $x^2 + y^2 = \sigma^2$ it follows that

$$\frac{x^2}{\sigma^2}\left(\frac{G}{\sigma^2} - 1\right) = g_{22} - 1 = \alpha x^2 + \cdots,$$

and hence that $G/\sigma^2 - 1$ behaves like $\alpha\sigma^2$ near $\sigma = 0$. Thus for G the development

$$G = \sigma^2 + \alpha\sigma^4 + \cdots$$

holds. Counterparts of the formulas (7.24) for geodetic polar coordinates now result from the following formulas:

$$\sqrt{G} = \sigma + \frac{\alpha}{2}\sigma^3 + \cdots,$$

$$\frac{\partial\sqrt{G}}{\partial\sigma} = 1 + \frac{3\alpha}{2}\sigma^2 + \cdots,$$

$$\frac{\partial^2\sqrt{G}}{\partial\sigma^2} = 3\alpha\sigma + \cdots.$$

The desired formulas are

$$\sqrt{G}\,\big|_{\sigma=0} = 0,$$

(7.24)′
$$\frac{\partial\sqrt{G}}{\partial\sigma}\bigg|_{\sigma=0} = 1.$$

Also, since (7.25) is generally valid for a line element of the form $ds^2 = du^2 + G\, dv^2$, it follows that

(7.25)'
$$\frac{\partial^2 \sqrt{G}}{\partial \sigma^2} + K\sqrt{G} = 0$$

holds for geodetic polar coordinates. Upon comparing this with the first and the last of the formulas a few lines above the constant α is seen to have a value given by

$$K(0,\, \theta) = -3\alpha.$$

Furthermore, the length $L(\sigma)$ of a geodetic circle of radius σ is given by

$$L(\sigma) = \int_0^{2\pi} \sqrt{G}\, d\theta = 2\pi\sigma - \frac{K}{3}\pi\sigma^3 + \cdots.$$

It follows that

$$\frac{\pi}{3} K(0,\, \theta) = \lim_{\sigma \to 0} \frac{2\pi\sigma - L(\sigma)}{\sigma^3},$$

which defines the Gaussian curvature K in terms of quantities that are isometric invariants. All of the above discussion was carried out with the aid of a special coordinate system; and hence the fact that the Gaussian curvature is independent of the choice of a coordinate system is not brought out by that discussion. However, the above formula, as one of the end results of the calculation, *does* bring out that fact, since the right-hand side of the equation clearly is invariant with respect to coordinate changes, since σ and $L(\sigma)$ have that property.

It is possible to give still another characterization of K, as follows. The area $A(\sigma)$ of a geodetic circle is given by

$$A(\sigma) = \int_0^{\sigma} \left[\int_0^{2\pi} \sqrt{G}(r,\, \theta)\, d\theta \right] dr$$

$$= \int_0^{\sigma} L(r)\, dr$$

in which $L(r)$ is the length of a geodetic circle, as defined above. Hence

$$A(\sigma) = \pi\sigma^2 - \frac{\pi}{12} K(0,\, \theta)\sigma^4 + \cdots,$$

and consequently

$$\frac{\pi}{12} K(0,\, \theta) = \lim_{\sigma \to 0} \frac{\pi\sigma^2 - A(\sigma)}{\sigma^4}.$$

It is to be noted that both $L(\sigma)$ and $A(\sigma)$ approach the values of the quantities for the circle in Euclidean geometry when $\sigma \to 0$ and that the line element $ds^2 = d\sigma^2 + G\, d\theta^2 = d\sigma^2 + (\sigma^2 + \alpha\sigma^4 + \cdots)\, d\theta^2$ differs from the line element of the plane in polar coordinates by a term of order σ^4 for σ small.

14.   Surfaces of Constant Gaussian Curvature

It is known that any two isometric surfaces, i.e., surfaces that are mapped in a one-to-one way on each other with preservation of lengths, have of necessity the same Gaussian curvature K at corresponding points.   However, the converse of this statement is not in general true, i.e., two surfaces that are mapped on each other in such a way that the values of K are preserved will not in general be isometric,—rather, the function K must satisfy certain conditions in order that this be true (cf. Eisenhart [E.6], p. 168, for example).   Nevertheless, in one very important special case the converse does happen to be true, i.e., in the case of surfaces for which the Gaussian curvature is *constant*.

With the aid of (7.24) and (7.25) it is to be shown that, in the small at least, two surfaces with the same constant Gaussian curvature can be put into isometric correspondence.   In fact, the differential equation (7.25) can be integrated explicitly in this case with the following results:

$$K = 0: \sqrt{G} = c_1 v_1 + c_2,$$

$$K > 0: \sqrt{G} = c_1 \sin \sqrt{K}\, v_1 + c_2 \cos \sqrt{K}\, v_1,$$

$$K < 0: \sqrt{G} = c_1 \sinh \sqrt{-K}\, v_1 + c_2 \cosh \sqrt{-K}\, v_1,$$

with c_1 and c_2 arbitrary functions of v_2.   In each case the functions $c_1(v_2)$ and $c_2(v_2)$ are uniquely determined by the initial conditions (7.24); and from them one finds easily the following values for $\sqrt{G}$:

$$K = 0: \sqrt{G} = 1,$$

$$K > 0: \sqrt{G} = \cos \sqrt{K}\, v_1,$$

$$K < 0: \sqrt{G} = \cosh \sqrt{-K}\, v_1.$$

In other words, the first fundamental form is uniquely determined by the constant value of K, and consequently *any two surfaces having the same constant value of K are isometric*.

A variety of conclusions can be drawn from the above results.   First of all it is seen that in the case $K = 0$ the line element is $ds^2 = dv_1^2 + dv_2^2$, i.e., it is the line element of the Euclidean plane.   A surface in three-dimensional space on which $K = 0$ is what is called a developable surface; it has now been proved quite generally that such surfaces are indeed isometric to the plane, in the small at least. In Chapter V this result was also achieved, but by a rather complicated discussion involving a classification of points into parabolic and planar points; such a classification plays no role here since no embedding of the surface in E^3 need be involved.

In the case $K > 0$ the line element resulting from $\sqrt{G} = \cos \sqrt{K}\, v_1$, as given above, is the line element of a sphere, with v_1 and v_2 parameters corresponding to polar coordinates on the sphere. In the case $K < 0$, the line element is said to be that of the *pseudo-sphere*; more will be said about it later. One of the problems at the end of this chapter requires the discussion of all surfaces of *revolution* for which K is a constant different from zero; it is readily shown, for instance, that the sphere is not the only such surface in the case of positive curvature.

15.   Parallel Fields from a New Point of View

The notion of a parallel field of vectors relative to a curve on a surface was not introduced originally by Levi-Civita in the way chosen earlier in this chapter. Instead, Levi-Civita proceeded by considering first the special case of developables in three-dimensional space. Directions at two different points P and Q on the developable could be reasonably defined as parallel if they would be parallel in the Euclidean sense after mapping the developable isometrically on a plane. This procedure of Levi-Civita for parallel fields on developables gives the same result as would be obtained by the methods explained previously since parallel fields are invariant under isometric mappings, and a parallel field in the Euclidean plane is parallel in the ordinary sense. For more general surfaces such a definition is not reasonable, since only surfaces with $K = 0$ can be mapped isometrically on a plane. However, Levi-Civita proceeded in the following way. First of all a developable surface is constructed that is tangent to the given surface along a curve C: this can be done by taking the envelope of the tangent planes of the surface along C. A parallel field with respect to the developable along C is constructed in the manner just indicated, and this field is in turn taken by definition to be a parallel field relative to the given surface. That this process also furnishes the same field as would be obtained by the methods described earlier is readily seen upon examining the formula (7.7) for the geodetic curvature. Evidently, $\kappa_g = d\theta(s)/ds$ is the same relative to both surfaces along the curve C since $\mathbf{X}$ and $\mathbf{X}_3$ are the same for both. It follows that the angle $\theta(s)$ between the tangent vector to C and the field vector is also the same at all points with respect to both surfaces if it was the same for both at any one point.

16. Models Provided by Differential Geometry for Non-Euclidean Geometries

It is of interest to pursue some further general consequences of the theorem that all surfaces with the same constant value of K are isometric. Since geodetic polar coordinates can be introduced in the neighborhood of any point of a surface, it follows that the line element of any two surfaces with $K = $ const. is given by $ds^2 = d\sigma^2 + G\, d\theta^2$ (cf. the discussion above) and G is the same function *independent of the point chosen as origin of the geodetic polar coordinate system.* Furthermore, since the initial line from which θ is measured is also arbitrary, it follows that *neighborhoods of two different points of the same surface having $K \equiv$ const. can be mapped isometrically on each other in such a way that any given direction through one of the points will correspond to any arbitrarily chosen direction through the other point.* A more intuitive way of expressing the last result is to say that a given geometrical figure on a surface of constant curvature can be moved about freely on the surface without changing the distances between any of its points nor of the angles between any lines on it. If such motions are called rigid motions, it is seen that rigid motions are as freely possible on surfaces with $K \equiv$ const. $\neq 0$ as they are for $K \equiv 0$, i.e., for the Euclidean plane. It is clear that this property holds on the sphere, for which K is a positive constant—on elementary geometrical grounds if the surface is regarded as lying in three-space. This in turn opens up the possibility of the existence of new geometries for $K > 0$ and $K < 0$ that might be analogous to Euclidean geometry without perhaps having the identical structure. This is indeed the case, and the geometries that arise are concrete realizations of the non-Euclidean geometries called elliptic ($K > 0$) and hyperbolic ($K < 0$) when the straight lines of these geometries are identified with the geodesics on the surfaces, and distances and angles are measured in terms of the metrics found for these cases. It turns out that nearly all of the axioms of Euclidean geometry hold in the two new geometries. The most important exception is the famous parallel axiom that states that there is exactly one parallel line to a given line through any given point; by *parallel* is meant that the two lines have no point in common. For centuries this axiom had been felt to be doubtful, in the sense that it was thought it might be a consequence of the others, and not an independent axiom. All attempts to prove such a thing failed. Finally, in the nineteenth century Bolyai and Lobachefsky put the matter to rest by exhibiting a consistent geometry, the hyperbolic geometry, in which the parallel axiom was replaced by a different one—thus making it clear that attempts to deduce this axiom from the others could not but fail. In this geometry, there are many parallels to a given line through a point not on the line.

The simpler of the two non-Euclidean geometries to describe is the elliptic geometry, for which a sphere can serve as a model, with the straight lines defined as the great circle arcs on the sphere, and with lengths and angles measured in the usual way. This geometry is mastered, essentially, by spherical trigonometry. In the book of Hilbert and Cohn-Vossen [H.10] there is a good description of this geometry and a discussion of the axioms that hold for it, together with comparisons with the axioms of Euclidean geometry. In particular, it is clear that there are *no* parallel lines to a given line in this geometry since all great circles on a sphere intersect.

From the point of view of differential geometry the hyperbolic geometry is more interesting than the elliptic geometry since its discussion requires a good deal of use of the tools provided by this subject, and in addition the discussion takes a quite amusing course. The starting point is the observation that the line element given by the formula

$$ds^2 = du^2 + e^{2u}\,dv^2, \qquad -\infty < u,v < \infty,$$

is a line element defined in the whole u,v-plane for which K has the value -1: this follows immediately—for example, from (7.25). It was proved in Section 11 that the lines $v = $ const. are geodesics. It is convenient to introduce new parameters (first used by Poincaré [P.4] in a history-making paper in analytic function theory) by setting $x = v$, $y = e^{-u}$, which are valid in the upper half-plane $y > 0$ (Poincaré's half-plane), since the Jacobian of the transformation has the value e^{-u} in this case. In fact, this parameter system is easily seen to be valid in the large, since the curves $x = $ const., $y = $ const. correspond to the curves $u = $ const., $v = $ const. in such a way that a one-to-one correspondence is established between these pairs of curves over their whole extent. In the new parameters the line element is given by

$$(7.26) \qquad ds^2 = \frac{dx^2 + dy^2}{y^2}, \qquad y > 0, \; -\infty < x < \infty.$$

It is perhaps worthwhile to prove that (7.26) holds by virtue of the necessity to preserve lengths, which should not depend on the choice of a parameter system. Early in Chapter IV it was seen that if $ds^2 = g_{ij}\,du^i\,du^j$ then the coefficients $\bar{g}_{ij}$ of an invariant line element after a parameter transformation are given by $\bar{g}_{ij} = g_{kl}\,(\partial u^k/\partial \bar{u}_i)\,(\partial u^l\,\partial \bar{u}_j)$. In the present case $u_2 = \bar{u}_1$, $u_1 = -\log \bar{u}_2$, and $g_{11} = 1$, $g_{12} = 0$, $g_{22} = e^{2u_1}$, from which (7.26) follows. However, in the present case it is much easier to observe simply that $dv = dx$, $du = -(1/y)\,dy$ and (7.26) follows at once.

It is relatively easy to find all geodesics in the upper half-plane. So far it is known only that the lines $x = $ const. are geodesics, since $x = v$ and the lines $v = $ const. are already known to be geodetic lines simply because of the form of the line element in the parameters u, v. It is convenient to introduce polar coordinates (ρ, ψ) in the upper half-plane, in the manner indicated by

Fig. 7.3. This system of coordinates is given by $x = x_0 + \rho \cos \psi$, $y = \rho \sin \psi$, $0 < \rho < \infty$, $0 < \psi < \pi$, the center of the circles $\rho = \text{const.}$ being on the x-axis. It follows that this coordinate system is regular in the entire open half-plane $y > 0$, since the origin of the polar coordinate system is on the

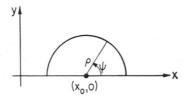

Fig. 7.3 Geodetic lines in the Poincaré half-plane.

x-axis. An easy calculation now gives the line element in these new coordinates in the form

$$(7.27) \qquad ds^2 = \frac{dx^2 + dy^2}{y^2} = \frac{d\rho^2 + \rho^2 \, d\psi^2}{\rho^2 \sin^2 \psi}$$

$$= \frac{1}{\rho^2 \sin^2 \psi} \, d\rho^2 + \frac{1}{\sin^2 \psi} \, d\psi^2.$$

Since new parameters $\rho_1 = \rho$, $\psi_1 = \int^\psi d\psi/(\sin \psi)$ can be introduced (since $\psi = 0$ and $\psi = \pi$ are excluded), in terms of which ds^2 would have the form $ds^2 = G(\rho_1, \psi_1) (d\rho_1)^2 + (d\psi_1)^2$, it is clear that the circles $\rho_1 = \rho = \text{const.}$ are geodesics. It can now be seen that the lines $\rho = \text{const.}$ and $x = \text{const.}$ furnish *all* geodesics, as follows. Consider any point P of the upper half-plane. It is known that a geodesic through P is uniquely determined by giving its direction. But there is a unique circle through P with its center on the x-axis and a given tangent at P, except for the vertical direction, which determines the geodesic $x = \text{const.}$ These curves fill out the upper half-plane without intersections other than those at P. Thus the geodesics are completely determined, and the totality of them through any point obviously fills the entire space without further intersections. In other words, a bundle of geodesics—that is, the straight lines in this geometry—through any point covers the space, and thus no two straight lines cut in more than one point— as in the Euclidean plane. Furthermore, any two points in the hyperbolic half-plane—as in Euclidean geometry—can be joined by a uniquely determined "straight line"; it is constructed by joining the points by the circle with center on the x-axis that is fixed by the (Euclidean!) perpendicular bisector of the (Euclidean) chord joining the two points.

In Euclidean geometry the straight lines are infinitely long in both directions. The same thing is true in the hyperbolic geometry. Consider first a geodesic $x = $ const. For it $ds = |dy/y|$, so that $L_{y_0}^{y_1} = \log(y_1/y_0)$, $y_1 > y_0$. It follows that $L \to \infty$ if $y_1 \to \infty$ or $y_0 \to 0$. Consider next a geodesic $\rho = $ const.; for it

$$ds = \frac{d\psi}{\sin\psi}, \qquad L_{\psi_0}^{\psi_1} = \int_{\psi_0}^{\psi_1} \frac{d\psi}{\sin\psi} = \log\frac{\tan\psi_1/2}{\tan\psi_0/2},$$

$\psi_1 > \psi_0$, and $L \to \infty$ if $\psi_1 \to \pi$ or $\psi_0 \to 0$.

However, *it is not true that parallels to a given line through a point not on it are uniquely determined*, as in Euclidean geometry. As was noted earlier, straight lines are called parallel lines if they have no points in common. Consider, for example, a straight line $x = $ const., and a point P not on it. Obviously, infinitely many half-circles with centers on $y = 0$ can be drawn through P that do not intersect the line $x = $ const., and such lines are parallel to it by definition.

It is of interest to consider angle measurements in the new geometry. It turns out that the angles are measured exactly as in Euclidean geometry, when the Poincaré half-plane is used as a model for this geometry. This is most readily seen by considering the line element in the form given by (7.26). If (dx_1, dy_1) and $(\delta x_1, \delta y_1)$ are direction numbers for line elements at a given point, then the angle ϕ between them is fixed (cf. Chapter IV) by

$$\cos\phi = \frac{E\, dx_1\, \delta x_1 + G\, dy_1\, \delta y_1}{\sqrt{E\, dx_1^2 + G\, dy_1^2} \cdot \sqrt{E\, \delta x_1^2 + G\, \delta y_1^2}},$$

when $F = 0$. In the present case $E = G = 1/y^2$, $F = 0$, so that E and G cancel out; thus for $\cos\phi$ the formula

$$\cos\phi = \frac{dx_1\, \delta x_1 + dy_1\, \delta y_1}{\sqrt{dx_1^2 + dy_1^2}\, \sqrt{\delta x_1^2 + \delta y_1^2}}$$

results—and it is exactly the same as for the Euclidean metric given by $ds^2 = dx^2 + dy^2$. In particular, orthogonal lines in this model of the hyperbolic geometry are orthogonal in the Euclidean sense. From this fact an interesting conclusion can be drawn. It is known from the study of geodetic polar coordinates in Section 11, that the orthogonal trajectories (orthogonal in the proper metric, of course) to the family of geodetic lines through a given point P are the loci obtained by marking off equal distances along the geodesics from that point. However, in the present case this family of geodesics is a one-parameter family of circles in the Euclidean plane with centers all on the x-axis, and it is a well-known fact that their (Euclidean!) orthogonal trajectories form a second set of circles with centers on the line orthogonal to the x-axis that passes through P, as indicated in Fig. 7.4. Since Euclidean and hyperbolic angles are the same, it is clear that these Euclidean circles orthogonal to the geodesics are at the same time geodetic circles in the hyperbolic geometry.

It was shown quite generally (see Section 14) for surfaces with $K =$ const. that they permit free "rigid" motions (in the small, at least), i.e., isometric mappings on themselves that take any one point and a direction

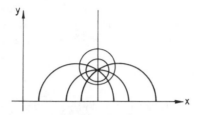

Fig. 7.4   Geodetic circles in the Poincaré half-plane.

through it into an arbitrary point and an arbitrary direction through the second point.   Thus in particular if the interior of a given geodetic circle is considered, it is clear that a rigid "rotation" of it through any angle about its center leaves all lengths preserved—although the shape of a sector of it will be deformed when looked at from the Euclidean point of view.   Actually, it is possible to describe explicitly all isometric mappings of the Poincaré half-plane on itself in a most elegant and amusing way (which presupposes some slight knowledge of complex function theory) and thus in effect to describe explicitly the rigid motions in this geometry.   This is done by describing the location of a point by the complex variable $z = x + iy, y > 0$. It is to be shown that the linear fractional transformation

$$(7.28) \qquad\qquad z^* = \frac{\alpha z + \beta}{\gamma z + \delta},$$

with $\alpha, \beta, \gamma, \delta$ all real numbers such that $\alpha\delta - \beta\gamma > 0$, transforms the upper half-plane into itself (that is, z^* is always a point such that $y^* > 0$ holds) and at the same time preserves the line element of the hyperbolic half-plane. Once this is proved it is to be expected—it will be proved a little later—that these transformations provide *all* isometric mappings since they depend on three independent real parameters (after eliminating by division one of the numbers $\alpha, \beta, \gamma, \delta$), and that is the correct number to fix a point (two coordinates) and a direction through it (an angle).   The proof is carried out by a direct calculation.   It is noted first that

$$z^* = x^* + iy^* = \frac{[\alpha(x + iy) + \beta][\gamma(x - iy) + \delta]}{|\gamma z + \delta|^2},$$

since $\overline{(\gamma z + \delta)} = \gamma\bar{z} + \delta$, the bar meaning that the conjugate complex value is to be taken; here the fact that γ and δ are real has been used. Thus y^* is given by

$$(7.29) \qquad y^* = (\alpha\delta - \beta\gamma) \frac{y}{|\gamma z + \delta|^2}.$$

It is therefore clear that the entire upper half-plane is indeed mapped on itself. As is well known, the mapping furnished by (7.28) is a *conformal* mapping, and it therefore preserves angles in the Euclidean sense—but then also in the hyperbolic geometry, as was proved above. It is shown next that the mapping is also *isometric*. This can be seen by computing dz^*, which is readily found to be given by

$$(7.30) \qquad dz^* = (\alpha\delta - \beta\gamma) \frac{dz}{(\gamma z + \delta)^2}.$$

From (7.27) $ds = |dz|/y$, and this in conjunction with (7.29) and (7.30) leads to the isometric invariance of the mapping, since $|dz^*|/y^* = |dz|/y$.

It is shown next that an arbitrary point and a direction through it can be carried by an appropriate mapping of the type (7.28) into any other point and any direction through that point. To this end it is noted first that the transformations (7.28) form a group, as can easily be verified. Any point $z_0 = x_0 + iy_0, y_0 > 0$, can be carried into the point $z^* = i$ by the transformation

$$z^* = \frac{1}{y_0} (z - x_0),$$

which belongs to the group. In addition, the half-plane $y > 0$ can be "turned" about the point $z = i$ through the angle ω by the transformation

$$(7.31) \qquad z^* = \frac{\tan \omega/2 + z}{1 - z \tan \omega/2},$$

since

$$dz^* = \frac{dz}{\cos^2 (\omega/2)(1 - i \tan \omega/2)^2} = e^{i\omega} \, dz$$

at $z = i$, and at the same time the point $z = i$ is easily found to remain fixed. Since the transformations (7.31) also belong to the group, it is clear that the above statement is proved, and, in fact, that the transformation needed is unique besides. In this way all isometric transformations which preserve the orientation of the plane have been found. To complete the set of isometric transformations it is necessary to add the "reflection" given by $x^* = -x$, $y^* = y$, which does not belong to the class defined by (7.28), since the determinant of this transformation is negative. In this case, of course, the orientation is not preserved. The situation in the Euclidean plane is analogous:

the orthogonal transformations with positive determinant yield the rigid motions, and once the reflections are added all isometric mappings of the plane itself are obtained.

It can now be seen that the geodetic circles are at the same time curves of constant geodetic curvature—in other words, points at constant distance

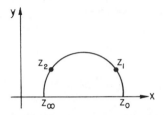

Fig. 7.5   Length as the logarithm of the cross-ratio of $(z_1, z_2, z_0, z_\infty)$.

from a fixed point in the hyperbolic plane fill out a closed curve whose geodetic curvature κ_g is constant. To prove this the given "center" is carried to the point $z^* = i$ as above; since the transformation (7.31) transforms geodetic circles into themselves (they are merely "rotated"), and since κ_g is an isometric invariant, it follows that it must be a constant on the geodetic circle. However, it is not in general true on an arbitrary surface that loci for which κ_g = const. are closed curves: if that happens to be true for all such loci, then it can be shown that the Gaussian curvature K must be constant.

It is left as an exercise to show that the length of a line in the hyperbolic plane is given by the logarithm of the cross-ratio of the points $(z_1, z_2, z_0, z_\infty$ shown in Fig. 7.5. Since the cross-ratio is the fundamental invariant in projective geometry, it is a plausible guess that there is a connection between the hyperbolic non-Euclidean geometry and projective geometry. In fact, there are very interesting connections of this kind (see, for a brief introduction to the subject, Courant-Robbins [C.13], p. 218, and Hilbert-Cohn-Vossen [H.10]).

17.   Parallel Transport of a Vector Around a Simple Closed Curve

As was pointed out earlier in this chapter, the transport of a vector once around a *closed* curve can be expected to lead to a final vector that does not in general coincide with the vector at the starting point unless the Gaussian curvature is everywhere zero. It is proved in this section that *the angle between the initial and final positions of the vectors imbedded in a parallel field along a simple closed curve is given by the integral of the Gaussian curvature over the area of the surface bounded by the curve.*

It is advantageous to consider a simple closed curve C that has vertices, i.e., it is assumed that C is continuous, but that it is composed of a finite number of regular arcs with finite discontinuities in the first derivative at the vertex points.   Along such curves the linear differential equations (7.5) will

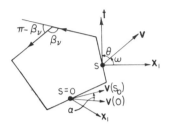

Fig. 7.6   Gauss-Bonnet formula.

define a parallel field $\mathbf{v}(s)$ uniquely along C as soon as the initial vector $\mathbf{v}(0)$ is prescribed, since the coefficients of the differential equations are continuous functions: the vertices on C thus cause no difficulties.   (It is implied here, of course, that the parallel displacement of $\mathbf{v}$ on any arc C_ν is to be carried up to its common "vertex" with $C_{\nu+1}$ and that this position of $\mathbf{v}$ is used as initial position for transport along $C_{\nu+1}$.)   As indicated in Fig. 7.6, the angle between the initial vector $\mathbf{v}(0)$ and the final vector $\mathbf{v}(s_0)$ after parallel displacement once around the curve is denoted by α.   By $\omega(s)$ is meant the angle through which the coordinate vector $\mathbf{X}_1(s)$ must be rotated in order to bring it into coincidence with $\mathbf{v}(s)$.   The angle α is then defined, quite reasonably, by the formula

$$\alpha = \omega(s_0) - \omega(0) = \int_0^{s_0} \dot\omega(s)\, ds,$$

since the vector $\mathbf{X}_1$ returns to its initial position.   This formula is valid since $\dot\omega(s)$ is a piecewise continuous function.   To calculate the angle ω it is convenient to assume that the parameter curves are orthogonal, so that $F = \mathbf{X}_1 \cdot \mathbf{X}_2 = 0$.   (That such orthogonal systems always exist in the small was shown in Chapter IV).   The field vector $\mathbf{v}(s)$ is expressed in the form

$$\mathbf{v} = v_1\mathbf{X}_1 + v_2\mathbf{X}_2 = \frac{\cos \omega}{\sqrt{E}}\,\mathbf{X}_1 + \frac{\sin \omega}{\sqrt{G}}\,\mathbf{X}_2:$$

in this way $\mathbf{v}$ is defined as a unit vector.   The derivative of $\mathbf{v}(s)$ is given by

$$\dot{\mathbf{v}} = \dot\omega\left(-\frac{\sin \omega}{\sqrt{E}}\,\mathbf{X}_1 + \frac{\cos \omega}{\sqrt{G}}\,\mathbf{X}_2\right) - \frac{\dot{E}\cos \omega}{2E^{3/2}}\,\mathbf{X}_1 - \frac{\dot{G}\sin \omega}{2G^{3/2}}\,\mathbf{X}_2$$

$$+ \frac{\cos \omega}{\sqrt{E}}\,\dot{\mathbf{X}}_1 + \frac{\sin \omega}{\sqrt{G}}\,\dot{\mathbf{X}}_2.$$

The condition that the field of vectors $\mathbf{v}(s)$ forms a parallel field is that $\dot{\mathbf{v}} \cdot \mathbf{X}_1 = \dot{\mathbf{v}} \cdot \mathbf{X}_2 = 0$ [cf. (7.21)]. The equation $\dot{\mathbf{v}} \cdot \mathbf{X}_1 = 0$, for example, leads to

$$\dot{\mathbf{v}} \cdot \mathbf{X}_1 = \sin \omega \left(-\dot{\omega}\sqrt{E} + \frac{1}{\sqrt{G}} \mathbf{X}_1 \cdot \dot{\mathbf{X}}_2 \right) = 0,$$

since $\mathbf{X}_1 \cdot \mathbf{X}_1 = E$ and $\mathbf{X}_1 \cdot \dot{\mathbf{X}}_1 = \frac{1}{2}(\mathbf{X}_1 \cdot \mathbf{X}_1)^{\cdot} = \frac{1}{2}\dot{E}$, while $\mathbf{X}_1 \cdot \mathbf{X}_2 = 0$. Similarly, $\dot{\mathbf{v}} \cdot \mathbf{X}_2$ is given by

$$\dot{\mathbf{v}} \cdot \mathbf{X}_2 = \cos \omega \left(\dot{\omega}\sqrt{G} + \frac{\mathbf{X}_2 \cdot \dot{\mathbf{X}}_1}{\sqrt{E}} \right) = 0.$$

The coefficients of $\sin \omega$ and $\cos \omega$ are essentially the same, since $\mathbf{X}_1 \cdot \dot{\mathbf{X}}_2 + \dot{\mathbf{X}}_1 \cdot \mathbf{X}_2 = 0$ from $\mathbf{X}_1 \cdot \mathbf{X}_2 = 0$. The field is therefore a parallel field if the following condition is satisfied:

$$\dot{\omega} = \frac{1}{\sqrt{EG}} \mathbf{X}_1 \cdot \dot{\mathbf{X}}_2.$$

Since $\dot{\mathbf{X}}_2 = \mathbf{X}_{21}\dot{u} + \mathbf{X}_{22}\dot{v}$, it follows that

$$\mathbf{X}_1 \cdot \dot{\mathbf{X}}_2 = \mathbf{X}_1 \cdot \mathbf{X}_{21}\dot{u} + \mathbf{X}_1 \cdot \mathbf{X}_{22}\dot{v}$$
$$= \frac{1}{2}(E_v\dot{u} - G_u\dot{v}).$$

This in turn results from the relation $\mathbf{X}_1 \cdot \mathbf{X}_{21} = \frac{1}{2}(\mathbf{X}_1 \cdot \mathbf{X}_1)_v = \frac{1}{2}E_v$, and the relation $G_u = \frac{1}{2}\mathbf{X}_2 \cdot \mathbf{X}_{12} = -\frac{1}{2}\mathbf{X}_1 \cdot \mathbf{X}_{22}$, since $(\mathbf{X}_1 \cdot \mathbf{X}_2)_v = \mathbf{X}_1 \cdot \mathbf{X}_{22} + \mathbf{X}_{12} \cdot \mathbf{X}_2 = 0$. Thus $\dot{\omega}$ can be expressed in the form

$$\dot{\omega} = \frac{1}{2\sqrt{EG}} (E_v\dot{u} - G_u\dot{v})$$

and this is to be integrated around C in order to determine α. But this is just the form that lends itself to an application of Gauss's theorem:

$$\oint_C F \, dv + G \, du = \iint_R (F_u - G_v) \, du \, dv,$$

in which the right-hand side is an integral over the region R bounded by C. As was remarked on other occasions, an orientation of the surface is defined locally by ordering the vectors $\mathbf{X}_1 \to \mathbf{X}_2$, and this in turn makes it possible to stipulate that the line integral is to be taken in the counter-clockwise sense around R. All of the conditions developed in the calculus that are needed for the validity of this formula are fulfilled, and consequently α can be given by the formula

(7.32)
$$\alpha = \oint_C \dot{\omega}(s) \, ds = -\frac{1}{2} \oint_C \left(-\frac{E_v}{\sqrt{EG}} \, du + \frac{G_u}{\sqrt{EG}} \, dv \right)$$

$$= -\frac{1}{2} \iint_R \left[\left(\frac{E_v}{\sqrt{EG}} \right)_v + \left(\frac{G_u}{\sqrt{EG}} \right)_u \right] du \, dv$$

$$= \iint_R K \sqrt{EG} \, du \, dv = \iint_R K \, dA$$

in view of the formula (6.17) for the Gaussian curvature K when an orthogonal coordinate system is used. The formula (7.32) establishes the result announced at the beginning of this section.

In Section 5 above it was shown that $K \equiv 0$ is a necessary condition for

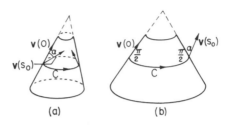

(a)   (b)

Fig. 7.7   Parallel transport on a cone.

the parallel transport of a vector along a curve on a surface to be independent of all pairs of points and all paths joining them. The formula (7.32) could also be used to establish this fact once more, since if K were not zero at some point, it would be different from zero and of one sign in some neighborhood of it; hence a simple closed curve in this neighborhood could be drawn and α would not be zero with respect to that curve: but this means that there would be paths for which the process of parallel transport of vectors would not be independent of the path.

It might be thought that $K \equiv 0$ would also be a *sufficient* condition for independence of the path in parallel transport, but that is not in general the case. In fact, (7.32) does not suffice to prove such a thing since it is a formula valid only when C is a simple closed curve that is the full boundary of a portion of the surface—and thus of a simply connected portion of it. In fact, Fig. 7.7 indicates an interesting counterexample. Shown there is a circular cone in three-dimensional space, together with its isometric development on the plane obtained by imagining the cone slit along a generator and then unwrapped on the plane. Angles and parallel fields are preserved in this process, and a parallel field in the plane is parallel in the ordinary sense. Thus $\mathbf{v}(s_0)$ is parallel to $\mathbf{v}(0)$ in Fig. 7.7b; and hence $\mathbf{v}(s_0)$ is turned at an angle $\alpha \neq 0$ to the image of the generator of the cone; but this means that $\mathbf{v}(s_0)$ is turned at the angle α to $\mathbf{v}(0)$, as indicated in Fig. 7.7a. Thus, even though the cone is a developable on which K is everywhere zero, it is not true that α is zero relative to the simple closed curve C—but then C does not bound a simply connected and regular portion of the surface and formula (7.32) does not apply. If a convex cap were to be put over the top of the conical piece of Fig. 7.7a, it is clear that (7.32) would then give for α a nonzero value. In fact, by considering the spherical image mapping as a means of determining the integral $\iint K\, da$, it is seen that α is determined by the vertex angle of the cone.   This is another example in which it is of advantage to consider a problem in three-space even though the question is one belonging to inner differential geometry.

The case shown in Fig. 7.2, in which the parallel transport of a vector around an octant of the unit sphere is in question, was discussed in Section 2. The result was the value $\pi/2$ for the angle α; this is as it should be since the total Gaussian curvature over the octant is the same as the area of the octant.

18.   Derivation of the Gauss-Bonnet Formula

One of the most important results of differential geometry is embodied in a formula called the Gauss-Bonnet formula, which is a formula involving the integral of the Gaussian curvature over a simply connected polygonal domain.

To derive this formula reference is made once more to Fig. 7.6.   It is convenient to introduce the angle τ by setting $\tau = \omega + \theta$, so that $\dot{\tau} = \dot{\omega} + \dot{\theta}$. Thus τ refers to the angle between $\mathbf{X}_1$ and the tangent vector $\mathbf{t}$ to any one of the regular arcs $\mathbf{C}_\nu$ that compose the curve C.   The first step is to calculate the total change in the angle τ upon transversing C once in the counterclockwise sense.   Along any regular arc C_ν of C the following relation holds:

$$\int\limits_{C_\nu} \dot{\tau}\, ds = \int\limits_{C_\nu} \dot{\omega}\, ds + \int\limits_{C_\nu} \dot{\theta}\, ds.$$

By summing all such equations for all regular arcs C_ν of C, the result is

$$\sum_\nu \int\limits_{C_\nu} \dot{\tau}\, ds = \sum_\nu \int\limits_{C_\nu} \dot{\omega}\, ds + \sum_\nu \int\limits_{C_\nu} \dot{\theta}\, ds.$$

Since the geodetic curvature κ_g is given by $\kappa_g = d\theta/ds$ [cf. (7.7)], it follows that

$$\sum_\nu \int\limits_{C_\nu} \dot{\theta}\, ds = \oint\limits_{C} \kappa_g\, ds,$$

while

$$\sum_\nu \int\limits_{C_\nu} \dot{\omega}\, ds = \iint\limits_{R} K\, dA$$

from (7.32).   Thus the following relation has been found:

$$\sum_\nu \int\limits_{C_\nu} \dot{\tau}\, ds = \oint\limits_{C} \kappa_g\, ds + \iint\limits_{R} K\, dA.$$

The calculation of the left-hand side, which is in some respects the most delicate step in the derivation, remains to be done.   It is noted first that

$$\sum_v \int_{C_v} \dot{\tau} \, ds + \sum_v (\pi - \beta_v) = n \cdot 2\pi,$$

in which the angles β_v, with $0 \le \beta_v \le 2\pi$, are the interior angles at the vertices of C and n is some integer.   This relation states that the total change in the angle between $\mathbf{X}_1$ and $\mathbf{t}$ along all regular arcs of C, plus the sum of the exterior angles at the vertices must be an integral multiple of 2π.   But this is evidently true since the final positions of both $\mathbf{X}_1$ and $\mathbf{t}$ are the same as the initial positions.   Of course it is to be kept in mind that all angles are to be measured by using the first fundamental form.   (It should be noted that all of the above statements are independent of the point chosen as starting point on C, as can readily be seen.)   The fact is that n *has the value one*. This statement is proved here by showing that n *has the same value for the plane as for any other regular surface*, and therefore, as was shown in Chapter II (Section 18), its value is one.   To prove this statement consider the following line element that depends on a parameter λ varying over the interval $0 \le \lambda \le 1$:

$$ds^2 = I(\lambda) = (1 - \lambda)(\dot{u}_i \dot{u}_i) + \lambda g_{ik} \dot{u}_i \dot{u}_k,$$

with $g_{ik}\dot{u}_i\dot{u}_k$, the first fundamental form of the surface.   Since $1 - \lambda$ and λ do not vanish simultaneously in $0 \le \lambda \le 1$ and each is greater than or equal to zero, it follows that $I(\lambda)$ is positive definite for all such λ, since it is a linear combination of positive definite forms with positive coefficients.   Thus $I(\lambda)$ (taken as a first fundamental form) may be used to measure angles.   Since all angle measurements made using $I(\lambda)$ would clearly be continuous in λ, it follows that the number n for it is continuous in λ.   (Since the quantities involved belong to inner differential geometry, it is legitimate to confine the discussion to the parameter plane.   Thus the question of the existence of a surface with such a line element in three-dimensional space is irrelevant.) But n is an integer, so that it must remain constant for all λ, and in particular for $\lambda = 0$, when $I(0) = \dot{u}_1^2 + \dot{u}_2^2$, i.e., when the geometry is the Euclidean geometry of the plane.   Thus the problem has been reduced to that of finding n in the Euclidean plane, and it is known that $n = 1$ in that case.

This step completes the derivation of the Gauss-Bonnet formula in the form

(7.33) $$\iint_R K \, dA + \oint_C \kappa_g \, ds = 2\pi - \sum_v (\pi - \beta_v).$$

19. Consequences of the Gauss-Bonnet Formula

The Gauss-Bonnet formula has many consequences.   Consider first the special case in which R is a triangular region bounded by three geodesics.   In that case $\kappa_g = 0$ on the boundary arcs of R, and the formula (7.33) takes the following form:

$$\sum_{v=1}^{3} \beta_v - \pi = \iint_A K \, dA.$$

Thus the sum of the interior angles of a geodesic triangle is greater than, less than, or equal to π according to whether K is positive, negative, or zero.

For a sphere of radius R, for example, $K = 1/R^2$, and

$$\sum_{v=1}^{3} \beta_v - \pi = \frac{A}{R^2},$$

which is the formula of spherical trigonometry for the so-called *spherical excess* of a triangle made up of great circle arcs. If $K = -1$—so that the surface is a so-called pseudosphere—the corresponding formula is

$$-A = \sum_{v=1}^{3} \beta_v - \pi \quad \text{or} \quad \sum \beta_v = \pi - A.$$

If K is everywhere zero $\sum \beta_v = \pi$, as it should be.

The general formula for geodetic triangles can be put in the form

$$\frac{\sum\limits_{v=1}^{3} \beta_v - \pi}{A} = K^*,$$

where K^* is the average Gaussian curvature over the triangle. If now the triangle shrinks down to a point, it is seen that the Gaussian curvature K can be defined as

$$K = \lim_{A \to 0} \frac{\sum \beta_v - \pi}{A}.$$

This brings out once more the fact that K is an isometric invariant, since the area A as well as the angles β_v are determined solely by the first fundamental form.

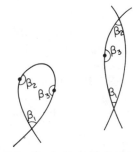

Fig. 7.8   *Geodesics with intersections.*

For $K < 0$, it is clear that $\sum_{\nu=1}^{3} \beta_\nu < \pi$. Hence a geodesic on a surface with $K < 0$ cannot have a double point, nor can two geodesics have more than one intersection, as is evident from Fig. 7.8 and the above inequality, since $\sum_{\nu=1}^{3} \beta_\nu > \pi$ in both cases after introducing the "angles" β_2 and β_3, which have the value π—while, on the other hand, $\sum \beta_\nu - \pi$ should be negative.

The Gauss-Bonnet formula has many more important consequences that belong to differential geometry in the large. Some of them are treated in the next chapter.

20. Tchebychef Nets

This chapter is brought to a close with another problem belonging to inner differential geometry, i.e., a mathematical problem that is said to correspond to the physical problem of applying a piece of cloth smoothly to a curved surface. This problem of "clothing" a given surface was first formulated and treated by Tchebychef [T.1]. It is desired to spread a net—a fishnet for example—over a surface in such a way that no cord is slack and all cords are unstretched. Instead of attacking this problem Tchebychef treats a different problem in which a net of discrete meshes is regarded as a part of a family of regular parameter curves that cover the surface completely. Thus he raises the question whether surface parameters v_1, v_2 can be introduced such that the line element takes the form

$$(7.34) \qquad ds^2 = dv_1^2 + 2F \, dv_1 \, dv_2 + dv_2^2,$$

since then any net formed by selecting curves $v_1 = nk_1$, $v_2 = nk_2$, ($n = 1, 2, \ldots$), k_1 and k_2 both constant, will have the desired property: all pairs of opposite sides of a parallelogram of the net will have the lengths nk_1 or nk_2, since the arc length is clearly the parameter on both sets of curves. The problem of clothing the surface is therefore regarded as solved if surface parameters $v_1(u_1, u_2)$, $v_2(u_1, u_2)$ can be introduced in such a way that the line element takes the form (7.34). Parameter curves with this property are called Tchebychef nets.

When the line element is in the form (7.34), it is clear that $F = \cos \omega$ (from $\cos \omega = F/\sqrt{EG}$), where ω is the angle through which the coordinate vector $\mathbf{X}_1$ must be turned to bring it into coincidence with $\mathbf{X}_2$. The theorema egregium applied to (7.34) must therefore yield a relation between ω and the Gaussian curvature K. It is interesting and useful to obtain this relation explicitly, e.g., by using the relations developed in Section 6 of the preceding chapter. Since $E = G = 1$, the formula is

$$K = \frac{1}{2\sqrt{1 - F^2}} \left[\left(\frac{F_1}{\sqrt{1 - F^2}} \right)_2 + \left(\frac{F_2}{\sqrt{1 - F^2}} \right)_1 \right], \qquad F_i = \frac{\partial F}{\partial v_i},$$

and since $F = \cos \omega$ this is easily seen to take the form

(7.35)
$$\frac{\partial^2 \omega}{\partial v_1 \, \partial v_2} = -K \sin \omega.$$

This is a second-order partial differential equation, of the type called hyperbolic, for the function ω, with $v_1 = $ const. and $v_2 = $ const. as the so-called characteristics.

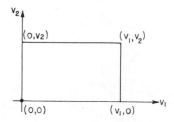

Fig. 7.9   Parallelogram from a Tchebychef net.

An interesting conclusion can be drawn by integrating (7.35) over any "parallelogram" of the net, the coordinates of whose vertices (in the v_1, v_2-plane) are indicated in Fig. 7.9.   In that case the result is

$$\iint \frac{\partial^2 \omega}{\partial v_1 \, \partial v_2} \, dv_1 \, dv_2 = - \iint K \sin \omega \, dv_1 \, dv_2 = - \iint K \, dA = -K_T,$$

where K_T is, of course, the curvature integrated over the area of the parallelogram.   The left-hand side of the equation can be integrated to yield the formula

(7.36) $-K_T = \omega(0, 0) + \omega(\bar{v}_1, \bar{v}_2) - \omega(\bar{v}_1, 0) - \omega(0, \bar{v}_2)$

$$= \omega_1 + \omega_3 - \omega_2 - \omega_4,$$

in which the angles ω_i (together with the interior angles α_i) of the parallelogram are as indicated in Fig. 7.10.   From this K_T is given by

(7.37)
$$K_T = 2\pi - \sum_{i=1}^{4} \alpha_i.$$

This formula is known as the formula of Hazzidakkis [H.6]. The angles α_i are subject to the condition $0 < \alpha_i < \pi$ since the angles ω_i must lie between 0 and π.   It therefore follows that the numerical value of K_T for any parallelogram in the net can never exceed 2π.   From this it is inferred that a surface may not be clothed without either tearing or wrinkling the cloth if its total Gaussian curvature is too large.   Tailors have learned this fact from experience.   They break up a surface to be clothed into pieces, and join the pieces along curves over which the net curves are not smoothly connected—as, for example, in the seam of a sleeve at the shoulder.

Tchebychef nets have a number of interesting properties, some of which are formulated among the problems at the end of this chapter. They also play an essential role in an important theorem in the large, due to Hilbert, to be treated in the next chapter.

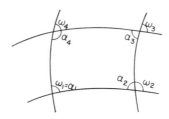

Fig. 7.10   Angles of a Tchebychef net.

It has already been hinted that the author is not convinced that the existence of a Tchebychef net is a sufficient condition for the existence of a fishnet covering a given surface. The reason is that such a net if stretched tightly over a smooth surface would of necessity have all of its cords take the form of segments of geodetic lines: this is readily proved since the equilibrium of such a cord under tension would require its principal normal to fall on the surface normal. The curves forming a Tchebychef net are, however, not geodesics in general. Also, there would in general be breaks in the tangent directions of the cords of a fishnet at the knots in it, whereas curves with continuous turning tangents are required for Tchebychef nets. (Some questions relevant to these remarks are formulated as Problem 12 at the end of the next chapter.)

PROBLEMS

1. In inner differential geometry show that the "straight" angle—i.e., the angle between two vectors opposite to each other—has the value π, no matter what the metric form may be. If the angle between two geodesics at a point p is less than π, show that a geodesic through their common point exists such that the points between them near p lie on one side of the geodesic. Prove also the converse. *Hint:* Consider a geodetic polar coordinate system (σ, θ) with pole at p and the map of a neighborhood of the pole on the Euclidean plane that results upon using these same numbers as polar coordinates in the plane. Are the same statements true for regular curves other than geodesics?

2. Show the geodetic curvature of a curve on a surface is the curvature of its orthogonal projection on the tangent plane.

3. Show with the use of (7.5) that the length of a vector is preserved in parallel transport along a curve.

4. Which two of the following properties of a regular curve on a surface imply the third: (a) asymptotic line, (b) geodetic line, (c) straight line?

5. Same question as in (4) but for (a) line of curvature, (b) geodetic line, (c) plane curve.

6. Justify the name "rectifying plane" for the plane determined by the tangent and binormal vectors of a regular space curve. *Hint:* Show that the curve is a geodesic on the developable surface that results as the envelope of the planes in question.

7. Show that the osculating developable along a regular curve C of a surface is a regular surface if C is not tangent to an asymptotic line of the surface.

8. Describe the possible surfaces of revolution in three-space for which the Gaussian curvature is constant.

9. Give explicit examples of surfaces of revolution mapped on each other in such a way that the Gaussian curvature is the same at corresponding points but such that the surfaces are not isometric.

10. Show that the area of the Poincaré half-plane is infinite.

11. Show that the logarithm of the cross-ratio of the points (z_1, z_2, z_0, z_{00}) (cf. Fig. 7.5) gives the length of the geodesic joining z_1 and z_2 in the Poincaré half-plane.

12. Prove that the tangents to one family of curves of a Tchebychef net along a curve of the other family form a parallel field in the sense of Levi-Civita, and conversely that a net is a Tchebychef net if all curves of the net have this property.

13. Consider the surface of revolution formed by rotating one branch of a hyperbola about its axis of symmetry to form a convex surface of revolution. Give a bound for the angle between its asymptotes above which no geodesics can intersect in more than one point.

14. A circular cone of semivertical angle β is given, together with a simple closed curve C on it which cuts all of its generators. Show that the angle α between the initial and final vectors of a parallel field obtained by parallel transport once around C is given by $\alpha = 2\pi(1 - \sin \beta)$.

15. Why does the angle α between the initial and final positions of a vector in parallel transport seem to be zero for a complete great circle arc on a sphere, although $\iint K \, dA$ over the half-sphere bounded by it is not zero?

16. Use the result of Problem 12 to derive the following pair of partial differential equations for the functions $u_1 = u_1(v_1, v_2)$, $u_2 = u_2(v_1, v_2)$, which

determine a Tchebychef net in terms of the original parameters (v_1, v_2) of the surface:

$$\frac{\partial^2 u_r}{\partial v_1 \, \partial v_2} + \frac{\partial u_i}{\partial v_1} \frac{\partial u_j}{\partial v_2} \, \Gamma_{ij}^r = 0, \qquad r = 1, 2.$$

Show that a Tchebychef net exists, for sufficiently smooth surfaces, in a neighborhood of a point p when two arbitrary intersecting regular curves through p are given as initial curves of the net. *Hint:* Show that integrals of the partial differential equations satisfying these initial conditions can be obtained by an iteration process that is exactly analogous to the standard method used for the corresponding problem for systems of ordinary differential equations.

17. If κ_{v_1} and κ_{v_2} denote the geodetic curvatures of two families of curves $v_1 = $ const., $v_2 = $ const. forming a Tchebychef net, and $\omega(v_1, v_2)$ is the angle between curves of the net, show that $\kappa_{v_1} = \partial \omega / \partial v_2$, $\kappa_{v_2} = -\partial \omega / \partial v_1$.

18. Derive the differential equations for geodetic lines as the Euler equations of the variational problem concerned with the length integral in the form

$$L = \int_{t_0}^{t_1} \sqrt{g_{ij} \dot{u}_i \dot{u}_j} \; dt.$$

19. The line element of a surface of negative curvature is given by $ds^2 = E \, du^2 + F \, du \, dv + G \, dv^2$ in terms of the asymptotic lines as parameter curves. Show that the line element of the spherical image in terms of the spherical images of the asymptotic lines as parameters is given by

$$d\sigma^2 = -K [E \, du^2 - 2F \, du \, dv + G \, dv^2].$$

From this show that if the asymptotic lines form a Tchebychef net their images on the sphere also form such a net on the sphere.

CHAPTER VIII

Differential Geometry in the Large

1. Introduction. Definition of n-Dimensional Manifolds

So far in this book, except for occasional easily described special cases, only problems of differential geometry in the small have been considered. This means that the geometric object considered was the topological image of the interior of a circle, or of an open segment of the real axis, i.e., a one-to-one mapping of these domains that is continuous in both directions. The image of the circle or the segment might or might not be regarded as lying in some other given higher dimensional space. Thus, in the first six chapters of this book the curves and surfaces studied were the topological images of a line segment or of the Euclidean disk in a two- or three-dimensional Euclidean space. In the preceding Chapter VII surfaces were studied from the point of view of their inner differential geometry, although they were regarded as being embedded in Euclidean space. In most of these cases it was assumed that a single curvilinear coordinate system without singularities covered the portion of the surface being studied. In problems in the large such a coordinate system that covers the whole surface rarely exists for reasons that are concerned with questions of topology. In fact, practically nothing of interest in differential geometry in the large can be done without some reference to topology, and a large proportion of the theorems in this field either require essential hypotheses regarding topological properties, or the theorems have a topological property as the main item to be proved.

It is therefore necessary to deal with certain portions of the subject of topology in order to make progress with the problems to be taken up in this chapter. Fortunately, only relatively simple ideas and facts from this field are needed; they will be outlined here and in the next sections without attempting anything like a complete discussion. For more complete treatments the little book of Patterson [P.1] and the book of Weyl [W.4] are recommended.

Questions of topology are formulated in what is called a topological space. In this chapter only a very special connected space of this kind is considered, i.e., what is called an *n-dimensional*[1] *differentiable manifold S.* That is, to begin with, a set of points S in which certain subsets, called *open sets*, are assumed to be defined. Such open sets are required to have the following properties:

(a) The null set and the space S are open sets.

(b) The union of any number of open sets is an open set.

(c) The intersection of any finite number of open sets is an open set.

(d) Every point of S is contained in some open set. In addition, if p_1 and p_2 are any distinct points of S, then an open set O_1 containing p_1 and an open set O_2 containing p_2 exist such that O_1 and O_2 have no points in common.

A set S covered by such open sets is called a *Hausdorff space.* For the purposes in view here an additional restriction is made so that the space becomes what is called a *separable* space:

(e) If any open set M in S is given, a *countable* collection C of open sets is assumed to exist such that the open set M is contained in the union of the sets of C.

With the aid of the open sets in a Hausdorff space it becomes possible to define what is meant by *continuous* and by *topological* mappings of such a space. To this end, suppose that U_1 is a subset of S_1 and that a *function f* is given that maps it into a subset U_2 of S_2, both S_1 and S_2 being Hausdorff spaces. This means, of course, that to each point p_1 of U_1 a point p_2 of S_2 is assigned by $f(p_1) = p_2$. The function f *is continuous at a point p_1 in U_1 if, for each open set O_2 in S_2 containing the point $p_2 = f(p_1)$, there exists an open set O_1 in S_1 containing p_1 such that $f(O_1 \cap U_1)$ is contained in O_2.* The function f is said to be continuous in U_1 if it is continuous at all points of U_1. The mapping f is said to be *topological* if it is a *one-to-one mapping of U_1 onto U_2 that is continuous together with the inverse mapping f^{-1}* (which evidently exists).

It is now possible to define what is called *an n-dimensional manifold M^n, differentiable of order r*: such manifolds are the principal objects of investigations in differential geometry. A manifold is characterized by the following properties:

---

[1] In all but the present section of this chapter only two-dimensional manifolds are treated. However, such a restriction is pointless for the discussion in this first section, which then serves a useful purpose in the next chapter as a preliminary to the discussion of tensor calculus in n-dimensional manifolds and its applications to Riemannian geometry and relativity. For more extended discussions of the early sections of this chapter, and a good part of the next chapter, the books by Sternberg [S.3], Auslander and MacKenzie [A.5], and by Kobayashi and Nomizu [K.4] might be consulted.

1. It is a separable Hausdorff space S covered by a countable collection D of open sets $D: \{N_j\}$ that are now called *patches*.

2. To each patch N_j in D there exists a *topological mapping* f_j of N_j into the *unit ball* S^n of n-dimensional Euclidean space E^n. (A unit ball is the interior of an n-dimensional unit sphere; E^n itself is, of course, a separable Hausdorff space. Its open sets can be defined as the sets containing only inner points, and these in turn are defined in the usual way as points that are the center of a ball contained in the set.)

3. If N_j and N_k are patches with a nonempty intersection, it is assumed that (a) $f_j(f_k^{-1})$ is a topological mapping between $f_k(N_j \cap N_k)$ and $f_j(N_j \cap N_k)$ and (b) the mapping $f_j(f_k^{-1})$ has a nonvanishing Jacobian and is differentiable of order r with respect to Cartesian coordinates $(x^1, x^2, \ldots, x^n)$ of E^n in $f_k(N_j \cap N_k)$. (It is advisable to make a sketch of this situation.)

4. If any two distinct points p and q of S are given, it is assumed that a finite number of pairwise overlapping patches $N_{i_1}, N_{i_2}, \ldots, N_{i_s}$ exists connecting p and q, i.e., such that $p \subset N_{i_1}$, $q \subset N_{i_s}$, and $N_{i_{j-1}} \cap N_{i_j}$ are nonempty for $j = 2, 3, \ldots, s$. This condition ensures that the space is what is called a *connected* space.

5. The mapping function $f_j(p)$ for all points p in a patch N_j of course yields Cartesian coordinates of the points $f_j(p)$ in an n-ball. These coordinates are said to furnish, by definition, a *local coordinate system*, or *coordinate patch* for the points of S in N_j. Two coverings D, D' of S furnish *equivalent* coordinate systems if their union is a covering of S satisfying the crucial properties given in (3). One way that is often used to obtain such an equivalent covering D' from a given covering D is to replace the mapping functions f_j by the functions $\phi_j(f_j)$, in which ϕ_j defines any topological mapping of the unit ball onto itself that is differentiable of order r and has a nonvanishing Jacobian. Only properties of manifolds that they possess for *all* coverings by systems of local coordinates of a given class are of interest. Or, in other words, the coordinate systems considered are assumed to be maximal in the sense that they cannot be enlarged by the addition of others.

As usual in this book, no great circumstance is made regarding smoothness properties, but it should be remarked that a manifold of class C^r may under some circumstances be reduced to one with fewer derivatives—for example, if new coordinates are introduced with the use of functions determined as solutions of differential equations defined in the manifold. In the more recent literature such a possibility is often avoided by working with manifolds of class C^∞.

In a manifold M^n the usual notions concerning point sets M in S, such as that of an accumulation point, closed sets, etc., are defined in the customary way. For example, a sequence p_n of points in M^n is said to converge to a point p in M^n if every open set containing p contains all but a finite number of points of the sequence; it is to be noted that the "separation requirement" (d) above ensures that the *limit point* p thus defined is uniquely determined.

The notion of a differentiable curve in M^n (or of differentiable submanifolds of higher dimensions) can now be introduced through use of the local coordinate systems. For example, a differentiable curve C in M^n is a set of points $p(t) \subset M^n$ defined for $0 \le t \le 1$, with the property that if $N(p(t_0))$ is a neighborhood with coordinates $f(N)$ of a point $p(t_0)$, then an interval $|t - t_0| < \epsilon, \epsilon > 0$, can be found such that the points $p(t)$ in it lie in $N(p(t_0))$ and $f(p(t))$ is a differentiable image in a unit ball of E^n of that segment of the real axis.

It is now possible to prove the equivalence of the property (4) and the following property:

4′. The manifold M^n is *arcwise connected*. By this is meant that any two points p, q in M^n can be joined by a continuous curve in M^n.

In fact, if the condition (4) is fulfilled it is easily seen that any pair of points can actually be connected by arcs which are piecewise differentiable. To this end it is noted that each point $p(t)$ of C is contained in a neighborhood of S in which a coordinate system is defined. Since the values of t range over a close bounded set of the t-axis it follows, by the Heine-Borel theorem, that a finite number of the neighborhoods suffices to cover C. It is therefore clear that piecewise differentiable curves can also be constructed in S which join p and q.

It is perhaps worthwhile to pause at this point to consider an interpretation of a coordinate patch N_j in the two-dimensional case. The points p of N can be identified with the points $(x^1, x^2) \equiv f(p)$ of the unit disk. But the coordinates $\phi^1 = \phi^1(x^1, x^2)$, $\phi^2 = \phi^2(x^1, x^2)$ defined by $\phi(f(p))$ are equally valid when ϕ has the properties prescribed above for a change of local coordinates. As a consequence such a patch is a rather formless thing since, for example, no measurements of length, or angle, no concept of a vector or vector field, nor of curvature, would be a priori available in the space. In fact, this is the parameter plane of surface theory, but without a metric form and the other concepts that give rise to the developments that lend to differential geometry its character and interest. The next task is to introduce such concepts in an intrinsic way.

2. Definition of a Riemannian Manifold

In the preceding chapter the intrinsic geometry of surfaces in the small was treated, but the treatment was carried out by regarding the surfaces as embedded in three-dimensional space. The concept of an n-dimensional manifold has now been introduced without making use of an embedding, and in the large rather than in the small. Even when it is desired to deal with problems concerning surfaces, or higher dimensional manifolds, when they are embedded in a given space (which might or might not be Euclidean) the reasonable way to treat them is to start with their local coordinate systems as parameter domains—as will be done later on in this chapter. A great many of the problems to be considered here—and practically all of the problems of the next chapter—are examined strictly from the intrinsic point of view, and it is therefore appropriate to introduce some at least of the essential concepts without reference to an embedding.

The two primitive notions of intrinsic differential geometry that must be defined are the concept of length and the concept of a vector or, better, of a field of vectors. These two concepts in turn can be defined in an intuitive and natural way, by analogy with what was done when an embedding in Euclidean space was assumed, once the notion of a differentiable *curve* in the manifold has been introduced. The proper way to define such curves is clear (it was in the essentials already done above): a *differentiable curve* is defined as *the locus of points $p(t)$ in the manifold that results when the local coordinates u^i of the points of the locus are prescribed to be differentiable functions $u^i = u^i(t)$ of a parameter t that ranges over an interval $0 \leq r \leq 1$*. (It is implied here that the curve might lie in many different coordinate patches, the functions $u^i(t)$ being differentiable in each patch.)

Since no embedding is available, the length of such a curve cannot be defined by making use of the concept of length that belongs to the containing space, as was done in the previous chapter. Instead, lengths of curves are determined by *assigning to the manifold*, once for all, a line element ds^2 that is by assumption independent of the choice of local coordinates. That is, in terms of a given set of local coordinates (the summation convention being implied) the line element is defined by

$$(8.1) \qquad\qquad ds^2 = g_{ik}\, du^i\, du^k.$$

Here, the functions $g_{ik} = g_{ik}(u^1, u^2, \ldots, u^n)$ are defined with respect to the coordinates u^i; they could be any sufficiently smooth functions such that the quadratic form in the differentials is positive definite.[1] However, it is evidently necessary to arrange matters so that the length measurements do not depend on any special choice of coordinates. This in turn is secured by *requiring* that the coefficients g_{ik} of the fundamental metric form defined by (8.1) transform in just such a way that the invariance of ds^2 is upheld. Already in Chapter IV the appropriate calculation was made to obtain the

---

[1] In the theory of relativity this last requirement must be given up; this special case will be dealt with at the appropriate place in the next chapter.

transformation formulas for two-dimensional surfaces; they are readily seen to be of the same form, whatever the dimension, and thus, if the coordinates are transformed by taking $u'^i = u'^i(u^1, u^2, \ldots, u^n)$, the new coefficients g'_{ik} are of necessity defined by the formulas

$$g'_{ik} = g_{lm} \frac{\partial u^l}{\partial u'_i} \frac{\partial u^m}{\partial u'^k}.$$

The length of a differentiable curve segment in the manifold that is defined by $u^i = u^i(t)$, $t_0 \le t \le t_1$, is then, in turn, defined by the integral

$$(8.2) \qquad s_{t_0}^{t_1} = \int_{t_0}^{t_1} \left(g_{ikj} \frac{du^i}{dt} \frac{du^j}{dt} \right)^{1/2} dt.$$

Evidently the length of the curve segment is now not only invariant with respect to a transformation of coordinates but also with respect to a transformation of the parameter t to another parameter τ fixed by giving $\tau = \tau(t)$, with $d\tau/dt \ne 0$, simply because the definite integral defining the length is invariant under such a change of the integration variable.

The next task is to introduce the concept of *vectors in a manifold*. In the preceding chapter vectors were defined in the surface as *tangent vectors* of the surface when embedded in Euclidean three-space; such tangent vectors, in turn, occurred as tangent vectors of differentiable curves in the surface. In manifolds a similar thing can be done without an embedding with the help of the notion of a differentiable curve, as introduced above; the basic idea is to identify the vectors in the manifold as the *tangent vectors* of curves. To this end, consider in local coordinates a curve $u^i = u^i(t)$ through a point p that corresponds to $t = 0$; by *definition* its tangent vector at p is *the ordered set of first derivatives* $\dot{u}^i(0): (\dot{u}^1(0), \dot{u}^2(0), \ldots, \dot{u}^n(0))$ *with respect to t of the coordinates of the curve when evaluated for $t = 0$*. Every differentiable curve through p thus defines a vector $\mathbf{v} = (\dot{u}^1, \dot{u}^2, \ldots, \dot{u}^n)$; conversely, every such vector is the tangent vector of a curve—for example, of the curve $u^i(t) = \dot{u}^i(0) \cdot t$. If these vectors are compounded in the way that is customary in linear algebra (cf. Appendix A of this book), it follows that the totality of them forms a *linear vector space* of dimension n, and this space is called *the tangent space of the given manifold at the point p*. The tangent space has what may very properly be called a *natural basis*, i.e., the basis formed by the tangent vectors to each of the coordinate curves of the patch. By this is meant tangent vectors $\mathbf{v}_i$ of the curves

$$C^i \colon \begin{cases} u^i(t) = u^i(0) + t, \\ u^j(t) = u^j(0), \qquad j \ne i, \end{cases}$$

defined for $1 \le i \le n$. The tangent vectors $\mathbf{v}_i = (\dot{u}^1_{(i)}, \dot{u}^2_{(i)}, \ldots, \dot{u}^n_{(i)})$ to these curves have components which are zero except for the ith component, which has the value 1; they therefore form a basis that spans the tangent vector space. At each point of the manifold an n-dimensional linear vector space is thus defined relative to a particular local coordinate system, and hence vectors could be defined at all points of the manifold as vectors in such spaces. However, the essence of the notion of a vector in a linear space lies in its invariance with respect to a change of basis vectors, and this in turn is coupled in a precise way in a manifold with a change of local coordinates in the manifold.

Some further stipulations, comments, and justifications regarding these important conventions are in order, since they form the basis for the employment of the tools of linear algebra to vectors (and tensors) in the tangent space. Consider a change of curvilinear coordinates in the manifold from u^i to u'^i given by $u'^i(u^1, u^2, \ldots, u^n)$ and its effect on a curve $u^i = u^i(t)$. Since

$$\frac{du'^i}{dt}\bigg|_0 = \frac{\partial u'^i}{\partial u^j}\bigg|_0 \frac{du^j}{dt}\bigg|_0 \quad \text{or} \quad \dot{u}^i(0) = \gamma^i_j \dot{u}^j(0) \quad \text{with} \quad \gamma^i_j = \frac{\partial u'^i}{\partial u^j}\bigg|_0,$$

it is seen that the basis vectors defined by tangent vectors to the original coordinates of the manifold are related to the basis vectors similarly defined with respect to the new coordinates at a given point by a *linear transformation with a matrix given by the elements of the Jacobian of the coordinate transformation in the manifold at that point of the space.* It is one of the basic assumptions regarding the space that this matrix should be nonsingular, so that a coordinate transformation of the space leads in a natural way to a unique nonsingular linear transformation in the tangent space. If constants $c^k_i = \partial u^k / \partial u'^i|_0$ are introduced it is then seen that the matrices (c^k_i) and (γ^i_j) play the same roles in the tangent space as the matrices given in the same notation do in the Appendix A on linear algebra. It is to be noted also that every nonsingular matrix may occur, since the linear transformations of coordinates given by $u'^i = a^i_k u^k$, a^i_k constant, belong to the group of admitted transformations of the manifold. It is stipulated always in what follows that *the transformations of the tangent space are defined by using the Jacobian of the transformations of coordinates in the manifold in this fashion.*[1] Thus the in-

---

[1] The procedure given here for defining vectors and vector fields is sometimes criticized because it starts from a very special case, i.e., the tangent vectors of curves. It is quite possible to proceed in an abstract fashion and thus avoid this criticism (which is not without point); such a treatment is given in the book of Willmore [W.7]. In the next chapter a somewhat different way of defining vectors and tensors in manifolds that is due to E. Cartan [C.2] will be taken. Cartan's approach is strongly intuitive, and follows closely Riemann's ideas in considering the geometry of a manifold as being approximately affine in a neighborhood of each of its points. Also, assuming the geometry in the first instance as affine rather than Euclidean—an idea due to Weyl [W.1]—makes it possible to define vector and tensor fields at the outset, without first introducing a metric.

variant quantities customarily defined in linear vector spaces become *by definition* intrinsic invariants of the more general spaces, and it is with respect to this type of invariance that the geometrical studies in all that follows are carried out.

For example, a metric is now defined in a natural way in the tangent vector space attached to a given point of the manifold by making use of the values of the coefficients g_{ik} of the line element in (8.1) at the point as coefficients of an invariant metric form in the tangent space.   As in Appendix A, such a form $g_{ik}\xi^i\eta^k$ in the tangent space serves to define a scalar product $\mathbf{x}\cdot\mathbf{y}$ and with it the lengths of vectors $\mathbf{x} = (\xi^1, \xi^2, \ldots, \xi^n)$ and the angle between pairs of them; in particular, the angle ϕ between the vector $\mathbf{x}$ and another $\mathbf{y} = (\eta^1, \eta^2, \ldots, \eta^n)$ is defined by the formula

$$\cos \phi = \frac{g_{ij}\xi^i\eta^k}{\sqrt{g_{ij}\xi^i\xi^k} \sqrt{g_{ij}\eta^i\eta^k}} = \frac{\mathbf{x}\cdot\mathbf{y}}{\sqrt{\mathbf{x}\cdot\mathbf{x}} \sqrt{\mathbf{y}\cdot\mathbf{y}}}.$$

It is clear also that the coefficients g_{ik} have the values $g_{ik} = \mathbf{v}_i\cdot\mathbf{v}_k$ obtained by taking the scalar products of pairs of basis vectors.   This same formula thus serves to define angles between the tangent vectors of two curves given by $u^i_{(1)}(t)$ and $u^i_{(2)}(t)$ in the manifold:

$$(8.3) \qquad \cos \phi = \frac{g_{ij}\dot{u}^i_{(1)}\dot{u}^j_{(2)}}{\sqrt{g_{ij}\dot{u}^i_{(1)}\dot{u}^j_{(1)}} \sqrt{g_{ij}\dot{u}^i_{(2)}\dot{u}^j_{(2)}}}.$$

Thus, with the aid of the tangent spaces, vectors and tensors are defined at any point of the Riemannian space; actually, such quantities are quite often defined at *all* points of the space and thus vector and tensor *fields* result.   Starting from this it is possible to go on to develop the whole of the intrinsic geometry of manifolds in an invariant way.   Thus the area and volume element, the notion of parallel fields along curves and the differential equations characterizing them, and the Gaussian curvature, for example, all could be dealt with systematically in this fashion.   That course is, however, not taken in the present chapter, but is reserved rather for the next chapter, where the central point at issue is in many ways the formal development of a calculus, the tensor calculus, that puts this notion of invariance in the foreground.   In the present chapter, in which only surfaces having two dimensions are considered, the point of view is that the invariant character of the geometric entities introduced has been shown in the preceding chapter by considering the surfaces in three-space, or else it should be studied in the next chapter.   It would also be legitimate to take the attitude that it would be possible, in each instance, to check by a formal calculation whether a given quantity or formula is properly invariant or not—but that would admittedly be somewhat tedious.   In particular, for example, use will be made later on of the definition of the element of area of a surface as $dA = \sqrt{g_{11}g_{22} - g_{12}^2} \cdot$ $du^1 du^2$ in terms of any local parameters u^1, u^2, whether or not the surface is regarded as embedded in E^3. The fact is that the material of the present chapter is such that the interest is concentrated on the discussion of specific geometrical situations and problems, and while it should be known that they have an invariant formulation, that circumstance is not of paramount interest.

3.   Facts from topology relating to two-dimensional manifolds

As was remarked earlier, questions concerning the topology of manifolds play a major role in differential geometry in the large. By definition, as was stated above, *a property of a manifold is said to be a topological property if it is invariant under transformations of the manifold that are one-to-one and continuous in both directions.* Two manifolds that can be put into relation with each other in this way are said to be topologically equivalent, or homeomorphic. Intuitively speaking, a topological property is one which is preserved when the manifold is continuously deformed without interpenetration or tearing— although, as every student of analysis knows, it is dangerous to lean too heavily on geometric intuition when the continuity of geometrical objects alone, without further restrictive smoothness properties, is assumed.

It would be very pleasant, if it were possible, to give at this point a means of classifying the n-dimensional manifolds into topological types by defining topological invariants, which if identical for two different manifolds would insure them to be homeomorphic. This has not up to now been done for the case $n = 3$, let alone for higher dimensions, and it seems to be an extraordinarily difficult problem to solve.

Although this general problem remains unsolved, many important necessary conditions for topological equivalence are known, and even both necessary and sufficient conditions are known for the case of greatest interest for the present chapter, i.e., that of *the compact two-dimensional manifolds.* One necessary condition for topological equivalence of two manifolds that is intuitively expected, but not easy to prove in general, is the *dimension* of the manifold: thus a two-dimensional manifold cannot be homeomorphic to a three-dimensional manifold. Another interesting and useful property that is preserved in a topological mapping is the property of compactness, already referred to above. A manifold *is said to be compact if out of any covering of it by a class of open sets a finite number of sets covering it can be selected.* That this property is a topological invariant of a manifold is readily seen upon using the definition of continuity of a mapping f of M_1^n into M_2^n which implies that the antecedent set $f^{-1}(N)$ is open in M_1^n if N is open in M_2^n.[1]

The compact two-dimensional manifolds are called also *closed surfaces*. Some special kinds of *open*, i.e., noncompact, surfaces are also considered in this chapter.

In the rest of this chapter the discussion is restricted to two-dimensional manifolds S (the symbol S, for *surface*, is preferred now instead of M^2), the topology of which is relatively simple to describe. The principal tool for this purpose is a *triangulation* of the manifold. This means that the entire two-dimensional manifold S, which will be called a *surface* from now on, is subdivided into a set of curvilinear triangles, each of which is the topological image of a Euclidean triangle, and such that the following conditions are satisfied:

1. If p is an inner point of a certain triangle, then it and some neighborhood of it belong to a unique triangle.

2. If p lies on an edge of a triangle Δ_0 but is not a vertex of it, then another triangle Δ_1 exists to which p belongs, Δ_0 and Δ_1 have exactly that edge in common, and neither p nor a neighborhood of it has points in any triangles of the set other than Δ_0 and Δ_1.

3. If p is a vertex point of a triangle Δ, there exist finitely many triangles $\Delta_0, \Delta_1, \ldots, \Delta_k$ with p as a common vertex point; and these triangles are arranged in cyclic order so that consecutive triangles have exactly one common side passing through p. The various triangles should have otherwise no points in common except p, and the last triangle Δ_k should be attached to the triangle Δ_0 along a common edge. Such a configuration is sometimes called a *star* of triangles associated with the vertex p.

These circumstances are illustrated in Fig. 8.1.

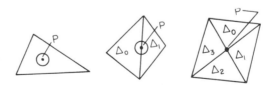

Fig. 8.1   Inner, edge, and vertex points.

[1] Eventually the manifolds to be studied here will be made into metric spaces and a covering of the manifold by neighborhoods equivalent to n-balls will be introduced, i.e., a covering by the sets of all points at a distance from a given one which is less than an arbitrary positive number. Once this has been done, all of the customary (ϵ, δ) statements of analysis can be expressed in the usual form by using the distance between points.

It ought to be mentioned also that the word "triangulation" is on occasion used in a figurative sense with reference to a subdivision of a surface that employs any simple closed polygons having any finite number of sides, rather than a subdivision restricted exclusively to triangles. The modifications that should then be made in the above three requirements are obvious.

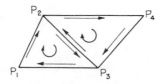

Fig. 8.2   Orientation of a surface.

It can be shown (though not without some difficulty) that every two-dimensional manifold permits of such a triangulation, but it is not known whether that is true for all manifolds of higher dimension. This is one of the reasons why two-dimensional manifolds are much easier to deal with than others. However, more decisive here is the fact that in two dimensions the topological structure of a manifold can be completely deciphered in terms of a single triangulation. This seems not to be the case in higher dimensions, so that even the restriction to triangulable cases for $n > 2$ does not lead to a complete topological classification of these manifolds.

As a preliminary to describing this process of classification for surfaces it is of advantage to deal first with an important and very useful notion that is readily introduced intuitively once a triangulation is available. That is the notion of an *orientation* of the surface. To this end consider any triangle (cf. Fig. 8.2) with vertices p_1, p_2, p_3 and with sides oriented in the sense p_1p_2, p_2p_3, p_3p_1, and also a triangle p_2, p_3, p_4 having the edge p_2p_3 in common with the previous triangle. The second triangle can be given, as it is customary to say, an orientation *coherent* with that of the first triangle by orienting the common side in the opposite sense p_3p_2 and continuing then in order with p_2p_4, as indicated in the figure. If this can be done for the entire triangulation in a way that is coherent throughout, the surface is said to be *orientable*; otherwise the surface is said to be *nonorientable*. (It is shown in topology that *one or the other of these properties must hold for all triangulations of the surface*, and that the property is a topological invariant.) In addition, it is important in differential geometry to know that the surface can be shown to be orientable if the Jacobians of the transformations induced by coincidence in the overlapping portions of the patches that cover the surface are all of one sign.

It is now possible to give simple necessary and sufficient conditions that characterize the topological structure of one of the most important classes of surfaces dealt with in differential geometry in the large, i.e., the class of *compact* surfaces referred to earlier. These were defined as the two-dimensional manifolds that could be covered by a finite number of the patches that figure

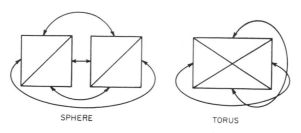

SPHERE   TORUS

Fig. 8.3   Abstract closed surfaces.

in the basic definition of a manifold. It is a fact, not proved here, that this condition is equivalent to the condition that a triangulation of the surface exists that consists of a *finite* number of triangles. It was remarked earlier that the compact surfaces are also commonly called *closed* surfaces, in contrast with *open* surfaces—which cannot be triangulated by a finite number of triangles.

To a closed, or compact, surface it is possible to assign a uniquely defined integer, the *Euler characteristic* $\chi(S) = F - E + V$ in which F, E, and V represent the number of faces, edges, and vertices of an arbitrary triangulation: it is, in fact, one of the more interesting theorems in topology that this number is the same for all triangulations and, in addition, it is a *topological invariant*, i.e., a number which is the same for any pair of compact surfaces that are topologically equivalent. But that still is not all, since the following remarkable theorem holds: *any two compact surfaces that have the same Euler characteristic are topologically equivalent if both are orientable or if both are not orientable.* Thus the value of the Euler characteristic together with the property of being orientable or not are *sufficient* conditions for fixing the topological structure of compact surfaces.

The above discussion evidently implies that prototypes of the various kinds of topologically equivalent abstract compact surfaces can be described in terms of finite collections of triangles in the plane that are joined in such a way that the properties postulated above for a triangulation hold. In this way of constructing examples of abstract surfaces, however, it is necessary to make use of the process of identification of edges in order to produce closed surfaces, in the fashion indicated in Fig. 8.3. The first example shows two replicas of a square, the sides of which are supposed to be joined at congruent points in the manner indicated; this yields a surface that is quite naturally called a *sphere*.   The second example is a rectangle with pairs of its own sides identified in the fashion shown; this yields a *torus*.   In the first case $F = 4$, $E = 6$, $V = 4$, so that $\chi = 2$.   In the second case $F = 2$, $E = 3$, $V = 1$, so that $\chi = 0$.   These two surfaces are therefore not topologically equivalent. (It is easily seen that both are orientable.)

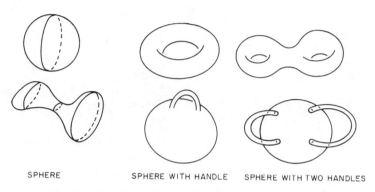

SPHERE   SPHERE WITH HANDLE   SPHERE WITH TWO HANDLES

Fig. 8.4   Topological forms of compact orientable surfaces.

In the special case of the compact, or closed, orientable surfaces it is important and interesting to know that all of them have realizations in E^3 that are topological images of their abstract prototypes, so that the latter might be used conveniently as parameter domains for them.   Such surfaces in E^3 would of necessity be free of double points and lines since such occurrences would mean that the mapping between a surface and its abstract prototype would not be one-to-one.   Such topological images of the abstract surfaces in E^3 are called *simple* surfaces.   All of the possibilities for simple surfaces in E^3 can be explained with reference to Fig. 8.4.   As the figure indicates, these surfaces are either the sphere, or the sphere with n handles.   It is perhaps helpful and of interest to consider how the torus, for example, is realized by deforming in E^3 its abstract prototype.   The steps to be taken are illustrated in Fig. 8.5; the rectangle is first rolled into a cylinder until the side AB coincides with CD, and the cylinder is then bent until the end circles coincide.

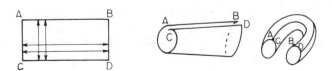

Fig. 8.5   The torus obtained by bending a strip.

One says that the sphere has *genus* g equal to zero, and that the other possible types of closed surfaces have a genus g equal to the number of handles. Thus the torus has genus $g = 1$. Also, it can be verified by using special triangulations that the genus is connected with the Euler characteristic $\chi(s)$ by the formula $\chi(s) = 2(1 - g)$.

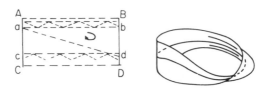

Fig. 8.6   Möbius strip and its orientation.

Although nonorientable surfaces will be dealt with only incidentally later on, it nevertheless is of interest to describe the Möbius strip—the best known example of such a surface. The Möbius strip is obtained from the rectangle by twisting it through a half-turn and then joining the ends so that A falls on D and C on B (cf. Fig. 8.6). It is easily seen that the resulting surface is not orientable. For example, if a clockwise orientation is chosen for the triangle *abd*, then a like orientation of the triangle *adc* would lead to the same direction of the sides *ac* and *bd* once they are made to coincide in the manner prescribed. (The figure also indicates, through the inclusion of a few triangles in the upper part of the strip, that this manifold, which is an open rather than a closed or compact manifold, cannot be triangulated with the use of a finite number of triangles.) It is interesting to observe that if a fly were to creep around the strip once that it would, so to say, pass from one side of the surface to the other without needing to break through it or to cross over its edge—thus it is said that the Möbius strip is a *one-sided* rather than a *two-sided* surface.

It is a general fact—which, like all of the statements being made here concerning the topology of surfaces, will not be proved—that the *simple* closed surfaces in E^3 (i.e., closed surfaces free of double points and self-intersections) are in all cases both orientable and two-sided. In the case of these orientable simple closed surfaces the property of being two-sided is often expressed by saying that the surface encloses an interior region of E^3 and that a passage from this region to the exterior of the surface can be achieved only by passing through the surface. This is, in fact, part of the content of the Jordan-Brouwer theorem that generalizes the Jordan curve theorem in the plane, and which states that a simple closed surface in E^3 is the full boundary of two disconnected open nonempty sets, each of which is connected; one of

them is bounded (and called the interior of the surface) and the other is unbounded (and called the exterior of the surface), such that these two sets plus
their boundary points fill out the whole of E^3.   Furthermore, an orientation of
the surface, now assumed to be a regular surface, which can be achieved by an
appropriate ordering of the local coordinates, leads to a normal vector of the
surface that either points everywhere into the exterior domain or everywhere into the interior domain bounded by the surface.

It would be possible also to give a means for the classification of noncompact, that is to say, open, surfaces, but since there will be no occasion to
use such a classification later on this subject hence will not be discussed here.

It is perhaps worthwhile to mention once more that the word "closed"
is synonymous with the word "compact" when referring to surfaces, and
that, on the other hand *noncompact* and *open* are also synonymous terms.
This is a little unfortunate since a closed, or compact, *surface* is an open, as
well as a closed, *point set*.   In German the language is clearer, since a closed
surface is called a *geschlossene Fläche*, but a closed point set is called an
abgeschlossene Punktmenge.   It is perhaps also worthwhile to observe that a
closed surface has no boundary points and can obviously not be extended to
a larger closed surface.   This is not in general true of open surfaces; for them
an additional condition (which is superfluous for closed surfaces) is usually
imposed in differential geometry in the large (see Section 6 of this chapter).

4.   Surfaces in Three-Dimensional space

In dealing with surfaces in E^3 in earlier chapters the surfaces were defined, with minor exceptions, in the small.   Many problems of differential
geometry in the large are also concerned with surfaces that are embedded in
E^3, and hence it becomes necessary to consider ways and means of dealing
with them in the large.   An abstract surface S is said to be embedded in E^3
if there exists a continuous mapping ϕ of S onto a regular surface S' in E^3,
such that all of the patches in a covering D of S are mapped topologically
by ϕ onto the patches of a covering D' of S', and in such a way that if f is the
mapping of the patch $N \in D$ on the unit ball and f' is the mapping of the
corresponding patch $N' \in D'$ then $f'(\phi(f^{-1}))$ has a nonvanishing Jacobian and
r derivatives.   Each patch of S' is thus given in E^3 by means of a vector
$\mathbf{X}(u, v) = (x_1(u, v), x_2(u, v), x_3(u, v))$ defined in terms of local coordinates
over the unit disk.   Or, put somewhat differently, the following requirements are introduced: If $N'(p')$ is a patch containing a point $p' \subset S'$, in
which (u, v) are any admissible local parameters (which can be regarded as
Cartesian coordinates in a u,v-plane), then $N'(p')$ is assumed to be the set of
points given by a vector

$$\mathbf{X}(u, v) = (x_1(u, v), x_2(u, v), x_3(u, v))$$

in which the mapping $\mathbf{X}$ is one-to-one, and the functions $x_i(u, v)$ are differentiable of a certain order. In addition the rank of the matrix

$$\begin{pmatrix} x_{1u} & x_{2u} & x_{3u} \\ x_{1v} & x_{2v} & x_{3v} \end{pmatrix}$$

should be two. Furthermore, only *isometric embeddings* are considered in this chapter, i.e., those in which the line element of S' as a surface in E^3 (as defined in Chapter IV, and in the preceding Chapter VII) agrees with the line element of S, as defined above.

If the surface S' were merely embedded in three-dimensional space it would acquire a metric from E^3, since the lengths of all curves on it would be determined; that is to say, a line element $ds^2 = \sum_{i,k} g_{ik} \, du_i \, du_k$ would be defined on S'. The functions $g_{ik}(u_1, u_2)$ could then be carried back to the surface S by using the local coordinates, so that S would acquire a Riemannian metric, and $S' \equiv \mathbf{X}(S)$ would be isometric. More often, perhaps, this procedure is reversed; i.e., a metric is defined at the outset in S, and $S' \equiv \mathbf{X}(S)$ is *required* to *exist* (or be in some way determined) so that S' is isometric to S. In all that follows it is assumed, without necessarily saying so explicitly, that S and its image $\mathbf{X}(S)$ are isometric. Thus the local geometry of a surface in E^3 is dealt with in the same way as was done in Chapters IV, V, VI, and VII.

The definition of surfaces $\mathbf{X}(S)$ in the large in E^3 requires that they should be *locally* in one-to-one correspondence with the surface S (which will on occasion be referred to as the *parameter surface*), but it does not require that $\mathbf{X}(S)$ should be in one-to-one correspondence with S *in the large*—in other words $\mathbf{X}$ need not be the topological image of S because of the possible occurrence of multiple points and self-intersections. If it is desired to consider the special case in which $\mathbf{X}$ is the *topological image* of the abstract surface S, then $\mathbf{X}$, as was remarked above, is called a *simple* surface. The possible types of the closed, or compact, simple surfaces were indicated in Fig. 8.4.

5. Abstract Surfaces as Metric Spaces

In studying intrinsic geometry in the large it is convenient to introduce the line element

$$ds^2 = g_{ik} \, du^i \, du^k$$

in terms of functions $g_{ik}(u^1, u^2)$ of the local parameters; in this way a *Riemannian metric* is defined, as was discussed in Section 2.

Experience has shown that it is of basic importance for the purposes of differential geometry in the large to introduce in a particular way a notion of *distance* between all pairs of points p,q of the Riemannian manifold. This is done in a rather natural way by making use of the Riemannian metric. Since the space is assumed to be arcwise connected (assumption (4)′ above), which implies the existence of piecewise regular curves joining all pairs of points, it follows that any such curve has a uniquely defined finite length that can be calculated by using the line element.   A distance function $\rho(p, q)$ is then defined for all pairs of points p, q as *the greatest lower bound of the lengths of all piecewise regular curves in the manifold joining p and q.*   It is clear that such a unique non-negative number ρ exists.   In addition, it is not difficult to verify that the function $\rho(p, q)$ has the usual properties needed in order that a manifold with this function as distance function is what is generally called a *metric space*.   To be shown is that (1) $\rho(p, q) \geq 0$ holds, with $\rho = 0$ only if p and q are the same point, and that (2) the *triangle inequality* $\rho(p, r) + \rho(r, q) \geq \rho(p, q)$ is valid for any points p, q, r.   That property (2) holds is clear from the definition of $\rho(p, q)$.   That $\rho(p, q) > 0$ for $p \neq q$ holds can be seen from the fact proved in Chapter VII (Section 11) that geodetic polar coordinates centered at any point p exist in the small, and that these arcs are the shortest arcs joining p with points in its neighborhood; since p and q are distinct they are contained in disjoint neighborhoods on S since that is a basic property postulated for a manifold, and hence p can be made the center of a geodetic circle, of radius r, say, which excludes q, so that $\rho(p, q)$ is necessarily at least equal to r.

Once the manifold has been converted into a metric space through the introduction of the distance function $\rho(p, q)$, it becomes possible, as was noted earlier, to define *spherical neighborhoods* of a point p of it: a neighborhood of this kind consists, quite naturally, of all points q such that $\rho(p, q) < r$ holds for some positive number r.   Furthermore, it is not hard to show, if a new system of open sets covering S is defined as the set of spherical neighborhoods of S, that the topology of S is unchanged.   To prove this is equivalent to showing that the limit processes are the same for both systems of open sets, or, put differently, that if $p = \lim p_i$ for the original system of open sets then $\lim \rho(p, p_i) = 0$ holds, and vice versa.   This is proved by making use of the properties in the small of the solution curves, emanating from a point, of the geodesic differential equations in the parameter disk.   If the sequence p_i does not converge to p, then infinitely many of the points p_i lie outside of some open set containing p and hence outside of some geodetic circle with sufficiently small radius r, so that $\rho(p_i, p) > r$ for these points p_i; it follows in such a case that $\lim \rho(p_i, p) \neq 0$.   On the other hand, if $p = \lim p_i$, then nearly all points p_i lie within a geodetic circle of arbitrarily small radius, and hence $\lim \rho(p_i, p) = 0$, since the geodetic radii are the arcs of shortest length. Thus the topology of the manifold is not changed by making use of spherical neighborhoods to define open sets.

It has some point to mention a few facts about point sets in metric spaces and to relate them to the corresponding notions introduced above in relation to Hausdorff spaces, since from now on only metric spaces with the metric just now introduced will be considered. For example, as was already indicated, the convergence of a sequence of points p_i to a point p is equivalent to saying that $\rho(p_i, p) \to 0$ as $i \to \infty$. It is also very important to consider the significance of the property of *compactness* in a metric space. A compact set in a Hausdorff space was defined in Section 3 as a set of points which when covered by any system of open sets, is already covered by a suitably chosen finite number of these open sets. (Strictly speaking, the concept was introduced earlier for the whole space rather than for a set contained in the space.) It is easily proved (cf. Patterson [P.1], p. 59) that *a compact set in a Hausdorff space is closed* and that *a compact set in a metric space is bounded* (i.e., all of its points are at a bounded distance from any one of them). Since a metric space is a special type of Hausdorff space it follows that *a compact set in a metric space is both closed and bounded*. In E^n, the converse holds; i.e., a closed and bounded set in E^n is compact.

It is worthwhile to note that the distance function $\rho(x, y)$ is a continuous function of position of the points x, y. The proof is simple. It follows from an application of the triangle inequality:

$$\rho(x, y) \leq \rho(x, x') + \rho(x', y') + \rho(y', y),$$

$$\rho(x', y') \leq \rho(x', x) + \rho(x, y) + \rho(y, y').$$

Appropriate subtractions and use of $\rho(a, b) = \rho(b, a)$ lead to

$$\left| \rho(x, y) - \rho(x', y') \right| \leq \rho(x, x') + \rho(y, y'),$$

and this clearly implies $\rho(x, y) \to \rho(x', y')$ when $x' \to x$ and $y' \to y$.

It is also important for later purposes to have in mind that a *real-valued continuous function defined over a compact set of a metric space takes on its maximum and minimum values at one or more points of the set.*

6. Complete Surfaces and the Existence of Shortest Arcs

Once the distance function $\rho(p, q)$ has been introduced, it becomes possible to deal with what has been described in earlier chapters as the most important single question in differential geometry in the large, i.e., the question of the existence of a shortest curve joining an arbitrary pair of points among all continuous piecewise regular[1] curves joining them, together with

---

[1] It might seem more reasonable to widen the class of curves among which a curve of shortest length should be found to the class of *rectifiable* curves; however, all purposes of this book are served by limiting the discussion to the narrower class. The definition of the metric distance given in the preceding section could also have been based on rectifiable curves, but the resulting distance function would be the same.

the confirmation that such a curve is a geodetic line.   (In particular, such a curve, which has continuous second derivatives, is therefore smoother than is required for the class of competing curves.)   It was, however, pointed out in the previous chapter on the basis of the special example of the Euclidean plane from which the origin had been removed, that such a theorem cannot hold without some restrictive hypothesis, which should at least insure that the manifold is without holes, or boundary points at finite distance from any given point.   (There is no need to rule out possible singular points at finite distance, it might be noted, since the definition of a manifold is such that every point has a neighborhood that is in all respects a regular surface.)   It is to be expected that the class of *compact*, or *closed*, manifolds would serve such a purpose since these manifolds are obviously free of holes and boundary points.   However, this is a narrower class than is necessary or desirable in differential geometry, since there are many interesting questions in differential geometry in the large that are concerned with *open*, that is *noncompact* manifolds.

Experience has proved that a restriction on open manifolds first introduced by M. Morse is quite generally appropriate and useful (i.e., the restriction to manifolds that are *finitely compact*).   This means simply that every infinite point set that is bounded in the metric $\rho(p, q)$ (i.e., is such that all points of it are at a bounded distance from any fixed point) is required to have at least one accumulation point in the manifold; clearly this is indeed a restriction only for open manifolds.   In Euclidean spaces—which are not compact—the Bolzano-Weierstrass principle states that every bounded set has at least one accumulation point.   Thus the assumption of finite compactness in open manifolds postulates in effect that a similar property is made available in such spaces.

It is possible to give a number of different but equivalent requirements for surfaces which may be used to replace the requirement of finite compactness, as has been done by Hopf and Rinow [H.16].   These requirements are

1. Every length can be measured off along every geodesic.

2. Every divergent curve [i.e., the rectifiable image $p(t)$ of a half-open interval $0 \le t \le t_0$ such that any sequence $p(t_n)$ with $t_n \to t_0$ has no accumulation point on the surface] is infinitely long.

3. Every Cauchy sequence [i.e., every sequence p_i such that $\rho(p_m, p_n) \to 0$ as $m, n \to \infty$] converges to a point in the manifold, so that the manifold is said to be *complete*.

4. Finite compactness.

It can be shown without much trouble that each of these requirements implies the validity of the preceding one, but it is not so easily shown that the statement holds in the reverse sense.   For example, while (4) is at once seen to be a more stringent condition than the completeness condition (3), since a Cauchy sequence is bounded, it is mostly not true in metric spaces that (3) implies (4), as the well-known example of the sequence $(1, 0, 0, \ldots)$, $(0, 1, 0, 0, \ldots)$, $(0, 0, 1, 0, 0, \ldots)$, $\ldots$ in Hilbert space shows, since this sequence of points forms a bounded set in a complete metric space with no accumulation point. It is therefore quite interesting that in the special case of surfaces with the metric distance $\rho(p, q)$ defined as above the four conditions just given are equivalent, and thus in particular that completeness implies finite compactness.

In this section it will be shown, following Hopf and Rinow, that the conditions (1), (3), and (4) are equivalent [the proof that (2) is also equivalent to the others is left as an exercise]. It has already been noted that (4) implies (3). [That (3) implies (1) is also readily seen, since if the length L could not be laid off from a fixed point on a geodesic, but every shorter length could, it follows easily that a divergent Cauchy sequence of points on the geodesic would exist.] Consequently, *the equivalence of* (1), (3), *and* (4) *will be proved if it is shown that* (1) *implies* (4). In what follows (and in the literature generally) the class of surfaces singled out by any of these conditions is called the class of *complete surfaces*.

The method used by Hopf and Rinow to prove the equivalence statement hinges on proving the more important theorem already mentioned, i.e., the theorem on the existence of a shortest geodetic arc joining any pair of points, when condition (1) is assumed to be satisfied. Thus two theorems will be proved in this section:

THEOREM I. *If x, y are arbitrary points on a surface satisfying condition* (1) *a geodetic arc joining them exists with a length equal to $\rho(x, y)$, the distance between the two points.*

THEOREM II. *The condition* (4) *follows from the condition* (1), *so that the four conditions are equivalent.*

The proofs of these theorems given below follow that given by de Rham [D.2], which is somewhat simpler in the details than the proof of Hopf and Rinow. The basic idea of de Rham is to work with a set of points G_r on S defined as follows: A fixed point $b \in S$ is chosen and G_r is defined as the set of points consisting of the closed arcs of length r, measured from b, of all geodesics issuing from b. Such a set[1] exists for all positive values r, because of the basic assumption (1).

The first step in the proofs of both theorems is to show that *the set G_r is*

[1] If a closed geodesic happened to exist, it might be noted that this curve would be covered many times in this process if r were to be taken large enough. This possibility, and also the possibilities brought about by geodesics with self-intersections, are irrelevant to what follows.

compact. This property of G_r is an almost immediate consequence of the theorem on the existence and on the continuous dependence of the solutions of the differential equation of the geodesics on variations in their initial directions, as follows.   Let x_n be any sequence in G_r, g_n a set of geodetic arcs joining b with these points, and $l(b, x_n)$ the lengths of the arcs g_n.   A subsequence of initial directions from those fixing the arcs g_n can be chosen that converges to a certain direction, and g is defined as the geodesic going out from b in that direction.   The lengths of a subsequence of arcs, again called g_n, from b to the points x_n, converge to a length l^*, which is laid off from b along g to a point x; that can be done because of condition (1).   Because of the theorem on differential equations just mentioned x is the limit of the sequence x_n, and since l^* is clearly at most equal to r it follows that x belongs to G_r, thus proving G_r to be compact.

The proof of Theorem I that follows is an indirect proof.   Thus it is assumed that a point y exists at distance $R = \rho(b, y)$ from b, which cannot be joined to b by a geodetic arc of length R; a contradiction will be established by using facts of a local character about geodesics together with the compactness of the sets G_r.   It is known (cf. Chapter VII, Section 11) that all points close enough to b can be joined to it by a geodetic arc of shortest length; in fact geodetic polar coordinates are valid in a circle of radius $\sigma > 0$.   Thus the set G_σ is identical in this case with the closed spherical neighborhood S_σ consisting of all points of S at distances $\leq \sigma$ from b.   Evidently, G_r is a subset of S_r for any value $r \geq 0$.   By hypothesis the sets G_R and S_R are not identical, and hence it follows that a least upper bound $d \leq R$ of values of r exists such that $G_r \equiv S_r$ holds for $0 \leq r < d$.   It is now to be shown that the existence of the least upper bound d involves a contradiction by proving that (1) if $G_r \equiv S_r$ holds for $r < d$, it holds for $r = d$, and (2) if $G_r \equiv S_r$ holds for $r = d$ it holds for $r = d + \delta$, with δ a properly chosen positive quantity. Thus d could not exist, and Theorem I would be proved.

To prove the statement (1) it suffices, since G_d is a subset of S_d, to show that a sequence x_n inside G_d can be found that converges to any given point y in S_d at distance d from b, since G_d is compact and hence y lies in it.   Such a set x_n can be defined, for instance, as a set such that the minimum distance from y to the compact sets $G_{d-1/n}$, $n = 1, 2, \ldots$, is obtained for a point x_n; such points exist; they lie in G_d, and they clearly converge to y.

It remains to prove statement (2).   To this end it is to be noted first that for all points $x \in G_d$ a uniform positive lower bound δ exists such that geodetic polar coordinates are valid for circles of radius δ when centered at any point of the compact set G_d.   In de Rham's discussion this fact is deduced from a theorem stating that the radius $\rho(x)$ of geodetic circles within which polar coordinates are valid can be defined as a continuous function of position of their centers x—and hence would have a positive lower bound δ on the compact set G_d. Another way to establish the same result is to observe that a lower estimate for $\delta(x)$ follows from the theory of ordinary differential equations, and from the implicit function theorem, when a quantitative formulation is given for the latter theorem. Such an estimate for the size of

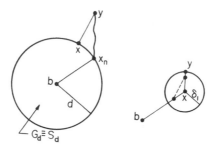

Fig. 8.7 The sets G_d and S_d.

$\delta(x)$ at any point of S depends only upon bounds for the coefficients of the differential equation and certain of their derivatives in a coordinate patch containing the point; this was shown at the end of Chapter IV. Since G_d is compact it can be covered by a finite number of coordinate patches, and hence a uniform lower estimate δ for the radius $r(x)$ of geodetic circles about all points $x \in G_d$ can be obtained.

Consider any point y such that $d < \rho(y, b) \leq d + \delta$ holds, when δ is the bound for the radii of geodetic circles just now introduced. Theorem I will clearly be established once it is shown that any such point lies in $G_{d+\delta}$, since then $G_{d+\delta} = S_{d+\delta}$ would hold for $\delta > 0$. As a first step, it is shown that a point x exists in G_d such that $\rho(b, x) = d$ and with the property that

$$\rho(b, y) = d + \rho(x, y).$$

To this end it is noted first that a so-called minimizing sequence of curves C_n joining b and y exists, such that their lengths l_n satisfy the inequality

$$l_n < \rho(b, y) + \frac{1}{n}, \qquad n = 1, 2, \ldots.$$

This is a direct consequence of the definition of the distance function. On these curves call x_n (see Fig. 8.7) the last points of C_n in G_d upon going from b toward y: these points exist because each curve C_n and the set G_d are compact. Thus $\rho(b, x_n) = d$ and $d + \rho(x_n, y) \leq l_n < \rho(b, y) + 1/n$ holds for all n. Choose, if necessary, a subsequence of the points x_n, again denoted by

x_n, that converges to a point x. Since $\rho(x_n, y) \to \rho(x, y)$ because of the continuity of the distance function it follows upon letting n tend to ∞ that

$$d + \rho(x, y) \le \rho(b, y).$$

However, the triangle inequality states on the other hand that $d + \rho(x, y) \ge \rho(b, y)$ holds; hence $d + \rho(x, y) = \rho(b, y)$ and the property desired for the point x is established. Thus the distance from b to y is the sum of the lengths of two geodetic arcs, one from b to x, the other from x to y, and each of these arcs have lengths equal to the distances between their end points. It follows that the two geodetic arcs do not have an angular point at x. To prove it, note first that a uniform bound δ_1 for the radii of geodetic circles exists for the points of $G_{d+\delta}$—by the same argument as for the case of G_d. On opposite sides of the angular point at x there exists, evidently, a pair of points on the curve under consideration at a distance less than δ_1 apart (see Fig. 8.7), and hence the geodetic arc joining them is unique and is of shortest length—and that means that this arc, if used to cut off the vertex point in the obvious way to form a new curve joining b and y, would have a length less than the distance $\rho(b, y)$—which is impossible. Thus Theorem I is proved.

The proof of Theorem II is almost immediate. Any bounded set of points of S lies in a spherical neighborhood S_r of bounded radius r, hence it lies in G_r, as was just shown, and since G_r is compact the theorem is proved.

It was noted by de Rham that this Theorem II is slightly more general than the equivalence theorem proved by Hopf and Rinow, since it suffices to require for its validity, instead of condition (1), that it should be possible to lay off all lengths on the geodesics going out from any *one* point.

Hopf and Rinow introduced still another class of surfaces which, at first sight, might seem to qualify as surfaces without holes or boundary points that might prevent the existence of shortest geodetic arcs joining all pairs of points. This class of surfaces is characterized by the property of *noncontinuability*, which is defined to mean that a surface S of the class cannot be mapped isometrically and in a one-to-one way on a proper subdomain of a larger surface S'. It is relatively easy to show that surfaces that satisfy condition (1), and hence belong to the class of complete surfaces, also belong to the class of noncontinuable surfaces. Hopf and Rinow go further in proving the additional interesting fact that *the class of noncontinuable surfaces is larger than the class of complete surfaces* by giving an example of a noncontinuable surface on which the theorem concerning the existence of geodetic shortest joins does not hold for all pairs of points—thus showing that this class of surfaces is probably not suitable for investigations in differential geometry in the large. The example is that of the Riemann surface of the many-valued analytic function log z of the complex variable z. On this infinitely sheeted surface with the origin as a branch point (which evidently does not belong to the surface) a Euclidean metric is introduced in the obvious way. Clearly no geodetic join exists for points properly located on a straight line passing through the origin. It is intuitively rather clear that this surface cannot be continued to a still larger surface, but the proof of this fact is not entirely obvious.

Some writers introduce a metric concept in the large for a surface S embedded in the Euclidean space E^3 that is not an intrinsic condition and also is not the same as completeness. The condition is that the surface S should be *closed in* E^3, and by this is meant that if p is a point of E^3 not in S that a neighborhood of p exists that is also not in S (cf., for example, the books of O'Neill [O.1] and Hicks [H.8]. This condition is more restrictive than the condition of completeness. An example in point is furnished by a cylinder erected over a spiral in the plane that tends to the origin in such a way that the length along the spiral in both directions is unbounded. If the generators of the cylinder are infinite straight lines the cylinder has, evidently, a complete metric, but it is not closed in E^3: the points on the line through the origin and parallel to the generators are points not on S, but they have no neighborhood that excludes points of S.

It is of interest to observe that the above two theorems hold for n-dimensional Riemannian manifolds, with proofs that are identical with those given above, once the existence of geodetic polar coordinates in the small has been established for such manifolds.

7. Angle Comparison Theorems for Geodetic Triangles[1]

It is a well-known fact that the interior angles of a spherical triangle are all larger than the corresponding angles of a Euclidean triangle having sides of the same length. A comparison in the same way of two triangles in the Euclidean geometry and the Lobachefsky geometry (i.e., the geometry on a surface with constant negative curvature, as treated in Chapter VII) leads to the result that the angles in the case of the latter geometry are smaller than those in the former. In 1946 a broad generalization of these facts was given by A. D. Alexandrov [A.3], who made use of it in an essential way to solve other problems in differential geometry in the large. This generalization is discussed in the present section.

A simply connected domain D on a two-dimensional surface S with an

---

[1] The inclusion of this section, and its form and content, were the result of suggestions and discussion with H. Karcher, to whom the author is also indebted for criticism and advice with respect to several parts of the present chapter.

arbitrary Riemannian metric is considered, and a distance function $\rho(p, q)$ is introduced, as in Section 5 of this chapter, so that the surface becomes a metric space. It is assumed that the closure $\bar{D}$ of D is simply connected and compact. As in previous sections, a number $\delta < 0$ can be chosen so that geodetic polar coordinates are valid when centered at an arbitrary point of $\bar{D}$ as long as their radii r satisfy the inequality $r \leq 2\delta$. The triangles on S to be considered are defined as follows. Three points $p_i \in D$ are chosen with the restriction that the distances between all pairs of them are $\leq \delta$ in value and that they do not lie on the same geodesic; it follows that the geodetic segments joining the points p_i ($i = 1, 2, 3$) are uniquely determined and their lengths are equal to the distances $\rho(p_i, p_j)$ between them. In addition no pair of these geodetic segments intersects except at a point p_i. Thus a triangle T is defined on S without ambiguity, and it shares with triangles in the Euclidean plane the property that it is a closed Jordan curve. The triangle T defines therefore a simply connected open interior domain that is separated by T from the rest of the simply connected domain $\bar{D}$. (It is perhaps proper to point out that T in what follows is regarded as made up of its three sides, not of them plus the interior.)

The theorem of Alexandrov is formulated with the aid of a certain compact domain $W(T)$ of $\bar{D}$, defined as the union of three closed geodetic circles D_i centred at the points p_i of T and each of radius 2δ [i.e., of all points q such that $\rho(p_i, q) \leq 2\delta$ holds for $i = 1, 2,$ or 3]. The maximum and minimum of the Gaussian curvature K in the compact set $W(T)$ are denoted by K^* and K_*. The complete surfaces of *constant* Gaussian curvatures K^* and K_* are denoted by E^* and E_*: thus E_* is the Euclidean plane if $K_* = 0$, E_* is the sphere of radius $1/\sqrt{K_*}$ if K_* is positive, and E_* is the Lobachefsky plane if K_* is negative.

Triangles T_* and T^* are defined in E_* and E^* *by choosing the lengths of their sides to be the same as those of T.* Such triangles exist for arbitrary lengths of the sides of T in the Euclidean and Lobachefsky geometries; for triangles on the sphere it is always assumed that the sum of the side lengths of T is less than the length of a great circle arc; i.e., that the following inequality holds:

$$\rho(p_1, p_2) + \rho(p_2, p_3) + \rho(p_3, p_1) < \frac{2\pi}{\sqrt{K^*}},$$

This assumption is evidently appropriate, since it is fulfilled for all spherical triangles with interior angles $< \pi$ in value.

Since the triangle T is the boundary of a well-defined interior domain, it is possible to specify its *interior angles* α_i at its vertices p_i in an obvious way. The corresponding angles of T_* and T^* are denoted by α_{i*} and α_i^*. A. D. Alexandrov's theorem then is:

THEOREM.   (a) $\alpha_{i*} \leq \alpha_i$ $(i = 1, 2, 3)$, (b) $\alpha_i \leq \alpha_i^*$ $(i = 1, 2, 3)$, *with equality holding in both cases and for any one index i only if $K = K^*$ holds in the interior of T.*

The proof of the theorem is based on the Sturm-Liouville comparison theorem.   (This theorem will also be used extensively in later sections of this chapter.)   Its statement is:

STURM-LIOUVILLE COMPARISON THEOREM.   *Two second-order ordinary differential equations for functions $y(x)$, $z(x)$ are considered*:

$$y'' + K(x)y = 0,$$

$$z'' + k(x)z = 0,$$

together with the following two different sets of initial conditions:

$$(a) \begin{cases} y(0) = z(0) = a > 0, \\ y'(0) = z'(0), \end{cases} \qquad (b) \begin{cases} y(0) = z(0) = 0, \\ y'(0) = z'(0) = b > 0. \end{cases}$$

It is assumed that $K(x) \geq k(x)$ holds.   The conclusion of the theorem refers to non-negative solutions of the differential equations satisfying either of the conditions (a), (b) and is:

If x_0 is the first zero of $y(x)$ to the right of the origin, then $z(x) \geq 0$ and $y(x) \leq z(x)$, hold in the interval $0 < x \leq x_0$.   In addition, equality holds at any point ξ in the interval only if $K(x) \equiv k(x)$ in $0 < x < \xi$.

Since a short and very simple proof of the theorem can be given, and since the theorem is used in important ways later on, the proof is given here. The first differential equation is multiplied by $z(x)$, the second by $y(x)$, and the second equation is subtracted from the first with the result

$$y''z + y'z' - y'z' - yz'' + (K(x) - k(x))zy = 0,$$

after adding and subtracting the terms $y'z'$.   Integration of the equation from 0 to ξ, $0 < \xi \leq x_0$ leads to

$$y'z - yz' \Big|_0^\xi + \int_0^\xi (K(x) - k(x))zy \, dx = 0.$$

The contribution at $\xi = 0$ of the first term vanishes for either of the initial condition (a) or (b).   Hence

$$y'z - yz' \Big|^\xi = \int_0^\xi (k(x) - K(x))zy \, dx.$$

Since $(d/dx)(y/z) = (y'z - yz')/z^2$ is valid for $\xi \neq 0$ since z is positive, it follows that $(d/dx)(y/z) \leq 0$ holds for $\xi \neq 0$ because $K - k > 0$ is assumed and

y and z are both assumed to be positive.   Thus y/z is a nonincreasing function of ξ.   In either of the cases (a) or (b) it is clear that $\lim_{\xi \to 0} y/z = 1$, and it therefore follows that $z \geq y$ holds in general for $\xi \neq 0$, with equality for any value of ξ less than x_0 only if $K(x)$ and $k(x)$ are identical for $0 \leq x \leq \xi$.

It suffices to prove Alexandrov's theorem for the case $i = 1$, since no vertex of the triangle is distinguished from the others in the statement of the theorem.   The proof begins by considering the closed geodetic circle D_1 of radius 2δ on S that is centered at p_1.   No point of the triangle T is at a distance greater than $\frac{3}{2}\delta$ from p_1, since its perimeter is at most equal to 3δ: this follows because the sides of T have at most the length δ.   The geodetic radii of length 2δ which go out from p_1 in the sector defined by the interior angle α_1 must therefore leave the interior of T by crossing the side p_2p_3 of T, since they can have no points other than p_1 in common with the sides p_1p_2, p_1p_3. The geodetic radii cut the side p_2p_3 in exactly one point since these points lie in the geodetic circle of radius 2δ about p_1; it follows that the side p_2p_3 has a representation in geodetic polar coordinates of the form $r = r(\theta)$, $\theta_2 \leq \theta \leq \theta_3$, with $\theta_3 - \theta_2 = \alpha_1$, the interior angle of T at p_1. It is then clear that the interior angles α_2 and α_3 must have values that are less than π; hence α_1 also has that property, since the above discussion could just as well have been carried out with respect to p_2 and α_2.

Geodetic polar coordinates are introduced on the surfaces E_* and E^* of constant curvatures K_* and K^*, with centers at the points p_{1*} and p_1^* (which, of course, can be chosen arbitrarily because of the homogeneity of the spaces E_* and E^*).   Denote by p_{2*}, p_{3*}, and p_2^*, p_3^* the points in E_* and E^* with polar coordinates $(r(\theta_2), \theta_2)$, $(r(\theta_3), \theta_3)$ in those spaces that are the same as those of the points p_2 and p_3 of T.   Thus the triangles $p_{1*}p_{2*}p_{3*}$ and $p_1^*p_2^*p_3^*$, called T_{**} and T^{**}, have two sides and the included angle (of value α_1) the same as those of T.

The statement (b) of Alexandrov's theorem is proved first.   To this end consider the curve C^* in E^* defined by $r = r(\theta)$, $\theta_2 \leq \theta \leq \theta_3$ that joins p_2^* and p_3^*, with $r = r(\theta)$ the same function of θ that serves to determine the side p_2p_3 of T on S.   The length $l(C^*)$ of C^* is of course not less than the distance $\rho^*(p_2^*, p_3^*)$ between the points p_2^*, p_3^* on E^*.   The line element in polar coordinates on S has the form (cf. Section 11 of Chapter VII)

$$ds^2 = dr^2 + G^2(r, \theta) \, d\theta^2,$$

with $G(0, \theta) = 0$, $G_r(0, \theta) = 1$, and the *theorema egregium* of Gauss has in these coordinates the form

$$G_{rr} + KG = 0.$$

In E^* the line element is defined in terms of the coefficients $G^*(r, \theta)$, and it is of course a solution of the differential equation

$$G^*_{rr} + K^*G^* = 0$$

with the same initial conditions as for G. Thus the initial conditions corresponding to case (b) in the Sturm-Liouville theorem apply, and that theorem therefore yields the result that $G(r, \theta) \geq G^*(r, \theta)$ holds, since $K^* \geq K$ is valid, with equality for a given value of θ only if $K = K^*$ holds along the geodetic radius fixed by θ. (Used here also is the fact that G^* is positive; this always holds because of the restriction that r is less than $\pi/\sqrt{K^*}$ in case K^* should happen to be positive, and is obviously true when $K^* \leq 0$ holds; thus G is always positive in the interior of T.) The following inequalities therefore are valid:

$$\rho(p_2, p_3) = \int_{\theta_2}^{\theta_3} \left[\left(\frac{dr}{d\theta}\right)^2 + G^2(r, \theta)\right]^{1/2} d\theta \geq \int_{\theta_2}^{\theta_3} \left[\left(\frac{dr}{d\theta}\right)^2 + G^{*2}(r, \theta)\right]^{1/2} d\theta$$

$$= l(C^*) \geq \rho^*(p_2^*, p_3^*),$$

since ρ^* is at most equal to the length of C^*. Since $G \geq G^*$ holds for every value of θ in the range $\theta_2 \leq \theta \leq \theta_3$, the form of the two integrals shows that they can be equal only if G and G^* are identical along the curve $r = r(\theta)$, and that in turn requires, as was just seen, that K^* and K must then be identical throughout the interior of T. Otherwise the following inequality holds:

$$\rho(p_2, p_3) > \rho^*(p_2^*, p_3^*).$$

This states that the triangle T has the side opposite the vertex p_1 of greater length than the corresponding side of the triangle T^{**}. The "law of cosines" for triangles in each of the three geometries of constant curvature is

$$\cos a = \cos b \cos c + \sin b \sin c \cos \alpha \qquad (K^* > 0),$$

$$a^2 = b^2 + c^2 - 2bc \cos \alpha \qquad (K^* = 0),$$

$$\cosh a = \cosh b \cosh c - \sinh b \sinh c \cos \alpha \qquad (K^* < 0).$$

The triangle T^{**} has sides of length b and c that include the angle α_1; clearly, if the opposite side of length a were to be increased the angle α_1 would increase. Since T^{**} has this side shorter than the corresponding side of T^* (unless $K \equiv K^*$ holds), it follows that the angle α_1^* of the triangle T^* is greater than α_1. Thus case (b) is disposed of.

The case (a) is disposed of by reversing the roles of S and E_*, although there is a slight additional remark, not necessary in case (b), that must be made. Thus the side $p_{2*}p_{3*}$ of T_{**} in E_* is assumed given in polar coordinates by $r = r_*(\theta)$, $\theta_2 \leq \theta \leq \theta_3$, $\theta_3 - \theta_2 = \alpha_1$. Since $r(\theta_2)$ and $r(\theta_3)$

have lengths $< \delta$, it follows that the triangle $p_{1*}p_{2*}p_{3*}$ exists in any of the three geometries, including that of the sphere, since in the latter case the triangle lies on a hemisphere because of the assumption on δ made in that case; hence the representation $r = r_*(\theta)$ of $p_{2*}p_{3*}$ exists. It is thus seen that the distance from p_{1*} to all points of the side $p_{2*}p_{3*}$ is less than 2δ, i.e., that $r_*(\theta) < 2\delta$ holds simply because T_{**} lies on a hemisphere bounded by a great circle of radius $< 2\delta$. It then follows that a regular curve joining p_2 with p_3 on S is defined by giving it the same representation $r = r_*(\theta)$—since these points lie in a geodetic circle about p_1 of radius $< 2\delta$. From this point on the argument goes in the same way as for case (b):

$$\rho_*(p_{2*}, p_{3*}) = \int_{\theta_2}^{\theta_3} \left[\left(\frac{dr_*}{d\theta} \right)^2 + G_*^2(r_*, \theta) \right]^{1/2} d\theta$$

$$\geq \int_{\theta_2}^{\theta_3} \left[\left(\frac{dr_*}{d\theta} \right)^2 + G^2(r_*, \theta) \right]^{1/2} d\theta \geq \rho(p_2, p_3),$$

again because of the Sturm-Liouville theorem. Thus $\rho_* > \rho$ holds unless $K \equiv K_*$ holds in the interior of T. Thus the side of T_{**} opposite p_{1*} has a length that is not less than the length of the corresponding side of T_*, and hence that $\alpha_{1*} < \alpha_1$ holds unless $K \equiv K_*$ holds in the interior of T. Thus case (a) is disposed of and Alexandrov's theorem is proved.

8. Geodetically Convex Domains

It has been very useful to know that to every point of a surface a geodetic circle exists centered at the point with the property that the geodetic radius is the shortest connection between it and all others in the circle. However, it was not proved that such a circle is *geodetically convex*, i.e., that *all* pairs of its points can be joined by an arc of shortest length that lies in the circle. In fact, such a statement is not in general true. For example, consider geodetic circles centered at a point p on a unit sphere: they have the basic property, whatever their radii may be, that p is joined to all points of any such circle by a shortest geodesic. However, if the radius of the geodetic circle is greater than $\pi/2$, it is clear that the geodetic circle is not geodetically convex. In fact, an exercise at the end of this chapter requires a proof of the fact that a geodetically convex domain on the sphere is either the whole sphere or it lies on a hemisphere.

Some further concrete examples are of interest, especially when they are compared with analogous cases in the Euclidean plane. Consider, as one such example, a lune on the unit sphere, i.e., the simply connected closed domain bounded by two great circle arcs having the same pair of antipodal points as end points.   Such a domain is geodetically convex if the two vertex angles are less than π, but the shortest connecting geodesic is not unique with respect to the two vertices.   If the vertex angle is π the domain is still convex (it is, evidently, a hemisphere) and all antipodal points on the boundary have nonunique shortest connections.   Thus, a geodetically convex region need not be such that the shortest connecting segments are *uniquely* determined. In the plane, of course, the shortest connection, being a straight line, is of necessity uniquely determined.   Such domains on surfaces also need not be simply connected, as they are of necessity in Euclidean spaces; a counter-example is furnished by the domain between two circular cross sections of a circular cylinder.

It is therefore clear that the discussion of geodetically convex domains on curved surfaces must involve more complexities than are encountered in the plane.   In this section some special cases are discussed, including a proof that *sufficiently small* geodetic circles are geodetically convex with *unique* geodetic connecting segments, as well as a proof that all geodetic triangles, if not too large, also have these properties.   These two theorems are corollaries of a general theorem, which is an exact analogue of the theorem of Erhard Schmidt that was treated in Chapter II, except that the domains to be identi-fied as convex must of necessity be restricted in size.

The domains considered in this section are open connected domains on surfaces provided with a complete metric.   The latter assumption is con-venient even when dealing with small domains, although the requirement of completeness could in such cases be dispensed with in all of the proofs by making rather obvious modifications.   Evidently, any complete surface as a whole is an example of a geodetically convex domain, since it was proved in Section 6 that all pairs of points on such a surface can be joined by a geodetic arc of shortest length lying in the surface—though not necessarily a uniquely determined one.   The definition of convexity employed here does not require a priori that the shortest joins be unique, although it is evidently of interest to know whether or not such a property holds in a given case.

The discussion begins with an arbitrary closed circular neighborhood N of radius R about a point p of a complete surface S: that is, N consists of all points q at distance $\rho(p, q) \leq R$ in the standard metric introduced in Section 5.   Since N is compact it follows, as in previous discussions, that a positive lower bound r exists for the radii of geodetic circles centered at all points of N within which geodetic polar coordinates are valid.   (If S happened to be compact, it would of course be appropriate to replace N by S and fix r accordingly.)   A geodetic circle C of radius $r/3$ is taken with any point p of N as its center; *the basic theorem to be proved refers to domains D lying in such a circle C.*   These domains are required to be open connected domains such that a *local supporting geodetic line exists at every boundary point.*   By this last property is meant, as in the case of the plane domains treated in Chapter II, that *a geodetic line exists through every boundary point of D such that all points of the domain in a neighborhood of the boundary point lie on one side of the geodetic line.* The analogue of Erhard Schmidt's theorem[1] then is:

THEOREM. *All such domains D are geodetically convex, with uniquely determined geodetic segments connecting all pairs of points.*

Proof. The first step in the proof is the observation that a geodetic circle of radius r centered at any point of C—the circle containing any domain D—contains all of C in its interior, since the radius of C is, by assumption, $r/3$. Thus every point of C can be joined to every other point by a uniquely determined shortest geodesic, which, however, need not be entirely in C. But once these facts are known, E. Schmidt's proof of convexity given in Chapter II can be carried out in exactly the same fashion as it was there for the plane. For the sake of completeness, it is repeated here; it is in any case quite brief. Since D is connected, it follows that any pair p,q of its points can be connected by a geodetic polygon in D, consisting say of the points $p, p_1, p_2, \ldots, p_{n-1}, q$ in sequence. Such a construction can be carried out in the following rather obvious way: p and q are connected first by a continuous closed arc in D, which in turn can be covered (by the Heine-Borel theorem) by a finite number of small overlapping geodetic circles of radius $\leq r/3$ all of which lie in D, after which the points p_i can be selected in an obvious way. If it were possible to join p with p_2 by a geodetic segment lying in D, then the vertex p_1 could be removed; the idea of E. Schmidt's proof is to show that all of the vertices between p and q can be removed one by one so that finally the segment pq is found to lie in D. To this end, suppose that all vertices up to p_i ($i = 1$ not excluded) could be removed, but p_{i+1} not. Consider the segment $p_i p_{i+1}$. Near enough to p_i the points of this segment can be joined to p by segments lying in D since D is open, the segment pp_i lies in it, and p is the center of a geodetic polar coordinate system valid for radii larger than $\rho(p, p_i)$. Since this is not the case, by hypothesis, for points on $p_i p_{i+1}$ all the

---

[1] J. H. C. Whitehead [W.5] proved the existence of certain types of geodetically convex domains contained in all sufficiently small neighborhoods of a given point. His method is different from the method employed here, and he also deals with n-dimensional spaces that are not necessarily provided with a Riemannian metric. The results here, though for much less general manifolds, permit more detailed statements to be made, and the proofs are simpler. Also, Whitehead limits his discussion to domains with quite smooth boundaries, and that is a rather unnatural limitation to impose when dealing with convex sets. A recent thesis of Karcher [K.1] deals in much greater generality with convex domains on surfaces than is done here; somewhat different proofs of a good many of the results of this section are given, but in a much more sophisticated context that includes a detailed discussion of such things as the locus of the *conjugate points* (and also of the so-called *cut locus points*) of a set of points.

way up to p_{i+1}, it follows that a first point p^* exists on going out from p_i along the segment $p_i p_{i+1}$ such that the segment pp^* contains a boundary point b of D; all geodetic segments drawn from p to points between p_i and p^* on $p_i p_{i+1}$ thus lie in D. This holds, of course, because the boundary of D is a compact set. Consider the geodetic triangle pp_ip^*. All points in its interior belong to D, and also points on the segment pp^* near enough to the point p, which lies in the interior of D. It follows that a first boundary point b^* of D must occur on the segment pp^* on going out from p along it. But now it is readily seen that the local support property could not hold at b^* since a geodetic segment along pp^* at b^* contains interior points (i.e., those on the side toward p), and any other segment, however short, penetrates into the interior of the triangle pp_ip^* which is made up entirely of inner points. Thus the vertex p_i can, after all, be removed. It follows that D is geodetically convex. The connecting geodetic segments in D are also unique since that is true for all pairs of points in the circle C of radius $r/3$ that contains D. Thus the theorem is proved.

The convexity of the open set D was proved above, but it is easy to show that D plus its boundary is also convex. This is a special case of a general theorem, left as a problem at the end of the chapter, according to which the closure of a sufficiently small convex set is convex.

The theorem has a number of corollaries. The first of them refers to *geodetic triangles* contained in a circle C. A geodetic triangle is defined by giving any three points a, b, c in C that are not "collinear," that is, not on the same geodesic, and joining these vertices pairwise by geodetic segments. It was shown in the previous section that such a triangle is the full boundary of a simply connected interior domain, with interior angles that are less than π in value. The theorem just proved shows therefore that the following corollary is valid, since the local support property holds at all boundary points:

COROLLARY 1.   *Every geodetic triangle T (regarded here as a closed point set) contained in a circle C of radius $r/3$ is geodetically convex.*

If it had been proved earlier that geodetic semicircles of radius r are convex, then the convexity of T would already result from the fact that it could be defined as the intersection of three properly chosen semicircles each containing a side of the triangle in a bounding diameter. Such a statement is not generally true; however, it is possible to show that *sufficiently small* geodetic circles are geodetically convex:

COROLLARY 2.   *Any sufficiently small geodetic circle $\sum$ inside C is geodetically convex.*

The proof is carried out by showing that the radius can be chosen small enough so that E. Schmidt's theorem can be applied, and this in turn requires a proof that the local support property holds on the circumference of the circle. To do so a calculation is made, as follows. The line element in $\sum$ has the form $ds^2 = dr^2 + G\, d\theta^2$ in geodetic polar coordinates (r, θ). It is known from Chapter VII, Section 13, that G has the development

$$G = r^2 - \frac{K}{3} r^4 + \cdots$$

valid near $r = 0$; the quantity K is the value of the Gaussian curvature at $r = 0$. It is also known that the geodetic curvature κ_g of a geodetic circle is given (cf. Section 11 of Chapter VII) in general by the formula

$$\kappa_g = \frac{(\sqrt{G})_r}{\sqrt{G}} = -\frac{1}{2} \frac{G_r}{G}.$$

Near $r = 0$, therefore, κ_g has the development

$$\kappa_g = -\frac{1}{r} + \cdots.$$

In effect, this gives the expected result that the curvature of small geodetic circles is approximately the reciprocal of the radius, as it is for Euclidean circles of any radius. This means that if r is small enough the geodetic circle has the local support property. In fact, if the circle is oriented positively, its points lie always to the left of the positively oriented geodetic tangent line in a neighborhood of the point of tangency, since r is positive: this is seen with reference to the definition of κ_g as $d\theta(s)/ds$, with s meaning arc length along the circle and $\theta(s)$ the angle through which the oriented tangent vector must be turned in order to bring it into coincidence with the vector of a parallel field defined along the circle; the negative sign obtained for κ_g is thus what is needed. In effect, the situation locally is qualitatively the same as it is in the Euclidean plane. (In fact, the easiest way to establish this result would be to consider the surface as embedded in E^3, and project it orthogonally on a tangent plane at a point of the geodetic circle C_j, in which case C_j projects locally into a plane curve with curvature κ equal to κ_g.)

Another corollary is of interest, since it gives, in a sense, a converse of the main theorem that is analogous to the familiar situation with respect to convex domains in Euclidean spaces, and which can be proved with the aid of Corollaries 1 and 2.

COROLLARY 3. *The local support property holds at all boundary points of any geodetically convex domain.*

This corollary differs in an important respect from the basic theorem and its corollaries 1 and 2 in that the size of the domain is not in question. To prove Corollary 3, consider a boundary point b of any such domain D (regarded as a closed set), and a small geodetically convex circle C centered at b: such circles exist, as was just seen. Take any pair of points p,q distinct from b in the intersection of D and C; the geodetic segment pq lies in D and C, and hence in their intersection, since both are convex and pq is unique. This means that the sector V of C formed by drawing geodetic radii from b through all points of D that lie in C, and then taking the closure of this set, is geodetically convex. (It has already been remarked that a proof of this fact is required in one of the problems formulated at the end of the chapter.) If the sector V is a semicircle, its bounding diameter is evidently a supporting geodesic of V and hence also a local supporting geodesic of D since V contains all points of D in a neighborhood of b. If V is not a semicircle, it is clear that a small geodetic triangle with b as vertex exists that contains all points of D in a neighborhood of b: for that, it suffices to select as vertices a point near to b on each of the boundary rays of the sector V. Since a small triangle with b and these two points as vertices is convex, as was proved in Corollary 1, it follows that a local supporting geodesic exists for the triangle at b, since the vertex angle is less than π, and hence a supporting geodesic for D also exists.

It is possible to make more detailed statements about geodetically convex domains if the Gaussian curvature K is restricted to have a fixed sign at all points of a complete surface containing such a domain. For example, the following corollary holds:

COROLLARY 4. *A simply connected complete surface S of nonpositive Gaussian curvature is considered. All geodetic circles* (these exist with centers at any point of S, and with arbitrary radii) *are convex, with unique geodetic joins between all pairs of points.*

The proof of Corollary 4 is given in Section 15; it requires a discussion involving Jacobi's theory concerning conjugate points, and the existence of geodetic coordinates in the large.

On surfaces with positive Gaussian curvature matters are very much more complex. For compact surfaces with $K > 0$ a discussion and a limitation for the size of convex geodetic circles is also given in Section 15; in the case of the unit sphere geodetic polar coordinates, which are valid for the whole sphere except for the diametrically opposite point to the center of the coordinate system, obviously give rise to convex geodetic circles only if their radii are at most $\pi/2$ in length.

Finally, this section is closed by proving that *every compact surface S can be triangulated by convex geodetic triangles*: a fact that will be used in the next section. To this end let $\delta > 0$ again be a constant such that geodetic polar coordinates with radii less than 2δ are valid when centered at any point of S. From Corollary 1 it follows that every point p of S lies in the interior of a convex geodetic triangle. Such a triangle can be constructed by taking three points on geodetic radii going out from p at $120°$ apart at a distance $\delta/3$, say, from p. By the Heine-Borel theorem a finite number of such open geodetic triangles can be selected such that S is covered by them. There would be in general many overlappings of such triangles. It is to be shown that the triangles can be subdivided in such a way that a triangulation of S results having the properties required for a triangulation of a manifold, i.e., that two triangles of the subdivision are either distinct, or have one vertex, or one side, in common. The proof proceeds by induction. It is assumed that the union of k triangles T_i, $i = 1, 2, \ldots, k$, of the covering has been subdivided so that the triangulation condition is satisfied. This assumption holds for $k = 1$. Add one more triangle T_{k+1} from the covering. Any parts of T_{k+1} that lie in the union $\bigcup T_i$ of the triangles T_i can be removed. The part of T_{k+1} not in $\bigcup T_i$ is a union of a finite number of polygons P_j. However, geodetic segments in T_{k+1} behave for present purposes in the same way as straight-line segments in a Euclidean plane. Thus the same arguments that prove that polygons in the Euclidean plane can be triangulated could be used for the polygons P_j (this is not carried out here, but a proof can be found in Knopp[K.3], p. 17). Finally, all triangles that have vertices of other triangles on their edges must be again subdivided in an obvious way by chords starting from their vertices. This completes the construction and establishes the desired result.

9.   The Gauss-Bonnet Formula Applied to Closed Surfaces

Once the facts about the topology of closed two-dimensional surfaces are known it becomes possible to make use of them in the study of a variety of highly interesting problems in differential geometry in the large. This section is concerned with a study based on the Gauss-Bonnet formula as applied to each of the triangles of a triangulation of a closed surface S, followed by a summation over all of the triangles. In Chapter VII the Gauss-Bonnet formula was derived for a triangle; it has the form

$$(8.4) \qquad \iint_R K \, dA + \oint_C \kappa_g \, ds = 2\pi - \sum_{v=1}^{3} (\pi - \beta_v).$$

Here κ_g is the geodetic curvature of each of the regular arcs bounding the triangle, and the quantities β_v represent the interior angles of the triangle. In the previous section it was shown that a triangulation of a compact, or closed, surface can be achieved with the use of small geodetic triangles each of which is covered by a valid local coordinate system—as was assumed in Chapter VII in deriving (8.4). Thus the line integrals $\oint_C \kappa_g \, ds$ may be assumed to vanish since $\kappa_g = 0$ on the boundary arcs. If all of the equations (8.4) are summed for all triangles that subdivide the surface it is to be shown that the result can be expressed as follows:

$$(8.5) \qquad \iint_S K \, dA = 2\pi(F - E + V)$$

$$= 2\pi \, \chi(S),$$

in which F, E, and V represent the number of faces, edges, and vertices that occur in the triangulation. The term $2\pi F$ obviously results from summing the numbers 2π on the right-hand side of (8.4). The summation of the quantities $\sum_{v=1}^{3} (\pi)$ leads to the number $2\pi E$ since each edge occurs twice, while the sum $\sum_{v=1}^{3} B_v$ of all the interior angles is $2\pi V$ because of the fact that their sum at any one vertex point is 2π. Thus the integral $\iint_S K \, dA$, often called the *curvatura integra or total curvature* κ_T, has the value $2\pi\chi(S)$ in which $\chi(S)$ is the Euler characteristic of the surface. Since, as was explained earlier, the number $\chi(A)$ is a topological invariant, this establishes the important theorem that *the curvatura integra is a topological invariant of closed surfaces that has the value* $K_T = 2\pi \, \chi(S) = 4\pi(1 - g)$, *in which g is the genus (equal to the number of handles) of S*. Thus it has been seen progressively that the Gaussian curvature, defined first as an invariant under orthogonal transformation in E^3, then identified as an isometric invariant, now is seen to yield a topological invariant when it is integrated over a closed surface. (It might be noted also that the proof of the theorem holds without reference to the orientability of the surfaces.)

Though, as was remarked earlier, it is not thought important in this chapter to dwell on the point, it is clear that K, dA, κ_g, etc., can be, and should all be defined for an abstract surface so as to be invariant under coordinate transformations, since the surface is defined with the use of overlapping patches, each with its own coordinate system. Only then can (8.4) and (8.5) be regarded as valid for the abstract closed surface.

The theorem in inner differential geometry in the large expressed by (8.5) has a number of important corollaries and applications. First of all, it shows that it is not possible to define a Riemannian metric in a perfectly arbitrary way (apart from smoothness properties and positive definite character of the line element) on closed surfaces. The reason is that the metric fixes the Gaussian curvature K (that is what the famous theorem of Gauss states) and is subject to the integral relation just derived. Thus, on a surface homeomorphic to a sphere it is not possible to have $K \leq 0$ everywhere, since from (8.5) $K_T = 4\pi$ because $\chi(S) = 2$. On a surface homeomorphic to the torus, for which $K_T = 0$ it is clear that K may not be of one sign.   It may, however, be identically zero.   This can be seen as follows.   Consider a rectangle in a Euclidean α,β-plane, with its edges identified in the manner indicated by Fig. 8.5.   This abstract model of a torus is, by definition, given the metric $ds^2 = d\alpha^2 + d\beta^2$, which is of course Euclidean, and hence $K \equiv 0$ holds.   However, this torus could not be embedded isometrically in E^3 since any regular closed surface in E^3 must have points where K is positive, as was noted in Chapter IV, and is rather easily seen in any case.   Another example of interest is furnished by the closed orientable surfaces on which the Gaussian curvature is everywhere positive; it follows that their Euler characteristic is positive, and hence the only possibility is that these surfaces are topologically equivalent to the sphere.   This same fact will be established later on, in Section 12, as a corollary of a theorem of Synge.

10.   Vector Fields on Surfaces and Their Singularities

The introduction of the notion of an abstract manifold and the discussion of the topology of the two-dimensional manifolds which followed, might be regarded as already amply justified as preliminaries for the study of differential geometry in the large by the above results concerning the total curvature and the insight gained concerning the possibility of introducing metrics in the large in a given closed, or compact, two-dimensional manifold.   However, another more primitive reason has already been given indicating why it is necessary to consider abstract manifolds to be a system of patches joined together properly; and that is, that it is rarely possible to introduce a single curvilinear coordinate system over a whole manifold without a singularity. In fact, it will be proved here that the only one of the orientable closed two-dimensional surfaces for which that is possible is the torus: here the "meridians" and their orthogonal trajectories form such a system for a torus of revolution, for example.   Consider, as another case in point, the sphere in E^3.   The commonest coordinate system used on it in practice is the curvilinear system consisting of all great circles through the north and south poles together with the circles orthogonal to them; evidently this system of coordinates has the poles as singular points.   In fact, a theorem will be proved in this section which says something precise about the totality of singularities that may occur in coordinate systems on a closed surface (at least in case there are not infinitely many of them) and puts them into relation with the genus of the surface.

The tools which make such a study possible—in particular, the concepts needed to define what is meant by a singularity of a coordinate system—are the notion of a differentiable vector field on the surface and of the *index of an isolated singularity* of a vector field.   The study of such vector fields and their singularities is of interest in itself—in the theory of vibrating systems, for example, and more generally in analytical dynamics—quite aside from the application to the special problem in view here.

It is convenient to begin the discussion with a special case, i.e., that of an open connected domain in the Euclidean plane in which two functions $P(x, y)$, $Q(x, y)$ with continuous first derivatives are defined.   These functions in turn define a vector field (P, Q) in the plane with integral curves that are solutions of the ordinary differential equations

$$(8.6) \qquad \begin{cases} \dfrac{dx}{dt} = P(x, y), \\[2mm] \dfrac{dy}{dt} = Q(x, y). \end{cases}$$

From the theory of ordinary differential equations it follows that these equations have a uniquely determined solution curve C: $x(t)$, $y(t)$ once a point $x(0)$, $y(0)$ is chosen as initial point.   From the point of view of the theory of differential equations there is thus nothing of a singular nature involved. From the geometrical point of view, however, it may well be that singular points occur where P and Q vanish simultaneously: that is, at points where no field direction is defined.   Without loss of generality such a point may be assumed to be at the origin, so that $P(0, 0) = Q(0, 0) = 0$ holds.   In that case, the uniquely determined solution of the system of differential equations, with the origin as initial point is $x = 0$, $y = 0$ for all t—and this solution is even analytic.   However, in such a case there is in general no unique *regular curve* in the sense of differential geometry through the origin that satisfies the differential equations.   For example, the pairs of differential equations $dx/dt = x$, $dy/dt = y$ and $dx/dt = y$, $dy/dt = -x$, which have the required property at the origin, have the solution curves shown in Fig. 8.8.   In the first case there are infinitely many solution curves that go into the singular point, in the second case none at all.

By definition, *a singular point of the vector field is an isolated point where $P = Q = 0$ and such that the vector field cannot be extended by continuity to this point.*   An integer j, called the *index* of the singularity, is now defined at such a point p in the following way:

1. Let C be a continuous simple closed curve containing p in its interior, but no other singular points, and such that no singularity occurs on C.

2. Consider the field of vectors on $C = C(t)$, $0 \le t \le 1$, with the parameter t so chosen that C is traversed once in the positive sense when t ranges over the interval.   Define $\theta(t)$ as the angle that the field vector makes with the positive x-axis, and in such a way that $\theta(t)$ is continuous.   (This can be

done using the Heine-Borel theorem in the same way as was done under similar circumstances in Chapter II.)   Call δ_C the total change in $\theta(t)$ in traversing C once in the positive direction—this number clearly would be the same for all starting points, so that δ_C can be defined uniquely as $\theta(1) - \theta(0)$. *The index j is then defined by the equation*

$$j = \frac{1}{2\pi}\,\delta_C.$$

It will now be shown that *j is an integer* (positive, negative, or zero) *that is independent of the special curve C which contains p in its interior—thus making it possible to assign the index j to p without ambiguity.*   That j is an integer follows from the fact that the field vector returns to the same position after a

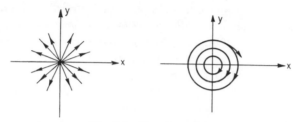

Fig. 8.8   *Singular points.*

circuit around C, and hence the number δ_C is an integer times 2π.   That j is independent of the special curve C used to define it follows from the fact that any two such curves could be deformed continuously into each other, and since $\theta(t)$ would vary continuously with C it follows that j would remain constant since its possible values are restricted to the integers.

It is useful to observe that the singular point in the two cases shown in Fig. 8.8 has the index $+1$, thus showing quite graphically that the value of the index is far from giving complete information about the nature of the field in its vicinity.   In Fig. 8.9 three examples are given in which j has the

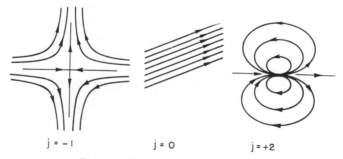

j = -1   j = 0   j = +2

Fig. 8.9   *Further types of singularities.*

values -1, 0, and $+2$. In particular, it is to be noted that $j = 0$ holds at a point that is not a singular point.

It is now in order to consider vector fields on a surface with a Riemannian metric, and to define the index of an isolated singularity on such a surface. The vector field is determined by means of the differential equations $du/d\tau = P(u, v)$, $dv/d\tau = Q(u, v)$, defined in the coordinate patches. The solution curves provide a unique system of curves in the surface; the vector field then is made up of tangent vectors to these curves. In Section 2 above it was explained how such vector fields are defined in an invariant way. (In the next chapter it will be seen that the pair of functions P,Q should transform in such a way when the surface parameters are changed that a covariant vector field results.) To determine the index of an isolated singularity at a point p consider again a continuous simple closed curve $C(t)$ encircling it, and measure the angle change δ_C *with respect to the field of vectors defined by the tangent vectors of one set of a pair of local parameter curves*, and set $j = (1/2\pi)$ δ_C as before. (The field of tangent vectors to a set of parameter curves is of course a vector field without a singularity since that is implied by the fact that the surface is assumed to be regular.) $C(t)$ might be taken in the first instance small enough to lie in a neighborhood of p in which local coordinates are valid. However, in the end the index is an integer independent of C by the same argument that was used above for a plane surface. The index must also be independent of the local coordinate system—evidently it is essential that this notion should have an invariant significance. To prove it, consider two different coordinate systems each valid in a neighborhood of p and call the angles between the field vector and a coordinate vector from each of the two fields θ_1 and θ_2, and the angle between coordinate vectors in each of the two fields θ; it is clear that $\theta_1 = \theta + \theta_2$. By choosing the curve $C(t)$ enclosing p small enough it is possible to make $\delta_C\theta$ arbitrarily small, since the two fields of coordinate vectors are continuous on C, and since $\delta_C\theta$ is at the same time an integer times 2π it follows that $\delta_C\theta = 0$. Hence $\delta_C\theta_1 = \delta_C\theta_2$ and the invariance of j is proved.

It is useful to know also that the index is *independent of the Riemannian metric which is used to measure angles*—that is, it depends only upon the functions $P(u, v)$ and $Q(u, v)$ that determine the vector field. To prove this statement, consider two metrics with g_{ik} and h_{ik} as components of the metric tensor and set $f_{ik} = (1 - \alpha)g_{ik} + \alpha h_{ik}$, with α a real number ranging over the interval $0 \leq \alpha \leq 1$. Since g_{ik} and h_{ik} both yield positive definite line elements, it is clear that f_{ik} does also, since $1 - \alpha$ and α are both positive. In addition f_{ik} depends continuously on α and hence any angle measured by it also depends continuously on α; the index measured with respect to it is thus a constant since it is an integer. Thus line elements defined by the functions g_{ik} and h_{ik} would both lead to the same value for the index. In particular, *the index of a singularity of a vector field defined in a Cartesian coordinate disk carries over to its image on the surface S.*

Before going on to apply this theory of indices of singularities of vector fields there are two remarks which should be made. First of all, it is of general interest, and also of practical value, e.g., in the theory of nonlinear

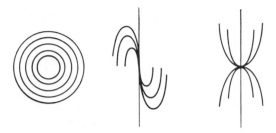

Fig. 8.10   Singularities with $j = +1$.

vibrations (cf. Stoker, Chapter III, for example), to be able to distinguish the various possible types of singularities by using the properties of the functions P and Q that define the field. Poincaré, who is the inventor of the theory described above, showed that the character of the singularity (supposed located at $u = v = 0$) can be deciphered from a study of the system

$$\frac{du}{d\tau} = au + bv + \cdots,$$

$$\frac{dv}{d\tau} = cu + dv + \cdots,$$

in which the dots refer to higher order terms in u and v and a, b, c, d are constants fixed by P and Q. If $ad - bc \neq 0$ Poincaré showed that these four constants determine (except for one very special case), by means of quite simple formulas, the nature of the singularity: in effect, the linear terms in the development of P and Q near the origin are sufficient to characterize it qualitatively. This description and classification of the various types of singularities can be done in an invariant way, so that enumeration of the qualitative possibilities in the Euclidean plane furnishes all of the possibilities on curved surfaces as well. In this connection it might be noted that the only type of singular points with $j = -1$ is the saddle singularity shown in Fig. 8.9, but that the possibilities in the case $j = +1$ are fairly numerous, since in addition to the two examples shown in Fig. 8.8 there can be added those in Fig. 8.10.

A second remark which should be made is that the theory described above can be developed in a more general way by considering not vector fields, but rather fields of *line elements* without a prescribed orientation.   For such fields there are more possible types of singularities than there are for vector fields, and the indices of such fields are not always integers, but rather numbers of the form $\pm n/2$, with n an integer.   For example, Fig. 8.11 illustrates fields of line elements with singularities for which $j = +\frac{1}{2}$ and $j = -\frac{1}{2}$,

$$j = +1/2 \qquad\qquad j = -1/2$$

Fig. 8.11   Singularities of fields of line elements.

and which thus cannot occur in vector fields.   H. Hopf (cf. his New York University lecture notes [H.15]) has made use of the theory of such fields in studying problems in the large concerning certain types of closed surfaces having constant mean curvature.   For the purposes in view here the slightly simpler theory of vector fields suffices.

11.   Poincaré's Theorem on the Sum of the Indices on Closed Surfaces

The following theorem, due to Poincaré, is to be proved: *on a closed surface S of genus g any continuous vector field defined everywhere on S and having a finite number of singularities with indices j_i is such that*

$$\sum_i j_i = 2(1 - g).$$

To begin with, it is easy to construct examples of vector fields on any of the closed orientable surfaces for which Poincaré's formula is readily seen to hold.   For example, the sphere in three-dimensional space, as was already noted above, can have on it two vector fields determined by the meridians and parallels, and these fields each have two singularities that are of the types shown in Fig. 8.8 and hence they have the index $+1$; thus the sum of the indices is 2 in each case—which is as it should be for $g = 0$.   Another way to define a continuous vector field on the sphere in three-space is to take the field of plane curves on it defined by the intersections of the sphere with all planes containing a fixed tangent line of the sphere.   The corresponding vector field has obviously one singularity of the type shown at the right of

Fig. 8.9, which has the index $+2$; again the count is correct. It has already been remarked that on the torus generated by rotating a circle about a line in its plane that does not cut it two vector fields without singularities can be defined: they correspond to the plane curves cut out of the torus by planes through the axis of rotation, and by planes orthogonal to it; here again the

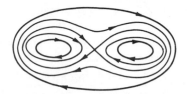

Fig. 8.12   Construction of a vector field on a surface with $g = 2$.

formula checks, since $g = 1$ for the torus. Still another way to define a vector field on the same torus of revolution is to think of it as resting on the x,y-plane with its axis parallel to that plane. Consider then the family of curves cut out of it by planes parallel to the x,y-plane. It is clear that there will be four singular points in the field of tangent vectors of the curves: two corresponding to the maximum and minimum of the z-coordinate on the torus (the latter is the point of contact with the x,y-plane) and two corresponding to the horizontal tangent planes at points of negative Gaussian curvature; there are thus four singular points, two with index $+1$ (the maximum and minimum) and two with index -1 (the saddle points, near which the singularity appears as in Fig. 8.9 at the left); thus the sum of the indices is zero. In similar fashion, a closed surface of genus 2 (doughnut with two holes) could be rested on the x,y-plane, and the curves cut out of it by planes parallel to it would determine a vector field with six singularities, corresponding to a maximum, a minimum, and four saddle points of the function $z(u, v)$; the sum of the indices would therefore be -2, which again checks since $g = 2$ in this case. Generalizing from these examples, it is seen that simple closed orientable surfaces in three-space can be found of all topological types and with vector fields such that Poincaré's formula holds.

Actually, there is no need to make the construction in three-space to find such examples—as should evidently be the case since the theorem belongs to inner differential geometry—nor is it necessary to require the surface to be orientable. To construct an example for an abstractly given surface S of genus 2 the procedure indicated in Fig. 8.12 can be used. Two distinct congruent disks with two holes are taken, together with a family of curves with a single saddle singularity; one of the two disks is shown in the figure. The Riemannian metric is assumed to be the same at congruent points of the two disks.   If the two exemplars are thought of as having their congruent boundary points identified it is clear that an abstract Riemannian surface of genus 2 results, and on it a field with two singularities each with index -1 is defined. It is clear that this process can be generalized for surfaces of any genus. However, the introduction of an appropriate Riemannian metric on such a disk as that of Fig. 8.12 presents a problem since it was proved in Section 9 above that the curvatura integra K_T, which is fixed by the metric form, has a value that is determined by the genus of the surface.

Poincaré's theorem is now proved by showing that the sum of the indices (in case only a finite number is present) is a *topological invariant*—in fact, that it has the value $(1/2\pi) \int_S K \, dA$—and therefore that it has a value determined by the genus g alone, i.e., $2(1 - g)$, in accordance with the discussion concerning the curvatura integra in Section 9.   It is therefore to be shown that $\sum_i j_i = (1/2\pi) \int_S K \, dA$.   To this end a triangulation of the surface is considered that has the following properties:

1. Each triangle contains at most one singular point in its interior, and none on its boundary.

2. Each pair of triangles having a side in common is covered by a valid local parameter system.

It was shown in Section 8 that a triangulation having property (2) exists —in fact, it can be constructed with geodetic triangles.   That condition (1) can be satisfied is clear, since the singularities are assumed to be isolated. Consider any one of the triangles T and let j be the index of the singularity in it (with the value zero, of course, if there is no such singularity).   Let $\mathbf{X}_1(t)$ be the set of tangent vectors of the u_1-system of coordinate curves along the boundary $C(t)$ of the triangle, and call $\mathbf{v}(t)$ the field vectors along the same curve.   Consider next a *parallel field* $\mathbf{w}(t)$ (in the sense of Levi-Civita) along $C(t)$; from (7.32) in Chapter VII it is known that

$$\iint_T K \, dA = \delta_C(\mathbf{X}_1, \mathbf{w}).$$

Here, the symbol $\delta_C(\mathbf{X}_1, \mathbf{w})$ means the change in angle between $\mathbf{X}_1$ and $\mathbf{w}$, measured from $\mathbf{X}_1$ to $\mathbf{w}$, in the sense of the orientation of the triangle, in making one circuit around it.   This relation can therefore be written, with a change of sign, as follows:

$$-\iint_T K \, dA = \delta_C(\mathbf{w}, \mathbf{X}_1).$$

The definition of the index gives us the relation

$$2\pi j = \delta_C(\mathbf{X}_1, \mathbf{v}).$$

Addition of the last two equations yields

$$2\pi j - \iint_T K \, dA = \delta_C(\mathbf{w}, \mathbf{X}_1) + \delta_C(\mathbf{X}_1, \mathbf{v}) = \delta_C(\mathbf{w}, \mathbf{v})$$

$$= \sum_{i=1}^{3} \phi(C_i),$$

in which $\phi(C_i) = \delta_{C_i}(\mathbf{w}, \mathbf{v})$ is a notation for the change in the angle between $\mathbf{w}$ and $\mathbf{v}$ along any one of the three sides C_i of T. The parallel field $\mathbf{w}$ along a curve C_i is the same, from its definition, no matter in which direction the curve is traversed, and $\mathbf{v}$ is also fixed once for all. The last equation above is now summed for all of the triangles that cover S; it is to be shown that the sum on the right-hand side must be zero. Thus the desired formula $\sum_i j_i = (1/2\pi) \iint_S K \, dA$ would be proved and with it Poincaré's theorem.

The reasons for the vanishing of the sum in question are as follows. First of all it is assumed that the surface is orientable; thus each triangle can be given an orientation coherent with that of a coordinate patch containing it. In case two triangles have common sides C, C', the desired cancellation clearly occurs, since $\mathbf{w}$ and $\mathbf{v}$ are fixed, but the two sides are traversed in opposite directions; hence $\phi(C) = -\phi(C')$. If the orientations of the two triangles are not coherent (which could occur if the surface is not orientable) $\mathbf{w}$ and $\mathbf{v}$ make angles θ, ϕ with the fixed coordinate vectors $\mathbf{X}_1$ of a coordinate patch covering them that are opposite in sign for the two sides, modulo an integral multiple of 2π in each case, but since the two sides are now traversed in the same sense it follows once more that $\phi(C) = -\phi(C')$. (The reader may find it useful to draw a figure illustrating this situation.) Thus cancellation occurs at all sides and Poincaré's theorem is proved.

Thus only on the torus, for which $g = 1$, can a coordinate system exist in the large without a singularity; all other closed surfaces must therefore be dissected into overlapping coordinate patches if singularities in the coordinates are to be avoided—as is usually the practice in differential geometry. In mathematical physics it is mostly the practice to accept singularities in the coordinate systems, but to deal with them explicitly—e.g., spherical polar coordinates are often used.

12. Conjugate Points. Jacobi's Conditions for Shortest Arcs

In Section 6 the existence of shortest arcs joining arbitrary pairs of points of a complete surface was proved, and it was shown that they are geodesics. In Chapter VII the geometry of geodesics on two-dimensional surfaces was studied in considerable detail by making deductive use of their definition as a solution of the Euler differential equation for the problem of finding an arc of shortest length between two points. In particular, it was shown that these solution curves of the Euler equation do indeed furnish the arcs of minimum length in the small and that such arcs are also uniquely determined in the small. More precisely, it was shown that a neighborhood of every point p of the surface exists within which *geodetic polar coordinates* are valid, i.e., coordinates such that one family of coordinate curves consists of all geodesics issuing from the point, while the other family consists of their orthogonal trajectories. The line element in these coordinates has the form

$$(8.7) \qquad ds^2 = dr^2 + G(r,\, \theta)\, d\theta^2.$$

Here r is the distance from p measured along the geodesics and θ is the angle from the $\mathbf{X}_1$-direction, say, to the tangent vector of a particular geodesic at p (which is determined uniquely in the small by these prescriptions). It was proved that a positive number r^* exists such that these coordinates, and (8.7), are valid for $0 \le r < r^*,\, 0 \le \theta \le 2\pi$. Once this is proved it is readily seen that the shortest rectifiable arcs joining p with any point in such a geodetic circle are the geodetic radii $\theta = \text{const.}$

The purpose of this section is to consider the problem of determining shortest arcs between two points of a surface (i.e., a two-dimensional manifold) not as a pure existence question, as was done in Section 6, but rather as a boundary value problem for the Euler equation of the relevant variational problem. Such a discussion is interesting for its own sake, and it also leads in some cases to a definite criterion for deciding whether or not a given geodetic join furnishes the shortest arc. In addition, the theory to which it leads furnishes a useful tool for various purposes, including the discussion of certain surfaces, called covering surfaces, which will be carried out in Section 15.

The principal tool needed for the present discussion is the so-called second variation of the minimum problem, from which a necessary condition for the minimum is obtained in addition to that furnished by the vanishing of the first variation. Consider a geodetic arc C connecting points p and q, and embed it in a geodetic parallel coordinate system. In Chapter VII, Section 11, it was shown that coordinates (u, v) could be introduced in a neighborhood of C such that $u = 0$ corresponds to C and such that the curves $v = \text{const.}$ are geodesics orthogonal to C. If v is chosen as arc length along C, so that v ranges over the interval $0 \le v \le L$, with L the length of C, the line element is given by

$$(8.8) \qquad ds^2 = du^2 + G(u, v)\, dv^2.$$

Since $u = 0$ corresponds to C, it follows that $G(0, v) = 1$, and in Chapter VII it was shown that $G_u(0, v) = 0$; thus G and its derivative with respect to u satisfy the equations

$$(8.9) \qquad G(0, v) = 1, \qquad G_u(0, v) = 0$$

on C. It is also seen more or less immediately that Gauss's theorema egregium for this line element takes the form

$$(8.10) \qquad K = -\frac{1}{2\sqrt{G}} \frac{\partial}{\partial u} \left(\frac{1}{\sqrt{G}} \frac{\partial G}{\partial u} \right).$$

The basic variational problem to be discussed in this section is concerned with what will be called *admissible arcs*; these arcs are assumed to lie in a neighborhood of C within which the geodetic parallel coordinate system is valid. By the term *admissible arc* is meant a continuous curve passing through the end points of C and having piecewise continuous first derivatives. For what follows it is useful to consider a special class of such curves, i.e., curves defined as the loci $u = \epsilon\eta(v)$ in the geodetic coordinate system, with $\eta(v)$ a continuous piecewise differentiable function defined for $0 \le v \le L$, and such that $\eta(0) = \eta(L) = 0$; clearly, if the constant ϵ is chosen sufficiently small in numerical value such an arc will lie close to C. The lengths $L(\epsilon)$ of such arcs in the neighborhood of C are given by

$$(8.11) \qquad L(\epsilon) = \int_0^L \sqrt{\epsilon^2 \eta'^2 + G(\epsilon\eta(v), v)} \, dv.$$

Since the curve C is assumed to be a shortest arc, it is clear that $dL/d\epsilon|_{\epsilon = 0} = 0$ must hold for it. That this is true follows from

$$\frac{dL(\epsilon)}{d\epsilon} = \int_0^L \frac{\epsilon\eta'^2 + \frac{1}{2}G_u(\epsilon\eta, v)\eta}{\sqrt{\epsilon^2 \eta'^2 + G(\epsilon\eta, v)}} \Bigg|_{\epsilon = 0} dv,$$

since C is characterized by $\epsilon = 0$, and $G(0, v) = 1$, $G_u(0, v) = 0$ on C [cf. (8.9)]. A further necessary condition results from consideration of the second derivative $d^2L(\epsilon)/d\epsilon^2|_{\epsilon = 0}$; as is well known, this must be non-negative if $L(0)$ is to furnish a minimum. This condition leads in the present case to what is called the *second variation* of the integral to be minimized. It is obtained from (8.11) by developing the integrand in the neighborhood of $\epsilon = 0$ up to terms of second order. Since the integrand is of the form $F(\epsilon) = \sqrt{f(\epsilon)}$, and since $f'(0) = 0$, $dL(\epsilon)/d\epsilon|_{\epsilon = 0} = 0$, it follows that $F''(0) = \frac{1}{2}f''(0)/\sqrt{f(0)}$. Also, $f(0) = F(0, v) = 1$, and the desired development therefore has the following form:

$$(8.12) \qquad L(\epsilon) = L + \int_0^L \frac{1}{2} \left[\eta'^2 + \frac{\eta^2}{2} G_{uu}(0, v) \right] \epsilon^2 \, dv + \cdots,$$

where the dots refers to terms that are of order ϵ^3 at least.   From (8.9) and (8.10) it follows that $G_{uu} = -2K$ for $u = 0$.   Thus the last formula can be written in the form

$$(8.13) \qquad L(\epsilon) - L = \frac{\epsilon^2}{2} \int_0^L [\eta'^2 - K(0, v)\eta^2] \, dv + \cdots$$

Clearly, if the length L of C is to furnish the minimum length, it is necessary that the *second variation* $\delta^2 L$,

$$(8.14) \qquad \delta^2 L = \int_0^L [\eta'^2 - K(0, v)\eta^2] \, dv,$$

should be ≥ 0 for all admissible arcs.   This is therefore another necessary condition that C should satisfy.

Since the minimizing curve C is assumed given in the present case by $\eta = 0$, it is at once seen from (8.14) that $\delta^2 L = 0$ with respect to it.   Thus the new necessary condition on C will be satisfied if the integral $\delta^2 L$ in (8.14) has $\delta^2 L = 0$ as its minimum for $\eta \equiv 0$ within the class of admissible functions introduced above.   This idea of introducing an auxiliary minimum problem based on the second variation is due to Jacobi.   A necessary condition for such a minimum is, of course, that the relevant Euler equation should be satisfied by a function $\eta(v)$ that vanishes at the end points; in this case the Euler equation is easily found to be

$$(8.15) \qquad \frac{d^2\eta}{dv^2} + K(0, v)\eta = 0.$$

Evidently, it is a *linear* differential equation of second order, and it is called Jacobi's equation.

It is now possible to introduce the important concept of a *conjugate point* on the geodetic arc C: a *conjugate point p^* on C is an interior point of the arc where a solution $\eta(v)$ of (8.15), which vanishes at p ($v = 0$) but not identically in v, also vanishes at p^* ($v = v^*$).*   With the aid of this concept the following theorem will be proved: *A necessary condition that the geodetic arc C should furnish the shortest arc among all admissible arcs in its neighborhood is that no conjugate point $p^* \neq q$ exists between p and q.*

The theorem is proved indirectly by showing that the existence of a conjugate point makes it impossible for C to yield an arc of shortest length. Thus it is assumed that p^* is a conjugate point on an arc of shortest length. It follows that an integral $\eta(v) \not\equiv 0$ of (8.15) exists which satisfies the initial conditions $\eta(0) = 0$, $\eta'(0) = 1$ [with no loss of generality in setting $\eta'(0) = 1$ since the differential equation is linear and homogeneous] and is such that

$\eta(v^*) = 0$ for a value v^* less than L. A special admissible variation $\bar{\eta}(v)$ is defined as follows:

$$\bar{\eta}(v) = \begin{cases} \eta(v) & \text{for} \quad 0 \le v \le v^*, \\ 0 & \text{for} \quad v^* \le v \le L. \end{cases}$$

For this variation $\delta^2 L(\bar{\eta})$ is given by

$$\delta^2 L(\bar{\eta}) = \int_0^L [\bar{\eta}'^2 - K(0, v)\bar{\eta}^2]\, dv$$

$$= \bar{\eta}\bar{\eta}'\Big|_0^{v^*} - \int_0^{v^*} \bar{\eta}[\bar{\eta}'' + K(0, v)\bar{\eta}]\, dv$$

$$= 0,$$

upon integrating the term $\bar{\eta}'\bar{\eta}'$ by parts, using the definition of $\bar{\eta}$, and noting that $\bar{\eta}'' + K(0, v)\bar{\eta} = 0$ in the interval $0 \le v \le v^*$. Thus $\bar{\eta}(v)$ furnishes a minimizing function for the integral $\delta^2 L$ in (8.14) since the integral has the value zero with respect to it. It will now be shown, however, that the existence of the variation $\bar{\eta}(v)$ implies the existence of another variation $\xi(v)$ for which $\delta^2 L(\xi)$ would be negative, and this contradiction would establish the correctness of the theorem.

The function $\xi(v)$ is determined with the aid of a solution of the differential equation

$$\zeta'' + (K - \alpha)\zeta = 0, \qquad \alpha > 0,$$

subject to the initial conditions $\zeta(0) = 0$, $\zeta'(0) = 1$, and for a proper choice of the constant α. Since $\eta(v)$ satisfies the same initial conditions and the same differential equation, except that $K - \alpha$ is replaced by K, it follows from the Sturm-Liouville comparison theorem proved in Section 7 above that the first zero v_0 of $\zeta(v)$ lies to the right of v^*, but will lie close to v^* if α is chosen small enough. (For this, the theorem on continuous dependence of solutions of ordinary differential equations on variations in a parameter is used.) In particular, if α is sufficiently small, v_0 will lie in the interval between v^* and L. An admissible variation $\xi(v)$ is now defined by the stipulations[1]

$$\xi(v) = \begin{cases} \zeta(v) & \text{for} \quad 0 \le v \le v_0, \\ 0 & \text{for} \quad v_0 \le v \le L. \end{cases}$$

[1] This way of defining the variation ξ was proposed to the author by H. Karcher.

The second variation $\delta^2 L(\bar{\xi})$ is therefore given by

$$\delta^2 L(\bar{\xi}) = \bar{\xi}\bar{\xi}'|_0^{v_0} - \int\limits_0^{v_0} \{\bar{\xi}[\bar{\xi}'' + K(0, v) - \alpha)\bar{\xi}] + \alpha\bar{\xi}^2\}\, dv,$$

after an integration by parts and addition and subtraction of the term $\alpha\bar{\xi}^2$. It follows that

$$\delta^2 L(\bar{\xi}) = -\alpha \int\limits_0^{v_0} \bar{\xi}^2\, dv,$$

and hence $\delta^2 L(\bar{\xi}) < 0$ holds. This contradiction proves the theorem.

It is perhaps of more interest for later purposes to develop a *sufficiency condition*, which guarantees that C minimizes the length integral in an appropriate sense if no conjugate point occurs.

The theorem to be proved next is that *if no conjugate point exists on C, then C minimizes the length integral with respect to admissible arcs in a certain neighborhood of it.*

The proof of this theorem is carried out in returning once more to the original variational problem of minimizing the length integral. It involves the construction of a *field* of geodesics within which the geodesic C is embedded, in the present case a field of geodesics issuing from the left end point $u = 0, v = 0$ of C at angles θ to the tangent of C in an interval $-\theta_0 < \theta < \theta_0$ with θ_0 a sufficiently small positive number. The same local coordinate system (u, v) is used as was introduced above in a neighborhood of C; the family of geodesics issuing from $u = 0$, $v = 0$ can therefore be put in the form $u = \eta(s, \theta)$, $v = \zeta(s, \theta)$, with s the arc length on these curves measured from the initial point on C. It is a fact from the theory of ordinary differential equations that such geodesics exist for θ_0 small enough and for values of s in a range that includes the range $0 \leq s \leq L$ over which v varies in describing C, and these geodesics remain in the coordinate strip in which C is embedded. The functions u and v of s and θ also have continuous derivatives of second order, since the original surface is assumed to be sufficiently smooth. The variational problem in the parametric form that characterizes these geodesics thus involves the integral

$$L(\theta) = \int\limits_0^{L_0} \sqrt{\eta_s^2 + G(u, v)\zeta_s^2}\, ds,$$

that arises from the first fundamental form (8.8). For the purposes in view here only the variational equation found by varying the function $\eta(s, \theta)$, while leaving $\zeta(s, \theta)$ unchanged, is needed; this equation is

$$\eta_{ss} - \tfrac{1}{2}G_u\zeta_s^2 = 0.$$

In it the quantity θ occurs as a parameter characterizing a given geodesic. (It is to be noted that $\eta \equiv 0$ is a solution of this differential equation for $\theta = 0$ since $G_u(0, v) = 0$ holds because of (8.9); this of course must be true, since C is the geodesic $\eta \equiv 0$.) The last equation is differentiated with respect to θ to obtain

$$\eta_{\theta ss} - G_u \zeta_s \zeta_{s\theta} - \tfrac{1}{2}(G_{uu}\eta_\theta + G_{uv}\zeta_\theta)\zeta_s^2 = 0,$$

and this in turn is seen to become for $\theta = 0$, i.e., along C (for which $u = 0$), the equation

$$\eta_{\theta ss} - \tfrac{1}{2}G_{uu}(0, s)\eta_\theta = 0,$$

since $v = \zeta(s, 0) = s$, $G_u(0, v) = 0$. This, however, means that η_θ is a solution of the Jacobi equation (8.15), since $K(0, v) = -\tfrac{1}{2}G_{uu}(0, v)$. The proof of the sufficiency condition is contained in the statement that s and θ *may be introduced as a coordinate system in a neighborhood of C*, since then C would be embedded in a *field* of geodesics within which the line element is given by (8.7) (but with r replaced by s). The argument used before in Chapter VII, Section 11, would then hold for all comparison arcs in a neighborhood of C and the theorem would be proved.

It is therefore to be proved that s and θ may be introduced as a valid curvilinear coordinate system.* This is done, of course, by considering the Jacobian determinant of $u = \eta(s, \theta)$, $v = \zeta(s, \theta)$ for $\theta = 0$, i.e., along C. This determinant is

$$\begin{vmatrix} \eta_s & \eta_\theta \\ \zeta_s & \zeta_\theta \end{vmatrix}_{\theta = 0} = \begin{vmatrix} 0 & \eta_\theta \\ 1 & \zeta_\theta \end{vmatrix} = -\eta_\theta.$$

The essential point in verifying this relation is that $\eta_s(s, 0) = 0$ and $\zeta_s(s, 0) = 1$, and these values result because $u = \eta(s, 0)$ is identically zero and $v_s = \zeta_s(s, 0)$ is identically 1 because C is given by $u = 0$ and v is by definition the arc length along C. As was remarked above, η_θ is a solution of Jacobi's equation, and the assumption of the theorem to be proved, i.e., that no conjugate point exists on C before the end point q, means that η_θ does not vanish along C. Thus the Jacobian does not vanish anywhere along C, and this suffices to prove the theorem (after a rather obvious application of the Heine-Borel theorem to a covering of C by neighborhoods within which the equations $u = \eta(s, \theta)$, $v = \zeta(s, \theta)$ can be inverted).

For the purposes of a later section it is of interest to make use of the last result to prove another theorem for special surfaces. The theorem states that *if the Gaussian curvature K is nonpositive, i.e., $K \leq 0$, then any geodetic*

---

* At the end of Chapter IV it was shown that this statement is valid in a neighborhood of the left-hand end point $u = 0$, $v = 0$; thus it needs to be established only at other points.

arc C of finite length can be embedded in a geodetic polar coordinate system (r, θ) with its origin at an end point of the arc and for some range of the angle θ. The proof of this theorem is almost immediate: if $K \leq 0$ holds it is seen at once from (8.14) that the second variation is never negative so that $\eta \equiv 0$ is the only admissible function that furnishes a minimum for $\delta^2 L$; thus no conjugate point occurs on C, the function η_θ introduced above does not vanish, and the theorem is therefore proved.   As a corollary it is clear that *any geodetic arc on a surface with $K \leq 0$ furnishes the arc of shortest length between any pair of its points with respect to all admissible arcs in a certain neighborhood of it.* In addition such an arc is the *unique shortest arc* in a certain neighborhood of it since two different arcs through the same end points could not exist, as was seen in Chapter VII, Section 19, where it was shown that no two geodesics in a simply connected region of a surface have more than one point in common when $K \leq 0$ holds—and the region under consideration here is simply connected.   It is, however, of great interest to note that these statements may not hold in the large: for example, two points on the same generator of a circular cylinder (K is zero everywhere, of course) can be connected by infinitely many geodesics (they are helices), but only the arc of the generator furnishes the arc of minimum length in the large, although it remains true that each helical arc furnishes a minimum for arcs in its neighborhood.   This remark foreshadows a later discussion to be found in Section 15.

It should be added that the discussion of this section will be carried out along much the same lines for n-dimensional manifolds in the next chapter. The difference is that the geodesic C is embedded first in a coordinate system consisting of C itself together with n mutually orthogonal geodesics, one of which is C, while the others are defined at all points of C by parallel transport along C of vectors that then furnish directions for the orthogonal geodesics. The calculations needed are straightforward, but rather intricate.

13.   The Theorem of Bonnet-Hopf-Rinow

In 1855 Bonnet proved that a closed regular convex surface in three-dimensional space E^3 for which the Gaussian curvature K is at least equal to the constant $1/k^2$ has, as a metric space in the sense of Section 5, a diameter $\leq \pi k$, i.e., that it has no pair of points at greater distance apart than πk as measured on the surface.   If $K \equiv 1/k^2 = $ const. the equality sign holds for a sphere of radius k, and hence the inequality cannot be made more precise. Hopf and Rinow [H.16] observed that this theorem is capable of a highly interesting generalization without any necessity to deviate from the main lines of the proof of Bonnet.   The theorem of Hopf and Rinow is as follows: *A complete surface S for which the curvature is $\geq 1/k^2$, $k = $ const., is such that the* distance $\rho(p, q)$ *of any pair of points in S is* $\leq \pi k$; *hence S is compact, i.e.,* *closed.*[1]

Thus the surface need not be assumed to be closed, as is done by Bonnet —rather, it is *proved* to be closed, and in addition the theorem belongs to inner differential geometry since no embedding in a Euclidean space need be involved.

The key point in the proof is the existence theorem concerning geodesics of shortest length, which was proved in Section 6 for complete surfaces. Consider any geodetic arc C of length L joining points p and q and furnishing the shortest arc. The second variation with respect to this arc is given, as was seen above in (8.14) by the integral $\int_0^L (\eta'^2 - K(0, v)\eta^2) \, dv$. Since K is by assumption positive it follows that the integral is not decreased in value by replacing $K(0, v)$ by its lower bound. The integral must be non-negative for any admissible function $\eta(v)$, and thus in particular for $\eta(v) = \epsilon \sin \pi v/L$, which is admissible for ϵ not too large since $\eta(0) = \eta(L) = 0$. The following inequality therefore holds:

$$0 \leq \int_0^L \left(\frac{\pi^2}{L^2} \cos^2 \frac{\pi v}{L} - \frac{1}{k^2} \sin^2 \frac{\pi v}{L} \right) dv = \frac{L}{2} \left(\frac{\pi^2}{L^2} - \frac{1}{k^2} \right).$$

From this it follows at once that L must have a value $\leq \pi k$. In other words, the distance between any pair of points of S is at most πk. The complete surface S is therefore such that all of its points are at a bounded distance from any one of them; hence every set of points has a point of accumulation; thus S is compact.

The theorem of Hopf and Rinow can be extended without much difficulty to n-dimensional Riemannian manifolds. This discussion is deferred to the next chapter, after the notion of curvature greater than a given positive constant, and other notions, such as that of parallel displacement in such manifolds, have been introduced. The proof of the extended Hopf-Rinow theorem will then be seen to be in essence the same as for two-dimensional manifolds.

14. Synge's Theorem in Two Dimensions

The discussion of the preceding section shows that the condition $K \geq 1/k^2$ on a complete surface makes it compact, i.e., closed. Synge's theorem goes further in showing that if $K \geq 1/k^2$ holds and the surface is in addition

---

[1].Reference is made here to Section 3 above, where the synonymous terms "compact" and "closed" are discussed. In a metric space it is known that compactness is equivalent to the property that every infinite point set should have a point of accumulation.

orientable, that it is simply connected. Actually, Synge proved the analogous theorem for complete n-dimensional Riemannian manifolds when n, the dimension, is an even number. After the introduction in the next chapter of the appropriate concepts of Riemannian geometry, it will be seen that the proof of this more general case can be reduced to the two-dimensional case and thus established by what is done in this section.[1]

The proof is carried out indirectly by supposing that a continuous piecewise regular closed curve C exists on the surface that cannot be deformed continuously into a point. It can then be shown that a shortest closed geodesic G exists within the homotopy class to which C belongs (i.e., within the class of those curves that can be deformed continuously into C). This is assumed for the moment without proof. Take, then, any point p of the closed minimizing geodesic G and consider a unit tangent vector to it as well as a unit vector in S orthogonal to it at point p. Upon parallel transport of both vectors around G these two vectors will come back to their same positions, since the process of parallel transport preserves angles, the tangent vector comes into coincidence with itself (since G is a geodesic), and thus the orthogonal vector does the same because of the assumed orientability of S. It follows that a neighborhood of the whole of G can be covered by a geodetic coordinate system consisting of geodesics orthogonal to G and their orthogonal trajectories. The parameter u can be chosen as the arc length on the geodesics orthogonal to G, with $u = 0$ on G. Thus G is embedded in a family of curves $u = \eta =$ const. and these curves are admissible variations of G in the usual sense employed in the calculus of variations (cf. Section 12, for example). For such varied curves the second variation relative to G, in the form analogous to (8.14), is given by $\delta^2 L = - \oint_C K\eta^2 \, dv$, and since K is everywhere positive $\delta^2 L$ is negative. Thus the curve G could not be a curve furnishing the minimum length; hence the original hypothesis that G is embedded in a homotopy class of curves not contractable to a point must be rejected, and S is therefore simply connected.

It remains to be proved that the geodesic G of minimum length exists. The proof to be given here would also serve, with minor changes, to prove the existence of a shortest arc in the form of a geodesic joining any *pair* of points on a finitely compact surface. In Section 6 above the proof of this fact was carried out differently since it was done in a context in which the property of finite compactness of S was to be deduced from a weaker assumption. Thus it has some point to give a proof for the existence of shortest connections that

---

[1] There is really no *need* to prove the theorem in two-dimensions since, as was pointed out earlier in Section 9 of this chapter, an orientable compact surface with positive Gaussian curvature is necessarily homeomorphic to the sphere because its Euler characteristic is positive—and hence of necessity the surface is simply connected.

makes direct use of the fact that S is actually compact. (Finite compactness is not a strong enough condition to guarantee in general the existence of a *closed* geodesic, but for the existence of a shortest join between a given *pair* of points the proof to follow can be easily modified in such a way as to hold when finite compactness is assumed.)

It is to be proved, in following essentially a method due to Hilbert, that a closed geodetic arc G exists with the desired properties. By assumption a piecewise regular closed curve C exists as one of a homotopy class of such curves that cannot be deformed continuously to a point on the surface. This means the following. The curve C is given in a parameter representation $p = p(t)$ with $0 \leq t \leq 1$, such that $p(0)$ and $p(1)$ coincide; this of course implies that the curve has been given an orientation. If C^* is a curve in the homotopy class of C this means by definition that C^* is embedded in a family $p = p(t, \tau)$ of piecewise regular closed arcs on the surface that depend continuously on τ for $0 \leq \tau \leq 1$, with t as parameter on each curve for $0 \leq t \leq 1$, and such that C^* is obtained for $\tau = 1$, and C, for $\tau = 0$. (For a discussion of the notion of homotopy see the book of Patterson [P.1].)

All piecewise regular curves C^* in the homotopy class of C are considered; it is to be shown first that the lengths of these curves have a greatest lower bound L that is different from zero because of the assumption that none of them can be deformed into a point. In fact, if L were zero, it follows that curves would exist such that all of their points would lie in a geodetic circle centered at a point q of the surface S within which geodetic polar coordinates are valid. This can be seen as follows: A sequence τ_n of τ-values would exist such that the lengths L_n of the curves $p(t, \tau_n)$ tend to zero as $n \to \infty$. Out of this sequence of curves consider for each n the points $p_n(t_0, \tau_n)$ fixed by $t = t_0$; since S is compact, a subsequence p'_n of these points could be found that converges to a point q on S. Since the lengths of the curves C'_n containing the points p'_n tend to zero as $n \to \infty$, it follows at once that for n sufficiently large all of these curves lie in a geodetic circle of sufficiently small radius centered at q within which geodetic polar coordinates are valid, since the length of a curve between any two of its points is at most equal to the distance between them. But in that case the curves could clearly be contracted to the point q; hence L is positive. (It is here that the compactness of S, which follows from the assumption that K has a positive lower bound, is needed. An open surface could have a "spine" going off to infinity, with closed curves encircling it, whose lengths tend to zero but which do not shrink down on a point of the surface.) Thus the lengths $l(C_i^*)$ of any curves C_i^* of the homotopy class satisfy the inequality $l(C_i^*) \geq L > 0$. It is very important to observe that this situation implies at once that *a sequence C_n^* of curves of the homotopy class exists such that* $\lim_{n \to \infty} l(C_n^*) = L$, since L was defined as the greatest lower bound of the lengths of all curves in the class C^*.

Since S is compact it follows—as has been noted several times in earlier sections—that a uniform positive lower bound r exists for the radius of geodetic circles about any point of S within which geodetic polar coordinates are valid. Consider any curve $C_i^* = C_i^*(t)$ of the sequence just introduced. This curve can be subdivided by a finite number k of points into arcs each of which has a length $\leq r/2$; the points of the subdivision are called $x_1, x_2, \ldots,$ x_k, x_{k+1} with $x_{k+1} = x_1$, and such that x_1 corresponds to the point of C_i^* for which the parameter t has the value 0 and the remaining points are arranged so that t does not decrease upon going from one to the next (two or more points may coincide, however). Since geodetic circles of radius r exist at all points of S, it follows that a closed geodetic polygon P_i can be inscribed in C_i^* with the points x_n as vertices, and the length of P_i is at most equal to the length $l_i = l(C_i^*)$ of C_i^*, since a geodetic segment $x_i x_{i+1}$ is a curve of shortest length between x_i and x_{i+1}. It can also be readily seen that P_i belongs to the homotopy class of curves C^*. In effect, a continuous deformation of each arc $x_i x_{i+1}$ of C_i^* into the geodetic segment $x_i x_{i+1}$ can be achieved, since the two arcs have the same end points and both of them lie inside a geodetic circle of S; hence they are clearly homotopic. Thus each arc of C_i^* can be deformed in turn into the geodetic segment having the same end points, and thus in the end all of C_i^* is deformed into P_i.

The set P_μ of all such geodetic polygons that are inscribed in the set C_i^* and which have k vertices is considered. The set P_μ is not empty. This can be seen by going out far enough in the sequence until the lengths $l(C_i^*)$ are close to L and then choosing $k > 2L/r$. Let $\bar{l}$ be the greatest lower bound of the lengths of all polygons P_μ; it is shown next that $\bar{l} = L$. It is clear that $\bar{l} \geq L$ holds, since the curves P_μ belong to the homotopy class C^* under consideration; it is to be shown indirectly that $\bar{l} > L$ is not possible. If such an inequality were to hold, it follows that a curve C_i^* of the sequence would exist with a length $l^* < \bar{l}$, because of the definition of L; but in that case a geodetic polygon P^* with k vertices could clearly be inscribed in C_i^* with a length at most equal to L^*, but since P^* is clearly in the set P_μ a contradiction results because of the definition of $\bar{l}$. As a consequence there exists in the set P_μ a subsequence of polygons, which is again denoted by P_μ, with lengths converging to L.

The next, and decisive, step to be taken is to show that a subsequence P_{μ_k} of the sequence P_μ can be selected that converges to a geodetic polygon P_∞ that has the minimum length L. For this purpose it is supposed that the polygons P_μ are oriented alike so that their vertex points are numbered $1, 2, \ldots, k$ in the proper sequence, and P_μ^j then denotes the jth vertex point on a polygon P_μ. Consider the set P_μ^1; this set has at least one accumulation point since S is compact, and consequently a subsequence P_{μ_1} of the polygons

P_μ can be selected such that the points $P^1_{\mu_1}$ converge to a point p_1. Out of the sequence P_{μ_1} a subsequence P_{μ_2} is in turn chosen so that the vertex points $P^2_{\mu_2}$ converge to a point p_2, and so on up to the sequence of geodetic polygons P_{μ_k} that yields a point p_k as limit point. In this way an ordered set of points $p_1, p_2, \ldots, p_k$ is determined, and this set is such that the lengths of all geodetic segments $\overline{p_i p_{i+1}}$ is at most equal to $r/2$: this follows from the continuity of the distance function and the fact that the lengths of all segments involved yield the distance between their end points. Thus the points $p_1, p_2, \ldots, p_k$ are the vertices of a closed geodetic polygon P_∞, with sides of length $\leq r/2$. It is easily seen that P_∞ belongs to the homotopy class of the set P_μ since for a fixed but sufficiently large value of μ the vertices P^i_μ lie arbitrarily near to the vertices P^i_∞ of P_∞ for each value of i from 1 to k. It is clear that the length of P_∞ is necessarily L since the sequence of polygons P_{μ_k} converges to P_∞, and hence their lengths converge to that of P_∞, and consequently that length is L. Thus P_∞ is a geodetic polygon in the homotopy class with the minimum length L.

It is now readily seen, finally, that P_∞ is a geodetic line G. This is proved by showing that any two consecutive sides of P_∞ have the same tangent at their common vertex. But this is clear from the fact that any three consecutive vertices p_{j-1}, p_j, p_{j+1} of P_∞ lie in a geodetic circle of radius r centered at p_{j-1}, by the triangle inequality (since the lengths of all sides is $\leq r/2$), and hence p_j must lie on the geodetic radius joining p_{j-1} with p_{j+1}— otherwise a geodetic polygon of length less than that of P_∞ could be constructed by eliminating the vertex p_j and connecting p_{j-1} and p_{j+1} with the geodetic radius. Thus G is a closed geodetic line, and this completes the proof of Synge's theorem.

15. Covering Surfaces of Complete Surfaces Having $K \leq 0$

In Section 16 it will be convenient to make use of the concept of a *covering surface* of a complete two-dimensional surface S having negative Gaussian curvature. Here such covering surfaces are constructed in a special way[1] for surfaces having Gaussian curvature $K \leq 0$.

For the purposes in view here a covering surface S' of S is defined as another surface with the following properties:

1. S' is mapped into S by functions that are regular and such that every

---

[1] The concept of a covering manifold is important in many connections. It is not at all limited to Riemannian manifolds of the special kinds considered here. A good presentation of the essential facts, which belong to topology rather than to differential geometry, can be found in the books of Weyl [W.4] and Pontrjagin [P.6].

point p' of S' has a neighborhood that is mapped in a one-to-one way on a neighborhood of its image point on S.

2. Each point of S' has only one image point in S.

3. Every point of S has at least one preimage in S', or, as it is also said, every point p of S is *covered* by a point of S'.

Under the hypothesis made here, i.e., that S *is complete with Gaussian curvature $K \leq 0$, it will be shown by an explicit construction that a complete covering surface S' exists that is homeomorphic with the Euclidean plane, and such that S' is mapped onto S with preservation of the local Riemannian metric of S.*

Before establishing these facts it is worthwhile to examine in an intuitive way a few special examples of covering surfaces. Consider first a circular cylinder embedded in E^3 with generators that are infinitely long in both directions. Such a surface is clearly complete. The mapping of a plane S' into the cylinder S could be achieved in an intuitive way by rolling the cylinder over the plane, thus dividing it into infinitely many congruent strips with parallel sides, each of which would be mapped in a one-to-one way on the cylinder (of course with an identification of congruent points on the two boundaries of the strip). Each point of the covering surface S' has exactly one point of S as image, but the mapping is evidently not a one-to-one mapping. Consider next a circular cone with straight generators that are rays of infinite length going out from its vertex (with the vertex excluded from the cone, however). This surface is not complete, since the vertex is a singularity at finite distance from any point in the surface. Its covering surface, again visualized by rolling it out over a plane, is seen to be in general an infinitely sheeted covering of the plane with the vertex corresponding to a branch point—like the Riemann surface for the analytic function $\log z$. These two examples are surfaces with $K \equiv 0$, but analogous ones could easily be given for surfaces with $K < 0$. If K is positive, however, the situation may be very different, since the plane cannot be the covering surface of a sphere, for example. Thus if the condition $K \leq 0$, or the condition of completeness, is violated a covering surface of the sort defined above need not exist.

A covering surface S' for complete surfaces with $K \leq 0$ is constructed in the following rather easy way. Any point p of S is selected together with a geodetic polar coordinate system which is valid for geodetic radii r that are not too large and is such that the line element has the form $ds^2 = dr^2 + G(r, \theta)\,d\theta^2$. Since the surface S is complete, it is known (cf. Section 6 of the present chapter) that any point q of S can be joined to p by a geodesic (in fact, even by one of the shortest length, though this fact plays no particular role here). It is known also from Section 6 that arbitrary lengths can be laid off on every geodesic issuing from p (that is one of the conditions proved by Hopf and Rinow to be equivalent to completeness). These facts make it possible to construct a mapping of the entire plane S' onto S in the following way. Let the origin O of S' correspond to p. At O introduce a polar co-ordinate system (r, θ) for S' and assign to the point $q'(r, \theta)$ of S' the point q on S which is at the distance r from p along the geodesic determined by the angle θ. This mapping has continuous derivatives because of the theorem on the differentiable dependence of solutions of ordinary differential equations on initial conditions. This prescription of course fixes a uniquely de-fined point q for any values of r and θ since the geodesics on S are uniquely defined by their initial conditions. Thus every point of S' maps into a uniquely determined point of S and all points of S are covered, since every point q of S can be joined to p by a geodesic (or, perhaps, by many of them). Thus the properties (1) and (2) given above hold on the surface S.

Thus far a restriction on the sign of the Gaussian curvature of S has not been needed. The restriction $K \leq 0$ is now used in order to show that prop-erty (1) holds for S' by showing that the coordinates (r, θ) on S' are legitimate parameters in some neighborhood of every point of S. This in turn follows at once from the discussion in Section 12 above that centered around the concept of a conjugate point. It was seen there that if $K \leq 0$ holds any arc of a geodesic of finite length can be embedded in a geodetic polar coordinate system, with the pole at one end of the arc, such that a neighborhood of the arc is covered by the geodesics. This evidently establishes the result wanted here. It is then also clear that the function $G(r, \theta)$ that serves to define the line element of S can be transferred from S back to S' by assigning its values on S' as those fixed by (r, θ) on S, thus making S' a locally isometric map of S. That S' has a complete metric is then clear. Actually, S' is called the universal covering surface of S, since its completeness means that it is not part of any larger covering surface.

It was shown by Cartan [C.2] that if S is, like S', simply connected, then S' and S are identical as Riemannian manifolds, i.e., they are topological images of one another and isometric in the large. In particular, then, if K were in addition everywhere zero it follows that S would be the Euclidean plane. Cartan makes no restriction to the dimension two, however. In fact, the discussion of this section follows much the same course for n-dimen-sional manifolds once the discussion of the theory of conjugate points of Section 12 has been carried out for manifolds of higher dimension.

In the context of the treatment presented here the result of Cartan can be established rather easily. It is thus assumed that S is simply connected. It is known that each point p' of S' has exactly one image point in S. Sup-pose that two points q'_1, q'_2 of S' have the same image point q in S. The shortest join G' of q'_1 and q'_2 in S' (a geodetic segment) would therefore correspond to a regular curve G in S that goes out from q and returns to it; here the fact that S and S' are mapped locally on each other in a one-to-one way is used. Since S is by assumption simply connected the curve G could be shrunk down continuously on a point q, while remaining a piecewise regular curve. Once such a curve lies in a sufficiently small neighborhood of q, it could be deformed continuously into a curve of arbitrarily small length; but the image of such a curve on S' would have the same length, and that is impossible because the distance between q_1' and q_2' has a fixed value and any curve joining these points would have a length at least as great. Thus the mapping of S' onto S is one-to-one, and the desired result is obtained.

It is of interest in itself and also useful for a later purpose to show that covering surfaces S' of the type just considered for surfaces having $K \leq 0$ *have infinite area.* This follows easily from the existence of the line element $ds^2 = dr^2 + G(r, \theta) \, d\theta^2$ in the large, with $G(r, \theta)$ defined for all positive values of r. In Chapter VII it was seen that

$$K = - \frac{1}{\sqrt{G}} (\sqrt{G})_{rr}$$

holds, with $(\sqrt{G})_r = 1$ for $r = 0$. Thus $(\sqrt{G})_r$ for $K \leq 0$ is a function that is nondecreasing, hence is not less than one for all $r \geq 0$ and therefore $\sqrt{G} \geq r$ holds uniformly in θ. Since the area $A(\rho)$ of a geodetic circle of radius ρ is given by $A(\rho) = \int_0^{2\pi} \int_0^{\rho} \sqrt{G} \, dr \, d\theta$ it follows that $A(\rho)$ is at least equal to $\pi\rho^2$, and the desired property of S' is established.

If the restriction $K \leq 0$ were to be dropped, it is easily seen that complete open surfaces S can be constructed that have finite area but have a simply connected covering surface with infinite area. For example, a surface of revolution with two "spikes" going off to infinity while approaching the axis of rotation rapidly enough would serve; such a surface is topologically a cylinder, and it would have points of positive curvature.

At the end of Section 8 it was stated that some results concerning geodetically convex domains would be established in the present section. The first of these was formulated in Section 8 as the following corollary:

COROLLARY 4. *A simply connected complete surface S of nonpositive Gaussian curvature is considered. All geodetic circles on S are convex, with unique geodetic segments joining pairs of their points.*

The necessity of the restriction to simply connected surfaces S is seen by considering what happens upon expanding a geodetic circle with a fixed center that lies on a circular cylinder or a hyperboloid of revolution of one sheet: such circles would eventually overlap to create nonconvex domains.

The proof of the corollary is as follows.   Since geodetic circles on S of *arbitrary* radius are covered by a valid geodetic polar coordinate system, it follows that any of them could serve as a circle C in the sense of the basic theorem of Section 8.   Consequently, the proof of Corollary 2 of Section 8 can be extended to this case if it can be shown that the geodetic curvature κ_g of any geodetic circle is always negative no matter what its radius may be.   This in turn follows from a calculation based on the use of the Gauss *theorema egregium* in the form

$$(\sqrt{G})_{rr} = -K\sqrt{G},$$

in terms of geodetic polar coordinates.   It is also known from Chapter VII, Section 13, that $\sqrt{G}$ satisfies the following conditions at the origin $r = 0$:

$$(\sqrt{G})_r = 1, \qquad \sqrt{G} = 0.$$

Since, as was noted in Section 8, the geodetic curvature of a geodetic circle is given by $\kappa_g = -(\sqrt{G})_r/\sqrt{G}$, it is seen easily that κ_g is always negative, as follows.   An integration of Gauss's equation yields

$$(\sqrt{G})_r = 1 - \int_0^r K\sqrt{G} \, d\sigma;$$

hence $(\sqrt{G})_r$ is always positive, since $K \leq 0$ is assumed and $\sqrt{G}$ is positive. It follows that $\kappa_g = -\frac{1}{2}[(\sqrt{G})_r/\sqrt{G}]$ is negative at all points of a geodetic circle.   This completes the proof of Corollary 4.

On complete surfaces with $K > 0$ matters are more complex with respect to geodetically convex domains, as is well known from the case of the sphere, in which any convex domain, if not the whole sphere, must lie on a hemisphere.   (This is easily proved by considering the sphere as embedded in three-dimensional space, but a proof can be given without an embedding.   A proof of these facts is required in Problem 1 at the end of the chapter.)   On open complete surfaces of positive curvature matters are rather complex, but on the closed or compact surfaces of positive curvature a number of statements can be made concerning the size of convex *circular disks*.   By a circular disk is meant the set of all points at distance $0 \leq r < r_0, r_0 > 0$, from some fixed point of the surface.

These statements result in part from the Sturm-Liouville comparison theorem (proved in Section 7) for solutions of the following pair of ordinary differential equations:

$$\frac{d^2y}{dr^2} + \lambda^2 y = 0, \qquad \lambda = \text{const.},$$

$$\frac{d^2z}{dr^2} + K(r)z = 0.$$

It is assumed that the quite special initial conditions $y(0) = z(0) = 0$, $y'(0) = z'(0) = m > 0$ are prescribed to be the same with respect to solutions of both equations. In addition, $K(r)$ is assumed to satisfy the inequality $0 < a \leq K \leq b$, with a and b both constant. Solutions $y(r)$ and $z(r)$ of the initial value problems that are *non-negative* are the only ones of interest here. The solutions $y(r; a)$ and $y(r; b)$ obtained by setting $\lambda^2 = a$ and $\lambda^2 = b$ are half-waves of sine curves that extend from $r = 0$ to $r = \pi/\sqrt{a}$ and $\pi/\sqrt{b}$. The comparison theorem shows that $z(r)$ *satisfies the inequality* $y(r; b) \leq z(r) \leq y(r; a)$ *for all values of r up to the first zero r^* of $z(r)$ or $y(r)$ after $r = 0$*; this zero satisfies the inequality $\pi/\sqrt{b} \leq r^* \leq \pi/\sqrt{a}$. (Here, the fact that all solutions satisfy the initial conditions, with m positive, is used in an essential way.)

On a compact surface K assumes its maximum b and minimum a. The Gauss *theorema egregium*, expressed in terms of geodetic polar coordinates, is expressed by the equation $(\sqrt{G})_{rr} + K\sqrt{G} = 0$, which therefore, with $\sqrt{G}$ as dependent function, coincides with the differential equation for z given above. It is known also that the initial conditions for $\sqrt{G}$ at the center of the geodetic circle are $\sqrt{G(0)} = 0$, $(\sqrt{G(0)})_r = 1$. Thus the first zero r^* of $\sqrt{G(r)}$ lies between $\pi/\sqrt{b}$ and $\pi/\sqrt{a}$. On the other hand, the differential equation for $\sqrt{G}$ is also Jacobi's differential equation with respect to any given geodetic radius [see equation (8.15) in Section 12], and hence r^* is the first conjugate point on it upon going out from the center of the circle; consequently the geodetic polar coordinates are valid for a neighborhood of the geodetic radius up to that point. Since all distances can be laid off on all geodesics, because the surface is complete, it follows from the comparison theorem that a conjugate point $r^* = r^*(\theta)$ exists on every geodetic radius. Since the solutions $\sqrt{G(r, \theta)}$ of the differential equation depend continuously on θ and r, it follows that r^* has a nonzero minimum r_m for some initial angle $\theta = \theta^*$. *A circular disk of radius r_m could not be geodetically convex.* To prove it consider the geodetic ray $\theta = \theta^*$. Along it the function $(\sqrt{G})_r$ decreases monotonically from the value 1 at $r = 0$, since $(\sqrt{G})_{rr} = -K\sqrt{G}$ and K and $\sqrt{G}$ are both positive. Hence $(\sqrt{G})_r$ has a zero between $r = 0$ and $r = r_m$, say at r^{**}, i.e., at the point where $\sqrt{G(r)}$ has a maximum. Thus $(\sqrt{G(r)})_r$ becomes negative at a point r^{**} before r_m is reached, and remains negative from r^{**} to r_m since it is a monotonically decreasing function of r. It follows, since the geodetic curvature κ_g of a geodetic circle is given by $\kappa_g = -(\sqrt{G})_r/\sqrt{G}$, that κ_g is positive for $r^{**} < r < r_m$ on the geodetic radius $\theta = \theta^*$. It follows that circular disks with radii in this interval do not have the local support property at all of their boundary points, and they are therefore not convex. (It should be noted, however, that geodetic circles of these radii might not exist with smooth boundaries.)

In case the curvature is a positive *constant* and S is a sphere, so that $K = a = b = 1$, say, it follows that $\sqrt{G(r)}$ is the function $\sin r$. In that case $r^* = \pi$, while $r^{**} = \pi/2$ on all geodetic radii; thus the hemispheres are the largest convex circular disks. They are, of course, also the interiors of geodetic circles. This result, which of course belongs to inner differential geometry, is very readily seen to hold if the surface—the unit sphere—is embedded in three-dimensional space; in this case it has also been mentioned repeatedly that *any* convex domain, if not the whole sphere, is contained in a hemisphere. That the latter statement also holds for inner differential geometry is formulated as a problem at the end of the chapter.

16. Hilbert's Theorem on Surfaces in E^3 with $K \equiv -1$

It has been mentioned on various occasions that any Riemannian manifold can be embedded in the small in a Euclidean space of high enough dimensions. It has also been proved by H. Whitney [W.6] that every sufficiently regular Riemannian manifold of dimension n can be embedded isometrically in the large in a Euclidean space of dimension $m \geq 2n$. However, this result leaves undecided whether a two-dimensional surface can be so embedded in the large in a Euclidean three-dimensional space E^3, and in fact it is not in general true. In this section two proofs of a famous theorem of Hilbert [H.9] are given; the theorem states that *a complete regular surface S of curvature $K = -1$ cannot be embedded isometrically in Euclidean three-space without a singularity.*

The proofs to be given use the same basic tool that was used by Hilbert, i.e., the fact that the asymptotic lines on the surface form at the same time a Tchebychef net. A Tchebychef net is by definition a set of parameter curves $p = $ const., $q = $ const. on a surface in terms of which the line element takes the form

(8.16) $$ds^2 = dp^2 + 2F \, dp \, dq + dq^2 ;$$

this means that p and q both represent lengths on the parameter curves. At the end of Chapter VII various properties of such nets were discussed. In terms of these parameters the equation for the *theorema egregium* of Gauss takes on a simple form if the angle $\omega(p, q)$, $0 < \omega < \pi$, between the coordinate curves is introduced. Clearly, $F = \cos \omega$ when $E = G = 1$, $\sqrt{EG - F^2} = \sin \omega$, and Gauss's formula

$$K = \frac{1}{2\sqrt{EG - F^2}} \left[\left(\frac{F_p}{\sqrt{EG - F^2}} \right)_p + \left(\frac{F_q}{\sqrt{EG - F^2}} \right)_q \right]$$

thus can be written, in case $K \equiv -1$ holds, as follows:

$$(8.17) \qquad\qquad \omega_{pq} = \sin \omega,$$

as can be easily seen.   By integrating over a parallelogram R formed by co-ordinate curves the formula of Hazzidakis (see Chapter VII, Section 20) is obtained:

$$(8.18) \qquad \iint\limits_R \sin \omega \, dp \, dq = A = 2\pi - (\alpha_1 + \alpha_2 + \alpha_3 + \alpha_4),$$

in which A is the area of R and the quantities α_i represent the interior angles at the vertices of R.   Since the inequality $0 < \alpha_i < \pi$ clearly holds, it follows that A cannot be larger than 2π.

Hilbert's proof consisted in showing that (a) the asymptotic lines on the isometric image surface $\mathbf{X}(S)$ of S in E^3 form a Tchebychef net, which (b) is in the large a regular coordinate system such that $\mathbf{X}(S)$ is in one-to-one correspondence with S and that both surfaces are simply connected.   The first is easy to show, but the second is bound to be a little awkward since quite a number of facts are to be proved for a surface of unknown topological structure at the outset, and which does not really exist.   Once these facts are established, however, it follows from what was done above that $\mathbf{X}(S)$ cannot be everywhere regular since it has infinite area (cf. Section 15), and that is in conflict with (8.18) for a sufficiently large parallelogram of the Tchebychef net.   A second proof, due to Holmgren [H.11], can be carried out by proving far less, i.e., that a parallelogram of net curves with area $A > 2\pi$ must always arise.   Both proofs are given here, following in the main the courses taken by Hilbert and Holmgren, but with quite a few differences in the details in the case of Hilbert's version.   In addition it will be shown, on the basis of Holmgren's work, how to estimate, from the information obtainable in a neighborhood of any given point of the surface the distance from the point to a singularity.   In other words, the theorem of Hilbert is proved in this way by showing that a singularity must occur at some finite distance from any fixed point.   A still stronger positive formulation of Hilbert's theorem has been given by Amsler [A.4], who showed that (for analytic surfaces) an analytic curve exists, made up of points where a principal curvature becomes infinite.

A proof of the fact that the asymptotic lines on the surface $\mathbf{X}(S)$ in E^3 form a Tchebychef net follows.   Suppose that $p = p(u, v)$, $q = q(u, v)$ are local transformations of coordinates to asymptotic lines $p = \text{const.}$, $q = \text{const.}$ Since $K \equiv -1$, these exist as regular local parameters in a neighborhood of the point $p = q = 0$, say.   It is always legitimate to set $E(p, 0) = 1$, $G(0, q) = 1$, i.e., to choose the arc length as parameter on any two intersecting parameter curves.   What is exceptional about the present case is that this then holds on all of them.   To prove it consider the unit normal vector $\mathbf{X}_3(p, q)$, and the following easily proved identity, based on the fact that $\mathbf{X}(p, q)$ has continuous third derivatives:

$$(\mathbf{X}_3 \times \mathbf{X}_{3q})_p - (\mathbf{X}_3 \times \mathbf{X}_{3p})_q = 2\mathbf{X}_{3p} \times \mathbf{X}_{3q} = 2K\sqrt{EG - F^2}\,\mathbf{X}_3,$$

which holds in any coordinates, as is known from Chapter IV [cf. (4.27) and (4.8)].   Consider also the identities

$$\mathbf{X}_3 \times \mathbf{X}_{3p} = \frac{1}{\sqrt{EG - F^2}} (\mathbf{X}_p \times \mathbf{X}_q) \times \mathbf{X}_{3p}$$

$$= \frac{1}{\sqrt{EG - F^2}} [(\mathbf{X}_p \cdot \mathbf{X}_{3p})\mathbf{X}_q - (\mathbf{X}_q \cdot \mathbf{X}_{3p})\mathbf{X}_p]$$

$$= \frac{1}{\sqrt{EG - F^2}} (M\mathbf{X}_p - L\mathbf{X}_q),$$

in which L and M are coefficients of the second fundamental form.   In the same way it is seen that the following also holds:

$$\mathbf{X}_3 \times \mathbf{X}_{3q} = \frac{1}{\sqrt{EG - F^2}} (N\mathbf{X}_p - M\mathbf{X}_q).$$

From these three identities the further identity follows:

$$2K\,\varDelta\mathbf{X}_3 = \left(\frac{N}{\varDelta}\mathbf{X}_p - \frac{M}{\varDelta}\mathbf{X}_q\right)_p - \left(\frac{M}{\varDelta}\mathbf{X}_p - \frac{L}{\varDelta}\mathbf{X}_q\right)_q,$$

with $\varDelta = \sqrt{EG - F^2}$.   This is specialized to the case of the asymptotic lines for which $L = N = 0$, $-M^2/\varDelta^2 = K$, to obtain for $K \equiv -1$ the result

$$\varDelta\mathbf{X}_3 = \pm\mathbf{X}_{pq}.$$

Since $\mathbf{X}_p$ and $\mathbf{X}_q$ are orthogonal to $\mathbf{X}_3$ it follows from this equation that $\mathbf{X}_p \cdot \mathbf{X}_{pq} = \mathbf{X}_q \cdot \mathbf{X}_{pq} = 0$, or that $E_q = (\mathbf{X}_p \cdot \mathbf{X}_p)_q$ and $G_p = (\mathbf{X}_q \cdot \mathbf{X}_q)_p$ are identically zero.   Since $E = 1$ for $q = 0$ and identically in p, while $G = 1$ for $p = 0$ and identically in q, it follows that E and G are everywhere equal to one.   The asymptotic lines thus form a Tchebychef net, since the line element is in the form given by (8.16).

It is very convenient to make use of the covering surface S' of the surface S which was introduced in Section 15; it is to be recalled that S' is locally isomorphic with S and is complete since S is complete.   The surface S' is topologically equivalent to the plane; on it geodetic polar coordinates exist in the large, and its area is infinite; these things were proved in Section 15.   Evidently, if $\mathbf{X}(S)$ exists as a regular surface in E^3, then $\mathbf{X}(S')$ would also be a regular surface in E^3.   Hilbert's theorem is to be proved by showing that

$\mathbf{X}(S')$ cannot exist in the large without a singularity. (Hilbert did not use the surface S', but showed rather that $\mathbf{X}(S)$ and S are homeomorphic and topologically equivalent to the plane by virtue of the special character of the asymptotic lines.) To this end it is to be noted first that the images of the direction fields of the asymptotic lines of $\mathbf{X}(S')$ on S' form two integrable fields of directions which are at nonzero angle to each other (since $K < 0$) at every point. Furthermore, the two fields can be distinguished from one another at every point by the value $\tau_\pm$ of the torsion of each since $\tau = \pm \sqrt{-K}$ (according to the formula of Beltrami-Enneper). Thus the asymptotic lines form a regular net in a certain neighborhood of an arbitrary point of S'. It will be shown that *this net can be extended as a regular coordinate net over the whole of S'*. Once this is shown Hilbert's theorem follows at once since S' has infinite area, and a contradiction results, as was already noted above, with the fact that no net parallelogram may have an area exceeding 2π.

It is therefore to be shown that the net of asymptotic lines forms a valid coordinate system in the large covering S'. At an arbitrary point O' of S' a local parameter system formed by the asymptotic lines is considered; these are two sets of curves $p(u, v) = $ const., $q(u, v) = $ const. forming a Tchebychef net, and hence such that p and q represent arc length on these curves, with $p = q = 0$ at O'. Thus, a "square" of a certain side length $2l$ centered at O' exists within which these curves form a regular parameter net. Our object is to extend such a square over S' to the maximum extent possible. It is clear that if such a square exists with a given length of side, all squares of smaller side length have the desired properties. The set of attainable values of l (it is clear what is to be understood by the word attainable) is thus a connected interval of the l-axis, and it is to be shown first that *all* values of l are attainable by showing that the existence of a least upper bound l^* for the attainable values would lead to a contradiction. It is assumed first that squares of side lengths $2l$ exist for all $l < l^*$, but not necessarily for $l = l^*$. Consider the union of all squares such that $l < l^*$ holds; the boundary points B of this set belong to S' since S' is complete, and B, since it is a closed set, is also compact since it is clearly bounded, in the usual metric (cf. Section 6), in S'. It is clear that the points of B could be obtained by measuring off the lengths l^* from the "axes" $p = 0$, $q = 0$ on the appropriate asymptotic lines. But each point $x \subset B$ is at the center of a square of net curves that form a locally valid coordinate system. Since B is compact it is covered by a finite number of such squares. Call $2s$ the length of the side of the smallest of these squares. It is clear that the set of squares for which the asymptotic lines are valid parameters can be extended to the open square of side length $l^* + s$, since every square at a point of B in effect gives a local continuation of the coordinate system over the boundary which extends the parameter curves an additional length s. Thus regular parameter squares exist with side lengths that are unbounded. These considerations lead also to a remark that will be of interest in a moment, i.e., that *any segment of any asymptotic line can always be embedded in a regular net, and, in fact, in such a way that an arbitrary length could be laid off on the asymptotic lines crossing it to obtain as locus a "parallel" net curve.*

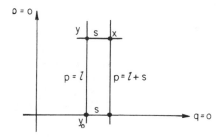

Fig. 8.13   Extending a net of squares.

It can now be seen, finally, that no points of S' are left uncovered in this process. Again we proceed indirectly. If there were such a point, it could be joined to O' by a geodesic—i.e., the geodetic ray R of the polar coordinate system that was used to define S' as a covering surface of S—and since a neighborhood of O' is covered by regular coordinate squares it follows that a point x on the geodetic ray R occurs such that regular coordinate squares exist covering all points of R between O' and x but not covering x. Thus there exists a nested sequence S_n of squares of increasing side length l_n, the union Q of which contains x in its boundary. If $\lim_{n \to \infty} l_n$ is finite, the situation dealt with above occurs and it is clear that the net of squares could be extended over x—in contradiction with the property assigned to it, and the desired result would be proved. However, the fact that the sequence l_n has a finite limit is easily proved.[1] The point x is the center of a certain regular coordinate square, and since x is a boundary point of Q it follows that points of Q lie in that square. Consider the asymptotic lines through x, and a point y on one of them that lies in the interior of Q (cf. Fig. 8.13), the line $p = l$, say, with s the length of the segment of the asymptotic line of the family $q = $ const. between y and x. This line, since it lies in the interior of Q, can

---

[1] It is perhaps of interest to observe that it is in the proof of this fact that the special property of the asymptotic lines, i.e., that they form a Tchebychef net, is used in an essential way. That net squares of infinite side length exist could have been shown without the use of this property, but if the nets are assumed to be no more than the integral curves of regular direction fields they would not in general cover the surface S'—counterexamples can easily be found.

be extended back to the "axis" $q = 0$ to its intersection point y_0; the segment yy_0 is of finite length.   As was remarked above, by laying off lengths s on the asymptotic lines crossing $p = l$ the locus obtained is an asymptotic line of the family $p =$ const.: evidently it joins x to the axis $q = 0$ by means of a segment of finite length L.   It is now clear that the square forming Q has sides of length $2L$.

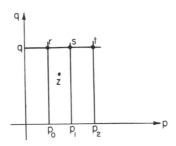

Fig. 8.14   Net parallelograms.

Thus S' is covered by a regular coordinate net, and the first proof of Hilbert's nonexistence theorem is complete.

A second proof of the theorem, due to Holmgren [H.11], will now be given. Holmgren's method consists in making a study of the solution $\omega(p, q)$ of the differential equation (8.17) over S' that goes beyond the mere use of the formula (8.18) of Hazzidakis.   Since $(\partial/\partial q)(\partial\omega/\partial p) \neq 0$, because of the basic fact that ω lies between o and π, it follows that $\partial\omega/\partial p$ is not constant in q. Arbitrarily near to any given point, there is therefore always a point O where $\omega_p \neq 0$, and ω_p can also be assumed to be positive at O since $\bar{\omega} = \omega(-p, -q)$ is a solution of (8.18) when ω satisfies the inequality $o < \omega < \pi$.   The point O is chosen as origin of coordinates in the S'-plane, and a number p_2 is fixed such that ω_p is positive for $o \leq p \leq p_2$, $q = 0$; ω_p is a positive increasing function of q for $q > 0$ and values of p in the interval $o \leq p \leq p_2$; in addition, it is clear that $\omega(p, q)$ is a positive increasing function of p for all $q > 0$ and values of p in the same interval.   The inequalities $\omega_1 = \omega(p_1, 0) - \omega(0, 0) > 0$ and $\omega_2 = \omega(p_2, 0) - \omega(p_1, 0) > 0$ hold, and since ω lies in any case between 0 and π it follows that the smaller of the two numbers ω_1, ω_2 is a positive number ϵ that is less than $\pi/2$:

$$\min\left[\omega(p_1, 0) - \omega(0, 0), \omega(p_2, 0) - \omega(p_1, 0)\right] = \epsilon < \frac{\pi}{2}.$$

Consider the points r, s, t on the line $q =$ const., $q > 0$, (cf. Fig. 8.14); their coordinates are $r:(p_0, q)$, $s:(p_1, q)$, $t:(p_2, q)$.   Since ω_p increases with q and is positive, it follows that

$$\omega(t) - \omega(s) = \int_{p_1}^{p_2} \omega_p(p, q)\, dp > \int_{p_1}^{p_2} \omega_p(p, 0)\, dp \geq \epsilon,$$

and, in the same way, that

$$\omega(r) - \omega(0, q) > \epsilon$$

holds.   These inequalities can be written in the form

$$\omega(r) > \epsilon + \omega(0, q),$$

$$\omega(s) < \omega(t) - \epsilon;$$

and hence $\epsilon < \omega(r)$, and $\omega(s) < \pi - \epsilon$ are valid inequalities since $o < \omega < \pi$ holds.   Since $\omega(s)$ is larger than $\omega(r)$ it is clear that both $\omega(s)$ and $\omega(r)$ lie in the interval between ϵ and $\pi - \epsilon$.   Since the line $q = $ const. was chosen arbitrarily, it follows for a point z in any rectangle over the segment $p_0 p_1$ that $\epsilon < \omega(z) < \pi - \epsilon$ holds since $\omega(r) < \omega(z) < \omega(s)$ holds.   Consequently the inequality

$$\sin \omega(z) > \sin \epsilon$$

is valid for all points in such a rectangle.   Since the area of a rectangle R of height q_2 is $< 2\pi$, by the formula of Hazzidakis, it follows that

$$2\pi > \iint_R \sin \omega \, dp \, dq > \sin \epsilon \int_{p_0}^{p_1} \int_0^{q_2} dp \, dq = q_2(p_1 - p_0) \sin \epsilon;$$

and hence the following inequality results:

$$q_2 < \frac{2\pi}{(p_1 - p_0) \sin \epsilon}$$

It follows that *the surface could not be regular over a parallelogram of the net with a side of length larger than* $2\pi/[(p_1 - p_0) \sin \epsilon]$. Since appropriate positive numbers p_0, p_1 could be chosen in any neighborhood of any point of the surface, it follows that an estimate of the distance from any point to a singularity of the surface has been found in terms of quantities that can be determined from the behavior of $\omega(p, q)$ in an arbitrary neighborhood of the point.   A *uniform* estimate for this distance for all points is, however, not possible in general since there are known surfaces of revolution in three-dimensional space with $K \equiv -1$ for which regular points exist at arbitrarily large distances from all singular points.   It might also be noted that Holmgren's proof does not require that the Tchebychef net should cover the whole surface, but only that it can be extended far enough over a short segment as a base.

It has been proved by Efimov [E.1] that the nonexistence theorem of Hilbert holds when K is not assumed to be constant but is negative and bounded uniformly away from zero.

17. The Form of Complete Surfaces of Positive Curvature in Three-Dimensional Space

In Chapter II it was proved that a *simple* closed plane curve with curvature of one sign throughout is the boundary of a bounded convex set.   It can also be shown (cf. Stoker [S.7]) that an open plane curve *free of double points*, with curvature of one sign, infinite in length in both directions along the curve, and such that the curve *goes to infinity in both directions*, is the boundary of an *unbounded* convex set.

The object of the present section is to prove analogous theorems for two-dimensional surfaces in three-dimensional space.   The first theorem of this kind was proved by Hadamard [H.2], who showed that a closed (i.e., compact) surface with $K > 0$ in three-dimensional space is the boundary of a bounded three-dimensional convex set and is therefore homeomorphic with the surface of a sphere.   Thus in this case the absence of double points or self-intersections is a *consequence* of the assumption of local convexity and not, as in the plane, a needed hypothesis.   In this section the theorem of Hadamard will be generalized to complete surfaces in three-dimensional space having $K > 0$ (Stoker [S.9]), and thus for surfaces which may be open surfaces.[1]   Again, unlike the analogous theorem for open plane curves, it is not necessary to postulate in advance that no double points or lines occur, nor is it necessary to stipulate that the surface should go to infinity in E^3 along divergent curves in the surface: these properties are again automatically satisfied for complete surfaces with $K > 0$.   Such surfaces are, in particular, *closed in E^3*, in the sense defined toward the end of Section 6.

The assumption that K is positive of course means that the surface in E^3 is locally convex.   The property of local convexity may hold if K is permitted to be zero (but, of course, not if K is permitted to be negative).   If K is permitted to be zero everywhere, however, the theorems partially formulated above will not hold in general.   For example, consider any curve in the plane with curvature of the same sign throughout, but with a double point, and which is infinite in length on going out from a point of it in both directions.

---

[1] S. Cohn-Vossen [C.10] dealt with questions similar to those treated in this section, but did not formulate the generalization of Hadamard's theorem that is proved below.   The methods used by Cohn-Vossen also have some similarity to those employed here.   However, the notion of completeness, for example, does not figure in Cohn-Vossen's discussion.

Erect over it a cylinder whose generators are infinite straight lines. This surface is evidently complete and locally convex, but it has a double line; however K is identically zero. It has been shown by van Heijenoort [V.1] that the generalized theorem of Hadamard remains true if local convexity is assumed, together with $K \geq 0$, provided that K is not identically zero. [Actually van Heijenoort considers more general surfaces than those treated here since he does not assume them to be differentiable, and hence they are without a Riemannian metric. He also considers n-dimensional surfaces in $(n + 1)$-dimensional space.] It is thus rather remarkable that Chern and Lashof [C.6] and Sacksteder [S.1] have proved such theorems for the case $K \geq 0$ *without the assumption of local convexity*, provided again that the necessary condition $K \not\equiv 0$ is fulfilled. Since all of these theorems are proved ultimately by showing that the surfaces in question form the boundary of a three-dimensional convex body, it follows that local convexity is a property which *follows* from the assumptions of completeness, $K \geq 0$, and $K \not\equiv 0$, and not a property which needs to be assumed. In what follows here, however, it will be assumed that K is everywhere positive, since the discussion is then much easier to carry out.

The theorem to be proved is formulated as follows. *Let $\bar{S}$ be a two-dimensional abstract Riemannian manifold that is complete in a metric derived from a metric fundamental form having everywhere positive Gaussian curvature. Let S be given by a vector $\mathbf{X}(\bar{S})$ as an isometric mapping of $\bar{S}$ into Euclidean three-dimensional space E^3, which is locally one-to-one and defines a regular surface in terms of local parameters.* It follows that:

1. *$\bar{S}$ and S are homeomorphic*, and hence S has no double points.

2. *$\bar{S}$ and S are homeomorphic to the Euclidean plane if $\bar{S}$ is open and to the sphere if $\bar{S}$ is closed.*

3. *S is the full boundary of a three-dimensional convex set in E^3, which is unbounded if S is open, bounded if it is closed.*

Before giving the details of the proof of the theorem it seems advisable to indicate first the general course of the proof and to describe the kind of geometrical constructions that will be made. An arbitrary point p of S is taken and a rectangular Cartesian coordinate system with this point as origin is chosen such that the x,y-plane is the tangent plane to S at p and the z-axis is taken so that the surface points in a neighborhood of p lie on that side of the tangent plane for which their z-coordinates are positive. This is possible, since K is assumed to be positive at all points. Near p the surface is assumed to be represented in the form $z = z(x, y)$. It is then intuitively clear, again because $K > 0$ is assumed, that the set of points above the surface $z = z(x, y)$ but below the plane $z = t$ for t a sufficiently small positive constant, fills out a three-dimensional open convex set K^t bounded below by the surface and

above by a convex set in the plane $z = t$ (see Fig. 8.15).   The essential idea
in the proof that follows is to extend such a convex body for the largest possible value t^* of t, which, it turns out, amounts to considering the maximal
such body.   (However, the convex sets K^t are not in general bounded by
surface points having a one-to-one projection on the x,y-plane, but rather

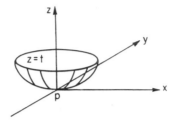

Fig. 8.15   *A convex body $K^{\dot{z}}$.*

appear as in Fig. 8.17.)   By virtue of the assumptions of the theorem it will
in fact be shown that if t^* is finite that S is compact, i.e., closed, and hence is
the full boundary of a bounded convex body, while if t^* is infinite it is then
obvious that S is open, and rather clear that S would be topologically equivalent to the plane.

  It has some point to begin filling in the details of this program by proving
the correctness of the first of the above assertions, since it foreshadows the
steps to be taken in proving the second assertion.   To this end consider
plane sections of S by normal planes at p, i.e., planes containing the z-axis;
these are all regular curves having nonvanishing curvature of the same sign.
They touch the plane $z = 0$ only at the origin p and hence are such that they
have a one-to-one projection on the x, y-plane if their z-coordinates are small
enough.   Thus the surface points for $0 \leq z \leq t$ in a neighborhood of p have
a one-to-one projection on the x, y-plane as long as t is small enough, and this
portion of S has a representation in the form $z = z(x, y)$.   The points such
that $z(x, y) < z < t$ evidently form the interior of a Jordan surface bounded
below by the surface $z = z(x, y)$ for $z < t$ and above by the plane convex
curve cut out of S by the plane $z = t$ plus its interior.   This curve bounds a
convex domain because it is a simple closed curve without inflections, since
$K > 0$ is assumed.   This fact was proved in Chapter II with the aid of a theorem of E. Schmidt.   This theorem, which is of basic importance for the
following, states that *a connected open set in E^n such that a local support plane*
(of dimension $n - 1$) *exists at every boundary point is convex.*   (By a local
support plane is meant a plane such that all points of the set in a neighborhood of the given boundary point lie on the same side of the plane.)   The
theorem of E. Schmidt is now applied in three dimensions to the set of in-

terior points of the Jordan surface just defined: at every boundary point of that set there is a local support plane, that is, the tangent plane at boundary points belonging to the surface S; otherwise the plane $z = t$. Since the set of interior points is a connected open set, the conclusion follows from E. Schmidt's theorem. Thus a convex body K^t exists for t small enough. This same theorem is used again in a decisive way to prove the second assertion of the preceding paragraph, which refers to the union of all of the convex sets K^t.

There are some advantages to numbering the steps to be taken in proving that assertion, which in turn proves the theorem.

1. The notion of a *convex part* $\bar{M}^t$ of $\bar{S}$ is introduced. By that is meant (a) a simply connected open set on the parameter surface $\bar{S}$ containing the point $\bar{p}$ whose image is the point p of S, which is such that (b) its image M^t given by $\mathbf{X}(\bar{M}^t)$ on the surface S in E^3 is homeomorphic under X to $\bar{M}^t$. In addition it is required that (c) a three-dimensional open convex set K^t exist which has $z = 0$ and $z = t$ as support planes and M^t as its boundary for $0 \le z < t$, while the remainder of its boundary points form a closed convex set in the plane $z = t$. The image M^t in S of $\bar{M}^t$ will also be called a convex part of height t. Because of what was done in the preceding paragraph it is clear that *a convex part $\bar{M}^t$ in $\bar{S}$ exists for small enough positive values t of z.*

2. The parameter surface $\bar{S}$ is converted into a metric space in the usual way (cf. Section 5) by defining the distance ρ between any pair of points as the g.l.b. of the lengths of all rectifiable curves of $\bar{S}$ joining them. It is proved next that (a) *every convex set K^t is bounded in E^3*, and that (b) *the set $\bar{M}^t$ is bounded in the metric ρ.* To prove the statement (a) consider three points of M^t near p, the origin of the Cartesian coordinate system in E^3, which lie in each of three planes containing the z-axis and that are $120°$ apart. The tangent planes to S at these points, which are therefore support planes of K^t, determine evidently a convex pyramid that contains K^t in its interior. Since K^t also lies below the plane $z = t$, which cuts off the pyramid above, it is clear that K^t is bounded in E^3. Thus statement (a) is proved. To prove statement (b) it need only be noted that since K^t is convex any two of its boundary points in M^t can be joined by a plane arc lying in M^t; all such arcs have lengths that are easily seen to be less in value than four times the diameter of K^t. The preimages in $\bar{M}^t$ of any such arcs (which are unique because M^t and $\bar{M}^t$ are by assumption homeomorphic) have the same lengths, and hence the distance ρ in the metric of $\bar{S}$ is uniformly bounded for all pairs of points in $\bar{M}^t$.

3. Some properties of the sets $\bar{M}^t$ and M^t, and of their boundaries $\bar{B}^t$ and B^t are next to be described. The boundary B^t of M^t evidently lies in the plane $z = t$. Take a sequence p_n of points of M^t which have as limit point a

point b of B^t: such points could, for example, be chosen on a plane section of M^t made by a plane through the z-axis and with z-coordinates tending monotonically with n toward b (cf. Fig. 8.17).   The points $\bar{p}_n$ in $\bar{M}^t$ (cf. Fig. 8.16) corresponding by the homeomorphism between $\bar{M}^t$ with M^t form a bounded

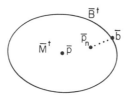

Fig. 8.16   The set $\bar{M}^t$.

set in the metric of $\bar{S}$ and hence have an accumulation point $\bar{b}$ in $\bar{S}$ since $\bar{S}$ has a complete metric.   (This is one of the decisive points at which the assumption of completeness is used.)   The point $\bar{b}$ is not a point of $\bar{M}^t$: if it were, its image b would be in M^t, which is not the case.   Thus $\bar{b}$ lies in the boundary $\bar{B}$ of $\bar{M}^t$, since there are points of $\bar{M}^t$ arbitrarily near to it.   Every point b of B^t is thus the image of some point $\bar{b}$ of $\bar{B}^t$, and thus belongs to S.

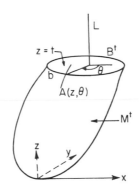

Fig. 8.17   Extension of a convex part.

4. Since K^t is a three-dimensional convex body, it is clear that every plane $z = \tau$ for $0 < \tau < t$ cuts K^t in a convex curve in M^t (which might be called B^τ) enclosing a two-dimensional convex set, and that these curves converge for $\tau \to t$ in an obvious sense to the boundary B^t of M^t.   In addition, since each such plane $z = \tau$ definitely *cuts* S (and is not tangent to it) at the points of B^τ, it is clear that B^τ is a regular curve, and hence its preimage $\bar{B}^\tau$ is also such a curve.

5. Two separate cases with respect to the boundary B^t of M^t are considered: (a) a point b of B^t, and thus a point of S, exists such that the plane $z = t$ is a tangent plane of S at b; (b) no such tangent plane exists at any point of B^t.

In case (a) it is clear that a convex part of S exists in a neighborhood of b—which by (3) is a point of S—that lies below the plane $z = t$—just as was found for the point p at the origin of the x,y,z-coordinate system. In view of (4) it is clear that values $z = \tau, 0 < \tau < t$, exist such that the two convex parts would have a plane convex section of S as their common boundary. Thus S would be the full boundary of a bounded convex body, and hence would be a compact surface topologically equivalent to a sphere. Its map $\bar{S}'$ in $\bar{S}$ is known to be one-to-one; hence the whole of $\bar{S}$ would be covered by $\bar{S}'$ since a manifold cannot contain a compact manifold as a proper part of it. This can also be seen by noting that $\bar{S}'$ is open in $\bar{S}$, but also closed in $\bar{S}$ since $\bar{S}$ has a complete metric; hence $\bar{S}'$ is the whole space $\bar{S}$. Thus the case of Hadamard's original theorem would be in question and would be proved in this case.

In case (b) the boundary points B^t of M^t are in the plane $z = t$ and they, again by (3), are points of S. Since the plane *cuts* S at all points of B^t it is a regular curve in a neighborhood of each of its points. These points are also the boundary points of a plane section of the closure of the convex set K^t, and hence form a regular closed convex curve having nonzero curvature everywhere, since the Gaussian curvature of S is positive. B^t is therefore the boundary of a two-dimensional convex set in the plane $z = t$.

6. It is now to be shown that in case (b) the convex parts $\bar{M}^t$ and M^t can be extended to convex parts $\bar{M}^{t_1}$ and M^{t_1} for a value t_1 larger than t. To this end it is first to be observed that the closures $\bar{M}^t + \bar{B}^t$ and $M^t + B^t$ of $\bar{M}^t$ and M^t are homeomorphic. This in turn, would be clear if it is once shown that the mapping of $\bar{B}^t$ on B^t is one-to-one; this however follows because B^t, as a plane closed convex curve, is free of double points and hence two different points $\bar{b}_1, \bar{b}_2$ of $\bar{B}^t$ cannot map into the same point of B^t. The extension of $\bar{M}^t$ to a larger convex part is then done as might be expected. Each point $\bar{b}$ in $\bar{B}^t$, the boundary of $\bar{M}^t$, has a neighborhood $\bar{N}$ in $\bar{S}$, part of which lies in $\bar{M}^t$, and the map $X(\bar{N})$ furnishes a local extension of M^t which contains the image point $b = X(\bar{b})$ of $\bar{b}$ in the boundary B of M^t in its interior. The Heine-Borel theorem is then used to select a suitable finite subset of such patches at the boundary B of M^t (which is a compact set) in such a way that their union leads to an extension of M^t which is the topological image of an extension $\bar{M}^{t_1}$ of $\bar{M}^t$ and also is such that the extension $\bar{M}^{t_1}$ has all of the properties required for a convex part. The circumstances are indicated in Fig. 8.17. The three-dimensional convex body K^t shown there has as boundary the union of the convex part M^t and a two-dimensional bounded plane

convex set having the curve B^t as its boundary. The line L shown in Fig. 8.17 is taken through a point in the interior of the convex set bounded by B^t and orthogonal to the plane $z = t$ that contains B^t. The arc $A(z, \theta)$ shown in the figure is an arc cut out of S by a plane through L and the point b of B^t; this plane is fixed by the angle θ. The arc $A(z, \theta)$ is a regular plane curve for values of z sufficiently near to $z = t$ for each value of θ in $0 \le \theta \le 2\pi$, since b is a point of S and the tangent plane to S at b does not coincide with the plane through b and L, as is readily seen. In addition, the coordinate z itself can be introduced as a parameter on the arc A for $t - \epsilon < z < t + \epsilon$ for $\epsilon > 0$ and not too large, again because the tangent plane of S at b is not horizontal and hence z is a monotonic function of the arc length s along the curve A. In this way an extension of M^t along such arcs for z-values larger than z and for $0 \le \theta \le 2\pi$ is achieved. In fact, it is clear that every point b has a neighborhood within which the curves $A(z, \theta)$ form one set of a family of curvilinear coordinates on S. By the Heine-Borel theorem it follows that B is covered by a finite number of such neighborhoods, and that a number $t_1 > t$ exists such that a strip ΔM^t of S exists above $z = t$ defined for $t \le z < t_1$, $0 \le \theta \le 2\pi$. This strip can have no double points if t_1 is not taken too large since such double points could result only if arcs $A(z, \theta)$ and $(A(z, \theta + \pi)$ had points in common for arbitrarily small values of t_1; and that is prevented by the fact that L is at a distance from B^t that is bounded below for all points of B^t. It follows that the inverse map $\Delta \bar{M}^t$ in $\bar{S}$ of ΔM^t is in one-to-one correspondence with ΔM^t, and hence that $\bar{M}^t + \Delta \bar{M}^t = \bar{M}^{t_1}$ and $M^t + \Delta M^t = M^{t_1}$ are homeomorphic open sets. By the same argument as in (5) above the boundary B^{t_1} of M^{t_1} is seen to be a simple regular plane convex closed curve, all of whose points are points of S. If the interior of the plane convex curve B^{t_1} is added to M^{t_1} the result is evidently a closed Jordan surface. At each point of this surface there exists a local support plane relative to the interior domain D since the Gaussian curvature K is positive at all points of $M^{t_1} + B^{t_1}$ and the remainder of the boundary is a portion of a plane below which all of D lies. Thus, the theorem of E. Schmidt can be applied to prove that the interior points D form a convex set. Thus a bounded convex body K^{t_1} exists and $\bar{M}^{t_1}$ is therefore seen to be a convex part. Thus *if the boundary B^t of M^t is a closed convex curve it is possible to extend the convex part $\bar{M}^t$ to larger values of t.*

The proof of the extension of Hadamard's theorem can now be completed by considering the union of all convex parts $\bar{M}^t$ and M^t, and the convex bodies K^t having M^t as the curved part of their boundaries; these sets are denoted by $\bar{M}^{t_*}$, M^{t_*}, and K^{t_*} with t_* defined as the l.u.b. of the t-values for which convex parts exist. This notation is reasonable since it is easily seen that $\bar{M}^{t_*}$ and M^{t_*} have the properties needed to identify them as convex parts: in fact, if t is any value less than t_*, a convex part M^t exists, and K^{t_*}

as the union of such nested convex bodies is convex. Two cases are to be considered: (a) t_* is finite, (b) t_* is infinite.

Consider case (a). The set K^{t_*} is the union of open convex bodies K^t for $0 < t < t_*$. It has the convex part M^{t_*} as the part of its boundary that lies under the plane $z = t^*$. The discussion above under (3) and (4) concerning M^{t_*} and its boundary B^{t_*} applies, since M^{t_*} is a convex part. It is to be noted that B^{t_*}, the points of which belong to S, cannot enclose a two-dimensional convex set in the plane $z = t^*$: if it did, the discussion under (6) shows that M^{t_*} could be extended to a convex part for a value of t larger than t^*—contrary to the definition of t_*. Thus B^{t_*} is of necessity either a point or a segment of a straight line. The latter case is not possible since a surface having positive Gaussian curvature contains no straight-line segment. Thus B^{t_*} would be a single point, with $z = t_*$ as a support plane of K^{t_*} there, and consequently $z = t_*$ is a tangent plane of S at the point. The discussion under (5) of case (a) then shows that S is a closed convex surface, with $\bar{S}$ homeomorphic to it, and Hadamard's original theorem is proved.

In case (b) it is clear that K^∞ (in an obvious notation) is an unbounded convex set in E^3, with S as its full boundary. Since every plane $z = t$, $0 < t < \infty$ cuts out of M^∞ a closed convex curve—they are level lines of S, in fact—it is clear that M^∞ is topologically equivalent to the plane.[1] The set $\bar{M}^\infty$ of $\bar{S}$, the topological image of M^∞, can now be seen to cover the whole of $\bar{S}$. If it did not, a boundary point $\bar{b}$ of $\bar{M}^\infty$ in $\bar{S}$ would exist. Such a point would have an image point b in S with the property that points of M^∞ arbitrarily near to it exist—since points $\bar{b}_i$ of $\bar{M}^\infty$ exist arbitrarily near to $\bar{b}$—and hence b belongs to M^∞ and therefore $\bar{b}$ to $\bar{M}^\infty$. Thus no boundary point $\bar{b}$ can exist, $\bar{S}$ is therefore identical with $\bar{M}^\infty$, and S is therefore identical with M^∞. Thus $\bar{S}$ like S, is homeomorphic with the plane if it is not compact. The extension of Hadamard's theorem to complete open surfaces is therefore established.

The open surfaces that figure in the theorem have a number of interesting properties: (a) their spherical images lie on a half-sphere, (b) they have infinite area, (c) there exists a tangent plane at some point of any such surface on which the orthogonal projection of it is convex and one-to-one, and hence a representation in the form $z = z(x, y)$ exists in the large. (This also shows again that the surface is homeomorphic with the plane.) The proofs of these and other facts are left as exercises among those formulated at the end of the chapter. It is perhaps worth mentioning also that any of these surfaces S is closed in E^3. This is readily seen since any surface S is the full boundary of

---

[1] It is known in any case (see Stoker [S.7]), and is easily proved, that the set of boundary points of a three-dimensional unbounded convex set is homeomorphic with the plane except when the convex set is a cylinder or the space between a pair of parallel planes.

a convex set, hence any point p of E^3 not on the surface either lies in its interior domain, which is an open set, or in its exterior, and in the latter case could be separated from S by a support plane of S: in either case it is clear that p would have a neighborhood free of points of S.

PROBLEMS

1. Prove that a geodetically convex set on the unit sphere, if not the entire sphere, lies on a hemisphere. *Hint:* Consider the three-dimensional cone obtained by drawing rays from the center of the sphere to the points of the set. Consider next an abstract compact surface not embedded in three-space on which $K \equiv +1$ holds and show that every geodetically convex set on it, if not the whole surface, lies in a geodetic circle of radius $\pi/2$.

2. In Chapter V it was shown that all developables satisfying a certain "completeness" condition (more general than that of Hopf-Rinow) are cylinders with infinitely long generators. Give an easier proof assuming completeness in the sense of Hopf-Rinow, and using the fact that the covering surface S' (in the notation of Section 15) is the entire Euclidean plane.

3. Construct examples of complete open surfaces with finite area. *Hint:* Seek them among surfaces of revolution having "spines."

4. Show that surfaces of revolution having $K = -1$ exist such that there are regular points on them at arbitrarily large distances from all singular points.

5. Consider a complete open surface S in E^3 having $K > 0$. From the results of Section 17 show the following: (a) The total Gaussian curvature of S is at most equal to 2π. *Hint:* Show first that the spherical image of S is convex on the sphere. (b) The area of S is infinite. (c) Every divergent curve on the surface goes to ∞ in E^3. (d) If a complete surface in E^3 with $K > 0$ has a pair of parallel tangent planes, it is a closed surface. (e) On complete open surfaces of revolution in E^3 with $K > 0$ find unbounded geodetically convex domains that are not the whole surface. (f) Show that the surfaces S have a valid representation in the large of the form $z = z(x, y)$, with x, y, z as rectangular Cartesian coordinates. Show that the x,y-plane can be chosen as a tangent plane of S and that the domain of definition of the function $z(x, y)$ is convex. *Hint:* Use the convexity of the spherical image.

6. Show that the closure of a not-too-large geodetically convex set is convex.

7. Show that no simple everywhere regular closed geodetic line can exist on a complete open surface with $K > 0$ that is embedded in three-space. *Hint:* Use the Gauss-Bonnet formula together with appropriate properties of such surfaces (see Problem 5).

8. A "parallelogram" of a Tchebychef net is defined by the net curves

$p = 0$, $q = 0$, $p = p_1$, $q = q_1$ and the lines $p = 0$, $q = 0$ are in addition assumed to be geodetic lines. Show that the interior of the parallelogram has the local support property at all boundary points with respect to geodesics if the Gaussian curvature of the surface is negative, and hence is geodetically convex. *Hint:* Use the formulas of Problem 17 of the previous chapter, together with the formula analogous to (8.17) that holds when K is not necessarily -1. (This result is due to Bieberbach [B.3].)

9. Show that all geodetic triangles on the covering surface S' (cf. Section 15) of a complete surface S having $K \leq 0$ are geodetically convex. Then show, by induction on the number of sides, that all simple closed polygons, on the same class of surfaces S', with geodesics as sides, and with interior angles $< \pi$, are geodetically convex.

10. In Section 6 it was not shown that the condition (2) is equivalent to the three other conditions for completeness formulated there. Show this to be true by proving that condition (1) implies condition (2).

11. Show that a compact surface of positive curvature cannot be triangulated by geodetic quadrangles such that exactly four such quadrangles occur at every vertex point. *Hint:* Sum the angles at all vertices, and also over all faces, and use the Gauss-Bonnet formula. Thus (cf. remarks at the end of Chapter VII), a fishnet of quadrangles cannot be stretched over a surface homeomorphic to the sphere with $K > 0$. On a torus in E^3 formed by rotating a circle show that such nets can be constructed, and in such a way that each mesh has its opposite sides of the same length.

CHAPTER IX

Intrinsic Differential Geometry of Manifolds.   Relativity

1. Introduction

In Chapter VII intrinsic geometry of two-dimensional surfaces in the small was treated, but the treatment was carried out initially by regarding the surfaces as embedded in three-dimensional space.   In Chapter VIII, the geometry of two-dimensional surfaces in the large was treated; this included both intrinsic problems and embedding problems.   An essential first step in Chapter VIII, as a preliminary to the discussion of differential geometry in the large, was the introduction of a variety of basic notions and conventions that make it possible to define in the large the geometric objects to be studied in such a way that the tools of analysis become available for this purpose. Thus in the first sections of Chapter VIII the notion of an n-dimensional Riemannian manifold was introduced; it is with such manifolds that the present chapter is concerned.

In addition to tools from analysis, certain tools furnished by linear algebra are also of great importance in dealing with differential geometry in manifolds, particularly when the point of view to be taken is strictly the intrinsic point of view—and that is the case in this present chapter.   It would be more correct to say that the mathematical ideas and techniques needed for the purposes of the present chapter, in addition to the elements of mathematical analysis, come from certain aspects of affine and Euclidean geometry, which in their turn are conveniently described in terms of linear algebra.   In order to avoid interruptions in the present chapter these matters have been summarized in Appendix A.

Thus as a preliminary to the reading of the present chapter it is recommended that the opening sections of Chapters VII and VIII, and the material of Appendix A be reviewed while having in mind one of the main objects of the present chapter: to introduce the basic concepts needed for the study of the differential geometry of n-dimensional manifolds in an intrinsic way. Such a review should make it fairly easy to grasp why the concepts to be introduced here are natural, even inevitable.

The key point in this process is the introduction of an n-dimensional linear vector space V, the so-called *tangent space*, at each point of the n-dimensional manifold; how this is done was explained in Section 2 of the preceding chapter. In two-dimensions, and with an embedding in three-space, this is nothing but the two-dimensional vector space spanned by the tangent vectors of any pair of parameter curves—in effect, the tangent plane. In Chapter VIII the analogous notion of an abstract tangent space formed by tangent vectors of the coordinate curves in an n-dimensional manifold was introduced, but without an embedding. Once this is done it becomes possible, with the aid of the vector space $\overline{V}$ dual to V, to define vectors and tensors of all types and ranks in the tangent space. Transformations of basis vectors in these spaces result through transformations of the coordinate curves in terms of which the manifold is parametrized; because of the fact that the Jacobian of the transformations is, by assumption, never zero the tangent vectors of the new parameter curves again span the tangent space. The invariance of vectors and tensors in the tangent space at any fixed point of the manifold is then understood with respect to such transformations; in this manner the tensors become by definition intrinsic invariants of the manifold: this is the only type of invariance considered in this chapter.

Thus vectors and tensors can be readily defined pointwise in the manifold in an almost purely algebraic way. However, for the purposes of *differential* geometry it is necessary to introduce differentiation processes with respect to such fields of vectors and tensors, which means in turn that these entities must be compared when located at *different* points. This also must be done in such a way as to preserve invariance of the objects obtained by differentiation in the sense just indicated. It can be done, following Weyl [W.1], by first introducing a notion of local parallelism of vectors in the manifold—so that the vectors may then be displaced parallel to themselves to a fixed point and then differentiated exactly as that is done for a system of vectors or tensors at a point when these objects depend upon scalar parameters. The way to do that was pointed out, really, by Riemann [R.3] in his famous *Über die Hypothesen, welche der Geometrie zu Grunde liegen*, in which he proposed to base the geometry of manifolds on the principle that it should be approximately *Euclidean* in the neighborhood of each point. Weyl modified this hypothesis somewhat by requiring as a preliminary step that the geometry should first of all be approximately *affine* near each point since that suffices to introduce the concept of vectors and of tensors, as well as a notion of parallelism of vectors near each point; and this makes a tensor *calculus* possible. Afterward an approximately Euclidean metric can be introduced in order to complete the definition of what is then called a Riemannian manifold.

Evidently, everything hinges on the interpretation of the phrase "approximately affine" and afterwards on the phrase "approximately Euclidean." The method of interpretation of these phrases employed here is modeled after that of Cartan [C.2],[1] in which the analogy with affine and Euclidean geometry is developed by treating these geometries themselves in an invariant way in terms of general curvilinear coordinates, which are the only sort available in manifolds. The appropriate definition of the terms "approximately affine" and "approximately Euclidean" in any manifold becomes then rather apparent, and the formal calculus of tensors in manifolds can be introduced in an invariant manner that seems very reasonable.[2] On this basis the differential geometry of manifolds can then be developed to yield a wide variety of interesting results.

In carrying out this program the present chapter is divided into three parts. In Part I affine and Euclidean spaces are discussed with the aid of tensor algebra and calculus. This is in itself of importance because these ideas are very much used in mechanics and physics, particularly when the employment of curvilinear coordinates in Euclidean space is made necessary or convenient. In Part II the ideas of Part I, following Cartan [C.2], are used as the motivation for the basic definitions and the treatment of the differential geometry of general manifolds, with emphasis on Riemannian manifolds. Finally, what is perhaps the most beautiful and striking of all the applications of these ideas, i.e., Einstein's general theory of relativity, is explained in Part III.

PART I. TENSOR CALCULUS IN AFFINE AND EUCLIDEAN SPACES

2. Affine Geometry in Curvilinear Coordinates

It has already been said that it is of interest to develop tensor calculus in affine spaces when the points of the space are located with the use of curvilinear coordinates. Thus it is assumed that the affine space, given originally by vectors X: $(x^1, x^2, \ldots, x_n)$, is covered in the large by a single system of curvi-linear coordinates u_i; i.e.,

$$(9.1) \qquad x^i = x^i(u^1, u^2, \ldots, u^n), \qquad i = 1, 2, \ldots, n;$$

---

[1] This book is very stimulating and interesting, particularly since it appeals strongly to the geometric intuition.

[2] In Chapter VII the local Riemannian geometry of two-dimensional surfaces was also introduced in making use of Euclidean geometry in such a way as to lead to the notion of parallel fields of vectors along curves. There, however, the surface was regarded as embedded in a higher dimensional Euclidean space, after which a notion of approximate parallelism could be introduced. Here, such an embedding is not made use of as a matter of principle; in its place the Euclidean geometry of a space of the same number of dimensions, but treated in general curvilinear coordinates, is used as the means to motivate the introduction of the appropriate concepts and definitions.

these functions are assumed to be defined in an open n-sphere, and to have derivatives of a certain finite order.   In these coordinates the position vector $\mathbf{X}$ becomes a function of the variables u^i.   The tangent vectors to the system of curves obtained by varying each of these in turn are, by definition, the vectors

(9.2)
$$\mathbf{v}_i = \frac{\partial \mathbf{X}}{\partial u^i}.$$

(Here, and in general in the following, it is understood that such a subscript or superscript as i in this equation takes on successively the values 1 to n.) Such a differentiation process is applied to vectors $\mathbf{X}$ located at the origin of the affine space; it therefore, in the first instance, produces vectors $\mathbf{v}_i$ that are also located at the origin.   However, it is a matter of *convention* that the vectors $\mathbf{v}_i$ are supposed located at the end point of the vector $\mathbf{X}$ and to be the result of a parallel displacement from the origin to that point in the affine space.   The vectors $\mathbf{v}_i$ are assumed to be everywhere linearly independent, and that in turn will evidently be the case if the Jacobian of the transformation (9.1) to curvilinear coordinates is not zero—an assumption that is always made.   In this fashion to every point of the space is attached a system of n linearly independent vectors, and the linear space spanned by such vectors at any one point is called the *tangent space* of the given affine space at that point.   The vectors $\mathbf{v}_i$ are called *basis* or *coordinate vectors* for the tangent space.   Of course in the present case such a tangent space also spans the whole space, but it is preferable not to think of it in that way since such a thing is just *not* possible in manifolds in general.

The crux of the present discussion lies in the introduction of differentiation processes for *vector fields* when these are defined with respect to the basis vectors of the tangent spaces.   As a preliminary step in this direction consider first the differentiation of the basis vectors themselves; the latter form a system of n sets of vector fields.   If the derivative is desired at a point P_0, the vectors $\mathbf{v}_i(P)$ in a neighborhood of P_0 are transported parallel to themselves to P_0 and then differentiated while regarding the parameter values u^i at P as being scalar parameters fixing the bundle of vectors at P_0; this leads, by definition, to the formulas

(9.3)
$$\frac{\partial v_i}{\partial u^k} = \frac{\partial^2 \mathbf{X}}{\partial u^i \, \partial u^k},$$

for the partial derivatives of the basis vectors upon using (9.2).   These are again vectors at P_0 and hence they can be represented in terms of the coordinate vectors of the tangent space at that point.   Thus all of the derivatives of the field vectors can be expressed in the form

(9.4)
$$\frac{\partial v_i}{\partial u^k} = \Gamma_{ik}^r \, \mathbf{v}_r;$$

clearly, a coefficient Γ^r_{ik} with three indices is necessary for this purpose. Thus, in the language of tensor algebra (cf. Appendix A), the quantities Γ^r_{ik} define, for fixed values of the indices i, k, the contravariant components of the vectors $\partial v_i/\partial u^k$ relative to the coordinate vectors at each point. The quantities Γ^r_{ik} thus are functions of the parameters defining the curvilinear coordinates; they vary in general from point to point. On the other hand if the coordinates are Cartesian it is clear that all of the quantities Γ^r_{ik} are zero in value. The differentials $d\mathbf{v}_i$ of the coordinate vectors $\mathbf{v}_i$ are of course given by the formulas

$$(9.5) \qquad\qquad d\mathbf{v}_i = \Gamma^k_{ir}\, du^r \mathbf{v}_k;$$

hence these differentials are vectors with contravariant components $\Gamma^k_{ir}\, du^r$ for each fixed index i. Cartan [C.2] introduces for these differential forms the notation

$$(9.6) \qquad\qquad \omega^k_{(i)} = \Gamma^k_{ir}\, du^r.$$

The subscript (i) in parentheses is used in order to underline the fact that this subscript has no tensor significance that would make $\omega^k_{(i)}$ a mixed tensor of second rank: the quantities $\omega^k_{(i)}$ are, however, the components of a contravariant vector for each fixed value of the index i.

Consider next a field of vectors $\mathbf{Y}(u^1, u^2, \ldots, u^n)$ in the affine space; this field is to be defined by giving the contravariant components ξ^i of the field at each point of the space with respect to the coordinate vectors at each point. The vector field is given by

$$(9.7) \qquad\qquad \mathbf{Y} = \xi^i \mathbf{v}_i.$$

The differential at a point P is taken (after a preliminary parallel transport of the fields $\mathbf{Y}$ and $\mathbf{v}_i$ to P); just as was done above [cf. (9.5)] in the case of the fields $\mathbf{v}_i$, the result is

$$(9.8) \qquad d\mathbf{Y} = d\xi^i \mathbf{v}_i + \xi^i \omega^k_{(i)} \mathbf{v}_k = (d\xi^i + \xi^k \omega^i_{(k)})\mathbf{v}_i,$$

upon using (9.5). Thus the differential $d\mathbf{Y}$ is a vector that has for components with respect to the basis $\mathbf{v}_i$ the quantities

$$(9.9) \qquad\qquad D\xi^i \equiv d\xi^i + \xi^k \omega^i_{(k)},$$

which are said to yield the *absolute differential* of the vector field; at the same time the operator symbol D is defined by implication.

It is useful to place alongside of (9.9) an analogous formula for the absolute differential of a vector field when it is defined by covariant components. To this end dual coordinate vectors $\mathbf{v}^j \equiv g^{ij}\mathbf{v}_i$ are introduced (cf. Appendix A, Section 6), and their differentials are denoted by $d\mathbf{v}^j \equiv \omega^{(j)}_k \mathbf{v}^k$ in terms of covariant components $\omega^{(j)}_k$. Since $\mathbf{v}_i \cdot \mathbf{v}^j = \delta^j_i$ and hence $d\mathbf{v}_i \cdot \mathbf{v}^j + \mathbf{v}_i \cdot d\mathbf{v}^j = 0$

it follows from (9.5) and (9.6) that $\omega_i^{(j)} = -\omega_{(i)}^j$. For a covariant field given by $\mathbf{Y} = \xi_i \mathbf{v}^i$, in place of the contravariant field of (9.7), it is now readily found by the same process that led to (9.9) that $d\mathbf{Y} = D\xi_i \mathbf{v}^i$ with

$$(9.9)_1 \qquad\qquad d\xi_j \equiv d\xi_j + \xi_k \omega_j^{(k)}$$
$$\equiv d\xi_j - \xi_k \omega_{(j)}^k$$
$$\equiv d\xi_j - \xi_k \Gamma_{jr}^k \, du^r.$$

Obviously, *if the vector field happens to be a constant field*, and thus also a parallel field, the absolute differentials are zero, and this fact is characteristic for such fields.   It is also important to observe that the absolute differentials reduce to the ordinary differentials in case the coordinate system is a Cartesian system, i.e., if $\Gamma_{ik}^r \equiv 0$ holds.

It would be possible to continue this process by considering *tensor fields* in an affine space, and their differentials.   However, since most of this chapter is devoted to geometries in spaces provided with a metric, all purposes are served for the moment by this brief discussion of the calculus of *vector* fields in affine spaces.   Tensor calculus in Euclidean spaces is treated in the next section.

3.   Tensor Calculus in Euclidean Spaces

An n-dimensional Euclidean space described in terms of curvilinear coordinates u^i is assumed given.   The elements of differential geometry in such spaces are to be introduced.

Any curve in the space is quite naturally defined by a vector $\mathbf{X}(t) = \mathbf{X}(u^1(t), u^2(t), \ldots, u^n(t))$, and its tangent vector $\mathbf{X}'(t)$ is given, by definition, by

$$(9.10) \qquad\qquad \mathbf{X}'(t) = \frac{\partial \mathbf{X}}{\partial u^i}\, \dot{u}^i = \dot{u}^i \mathbf{v}_i,$$

in terms of the coordinate vectors of the tangent space at some given point P. Once more, this vector, which in the first instance is located at the origin, is considered as having been displaced parallel to itself to point P.   The line element ds of the curve is by definition given, in terms of the scalar product defined in that space, by

$$(9.11) \qquad ds^2 = \mathbf{X}' \cdot \mathbf{X}' \, dt^2 = \mathbf{v}_i \cdot \mathbf{v}_k \dot{u}^i \dot{u}^k \, dt^2 = \mathbf{v}_i \cdot \mathbf{v}_k \, du^i \, du^k.$$

Quantities $g_{ik} = g_{ki} = \mathbf{v}_i \cdot \mathbf{v}_k$ are introduced, in accordance with the discussion in Section 6 of Appendix A concerning tensor algebra in Euclidean spaces, and the line element of the space takes the form

$$(9.12) \qquad\qquad ds^2 = g_{ik} \, du^i \, du^k.$$

It is an invariant in the tangent space, in the sense of tensor algebra, defined in terms of the covariant tensor g_{ik} and the contravariant tensor fixed by the differentials taken along the curve. Of course, it would be possible to identify this tangent space with the original Euclidean space, but it is preferable here also, as in the previous section, not to do that because of the later generalization to other spaces. Lengths of vectors, and angles between vectors, in the tangent space are now defined by using this metric form. Two directions $du^i_{(1)}$, $du^i_{(2)}$, for example, are at the angle ϕ to one another given, by definition, by the formula

$$(9.13) \qquad \cos \phi = \frac{g_{ij}\, du^i_{(1)}\, du^j_{(2)}}{\sqrt{g_{ij}\, du^i_{(1)}\, du^j_{(1)}} \, \sqrt{g_{ij}\, du^i_{(2)}\, du^j_{(2)}}}.$$

Obviously, it is a scalar invariant.

In the preceding section the derivatives and differentials of the coordinate vectors $\mathbf{v}_i$ were derived for an affine space: they continue to hold for Euclidean spaces since these are special cases of affine spaces. However, *it now becomes possible to express the differential forms of Cartan given by (9.5) in terms of the coefficients g_{ik} of the line element by calculating the coefficients Γ^k_{ir} in terms of these functions.* From $g_{ik} = \mathbf{v}_i \cdot \mathbf{v}_k$ the differentials dg_{ij} can be expressed as follows:

$$(9.14) \qquad g_{ki}\omega^k_{(i)} + g_{ik}\omega^k_{(j)} = dg_{ij};$$

they are a consequence of (9.5) and (9.6). Upon introducing the new symbols $\omega_{(i)j}$ by the formulas

$$(9.15) \qquad \omega_{(i)j} = g_{jk}\omega^k_{(i)} = g_{jk}\Gamma^k_{ir}\, du^r,$$

thus lowering the index and passing to a covariant representation, (9.14) obviously takes the form

$$(9.16) \qquad \omega_{(i)j} + \omega_{(j)i} = dg_{ij}.$$

These relations are at first sight too few in number to determine the n^3 functions Γ^k_{ir}; however, if it is noted that these symbols are symmetric in the indices i and r:

$$(9.17) \qquad \Gamma^k_{ir} = \Gamma^k_{ri},$$

since $\partial \mathbf{v}_i/\partial u^j = \Gamma^k_{ij}\mathbf{v}_k$ and $\partial v_i/\partial u^j = \partial \mathbf{v}_j/\partial u^i$ from (9.4), the relations (9.16) do suffice to fix them. To carry this out explicitly it is convenient to set

$$(9.18) \qquad \Gamma_{ijr} = g_{jk}\Gamma^k_{ir} = \Gamma_{rji},$$

and express (9.17) in the form

$$(9.19) \qquad \Gamma_{ijk} + \Gamma_{jik} = \frac{\partial g_{ij}}{\partial u^k},$$

which results from (9.15) and (9.16).   The symmetry property expressed by (9.18) permits (9.19) to be written as follows:

$$(9.20) \qquad \Gamma_{kij} + \Gamma_{ijk} = \frac{\partial g_{ij}}{\partial u^k}.$$

Cyclic permutation of the indices i, j, k in that order leads to two further relations which by addition and subtraction yield the values of the quantities Γ_{jki}:

$$(9.21) \qquad \Gamma_{jki} = \frac{1}{2}\left(\frac{\partial g_{jk}}{\partial u^i} + \frac{\partial g_{ik}}{\partial u^j} - \frac{\partial g_{ij}}{\partial u^k}\right).$$

Evidently, these quantities are linear combinations of first derivatives of the coefficients g_{ik} of the fundamental form.   By solving the linear equations (9.18) for the quantities Γ_{ir}^k these functions can be expressed also in terms of first derivatives of the coefficients g_{ik}, though not as linear functions of them; this is readily done by making use of the contravariant form g^{ik} of the metric coefficients [cf. equation (31) in Appendix A]:

$$(9.22) \qquad \Gamma_{ij}^k = \Gamma_{ji}^k = g^{kh}\Gamma_{ihj}.$$

It is worthwhile for later purposes to write down explicitly the formula for the symbols of the first kind:

$$(9.23) \qquad \Gamma_{ikj} = \frac{1}{2}\left(\frac{\partial g_{ik}}{\partial u^j} + \frac{\partial g_{jk}}{\partial u^i} - \frac{\partial g_{ij}}{\partial u^k}\right).$$

It is to be noted that there are as many quantities Γ_{ijk}, or Γ_{ij}^k as there are derivatives $\partial g_{ij}/\partial u^k$, since these derivatives are also symmetric in i and j.   It is also important to have in mind that *each one of these three sets of quantities determines the other set uniquely.*   Clearly, to establish this fact it suffices to prove that the derivatives $\partial g_{ij}/\partial u^k$ are determined by the quantities Γ_{ikj} and this does indeed result by reversing the process which led to (9.21) through the use of equations of the type (9.20).

An interesting fact has now emerged: *if the coordinates happen to be Cartesian, then all of the Cartan symbols have the value zero since the quantities g_{ik} are constants.*   If the functions g_{ik} are not given as constants, it is clear that they must satisfy conditions of some sort if the space is to be Euclidean, as might reasonably be inferred from the study of the analogous problem in the two-dimensional case.   In fact the Cartan symbols must satisfy certain integrability conditions if the space is to be Euclidean, which in turn impose conditions on the functions g_{ik} and certain of their first and second derivatives.   It is a very important fact that these integrability, or compatibility, conditions, which are necessary to ensure that the space should be Euclidean, are also sufficient conditions (at least in the small). This result is not obvious, and its proof will be deferred until a later occasion when a proof of it will be given in a more general context.

Once the three-index symbols of Cartan and Christoffel have been expressed in terms of the components g_{ik} of the metric tensor of the Euclidean space they are of course available for the calculation of the absolute differential of a vector field defined in the space, as given by formulas (9.9) and $(9.9)_1$. Thus the following formulas result:

$$(9.24) \qquad D\xi^i = d\xi^i + \xi^k \Gamma_{kr}^i \, du^r,$$

and

$$(9.25) \qquad D\xi_i = d\xi_i - \xi_k \Gamma_{ir}^k \, du^r.$$

It has point, perhaps, to dwell for a moment on a particular way of interpreting these formulas. If the differentials du^r are regarded as small quantities of first order, then $D\xi^i$ furnishes the changes in the components of the vector field at points near P located by the parameter values $u^r + du^r$ with an approximation that is correct up to second order terms in the quantities du^r. To achieve this, a correction to the differentials $d\xi^i$ is clearly needed that involves the curvature of the coordinate system through the quantities Γ_{kr}^i, and these in turn are fixed by the values of the first derivatives of the metric coefficients g_{ik}.

The formulas (9.24) and (9.25) can be generalized to furnish reasonable definitions for the absolute differential of any tensor field in a Euclidean space. Consider as a typical example a tensor field of the type a_i^j. Two *constant* fields of vectors of types ξ^i and η_i are introduced, and the invariant $a_i^j \xi^i \eta_j$ is then defined. By the absolute differential Da_i^j would be meant the tensor at a point P whose components are the differentials of the functions a_i^j at that point, on the assumption that this is done in defining the components of a_i^j at all points near P with respect to a Cartesian coordinate system at P. It follows that the ordinary differential $d(a_i^j \xi^i \eta_j) = da_i^j \xi^i \eta_j + a_i^j d\xi^i \eta_j + a_i^j \xi^i d\eta_j$ is also given at point P by $(Da_i^j)\xi^i \eta_j$ in the present special case since the two fields ξ^i and η_j represent constant vectors. Since the fields given by ξ^i and η_j are constant fields it follows that $D\xi^i = D\eta_j = 0$ and the formulas (9.24) and (9.25) can be used to determine $d\xi^i$ and $d\eta_j$; it follows that $(Da_i^j)\xi^i \eta_j$ is given by the formula

$$(Da_i^j)\xi^i \eta_j = (da_i^j - a_k^j \omega_{(i)}^k + a_i^k \omega_{(k)}^j)\xi^i \eta_j.$$

Since ξ^i and η_j are tensors that can be chosen arbitrarily, it follows that

$$(9.26) \qquad Da_i^j = da_i^j - a_k^j \omega_{(i)}^k + a_i^k \omega_{(k)}^j$$

is a tensor and hence yields an appropriate definition for the absolute differential of the tensor a_i^j in terms of curvilinear coordinates.

It is of interest to apply the rule analogous to (9.26) to the metric tensor g_{ij} to verify the *theorem of Ricci*, i.e., that the absolute differential of this tensor is always zero. This is really obvious since $g_{ij}\xi^i\eta^j$ is of necessity a constant for arbitrary constant fields ξ^i, η^j since it represents the scalar product of the vectors at each point; but it can also be verified from the formula $Dg_j^i = dg_{ij} - g_{kj}\omega_{(i)}^k - g_{ik}\omega_{(j)}^k$ upon making use of the formula (9.14).

From the formulas for the absolute differential formulas are obtained next for what are called the *covariant derivatives* of vectors and tensors. These result, in the case of (9.25) for example, by regarding the differentials $d\xi_i = (\partial\xi_i/\partial u^r)\,du^r$ as being taken with reference to the coordinates u^i; it follows that

$$(9.27) \qquad d\xi_i = \left(\frac{\partial\xi_i}{\partial u^r} - \Gamma_{ir}^k\,\xi_k\right) du^r,$$

when the equations defining the symbols $\omega_{(k)}^i$ in terms of the Christoffel symbols have been used. The *covariant derivative of the covariant vector* ξ_i is then defined by the formula

$$(9.28) \qquad \frac{D\xi_i}{du^j} = \xi_{i,\,j} = \frac{\partial\xi_i}{\partial u^j} - \Gamma_{ij}^k\,\xi_k.$$

The comma before the subscript j thus is a sign denoting a covariant derivative with respect to the jth coordinate. That the quantities $\xi_{i,\,j}$ thus defined form a covariant tensor of rank two is to be inferred from (9.27), since the absolute differential $D\xi_i$ was constructed as a vector and the contravariant vector du^r can be chosen arbitrarily. It is also entirely possible (and it should be carried out as an exercise) to prove that the quantities $\xi_{i,\,j}$ define a tensor in accordance with the basic transformation rule of tensor algebra when a change of basis vectors from $\mathbf{v}_i$ to $\mathbf{v}_i'$ is made in using (9.2) in conjunction with the change of curvilinear coordinates from u^i to u'^i. It turns out, in other words, that the quantities Γ_{ij}^k transform in just the correct way to compensate for the fact that the partial derivatives $\partial\xi_i/\partial u^j$ do *not* define a tensor (to prove it is a straightforward exercise) unless a Cartesian coordinate system is used. It is also easy to see, for example from (9.28) and the fact that $\partial\xi_i/\partial u^j$ is not a tensor, that the symbols Γ_{ij}^k—and with them all bracket symbols—are also not in general tensors.

The covariant derivative of a contravariant vector ξ^i and of a tensor a_i^k are in like manner defined with the aid of (9.25) and (9.26) by the formulas

$$(9.29) \qquad \xi_{,j}^i = \frac{\partial\xi^i}{\partial u^j} + \Gamma_{kj}^i\,\xi^k,$$

$$(9.30) \qquad a_{j,\,k}^i = \frac{\partial a_j^i}{\partial u^k} + \Gamma_{hk}^i\,a_j^h - \Gamma_{jk}^h\,a_h^i.$$

From these examples it is not difficult to see by analogy how the covariant derivative of any tensor is formed.

4. Tensor Calculus in Mechanics and Physics

It is now possible to use the calculus of tensors developed above for Euclidean spaces in interesting and important ways for various purposes in mechanics and physics. Consider, for example, a scalar field $\phi(u^1, u^2, \ldots, u^n)$ defined with respect to a curvilinear coordinate system. At any fixed point it is readily seen that the partial derivatives $a_i = \partial\phi/\partial u^i$ define a vector field, as follows. The derivative $d\phi/ds$ is taken along any curve $u^i = u^i(s)$, with the arc lengths as parameters, to obtain $d\phi/ds = (\partial\phi/\partial u^i)(du^i/ds) = t \cdot \text{grad }\phi$ in terms of the vectors $t = (\dot{u}^1, \dot{u}^2, \ldots)$ and $\text{grad }\phi = (\partial\phi/\partial u^1, \partial\phi/\partial u^2, \ldots)$ but for a given curve $d\phi/ds$ is independent of the coordinate representation of the curve, as is the unit tangent vector t. It follows that $\text{grad }\phi$ at P is also independent of the coordinate representation since $d\phi/ds$ has that property for any curve through P. In addition, since the contravariant vector t could be taken as an arbitrary unit vector, it follows that $\text{grad }\Phi$ is a *covariant* vector. The notation $\Phi_{,i}$ is sometimes used for the components of the gradient in evident analogy with the convention for covariant differentiation of vectors and tensors. Of course, the fact that $\text{grad }\Phi$ is a covariant vector can also be easily verified by computing the gradient in a new coordinate system (u'^i).

The square of the length of the gradient is, by definition, the quantity

(9.31) $$g^{ij}\phi_{,i}\phi_{,j}.$$

It is obviously an invariant, and is sometimes denoted by $\Delta_1\phi$ and called Beltrami's *differential parameter of first order*. In orthogonal Cartesian coordinates this quantity is given, obviously, by $\sum_i (\partial\phi/\partial x^i)^2$; consequently it determines the magnitude of the gradient of ϕ in any other coordinate system simply because it is an invariant.

Another tensor of interest is derived from a vector field ξ^i by taking its covariant derivative $\xi^i_{,j}$ and then contracting to obtain the scalar $\xi^i_{,i}$. This scalar is called the *divergence* of ξ^i and denoted by $\text{div }\xi^i$; thus

(9.32) $$\text{div }\xi^i = \frac{\partial\xi^i}{\partial u^i} + \xi^k\Gamma^i_{ki}.$$

The familiar expression $\sum_i \partial\xi^i/\partial x^i$ for the divergence results in Cartesian coordinates, since $\Gamma^i_{kj} = 0$ in that case; thus it is clear that (9.32) furnishes the value of the divergence in any coordinate system in view of the known invariant significance of the divergence in Cartesian coordinates as the flux of a vector field at a point—which can, of course, also be proved by a formal calculation.

Another scalar of very great importance is obtained by taking the divergence of the gradient of ϕ to obtain what is called the *differential parameter* $\Delta_2\phi$ *of second order* of Beltrami. The divergence was defined for a contravariant field; consequently it is convenient to take $\xi^i = g^{ik}(\partial\phi/\partial u^k)$ and use (9.32) to obtain

$$(9.33) \qquad \operatorname{div}(g^{ik}\phi_{,k}) = \frac{\partial}{\partial u^i}\left(g^{ik}\frac{\partial\phi}{\partial u^k}\right) + g^{kj}\frac{\partial\phi}{\partial u^j}\,\Gamma^i_{ki}.$$

This expression could be given a more elegant form, but since no use is made of it later on the calculation will not be made. It is to be noted, however, that it reduces to the *Laplacian* of ϕ in orthogonal Cartesian coordinates, i.e., to $\sum_i \partial^2\phi/\partial x^{i2}$. The significance of the quantity given by (9.33) in physical problems is thus made clear.

Finally, since the covariant derivative of a covariant vector ξ_i is given by $\xi_{i,j} = \partial\xi_i/\partial u^j - \xi_k\Gamma^k_{ij}$, it follows that $\xi_{j,i} - \xi_{i,j} = \partial\xi_j/\partial u^i - \partial\xi_i/\partial u^j$ because of the fact that $\Gamma^k_{ij} = \Gamma^k_{ji}$. This tensor of second rank is called the *rotation* of the vector field. That this name has been correctly chosen is once again seen by going over to orthogonal Cartesian coordinates; in that case the distinction between covariant and contravariant vectors disappears, the tensor has the components $\partial\xi_j/\partial x^i - \partial\xi_i/\partial x^j$, and thus represents the rotation, or curl, of the vector with components ξ_i. In three dimensions, there would be only three components of the rotation different from zero, and this makes it possible to regard the rotation in three-space as a *vector*, although it is really a so-called skew-symmetric tensor of second rank.

Thus the tensor calculus provides a very convenient means to obtain important invariant differential operators coming from mathematical physics when arbitrary coordinate systems are used. All that is necessary to obtain them explicitly is to calculate the coefficients g_{ik} of the line element for any coordinate system—e.g., $ds^2 = dr^2 + r^2\,d\phi^2$ for polar coordinates (r, ϕ) in the plane—and then substitute in the formulas. The *concept* of a tensor, and the calculus of tensors thus are of basic importance. However the tensor *notation* oftentimes does more to obscure than to illuminate a given subject. This is noticeably the case sometimes in continuum mechanics, where a notation employing vectors or matrices is often quite sufficient for the purposes in view, since there may be no necessity to operate with any coordinate system other than a single orthogonal Cartesian system, particularly when deriving the basic general theory. In the early chapters of this book, a vector rather than a tensor notation was employed for this very reason. In Riemannian geometry and in the theory of relativity, however, the situation is different, since the heart of the matter then largely centers around invariance properties of all sorts of entities with respect to any coordinate systems: but even there the notation employing differential forms and the wedge product together with the exterior derivative (cf. the next chapter) has to a considerable degree superseded the tensor notation (but of course not the tensor concept).

PART II. TENSOR CALCULUS AND DIFFERENTIAL GEOMETRY IN GENERAL MANIFOLDS

5.   Tensors in a Riemannian Space

This section is concerned with the discussion of tensor calculus in an n-dimensional Riemannian space.   As has been said a number of times, the first step to be taken is the introduction at each point in the n-dimensional space of an n-dimensional linear vector space V, its tangent space—the exact analogue of the two-dimensional linear space spanned by the tangents to the coordinate curves in the study of inner differential geometry in Chapter VII, and to the tangent spaces introduced above in n-dimensional affine and Euclidean spaces.   It is to this space that tensor algebra is applied, and from it stem all notions of invariance that are used in Riemannian geometry.

In Chapter VIII the spaces to be dealt with were introduced as abstract n-dimensional manifolds defined locally as topological images of an open coordinate sphere $\sum_{i=1}^{n} (x^i)^2 < \rho^2$ of an n-dimensional linear space.   The coordinate sphere is parametrized by differentiable functions $x^i = x^i(u^1, u^2, \ldots, u^n)$.   The space is required to be the same when any new coordinates $(u^1, u^2, \ldots, u^n)$ are introduced locally such that the transformation is one-to-one and the Jacobian is nowhere zero.   At the same time, the notion of the tangent space at a point was introduced; this is an n-dimensional linear vector space with the tangent vectors of a given set of coordinate curves as a basis.   It was Riemann's idea, as has been remarked before, to give this rather formless continuum a geometrical structure by postulating that it should be approximately Euclidean in a small neighborhood of each point.

However, as Weyl [W.1] first observed, it is possible to go rather far with a geometry of manifolds without introducing a Euclidean metric in the linear tangent space.   In fact, two important special cases of tensor fields can be at once introduced pointwise without the aid of a metric.

The first is that of the field of differentials du^i of the coordinates.   Since the transformed differentials du'^i are naturally assumed to be given by the usual rule of the calculus from the formulas $u'^i = u'^i(u^1, u^2, \ldots, u^n)$, i.e., by

$$(9.34) \qquad\qquad du'^i = \frac{\partial u'^i}{\partial u^j}\, du^j = \gamma_j^i\, du^j,$$

it follows that such differentials do indeed define a contravariant vector. Of course, the coefficients γ_j^i are now functions of the coordinates u^i and so vary from point to point in the manifold.

Another important tensor is given by any scalar field, which is simply a single function $\phi(u^i)$ defined over the space, and by definition invariant under coordinate transformations.   This leads at once to still another very important type of vector field, obtained by differentiation.   Consider the partial derivatives $\partial\phi/\partial u^i$ of a scalar field $\phi(u^i)$; when new coordinates are introduced the derivatives are given by

$$(9.35) \qquad \frac{\partial\phi}{\partial u'^j} = \frac{\partial\phi}{\partial u^k}\frac{\partial u^k}{\partial u'^j} = c_j^k\frac{\partial\phi}{\partial u^k}.$$

and it follows that this so-called *gradient of ϕ*, given by $\partial\phi/\partial u^i$, is a *covariant vector*.

Even though it amounts to a repetition, it is useful to write down explicitly the transformation equations which serve to define a tensor $A_{p_1p_2\cdots p_q}^{r_1r_2\cdots r_m}$ which is *contravariant of order m and covariant of order q*.   In fact, the following rule for the transformation of such tensors results at once from the corresponding rule in Appendix A:

$$(9.36) \qquad A_{p_1p_2\cdots p_q}^{'r_1r_2\cdots r_m} = \frac{\partial u'^{r_1}}{\partial u^{s_1}}\cdots\frac{\partial u'^{r_m}}{\partial u^{s_m}}\frac{\partial u^{t_1}}{\partial u'^{p_1}}\cdots\frac{\partial u^{t_q}}{\partial u'^{p_q}}A_{t_1t_2\cdots t_q}^{s_1s_2\cdots s_m}.$$

So far, no mention has been made of a metric in the space.   It was noted above that Weyl [W.1] and others go on to study the geometry in such spaces without a metric, called *spaces with an affine connection*.   To do so it seems necessary in order to develop a theory with geometrical interest to introduce a means for defining *local parallelism* of vectors so that the geometry at one point will have some connection with the geometry at neighboring points, as in Euclidean spaces.   This is done by introducing functions Γ_{ij}^k which are to a certain extent arbitrary; they then play the role of those introduced above, and thus make it possible to define the covariant derivative and a reasonable notion of parallelism.   This kind of geometry, which is very actively studied at the present time, will not be treated here; instead, the special case of Riemannian spaces, which result when a metric is introduced, is taken up at once.   By this is meant, as was described in the previous chapter, that a metric form is defined at all points of the space by prescribing functions $g_{ij}(u^1, u^2, \ldots, u^n)$, with $g_{ij} = g_{ji}$, and requiring these functions to have everywhere a positive definite matrix.   A natural basis $\mathbf{v}_i$ is defined in the tangent space at a point u_0^i by the tangent vectors of the coordinate curves there in the fashion also described in Chapter VIII.   In terms of the Cartesian coordinates defined by the basis vectors $\mathbf{v}_i$ at u_0^i, a Euclidean metric can be introduced in the tangent space so that the line element in it is given by

$$(9.37) \qquad ds^2 = \gamma_{ij} \, du^i \, du^j,$$

and it is assumed that *the constant values of the functions* γ_{ij} *in the tangent space are the same as those of the functions* g_{ij} *at the point* u_0^i. Thus at the point u_0^i the coefficients γ_{ij} are given by definition by

$$(9.38) \qquad \gamma_{ij} = \mathbf{v}_i \cdot \mathbf{v}_j$$

in terms of the natural basis $\mathbf{v}_i$ resulting from a given coordinate system u^i of the space. Transformations of these coordinates of course lead to different sets of basic vectors and these in turn to new values for the functions γ_{ij} at the origin of the tangent space. Since the line element ds^2 is naturally required to be an invariant in the manifold it follows that the quantities γ_{ij} form the components of a covariant tensor, since the functions g_{ij} must at each point be assumed to transform in accordance with the general rule (9.36). Once these stipulations have been made the space is said to be a Riemannian space with the metric or fundamental form

$$(9.39) \qquad ds^2 = g_{ij} \, du^i \, du^j.$$

6. Basic Concepts of Riemannian Geometry

Enough tools are now available to develop satisfactorily some of the concepts of Riemannian geometry. For example, it is now clear that the angle ϕ between the tangents $\dot{u}_{(1)}^i$ and $\dot{u}_{(2)}^i$ of two curves intersecting at a point can be defined by the following formula [cf. (9.13)] in terms of the differentials $du_{(j)}^i = \dot{u}_{(j)}^i \, dt$:

$$(9.40) \qquad \cos \phi = \frac{g_{ij} \, du_{(1)}^i \, du_{(2)}^j}{\sqrt{g_{ij} \, du_{(1)}^i \, du_{(1)}^j} \cdot \sqrt{g_{.j} \, du_{(2)}^i \, du_{(2)}^j}},$$

with the assurance that the right-hand side has a value independent of the choice of coordinates.

To these possibilities at a single point can also be added easily a few of a global character. For example, the length of a curve segment $u^i(t)$, $t_0 \le t \le t_1$, can be defined by

$$(9.41) \qquad s_{t_0}^{t_1} = \int_{t_0}^{t_1} \left(g_{ij} \frac{du^i}{dt} \frac{du^j}{dt} \right)^{1/2} dt,$$

and the invariance of this quantity with respect to coordinate transformations is clear (as is also the invariance with respect to parameter transformations $t = t(\tau)$, $dt/d\tau \neq 0$, since $s_{t_0}^{t_1}$ is defined by a definite integral). In the tangent space it is assumed known that the volume element is given by

$d\tau = \sqrt{g}\, du^1\, du^2 \cdots du^n$, with g defined as the determinant $|g_{ik}|$ of the coefficients of the fundamental form. Consequently, the volume of a portion of a Riemannian space can be defined reasonably as the definite integral $\int \sqrt{g}\, du^1\, du^2 \cdots du^n$.

In general, any notions which depend only on scalars, or upon vectors and tensors located at the same point, generalize at once from Euclidean to Riemannian spaces. However, as was remarked earlier on a number of occasions, such a purely local and, in a sense, disconnected kind of geometry needs to be supplemented by constructions which permit comparisons to be made at different points, and in particular to permit differentiation of tensor fields. In Euclidean geometry this was done in Section 2 for general coordinate systems by introducing the notion of the absolute differential and the covariant derivative, and these in turn depended upon the fact that parallel transport of vectors in such spaces is possible. As a preparation for generalizing this process it is convenient to introduce, following the terminology of Cartan, a special kind of Euclidean tangent space called an *osculating space*.

In the preceding section the tangent space of R at a point P was provided with a Euclidean metric

(9.42) $$ds^2 = \gamma_{ij}\, du^i\, du^j,$$

with γ_{ij} constants equal to the values of the functions g_{ij} at P. It is convenient to assume that $u^i = 0$ at P. Since the quantities γ_{ij} are assumed constant, the coordinates are Cartesian in this space. The equation (9.42) can be interpreted to mean that points near P in R characterized by the parameter values du^i have the coordinates du^i in the tangent space V, and this mapping of R into V furnishes an approximation for which length measurements near P would differ in the two spaces by quantities of higher than first order in the differentials du^i.

This, however, is not a sufficiently accurate approximation for all purposes. Again following Cartan the approximation is refined by introducing special coordinate systems in R and V with respect to which the *first derivatives* of the coefficients of the line element (9.42) in the tangent space should be the same as those of R at the point P, in addition to the values of those functions themselves. (It is assumed, as before, that P is given by $u'^i = 0$.) Such curvilinear coordinate systems in V are said to yield an *osculating metric* of R at P. It is easy to see that an infinity of such coordinate systems exists, as follows. Suppose that the quantities u^i yield the Cartesian tangent metric and introduce new coordinates u'^i in R and in the tangent space by the equations

(9.43) $$\begin{cases} u'^i = u^i + \tfrac{1}{2}(\Gamma^i_{rs})_0\, u^r u^s + \cdots, \\ u^i = u'^i - \tfrac{1}{2}(\Gamma^i_{rs})_0\, u'^r u'^s + \cdots. \end{cases}$$

The coefficients of the metric tensor in the tangent space are once more denoted by γ_{ij}, but they are now regarded as functions of the coordinates u'^i.
It is clear that the new coordinate system belongs among the admissible systems since the Jacobian of the transformation has the value one at P.   It is
customary sometimes to refer to the coordinate systems u'^i as *geodetic* coordinate systems (for reasons that will be apparent later on) because of the
fact that the transformed coefficients of the metric form have first derivatives
that have the value zero at the point P for both R and V.   In this way an
infinity of special osculating tangent spaces result through transformations
of the sort given by (9.43).   The proof is the result of a straightforward calculation.   For this purpose the symbols Γ_{jki} and Γ_{ji}^k, as defined with the aid
of (9.18) and (9.21), together with the general law $g'_{ij} = g_{rs} (\partial u^r/\partial u'^i)(\partial u^s/\partial u'^j)$
for the transformation of the coefficients of the metric form, are used.   In
the present case

$$g'_{ij} = g_{ij}(0) + \left(\frac{\partial g_{ij}}{\partial u_k} - \Gamma_{ijk} - \Gamma_{jik}\right) u'^k + \cdots,$$

since $\partial u^r/\partial u'^i = \delta_i^r - \Gamma_{ik}^r u'^k$.   The coefficients of u'^k are, of course, evaluated
at point P.   But (9.19) then implies that $\partial g'_{ij}/\partial u'^k = 0$ at P, as was to be
shown.

The totality of the osculating metrics at a point is independent of
changes of variables in R in the sense that for a fixed such transformation
new values of the functions g_{ij} and their first derivatives are uniquely fixed
by the original values of these same quantities.   In addition, only properties
which are common to all such osculating metrics arise in the discussion to
follow, and hence it can cause no confusion to refer to *the* osculating space.
It is also evident that such properties can be regarded as intrinsic properties
of the Riemannian space.   Thus all of the invariants studied in Euclidean
geometry which are expressed solely in terms of the tensor g_{ij} and its first
derivatives (or, what is the same thing, in terms of the bracket symbols) can
be at once defined in a reasonable way as invariants in the Riemannian
space.   In particular, this therefore holds for the absolute differential and
the covariant derivative of a tensor.

Before pursuing the possibilities opened up by these remarks it is of interest to look at a familiar special case and to interpret it in relation to the
notion of the osculating space—in much the same way as was done in Chapter VII.   The example is that of a surface in Euclidean space defined by an
equation $z = f(x, y)$ in orthogonal Cartesian coordinates so chosen that the
x,y-plane is a tangent plane of it at the origin.   The line element is given by

$$ds^2 = dx^2 + dy^2 + (f_x \, dx + f_y \, dy)^2.$$

The osculating space is in this case the tangent plane at the origin with the line element $ds^2 = dx^2 + dy^2$; it is clear that the two line elements are the same at the origin ($f_x = f_y = 0$ there), and also that their coefficients have the same first derivative there: these derivatives have, in fact, the value zero in both cases.   The osculating coordinate system is thus an orthogonal Cartesian system in this special case and the Cartan symbols are all zero at the origin.   A neighborhood of the surface about the origin is mapped onto its tangent space by an orthogonal projection in three space, and the map in the tangent space is an approximation to the surface in a precise sense, i.e., that corresponding points have coordinates in the three-dimensional space that differ near the origin by terms which are of higher than second order in these coordinates.   Two tangent vectors of the surface at points near the origin would have as images in the tangent space their orthogonal projections; if the latter were parallel and of equal length in the Euclidean sense it is not difficult to see that the corresponding vectors in the surface would differ in Euclidean three-dimensional space by a vector whose length would be quadratic at least in the coordinates of the base points of the vectors.   Thus it would be possible to define parallel fields on the surface by projection of Euclidean parallel fields in the tangent plane onto it, and these fields would form approximately constant fields in the Euclidean three-dimensional space containing the surface.   In fact, any two such vectors would have a difference which would be of at least second order in the coordinates x and y.   One of the ways chosen in Chapter VII to define a parallel field along a curve on the surface was to make use of this remark as a means to derive a *differential equation* for the purpose of defining parallelism by calculating the rate of change of a parallel field along a given curve by the appropriate passage to the limit.

    The situation in the general case is similar to this, though an embedding in a Euclidean space is not in question.[1]   The introduction of the osculating metric at a point P_0 makes it possible to define parallelism through the direct application of the formulas of Section 2 simply by identifying the quantities Γ^r_{ik} there with the same quantities calculated from the given coefficients g_{ik} of the metric tensor in R at the origin of the osculating space.   These formulas gave the position of the natural basis vectors of the curvilinear coordinate system at all points near P_0 with an error greater than first order in the differentials du^i.   Riemann's idea of developing geometry in a continuum

---

[1] One way to introduce the ideas of Riemannian geometry would be to make use of the fact that an n-dimensional Riemannian space can be embedded isometrically and smoothly in the small in a Euclidean space of dimension $\frac{1}{2}n(n + 1)$.   Parallelism can then be defined in a manner analogous to that of Chapter VII for the case of R^2 embedded in E^3.   The literature concerning this so-called embedding problem is quite extensive. A summary of it is given by Kuiper [K.5], where, in particular, references to the work of Nash and himself on this problem are given.

by making it "Euklidisch in den kleinsten Theilen," as he said, is thus carried out concretely by Cartan in assuming that the formulas of Section 3 for the differentials $d\mathbf{v}_i$ of the coordinate vectors and for the absolute differential $D\xi^i$ of a vector field are also valid in a Riemannian space with errors that are of higher than first order in the differentials du^i of the coordinates. Thus the formulas of Section 3 can be taken over bodily. This rather natural way of motivating the basic assumptions of Riemannian geometry is at the same time made reasonable since the invariants are those of linear algebra.

7. Parallel Displacement.   Necessary Condition for Euclidean Metrics

It might have turned out that nothing new is gained in proceeding in the manner just explained, or that the new geometry would lack interest. Neither of these things happens—as was seen in Chapter VII in the two-dimensional case. The new geometry is not Euclidean in general, since arbitrary metric forms $g_{ij}\, du^i\, du^j$ are permitted and, as has yet to be shown (but already known to be true in two dimensions), it is in general not possible to reduce the line element to the Cartesian form by a change of coordinates. The reason for that was indicated earlier, but it is now important to emphasize it.

A fact of basic importance is that *the formulas*

$$(9.44) \qquad d\mathbf{v}_i = \Gamma_{ir}^k\, du^r \mathbf{v}_k = \omega_{(i)}^k \mathbf{v}_k$$

do not define in general exact differentials $d\mathbf{v}_i$, *as they do in the Euclidean case*: the vectors $\mathbf{v}_i$ can no longer be derived from a vector $\mathbf{X}(u^1, u^2, \ldots, u^n)$ as was done in Section 2. The working out of the consequences of this fact is one of the next tasks to be undertaken. Before doing so it has point to recapitulate other facts and formulas which are now to be used.

The formula (9.40), which defines the angle between two curves, hardly needs to be written down again. The formulas for the Cartan (and bracket) symbols are

$$(9.45) \qquad \Gamma_{ijr} = \Gamma_{rji} = g_{ij}\Gamma_{ir}^k,$$

$$(9.46) \qquad \Gamma_{jki} = \frac{1}{2}\left(\frac{\partial g_{jk}}{\partial u^i} + \frac{\partial g_{ik}}{\partial u^j} - \frac{\partial g_{ij}}{\partial u^k}\right),$$

$$(9.47) \qquad \Gamma_{ij}^k = \Gamma_{ji}^k = g^{kh}\Gamma_{ihj} = g^{kh}[{}^i{}_h{}^j] = \{{}^i{}_k{}^j\}.$$

The formulas for the absolute differentials of vector fields and of tensor fields are

$$(9.48) \qquad D\xi^i = d\xi^i + \xi^k \omega_{(k)}^i,$$

$$(9.49) \qquad D\xi_i = d\xi_i - \xi_k \omega_{(i)}^k,$$

$$(9.50) \qquad Da_i^j = da_i^j - a_k^j \omega_{(i)}^k + a_i^k \omega_{(k)}^j.$$

The corresponding formulas for covariant derivatives are

(9.51)
$$\xi_{i,j} = \frac{\partial \xi_i}{\partial u^j} - \Gamma_{ij}^k \, \xi_k,$$

(9.52)
$$\xi_{,j}^i = \frac{\partial \xi^i}{\partial u^j} + \Gamma_{kj}^i \, \xi^k,$$

(9.53)
$$a_{j,k}^i = \frac{\partial a_j^i}{\partial u^k} + \Gamma_{hk}^i \, a_j^h - \Gamma_{jk}^h \, a_h^i.$$

The last three equations, it is to be recalled, define tensors of one covariant order higher than those differentiated.

The formulas of Section 3 could also be taken over just as they stand as the reasonable generalizations of the gradient, rotation, divergence, and Laplace operator in Riemannian spaces.

One of the most important concepts to be introduced first in Riemannian geometry is that of a *parallel field*[1] of vectors. In Euclidean space it was shown that such fields can be characterized in the small by the vanishing of the absolute differential. They can also be obtained in the large in Euclidean geometry—as will be seen shortly—by integrating the partial differential equations $\xi_{i,j} = 0$ or $\xi_{,j}^i = 0$. These partial differential equations are over-determined systems of first order, but in the Euclidean case the compatibility or integrability conditions are satisfied, and Theorem V of Appendix B then makes it possible to prove the existence of unique vector fields ξ^i or ξ_i as solutions of these differential equations once an initial vector is given at one point. In Riemannian geometry it is natural and reasonable to characterize local parallelism by the vanishing of the absolute differential, but it is no longer possible to define parallel fields in the large in a unique way. Nevertheless, this seeming defect in Riemannian geometry is really a virtue since it is a manifestation of properties of great geometrical interest, and also it can be overcome at least partially by a device already employed in Chapter VII in the two-dimensional case. That device is to restrict the definition of parallel fields to the case of fields *defined along curves* $u^i = u^i(t)$. In that case the vanishing of the absolute differentials $D\xi^i$ and $D\xi_i$ along a curve leads to *linear ordinary differential equations* defined along it [cf. (9.24) and (9.25)]:

(9.54)
$$\frac{d\xi^i}{dt} + \Gamma_{kj}^i \, \dot{u}^j(t) \xi^k = 0,$$

(9.55)
$$\frac{d\xi_i}{dt} - \Gamma_{ij}^k \, \dot{u}^j(t) \xi_k = 0.$$

The functions defined by the symbols Γ_{kj}^i are known along the curve since

---

[1] It would be better to say *constant field*, since vectors are always taken which are not only parallel but are also of the same length, but this terminology has been standard for a long time.

they are defined over the whole manifold, and the functions $\dot{u}^i(t) = du^i/dt$ are prescribed; thus if the curve has a continuous first derivative the standard existence and uniqueness theorems apply and they give by integration a uniquely determined vector field ξ^i or ξ_i along the curve once an initial vector has been chosen at one point. Such parallel fields thus satisfy differential equations identical (except for the number of dimensions) with those derived in Chapter VII for parallel displacement in the sense of Levi-Civita in two-dimensional surfaces. (It is possible, as some writers do, to use these differential equations at the outset as a basis for the development of Riemannian geometry, since they are properly invariant and go rather to the heart of the matter.)

It is to be expected that parallel fields along curves will have the same basic property here that such fields were found to have in the two-dimensional case, i.e., that *two such fields* $\xi_{(1)}^i(t)$, $\xi_{(2)}^i(t)$, say, *defined along the same curve have at each point a scalar product* $g_{ij}\xi^i\xi^j$ *with a value independent of t.* A formal calculation (which is left as an exercise) proves it, but it clearly must be correct since the derivative of the scalar product with respect to t involves only the values of the functions g_{ij} and their first derivatives, and hence the verification, because of the existence of an osculating tangent space, takes the same course as if the space were Euclidean. In particular, if $\xi_{(1)}^i = \xi_{(2)}^i$ is assumed it follows that the length of the field vectors is preserved. Thus the angle between the vectors of two parallel fields is also preserved in going along the curve.

Consider now any pair of points P_1, P_2 of R and connect them with a regular curve $u^i(t)$. A set of n linearly independent vectors $_{(1)}\xi_{(k)}^i$ in the tangent space at P_1 is chosen and each of these vectors is taken as initial vector of a parallel field along the curve. In the course of parallel propagation along the curve the n vectors retain their lengths and their mutual angles, and hence define a set $_{(2)}\xi_{(k)}^i$ of linearly independent vectors in the tangent space at P_2. It is therefore possible to make use of these two sets of vectors to define in the obvious way a mapping of the two tangent spaces on one another through identification of points having the same coordinate values. The introduction of the notion of parallelism at a distance, but along a specified curve, therefore leads in a quite natural way to an important possibility for connecting the geometry at two different points of the space.

One possible way of characterizing the straight lines in Euclidean geometry is to define them as the regular curves $u^i = u^i(t)$ having *tangent vectors that form a parallel field*;—that is, these vectors are assumed to be parallel in the ordinary sense. In curvilinear coordinates they would therefore be characterized by replacing ξ^k in (9.54) by $\dot{u}^k$ to obtain

$$(9.56) \qquad \ddot{u}^k + \Gamma_{sr}^k \dot{u}^r \dot{u}^s = 0, \qquad k = 1, 2, \ldots, n,$$

as a system of differential equations for them.   In other Riemannian spaces the general point of view assumed here makes it obvious that (9.56) is to be taken over as the means of defining the straight lines, or *geodetic lines*, in these spaces.   Since (9.56) is a system of n second-order ordinary differential equations for the functions $u^r(t)$, they will have a uniquely determined solution once a point $u^r(0)$ and a direction $\dot{u}^r(0)$ at $t = 0$ have been prescribed.   In Chapter VII the differential equation (9.56) was derived for the case $n = 2$, but a rather different derivation was given.

It has already been remarked that the parallel transport of a vector from one point to another in general yields a different vector if the parallel displacement is carried out along different curves joining the two points.   As in two dimensions it is an interesting problem to investigate conditions under which the parallel displacement of vectors is independent of the path—as it is in Euclidean spaces.   Evidently, such a condition prevails if, and only if, the differential equations (9.56) are satisfied by vector *fields* $w^r(u^i, u^2, \ldots, u^n)$ which originate with a vector at an arbitrary point and result through parallel displacement of it along arbitrary curves to other points.   Thus the equations (9.56) would be satisfied by $\dot{w}^r = (\partial w^r / \partial u^k)\dot{u}^k$; hence equations (9.56) would yield the partial differential equations

$$(9.57) \qquad \frac{\partial w^i}{\partial u^k} + \Gamma^i_{kj}\, w_j = 0,$$

after noting that the common factor $\dot{u}^k$ that appears in both terms has, by hypothesis, an arbitrary value because of the independence of the path assumed for the parallel displacement.   As one sees, the equations (9.57) form an overdetermined system of linear first-order partial differential equations —there are, in fact, n^2 equations for the n functions w^r.   In Appendix B it is proved that such a system has a uniquely determined solution once the values of the functions w^r are prescribed at a point, provided that the *compatibility conditions* are satisfied, i.e., provided that the conditions resulting from the equality of the mixed second derivatives $\partial^2 w^r / \partial u^k\, \partial u^l = \partial^2 w^r / \partial u^l\, \partial u^k$ are satisfied identically.   In the present case the compatibility conditions are seen with little trouble to be given by

$$(9.58) \qquad R^i_{.jlk} = \frac{\partial \Gamma^i_{jk}}{\partial u^l} - \frac{\partial \Gamma^i_{jl}}{\partial u^k} - (\Gamma^i_{sk}\Gamma^s_{jl} - \Gamma^i_{sl}\Gamma^s_{jk}) = 0.$$

These equations define a set of functions $R^i_{.jlk}$ depending on four indices, the vanishing of which ensures that parallel displacement of vectors from any given point is uniquely determined independently of paths.

The highly important set of quantities $R^i_{.jlk}$ was written purposely in that form because they are components of a tensor, as is to be shown, called the *Riemann* or *curvature tensor*, which is covariant in three indices and con-travariant in a fourth index. To this end it is observed first of all that the left-hand side of (9.57) is, in view of (9.52), the covariant derivative $w^i_{.k}$ of the vector field w^i. It follows that the expression $w^i_{.k,l} - w^i_{.l,k}$ obtained by tak-ing a further covariant derivative of $w^i_{.k}$, is a tensor; it is given by

$$(9.59) \qquad w^i_{.k,l} - w^i_{.l,k} = w^j R^i_{.jlk},$$

as is seen after a little calculation. Since the vector field w^j can be chosen arbitrarily, by hypothesis, at any one point, it follows from the so-called quotient rule for tensors (cf. Appendix A) that $R^i_{.jlk}$ must be a tensor, since the left-hand side of (9.59) is a tensor.

In two dimensions it was shown in Chapter VI that the Riemann tensor has essentially only one nonvanishing component, which in turn was the Gaussian curvature of the surface within a scalar multiplying factor. It was then shown, in Chapter VII, that if the Gaussian curvature vanishes the sur-face is necessarily isometric with the plane, i.e., that its line element could be transformed by a real nonsingular transformation of coordinates into the Euclidean form $ds^2 = (dx^1)^2 + (dx^2)^2 = dx^i \cdot dx^i$, at least in the small. In an n-dimensional Riemannian space an analogous result holds; i.e., that the *vanishing of all components of the Riemann tensor is a necessary and sufficient condition for the existence of real nonsingular coordinate transformations which will transform the line element into one with constant coefficients.* This import-ant fact is proved as follows. It is known from the discussion above that the vanishing of the tensor $R^i_{.jlk}$ is the necessary and sufficient condition (in the small, at least) that parallel transport of vectors should be independent of the path. Thus if a set of n linearly independent covariant vectors $_{(j)}w_i(0, 0, \ldots, 0)$, $(j) = 1, 2, \ldots, n$, is given at one point, it leads by parallel displacement to a uniquely determined set $_{(j)}w_i(u^1, u^2, \ldots, u^n)$ of linearly independent vectors in a whole neighborhood of the point. Upon setting $\partial v^j / \partial u^i = {}_{(j)}w_i$, it follows that this overdetermined system of linear partial differential equations has a uniquely determined solution $v^j(u^1, u^2, \ldots, u^n)$ once its value is prescribed at $(0, 0, \ldots, 0)$. This holds because compatibil-ity conditions $\partial_{(j)}w_i / \partial u^k = \partial_{(j)}w_k / \partial u^i$ are satisfied, since the counterpart of (9.57) is obtained from (9.55) and it holds for $\xi_i \equiv w_i$; the symmetry rela-tions $\Gamma^i_{kj} = \Gamma^i_{jk}$ are also used. Hence n functions $v^1, v^2, \ldots, v^n$ are deter-mined, and they have a nonvanishing functional determinant, since $\partial v^j / \partial u^i = {}_{(j)}w_i$ and the vectors $_{(j)}w_i$ are linearly independent. Thus the quantities v^i can be introduced in the small as coordinates, and the corresponding coordin-ate curves have the vectors $_{(j)}w_i$ as tangent vectors. In the coordinates v^j it is then clear that *all* parallel fields have components y_i which are constant in the coordinates v^j. Since any parallel field $y_i(t)$ satisfies the equations $\dot{y}_i - \Gamma^k_{ir} y_k v^r = 0$ for all curves $v^r = v^r(t)$ it follows from $\dot{y}_i = 0$ that $\Gamma^r_{ik} =$

$g^{rl}\Gamma_{ilk} = 0$, hence that all first derivatives of the functions g_{ik} vanish identically; thus the desired result is obtained. This discussion was based on the study of a field of covariant vectors, but the same result would follow for a field of contravariant vectors.

If consideration is further limited to spaces with a positive definite line element it is then easily seen that the space must be Euclidean. For the discussion of the theory of relativity later on it is important to remark that the above result is valid in spaces having a metric that is not necessarily positive definite provided only that the determinant $|g_{ik}|$ is different from zero. Again if the Riemann tensor vanishes a transformation exists relative to which the line element has constant coefficients; such spaces are called *flat spaces*. The special theory of relativity involves a four-dimensional flat space with a metric that is not positive definite, and this in turn is a special case of the space with which the general theory of relativity is concerned.

If the covariant curvature tensor is defined by

(9.60) $$R_{rjnp} = g_{rl}R^{l}_{.jnp}$$

it can be shown without too great difficulty that the following formula holds:

$$R_{rjnp} = \frac{1}{2}\left(\frac{\partial^2 g_{rp}}{\partial x^j\,\partial x^n} + \frac{\partial^2 g_{jn}}{\partial x^r\,\partial x^p} - \frac{\partial^2 g_{rn}}{\partial x^j\,\partial x^p} - \frac{\partial^2 g_{jp}}{\partial x^r\,\partial x^n}\right)$$
$$+ g^{ts}(\Gamma_{jsn}\Gamma_{rtp} - \Gamma_{jsp}\Gamma_{rtn}).$$

From this a number of symmetry relations follow:

$$R_{rjnp} = -R_{jrnp},$$
$$R_{rjnp} = -R_{rjpn},$$
$$R_{rjnp} = R_{nprj},$$

and also the relations

$$R_{rjnp} + R_{rnpj} + R_{rpjn} = 0.$$

Because of these symmetry relations the number of essentially distinct non-vanishing components of the tensor R_{rjnp} is a fraction of the total number. In fact it is given by $\frac{1}{12}n^2(n^2 - 1)$. For $n = 2$ this reduces to one, as is known from Chapter VI.

It has importance for Einstein's general theory of relativity to observe that though the tensor $R^{l}_{.jnp}$ can be contracted in three ways only one of them leads to a tensor with nonzero components; it is the tensor

(9.61) $$R_{jn} = R^{l}_{.jnl} = g^{ls}R_{sjnl},$$

and this tensor is found to be symmetrical; these facts are simple consequences of the symmetry relations.

The above treatment of some of the elements of Riemannian geometry was confined almost exclusively to *inner* differential geometry. It is, however, entirely possible and quite interesting to consider a Riemann space R_n immersed in another space R_m, with $m > n$, just as was done in earlier chapters where the cases $n = 1$, $m = 2$ or 3 (curves in the plane and in space), and the case $n = 2$, $m = 3$ (surfaces in space) were studied. It is, however, not at all necessary to restrict the containing space R_m to be a Euclidean space, as was always done in earlier chapters.

For their own sake, and also for the sake of an application to differential geometry in the large to be made later, it is of interest to consider two special cases of two-dimensional subspaces of a Riemannian space R_n of dimension $n > 2$.

The first case is that of an arbitrary two-dimensional surface in R_n which is supposed defined by equations $x^k = x^k(u, v)$ in terms of two parameters u and v; or, as it could also be put, these equations define a two-parameter family of curves in R_n. It is of interest to consider a small "parallelogram" formed from these curves which begins at a point (u, v), passes to $(u + \Delta u, v)$, then to $(u + \Delta u, v + \Delta v)$, then to $(u, v + \Delta v)$ and back to (u, v), always in holding alternately u and v constant. Consider next a vector ξ^i at (u, v) as initial vector of a parallel field to be constructed in going once around the parallelogram[1]; the end result will be a vector ξ^i_* at (u, v) which will be different from ξ^i. In fact, a familiar argument shows that ξ^i and ξ^i_* will, or will not, be the same for all closed paths, according to whether parallel propagation of vectors is, or is not, independent of the paths joining arbitrary *pairs* of points. Consider now the difference $\Delta \xi^h = \xi^h_* - \xi^h$. This difference must be of the order $\Delta u \, \Delta v$, since it vanishes if either Δu or Δv vanishes, and hence $\lim \Delta \xi^h / \Delta u \, \Delta v$ exists when Δu, $\Delta v \to 0$ and can be calculated by making use of (9.54). The result is seen without difficulty to be

$$\text{(9.62)} \qquad \lim \frac{\Delta \xi^h}{\Delta u \, \Delta v} = - R^h_{ijk} \xi^i \frac{dx^j}{du} \frac{dx^k}{dv},$$

with

$$\text{(9.63)} \qquad R^h_{ijk} = \frac{\partial \Gamma^h_{ik}}{\partial x^j} - \frac{\partial \Gamma^h_{ij}}{\partial x^k} + \Gamma^h_{j\alpha} \Gamma^\alpha_{ik} - \Gamma^h_{k\alpha} \Gamma^\alpha_{ij}.$$

Upon comparison with (9.58) it is seen that the quantities R^h_{ijk} are the components of the Riemann tensor. The equation (9.62) proves again that the quantities R^h_{ijk} define a tensor since $\Delta \xi^h$, as the difference of two vectors *at the*

---

[1] This curve does not have a continuous derivative throughout, but the parallel field is uniquely defined by taking as initial vector at each vertex of the parallelogram the field vector which resulted at the end of the preceding side through integration of the differential equations.

same point, is a vector and $\xi^i(dx^j/du)(dx^k/dv)$ is an arbitrary tensor. In addition (9.62) states that the Riemann tensor determines, to lowest order in Δu and Δv, the change at (u, v) resulting from parallel transport of a vector ξ^i around any parallelogram with dx^i/du, dx^i/dv as tangent vectors to its sides at (u, v). If the space is Euclidean the Riemann tensor vanishes and $\Delta \xi^h$ also, as it should; otherwise the deviation from Euclidean character is in some sense measured by this tensor.

8. Normal Coordinates.   Curvature in Riemannian Geometry

The Riemann tensor is often called the curvature tensor for reasons which it is of considerable interest to describe. Here, as on many previous occasions, it is advantageous to introduce a special type of coordinate system as a basis for the investigation. The coordinate system in question is obtained by considering the totality of geodesics, i.e., integral curves of the differential equations (9.56), which issue from a given point $P(0, 0, \ldots, 0)$ of R. The first object is to prove that these curves fill out a whole neighborhood N of P in such a way that through each point of N exactly one geodesic passes which joins it with P. Since such geodesics $u^i = u^i(s)$ are in turn uniquely determined by giving an initial tangent vector $\dot{u}^i(0)$ at P, it follows that a point Q of N would be determined uniquely by giving such an initial vector together with the arc length s of the geodesic containing Q—except for P itself, which corresponds to $s = 0$ on all geodesics. In effect, the situation is analogous to that of polar coordinates in a Euclidean space or of the geodetic polar coordinates treated in Chapter VII.

The proof that the situation is as described is as follows. The geodesics through P, as solutions of (9.56), are given by $u^i = f^i(t; \dot{u}^1(0), \dot{u}^2(0), \ldots, \dot{u}^n(0))$ and these functions are differentiable several times with respect to all arguments if R and the metric tensor g_{ij} are sufficiently regular. It is useful to observe that the parameter t in (9.56) with respect to which the derivatives are taken is proportional to the arc length of the geodesic since the tangent vectors $\dot{u}^i(t)$ have constant length because they form a parallel field. In fact (9.56) is easily seen to be invariant if t is replaced by $t' = t/k$, $k = \text{const.} \neq 0$. Consequently, if the initial vector $\dot{u}^i(0)$ is changed in length by the factor k while t is replaced by t/k it follows that the values u^i fixing the points of the geodesic would not be changed. As a consequence the following identities hold for $k \neq 0$:

$$(9.64) \qquad f^i\left(\frac{t}{k}; k\dot{u}^1(0), k\dot{u}^2(0), \ldots\right) = f^i(t; \dot{u}^1(0), \dot{u}^2(0), \ldots).$$

It is therefore permissible to set $k = t$ and thus obtain the equations of the geodesics through P in the form

$$(9.65) \qquad u^i(t) = f^i(1; t\dot{u}^1(0), t\dot{u}^2(0), \ldots, t\dot{u}^n(0)).$$

These equations hold also for $t = 0$ since the differential equations (9.56) are evidently satisfied for $u^i(t) = a^i = \text{const.}$, so that $\dot{u}^i = 0$; thus the following identities are valid:

$$(9.66) \qquad u^i(0) = 0 = f^i(1; 0, 0, \ldots, 0).$$

It is convenient to introduce the quantities

$$(9.67) \qquad \xi^i = t\dot{u}^i(0), \qquad f^i(1; \xi^1, \xi^2, \ldots, \xi^n) = h^i(\xi^1, \xi^2, \ldots, \xi^n),$$

so that the geodesics going out from P are given by

$$(9.68) \qquad u^i = h^i(\xi^1, \xi^2, \ldots, \xi^n), \qquad 0 = h^i(0, 0, \ldots, 0)$$

in a certain neighborhood of $\xi^i = 0$. In effect, the vector ξ^i determines the initial direction of a geodesic, and at the same time fixed a point on it at the distance $\sqrt{g_{ik}(0)\xi^i\xi^k}$ equal to its length, since $ds/dt = \sqrt{g_{ik}\dot{u}^i\dot{u}^k} = \sqrt{g_{ik}(0)\dot{u}^i(0)\dot{u}^k(0)}$ is constant in t and hence $s = t\sqrt{g_{ik}(0)\dot{u}^i(0)\dot{u}^k(0)} = \sqrt{g_{ik}(0)\xi^i\xi^k}$.

The equations (9.68) furnish a mapping of a neighborhood of P in the tangent space into the Riemann space R such that each point ξ^i has a unique image u^i in R. It is now to be shown—and it is the main point of the discussion—that, conversely, to each point Q in R in a certain neighborhood of P a unique point in the tangent space is determined by (9.68). Or, in other words, a neighborhood N of P exists within which *every point Q can be joined to P by a uniquely determined geodesic*. This results, clearly, if it is shown that the equations (9.68) have unique solutions ξ^i for arbitrarily given values u^i in N. To prove this it is sufficient to show that the Jacobian $|\partial h^i/\partial \xi^k|$ is not zero at P, since the functions $h^i(\xi^1, \xi^2, \ldots, \xi^k)$ have continuous derivatives and hence the implicit function theorem can be applied. In fact, this determinant has the value one, which is seen as follows. In (9.68) replace ξ^i by $t\dot{u}^i(0)$ and differentiate with respect to t:

$$\dot{u}^i = \frac{\partial h^i}{\partial \xi^k} \dot{u}^k(0).$$

If in this $t = 0$ is taken, the result is

$$\dot{u}^i(0) = \frac{\partial h^i}{\partial \xi^k}\bigg|_0 \dot{u}^k(0),$$

and these linear equations hold for arbitrary sets of values $\dot{u}^i(0)$; thus

$$\frac{\partial h^i}{\partial \xi^k}\bigg|_0 = \delta^i_k.$$

Thus a neighborhood of the point P in R exists which is in one-to-one correspondence with an open sphere S of a Euclidean space and in such a way that each geodesic from P is mapped isometrically on a unique radius of S. For a later application it has point to remark that the radius of S can be bounded below by a number which depends only upon bounds for derivatives of the functions $h^i(\xi^1, \xi^2, \ldots, \xi^n)$, and thus ultimately upon bounds for various derivatives of the functions involved in defining R and its metric: this results from a quantitative proof of the implicit function theorem (cf., e.g., Goursat [G.3], p. 81).

The coordinates ξ^i, which have now been proved to be legitimate, are called Riemann *normal coordinates*.[1] Their use often simplifies calculations of various kinds in Riemannian geometry. A first such application of them is now made which has as its purpose an analysis of certain curvature properties of a Riemannian space that are fixed by the Riemann tensor and which therefore justify calling this tensor the *curvature tensor*. Consider any pair of linearly independent vectors ξ^i, η^i at a point P of R. Normal coordinates, which are now called u^i, are introduced at P in such a way that ξ^i and η^i fall in the plane of the tangent vectors to the u^1 and u^2 curves. (It is clear that such a construction is always possible.) All geodetic lines which issue from P in the directions given by $a\xi^i + b\eta^i$, a, b arbitrary constants are considered; that is to say, all geodesics tangent to the plane of ξ^i and η^i are in question. These points satisfy the conditions $u^i = h^i(u^1, u^2, 0, 0, \ldots, 0)$ obtained from (9.68), with ξ^i replaced by u^i, by setting all $u^i = 0$ except u^1 and u^2. The result is a two-dimensional subspace of R which is composed of geodesics issuing from P. Such a surface has all of the properties necessary in order to identify it as a two-dimensional Riemann space when the line element is regarded as defined by $g_{11}(u^1, u^2, 0, 0, \ldots)$, $g_{12}(u^1, u^2, 0, 0, \ldots)$, $g_{22}(u^1, u^2, 0, 0, \ldots, 0)$, the main point being that the subspace near P has a neighborhood that is a differentiable and topological image of a Euclidean disk. At the same time the Riemann tensor for this surface that was employed in Chapter VI is obtained also by restricting the Riemann tensor in the space R to the indices 1 and 2: the definition given by (9.58) is identical with that of Chapter VI [cf. (6.23)]. It follows that the Gaussian curvature K of the geodetic surface is appropriately defined by the formula [cf. (6.27)]:

$$K = \frac{R_{1212}}{(g_{11}g_{12} - g_{12}g_{12})},$$

---

[1] The tangent space corresponding to these coordinates is an osculating space in the sense of Cartan. The proof is left as an exercise. It might also be noted that these coordinates are analogous to orthogonal Cartesian coordinates in a Euclidean space if the coordinate vectors are chosen at P to be an orthonormal set since then the quantities $\dot{u}^i(0)$ could be interpreted as the direction cosines of the initial direction of a given geodesic. However, the coordinates ξ^i would not be orthogonal in general at points other than P.

and this can be given the form

$$K = \frac{R_{1212}\xi^1\eta^2\xi^1\eta^2}{(g_{11}g_{22} - g_{12}g_{12})\xi^1\eta^2\xi^1\eta^2}$$

in which ξ^1 and η^2 are the only nonvanishing components at P of the vectors ξ^i and η^i which determine the orientation in R of the geodetic surface. It follows therefore that in *any* coordinates, and for any pair of vectors ξ^i, η^i, this quantity is given by

(9.69)
$$K = \frac{R_{ijkl}\xi^i\eta^j\xi^k\eta^l}{(g_{ik}g_{jl} - g_{il}g_{jk})\xi^i\eta^j\xi^k\eta^l}$$

since it is obviously an invariant which reduces to the preceding when the special coordinates are chosen. Thus K in (9.69) measures the Gaussian curvature of any geodetic surface tangent at P to the plane determined by ξ^i, η^i, and this in turn is fixed solely in terms of the components of the Riemann tensor. This tensor thus merits the name curvature tensor since it fixes the Gaussian curvature of all geodetic surfaces at a point however they are oriented.

9.   Geodetic Lines as Shortest Connections in the Small

Another very important fact can be proved with the aid of the above facts concerning the "geodetic polar coordinates." Just as in the two-dimensional case, the fact that the geodetic lines issuing from a point P fill out a full neighborhood N of the point in such a way that all points Q of N are joined to P by a unique geodetic arc leads to a proof that these arcs are the shortest among all piecewise differentiable curves in R joining P with the points of N. This theorem can be proved as follows. Consider first a point Q with coordinates $\bar{u}^i$ in N and the unique geodetic line g joining it with P; on this arc it is convenient to introduce a parameter t ranging over $0 \leq t \leq 1$ with P corresponding to $t = 0$, Q to $t = 1$. The normal coordinates of Q are $\bar{\zeta}^i = \bar{\zeta}^i(\bar{u}^1, \bar{u}^2, \ldots, \bar{u}^n)$, i.e., the appropriate solution of (9.68) is taken. For points $u^i(t)$ on g the normal coordinates $\xi^i(t)$ are given by $t\bar{\zeta}^i$ [cf. (9.67)], and hence from (9.68) the coordinates u^i of points on g are given by

$$u^i = h^i(t\bar{\zeta}^1, t\bar{\zeta}^2, \ldots, t\bar{\zeta}^n)$$
$$= h^i(t\bar{\zeta}^1(\bar{u}^1, \bar{u}^2, \ldots), t\bar{\zeta}^2(\bar{u}^1, \bar{u}^2, \ldots), \ldots, t\bar{\zeta}^n(\bar{u}^1, \bar{u}^2, \ldots));$$

they therefore depend in a differentiable way upon t and the end point $\bar{u}^i$ of the geodetic segment. Consider now any differentiable curve $C(s)$: $\bar{u}^i = \bar{u}^i(s)$ joining P with Q, lying in N, and for which s is the arc length, i.e., $g_{ij}(d\bar{u}^i/ds)(d\bar{u}^j/ds) = 1$. Consider also the unique geodetic segments

$x^i(t, \overline{u}^i(s))$ joining P with the points of C and such that $t = 1$ corresponds to points on C for all s; these functions $x^i(t, s)$ have continuous derivatives in both arguments. Let $L(s)$ represent the length of the geodetic segments joining P to the points $C(s)$ on C; it is given by

$$L(s) = \int_0^1 \sqrt{g_{ij}\dot{x}^i\dot{x}^j}\, dt = \int_0^1 \phi(x^i, \dot{x}^i)\, dt, \qquad \dot{x}^i = \frac{\partial x^i}{\partial t}.$$

The derivative dL/ds is now computed:

(9.70)
$$\frac{dL}{ds} = \int_0^1 \left(\phi_{x^i} \frac{\partial x^i}{\partial s} + \phi_{\dot{x}^i} \frac{\partial^2 x^i}{\partial t\, \partial s} \right) dt$$

$$= \int_0^1 \left(\phi_{x^i} - \frac{d}{dt} \phi_{\dot{x}^i} \right) \frac{\partial x^i}{\partial s}\, dt + \left[\phi_{\dot{x}^i} \frac{\partial x^i}{\partial s} \right]_0^1,$$

after an intergration by parts. The integral vanishes because the parentheses in it are the left-hand sides of the differential equations (9.56) for the geodesics $x^i(t)$ (the calculation is left as an exercise). Also, $\partial x^i/\partial s$ vanishes for $t = 0$ since $x^i(o, s)$ is the fixed point P for all s. It follows therefore that

$$\frac{dL}{ds} = \left. \frac{g_{ij}\dot{x}^i(\partial x^j/\partial s)}{\sqrt{g_{ij}\dot{x}^i\dot{x}^j}} \right|_{t=1} = \cos\theta(s),$$

in which $\theta(s)$ is the angle of intersection of the geodetic segments with C; in this calculation it is to be noted that $\partial x^j/\partial s|_{t=1}$ is a unit vector. The integral $\int_C dL = \int_C (dL/ds)\, ds$ along C is taken; the result is the length ρ of the geodetic segment joining P with Q, since $L(s)$ is uniquely defined along C and varies from o at P to ρ at Q. The following relations therefore hold:

$$\rho = \int_C \frac{dL}{ds}\, ds = \int_C \cos(\theta(s))\, ds \le \int_C ds.$$

Evidently, the sign of equality can hold only if $\theta(s) = 0$ holds everywhere on C, i.e., if C coincides with the geodetic segment. Hence the latter is the shortest arc among all differentiable arcs joining P with Q and lying in N. It can be seen that the above argument permits an easy modification so that it holds if C is assumed to be a continuous arc with a piecewise continuous first derivative. Also, if C did not lie entirely in N it would of necessity have a length $> \rho$ since it would leave a sphere about P of geodetic radius $\rho^* > \rho$ at some first point on going out from P and thus would have a length greater than ρ^*. The theorem is thus proved in general.

10.   Geodetic Lines as Shortest Connections in the Large

In dealing with geodesics as shortest connections in the large in the two-dimensional case in Section 10 of Chapter VIII, the theory of the second variation due to Jacobi, and the accompanying concept of the conjugate point, played a central role. In principle, this theory is restricted to two dimensions, but it can nevertheless be extended to n-dimensional Riemannian manifolds R, in a variety of circumstances of interest, by embedding a given geodesic in a two-dimensional submanifold and then applying the theory described in Chapter VIII. The presentation of this material in the present section follows closely a paper by Preissman [P.7].

A very useful tool in this process is furnished by a special system of coordinates called *Fermi coordinates*, so chosen that a given closed segment g_0 of a geodesic is one of the coordinate curves and the others form at each point of g_0 a system of Riemann normal coordinates in the $(n - 1)$-dimensional subspace of R orthogonal to g_0. More specifically, these coordinates are defined as follows. At one end point of the arc g_0 (assumed to be without double points), a set of mutually orthogonal vectors $\xi_{(1)}^i, \xi_{(2)}^i, \ldots, \xi_{(n-1)}^i$ orthogonal also to g_0 is chosen; these vectors are then displaced parallel to themselves along g_0 and thus remain mutually orthogonal to themselves and to g_0. At each point of g_0 a system of Riemann normal coordinates $x^1, x^2, \ldots, x^{n-1}$ is introduced in the $(n - 1)$-dimensional subspace orthogonal to g_0 with the geodesics in the directions $\xi_{(1)}^i, \xi_{(2)}^i, \ldots, \xi_{(n-1)}^i$ as "axes." A neighborhood N of each point of the segment g_0 exists, as is known from the preceding section, within which such a system of coordinates is valid. If P is a point in N not on g_0, there exists a unique geodetic arc through it in N which cuts g_0 orthogonally at a point P_0. (The proof of this is the subject of an exercise at the end of the chapter.) The point P has the coordinates $x^1, x^2, \ldots, x^{n-1}$, say, in the manifold orthogonal to g_0; its position is then uniquely fixed by giving in addition the distance x^n to P_0 as measured from an end point of g_0. The coordinates $x^1, x^2, \ldots, x^n$ are called Fermi coordinates in N. From the manner of their construction it follows that the coefficients g_{ij} of the line element defined with respect to them satisfy the following conditions at every point P_0 of g_0:

$$(9.71) \qquad\qquad g_{ij}(P_0) = \delta_{ij},$$

$$(9.72) \qquad \frac{\partial g_{ij}}{\partial u^k}(P_0) = 0, \qquad \Gamma_l^{ij}(P_0) = 0, \qquad \Gamma_{ijk}(P_0) = 0.$$

The first condition is obvious, and the second set results because the Riemann normal coordinates form an osculating coordinate system (the proof of this fact is given as an exercise at the end of the chapter). Since these conditions hold all along g_0, i.e., independent of the value of the coordinate x^n measuring distance along it, the following conditions also hold:

$$(9.73) \qquad \frac{\partial}{\partial x^n}\left(\frac{\partial g_{ij}}{\partial x^k}\right)(P_0) = 0, \qquad \frac{\partial \Gamma_k^{ij}}{\partial x^n}(P_0) = \frac{\partial \Gamma_{ijk}}{\partial x^n}(P_0) = 0.$$

It is now possible to construct a special set of "variations" of g_0, consisting of certain one-parameter families of curves lying near g_0. To this end consider any surface, or two-dimensional submanifold V_2, of R that contains g_0 and is given as a locus in R by equations $x^i = x^i(u, v)$ in terms of two parameters (u, v); the coordinates x^i are assumed to be Fermi coordinates: V_2 is assumed to have all of the properties of a manifold, and is given a metric that is derived from the metric of the space R in which it is embedded. (Concrete examples of how this is done will be introduced later.) It is assumed further that

(a) $x^i(u, 0) = 0$ if $i \neq n$ and $x^n(u, 0) = u$; i.e., that the curve $v = 0$ is the geodesic g_0 with u as arc length on it, and

(b) $(\partial/\partial v)\, x^n(u, v)|_{v=0} = 0$, i.e., that the curves $u = $ const. cut the geodesic g_0 orthogonally.

The curves $x^i = x^i(u, v_0)$, for a constant value v_0 of v, are defined as *variations* or *varied curves* of g_0; their lengths $L(v)$ are to be compared with the length $L(0)$ of g_0 in some interval $u_1 \leq u \leq u_2$. The lengths $L(v)$ are given by

$$L(v) = \int_{u_1}^{u_2} \left(g_{ij} \frac{\partial x^i}{\partial u} \frac{\partial x^j}{\partial v} \right)^{\!\frac{1}{2}} du.$$

In order to calculate $L'(0)$ and $L''(0)$, which are by definition the first and second variations in length, it is useful to introduce the following notation with respect to the derivatives of any function $F(u, v)$ defined on V_2:

$$\frac{\partial F}{\partial u} = \dot{F}, \qquad \frac{\partial F}{\partial v} = F'.$$

In particular, the partial derivatives of the coordinates $x^i(u, v)$ of points on V_2 are denoted by $\dot{x}^i$ and x'^i and also by

$$\xi^i = \dot{x}^i, \qquad \eta^i = x'^i$$

with the consequence that

$$\xi'^i = \dot{\eta}^i.$$

Since Fermi coordinates are being used it follows that

$$\xi^i(u, 0) = 0 \quad \text{for} \quad i \neq n, \qquad \xi^n(u, 0) = 1,$$

$$\eta^n(u, 0) = 0.$$

The quantity $(g_{ij}\xi^i\xi^j)^{\frac{1}{2}}$ in the integral for $L(v)$ is denoted by $F(u, v)$, with

$F(u, 0) = 1$ since the arc length is the parameter on g_0. F' may be written in the form $(F^2)'/2F$, from which

$$F' = \frac{1}{2F} \left(\frac{\partial g_{ij}}{\partial x^h} n^h \xi^i \xi^j + 2g_{ij} \xi^i \xi'^j \right)$$

$$= \frac{1}{2F} \left(\frac{\partial g_{ij}}{\partial x^h} \eta^h \xi^i \xi^j + 2g_{ij} \xi^i \dot\eta^j \right).$$

It is readily seen that $F'(u, 0) = 0$ independent of u, since $\partial g_{ij}/\partial x^h = 0$ on g_0, while the expression $g_{ij}\xi^i\dot\eta^j$ reduces to $\dot\eta^n$ for $v = 0$, and therefore also is zero since $\eta^n = 0$ at all points of g_0.   It follows that $L'(0) = 0$—a result that must of necessity be obtained since g_0 is assumed to be a geodesic.

The calculation of the second variation proceeds as follows:

$$L''(v) = \int_{u_1}^{u_2} F''(u, v) \, du = \int_{u_1}^{n_2} \frac{1}{2F} (F^2)'' \, du - \int_{u_1}^{u_2} \frac{(F')^2}{F} \, du.$$

The second integral has the value zero for $v = 0$.   The first integral can be written as follows:

$$\int_{u_1}^{u_2} \frac{1}{2F} \left\{ \frac{\partial^2 g_{ij}}{\partial x^h \, \partial x^k} \eta^h \eta^k \xi^i \xi^j + \frac{\partial g_{ij}}{\partial x^h} (\cdots) + 2g_{ij}(\dot\eta\dot\eta^j + \xi^i \dot\eta'^j) \right\} du.$$

The second variation for $v = 0$ reduces to the following, since then $F = 1$, $\partial g_{ij}/\partial x^h = 0$, and the quantities g_{ij}, ξ^i, and η^i satisfy for $v = 0$ various relations given above:

$$L''(0) = \int_{u_1}^{u_2} \left\{ \frac{1}{2} \frac{\partial^2 g_{ij}}{\partial x^h \, \partial x^k} \eta^h \eta^k \xi^i \xi^j + \sum_{i=1}^{n-1} (\dot\eta^i)^2 + \dot\eta'^n \right\} du.$$

The following conditions hold:

(9.74) $\xi^i(u, 0) = 0, \quad i \neq n, \quad \xi^n(u, 0) = 1, \quad \eta^n(u, 0) = 0,$

and hence the first sum in the integrand reduces to

$$\frac{1}{2} \frac{\partial^2 g_{nn}}{\partial x^h \, \partial x^k} \eta^h \eta^k.$$

Consider next the covariant curvature tensor R_{hnkn} of Riemann as evaluated along g_0 [see the equation following (9.60)]; it is given in this case by

$$R_{hnkn} = -\frac{1}{2} \left\{ \frac{\partial^2 g_{nn}}{\partial x^h \, \partial x^k} + \frac{\partial^2 g_k^h}{\partial x^n \, \partial x^n} - \frac{\partial^2 g_{hn}}{\partial x^k \, \partial x^n} - \frac{\partial^2 g_{kn}}{\partial x^h \, \partial x^n} \right\}.$$

since terms involving first derivatives of the coefficients g_{ij} are zero along g_0.

From (9.73), however, it follows that the terms involving derivatives with respect to x^n are also zero along g_0, and consequently the tensor reduces to

$$R_{hnkn} = -\frac{1}{2}\frac{\partial^2 g_{nn}}{\partial x^h \, \partial x^k},$$

This in turn makes it possible to introduce the following identity:

$$-\frac{1}{2}\frac{\partial^2 g_{ij}}{\partial x^h \, \partial x^k}\,\eta^h\eta^k\xi^i\xi^j = -R_{hikj}\eta^h\xi^i\eta^k\xi^j,$$

since from (9.74) the only nonvanishing terms occur for $i = j = n$, in which case $\xi^n = 1$. (An antisymmetry property of the covariant Riemann tensor has also been used.) Since the vectors ξ^i and η^i are orthogonal, the second member of the last equation, by (9.69), gives the value of the curvature K of the geodetic surface having the orientation fixed by ξ^i and η^i multiplied by the lengths of the vectors ξ^i and η^i. Since ξ^i is a unit vector the second variation $L''(0)$ given above can therefore be written as follows:

$$(9.75) \qquad L''(0) = \int_{u_1}^{u_2} \left(-K\eta^2 + \sum_{i=1}^{n-1} (\dot{\eta}^i)^2 + \ddot{\eta}^{\prime n} \right) du$$

$$= \int_{u_1}^{u_2} \left(\sum_{i=1}^{n-1} (\dot{\eta}^i)^2 - K\eta^2 \right) du + \eta'^n \Big|_{u_1}^{u_2}$$

Here η^2 represents the square of the length of the vector η^i. It is convenient to introduce a unit vector μ^i along g_0 that falls along η^i, so that $\eta^i = \eta\mu^i$, $\sum_{i=1}^{n-1} (\mu^i)^2 = 1$, $\sum_{i=1}^{n-1} \mu^i\dot{\mu}^i = 0$, $\dot{\eta}^i = \dot{\eta}\mu^i + \eta\dot{\mu}^i$. The equation (9.75) then takes the form

$$(9.76) \qquad L''(0) = \int_{u_1}^{u_2} \left[\dot{\eta}^2 + \eta^2\left(\sum_{i=1}^{n-1} (\dot{\mu}^i)^2 - K \right) \right] du + \eta'^n \Big|_{u_1}^{u_2}.$$

If $\dot{\mu}^i$ is zero all along g_0, it follows from (9.54) that the vectors μ^i form a parallel field along g_0, since all of the Cartan symbols vanish because of the choice of coordinates. In that special case the second variation takes the form

$$(9.77) \qquad L''(0) = \int_{u_1}^{u_2} (\dot{\eta}^2 - K\eta^2) \, du + \eta'^n \big|_{u_1}^{u_2}.$$

Another important special case occurs when all of the varied curves pass through the end points of g_0; such a choice is clearly feasible, and the parameter v could be chosen to represent the inclination angle of the varied curves with respect to g_0 at one of its end points so that $v = v_0 = $ const. would also in that case fix a particular one of the curves. If this is done it follows that

η^n is zero at both end points independent of v and that η'^n vanishes at these points. The second variation is then given by

$$(9.78) \qquad L''(0) = \int_{u_1}^{u_2} \left\{ \dot{\eta}^2 - \left[K - \sum_{i=1}^{n-1} (\dot{\mu}^i)^2 \right] \eta^2 \right\} du.$$

This formula can be used to derive a very useful lemma of Synge [S.13]. For this purpose consider once more the two-dimensional submanifold V_2 that figures in the above discussion. In this surface it is of course possible to introduce Fermi coordinates containing the geodesic g_0, and to calculate the second variation of g_0 with respect to variations passing through both of its end points, in accordance with (9.78). In that case the vector $\dot{\mu}^i$ vanishes since the vectors μ^i would define a parallel field simply because they are orthogonal to the geodesic g_0 on a two-dimensional surface. The quantity K in (9.78) then represents the Gaussian curvature of V_2, and is denoted by K_0. Thus the following formula holds:

$$L''(0) = \int_{u_1}^{u_2} (\dot{\eta}^2 - K_0 \eta^2) \, du.$$

Upon comparing this with (9.78) it is clear, because of the possibility of choosing the function $\eta(u)$ with a great deal of freedom, that the following identity holds along g_0:

$$(9.79) \qquad K - \sum_{i=1}^{n-1} (\dot{\mu}^i)^2 = K_0.$$

The formula (9.79) is called Synge's lemma. From it follows at once the inequality $K_0 \le K$, and this in turn means that *the Gaussian curvature of V_2 at any point of the geodesic g_0 is at most equal to the Riemannian curvature of the space R for a geodetic surface tangent to V_2 at the same point.* In addition, *the equality obviously holds only if $\dot{\mu}^i$ is zero,* i.e., only if the tangent plane of V_2 along g_0 results by parallel displacement (in the space R) of a vector μ^i along g_0.

Once Synge's lemma is available it becomes possible to generalize to n-dimensional Riemannian manifolds much that was done in the previous chapter for two-dimensional surfaces with respect to geodesics in the large. Before doing that, however, it should be noted that the facts derived in Section 6 of Chapter VIII concerning the Hopf-Rinow criteria for "completeness" of a surface carry over to n-dimensions more or less immediately once the same notion of distance in the space has been introduced, i.e., that the distance $\rho(x, y)$ between any pair of points is the g.l.b. of lengths of all rectifiable curves joining the two points. The discussion of the equivalence of the four different criteria for completeness, which includes as an essential step the proof of the existence of a geodetic arc between any pair of points having a length equal to the distance between them can then be carried out in the same way in n dimensions.

Synge's lemma comes into play when it is desired to generalize theorems about geodesics that make use of the second variation and the notion of the conjugate point. For example, a generalization of the theorem of Bonnet-Hopf-Rinow will now be given. In question is an n-dimensional Riemannian manifold R with a complete metric. It is assumed that *the curvature K of all geodetic surfaces is bounded below by a certain positive constant k uniformly for all points of the manifold and all orientations of the two-dimensional geodetic surface elements.* It is to be shown that *R is compact, with a diameter d satisfying the inequality $d \leq \pi/\sqrt{k}$, just as in two dimensions.*

The proof is almost immediate, as follows. Take any pair of points p, q in R; a geodetic line g_0 of shortest length joining them exists. Construct a two-dimensional submanifold V_2 containing g_0 and such that the tangent planes along g_0 result through parallel propagation (relative to R) of a vector along g_0. In that case the Gaussian curvature of V_2 along g_0 has the same value as the Riemannian curvature of R for the correspondingly oriented geodetic surface elements. The entire discussion of Chapter VIII then applies to V_2; it shows that the length of the segment g_0 is at most equal to $\pi/\sqrt{k}$ because otherwise g_0 could not furnish an arc of shortest length even with respect to the restricted variations that are confined to V_2—and hence still less in R.

In Chapter VIII (Section 14) a theorem of Synge was proved for the special case of two dimensions. It will now be proved in the general case. It states that *an orientable space R of even dimension having Riemannian curvatures always positive and bounded away from zero is simply connected.* The proof follows the same course as in the two-dimensional case, but with an amusing new twist at the very end. The proof is indirect. It is assumed that a closed geodesic g of shortest length exists relative to closed curves in its neighborhood. (It is true, but is not proved here, that such a geodesic exists in the homotopy class of any given closed curve that cannot be shrunk to a point in R.) Take any set $\mathbf{v}_i$ of $(n-1)$ mutually orthogonal vectors at a point p of g that lies in an $(n-1)$-dimensional subspace of the tangent space at p that is orthogonal to the tangent vector $\mathbf{v}_n$ of g at p. Transport by parallelism the set $\mathbf{v}_i$ once around g; upon returning to p these vectors again form a mutually orthogonal set. Thus in the $(n-1)$-dimensional subspace of the tangent space at p an orthogonal transformation is induced in the obvious way, and since the dimension of this space is odd and since the orientation is preserved it follows that the orthogonal transformation is a rotation. But a rotation about a fixed point p in an odd dimensional Euclidean space is well known to have at least one direction through p, fixed by a vector $\mathbf{v}_0$, say, that is invariant. The vector $\mathbf{v}_0$ is now displaced by parallelism once around g and a manifold V_2 is constructed tangent to the "band" thus determined. Since the Gaussian curvature of V_2 is bounded below by a positive constant by Synge's lemma, the same argument as was used in Chapter VIII shows that g could not furnish a closed geodesic of shortest length, and this contradiction proves the theorem.

It would be quite possible to continue this process of extending results of the previous chapter to manifolds of higher dimensions. A few of them are formulated as problems at the end of this chapter. For more extended treatments of the subject the books devoted to this specialty should be consulted. The remainder of the chapter is devoted to what is probably the most striking of the applications of Riemannian geometry, i.e., Einstein's general theory of relativity.

PART III.   THEORY OF RELATIVITY

11.   Special Theory of Relativity

In mechanics as it was developed in the seventeenth century, and especially in the form given to it by Newton in 1687, there are involved fundamental hypotheses concerning space and time that are tacitly assumed to be valid in the physical world. Toward the end of the nineteenth century these hypotheses were put into doubt with respect to some classes of phenomena, and as a result a series of new concepts and ways of considering space and time evolved very rapidly. They culminated in Einstein's general theory of relativity, which he published in definitive form in 1916 [E.5]. It seems reasonable to explain this development here since it belongs from the mathematical point of view to Riemannian geometry, and is indeed one of the beautiful and striking applications of it. However, it is not possible to deal with this subject here in detail. Only a brief introduction to the theory of relativity can be given, with the hope of arousing a desire to study it more thoroughly. Aside from the papers of Einstein of 1905 and 1916, the books of Weyl [W.1], Pauli [P.2], and Synge [S.11] are recommended for further study.

The general theory of relativity can be explained, however, only on the basis of what is called the special theory of relativity, which is outlined here in following the path traced out by Einstein [E.4] in 1905.[1] This requires first of all an examination of the basic assumptions of Newtonian mechanics. These are contained implicitly in Newton's law of motion: $\mathbf{F} = m(d^2\mathbf{x}/dt^2)$,

---

[1] The two papers of Einstein just cited are highly recommended. They are clearly and very well written.

in which $\mathbf{F}$ represents the force on a particle of mass m, and $\mathbf{x}(t)$ the position vector of the particle in terms of the time t as parameter. Evidently, this law implies the existence of a coordinate system with respect to which the law should hold. However, once such a system is known—which is then called an inertial system—it is almost immediately obvious that the law will be unchanged in form with respect to any other Cartesian coordinate systems moving relative to it with constant velocity, since from

$$(9.80) \qquad \mathbf{x}' = \mathbf{x} + t\mathbf{v}, \qquad \mathbf{v} = \text{const.}$$

follows $\ddot{\mathbf{x}}' = \ddot{\mathbf{x}}$. Such a transformation is called a *Galilean transformation*. On the other hand, it is clear that the law will not retain its form if the moving system has a nonzero acceleration relative to the fixed system.

A moment's reflection only is needed to make it obvious that a fixed coordinate system in the universe is something which simply cannot be conceived. Newton evaded this difficulty by restricting his mechanics to the solar system and making the provisional hypothesis that a coordinate system having the sun as origin and with axes through three fixed stars would lead, by the integration of his equations of motion, to predicted motions in accord with actual observations. A still more difficult and indeed a rather enigmatic concept is also involved, i.e., the concept of time. Newton simply assumed, without saying much if anything about it (though the inherent conceptual difficulties could hardly have escaped his notice) that time is an entity which flows on at a uniform rate under all circumstances—in particular, that time measurements in any two coordinate systems differ at most in the determination of the zero point[1]:

$$(9.81) \qquad t = t + t_0, \qquad t_0 = \text{const.}$$

Implied also is the assumption that quantitative time measurements can be achieved, for example, by making use of periodic processes in nature such as the period of the earth in its orbit around the sun.

It hardly needs to be said that the mechanics of Newton, based on these assumptions, was found to hold with accuracy for a great variety of phenomena, and that it had the most wonderful and remarkable successes in the centuries following its introduction (and still continues to have them, it should be added). However, toward the close of the nineteenth century, as was remarked above, some doubts concerning its validity under all circumstances began to arise, and these doubts were occasioned both by experimental observations and by theoretical considerations. On the theoretical side it was observed by Lorentz that the differential equations of Maxwell for the

---

[1] This further stipulation is implied in saying that (9.80) defines a Galilean transformation.

propagation of electromagnetic waves were *not* invariant under the Galilean transformations but rather were invariant under what are now called the *Lorentz transformations*. These are linear transformations which leave the expression $(x^1)^2 + (x^2)^2 + (x^3)^2 - c^2 t^2$ invariant; much more will be said about such Lorentz transformations shortly. On the other hand there was much evidence that Maxwell's equations represent large classes of phenomena with great accuracy, and hence doubt was thus thrown on the general validity of Newton's basic assumptions.

It was thought that the hypothetical medium, the ether, through which electromagnetic waves propagate, might possibly be such that motions relative to it could be detected, and thus this medium might furnish a fixed space of reference. Michelson and Morley devised an experiment to test out such a hypothesis. The experiment used the motion of the earth in such a fashion that the experimental errors could be expected to be smaller than the effects to be detected, but the result was negative in the sense that the speed of light turned out to be independent of the motion of the earth. Fitzgerald and Lorentz proposed to interpret this result physically by assuming that lengths of bodies are shortened in a moving system by an amount that depends on the relative velocity. The result can also be interpreted as suggesting that the speed of propagation of electromagnetic waves, in other words the speed of light, is independent of the motion of coordinate systems moving with constant velocity relative to each other. Einstein in 1905 gave the whole affair a new aspect by (a) accepting the constancy of the light speed as a physical fact and (b) adding to it the basic Newtonian hypothesis that all physical laws should have the same form with respect to Cartesian coordinate systems moving with constant velocity relative to each other.

The first assumption implies that if light rays go out from the origin O of a coordinate system S with coordinates x^1, x^2, x^3 at time $t = 0$ and reach the surface of a sphere of radius ct (with c, the speed of light) at time t, then in an S' system with coordinates x'^1, x'^2, x'^3, which is moving with a speed v at the time $t = 0$ and has the same origin momentarily as the first system, the light rays will also lie on a certain sphere which is attained by them in traveling with the same speed c. Furthermore, the sphere in the original system satisfies the equation $(x^1)^2 + (x^2)^2 + (x^2)^2 + (x^3)^2 - c^2 t^2 = 0$, and this "law" for the propagation of light should by Einstein's second assumption have the same form in both systems. To achieve this *it is essential to regard the time t as being transformed together with the space variables* in such a fashion that the equation $(x^1)^2 + (x^2)^2 + (x^3)^2 - c^2 t^2 = 0$ holds in the same form in the second system. Evidently, this condition will hold if the quadratic form $(x^1)^2 + (x^2)^2 + (x^3)^2 - c^2 t^2$ in the four variables x^i, t is invariant under transformation of these variables to new variables x'^i, t', i.e., if the transformations are what was called above Lorentz transformations. We proceed to find these linear[1] transformations in a special case. It is useful to begin by setting $x^4 = ict$, $x'^4 = ict'$ and thus to base the discussion on the invariance of the form $(x^1)^2 + (x^2)^2 + (x^3)^2 + (x^4)^2$; this procedure, which is due to Minkowski, thus leads to the orthogonal transformations in a four-dimensional Euclidean space. The special cases considered here are those in which two of the four coordinates are held fixed while the others vary. For example, if x^3 and x^4 are held fixed, all Lorentz transformations could be shown to have the form

$$(9.82) \quad \begin{cases} x'^1 = x^1 \cos\phi - x^2 \sin\phi, \\ x'^2 = x^1 \sin\phi + x^2 \cos\phi, \\ x'^3 = x^3, \qquad x'^4 = x^4. \end{cases}$$

These transformations represent a rotation in Euclidean space of three dimensions about the x^3-axis when ϕ is an arbitrary real angle; thus these classical rotations in Euclidean space are included among the Lorentz transformations. It is more interesting for present purposes to consider the transformations which arise when x^2 and x^3 are regarded as fixed, while x^1 and x^4 vary; this case could be expected to correspond to that in which the system S' moves with a speed v along the x^1-axis. As in (9.82) the following relations would hold:

$$(9.83) \quad \begin{cases} x'^1 = x^1 \cos\psi - x^4 \sin\psi, \\ x'^4 = x^1 \sin\psi + x^4 \cos\psi, \\ x'^2 = x^2, \qquad x'^3 = x^3. \end{cases}$$

However, the angle ψ can no longer be regarded as a real angle; in fact, upon reintroducing the t and t' coordinates the following equations result:

$$x'^1 = x^1 \cos\psi - ict \sin\psi,$$

$$ict' = x^1 \sin\psi + ict \cos\psi.$$

The origin of S' coincides with that of S at $t = 0$, but is moving with velocity v along the x'-axis; hence $x' = vt$ for $x'^1 = 0$ so that the first of the above equations becomes

$$0 = vt \cos\psi - ict \sin\psi.$$

---

[1] That the physical laws should be assumed to be invariant under *linear* transformations is reasonable on grounds other than mere simplicity. For example, it is to be expected that the Galilean transformations, which are obviously linear, should be correct in some simple limit situations of those now under investigation, and give good approximations in the wide class of phenomena for which Newtonian mechanics is sufficiently accurate. Also, nonlinear transformations would imply in some sense a lack of homogeneity of the space.

From this it follows that $v/c = i \tan \psi$, and hence $\sin \psi$ and $\cos \psi$ are given by

$$\sin \psi = \frac{-i \, v/c}{\sqrt{1 - (v/c)^2}}, \qquad \cos \psi = \frac{1}{\sqrt{1 - (v/c)^2}}.$$

Thus the transformations (9.83) take the form

(9.84)
$$\begin{cases} x'^1 = \dfrac{x^1 - vt}{\sqrt{1 - (v/c)^2}}, & t' = \dfrac{t - (v/c^2) \, x^1}{\sqrt{1 - (v/c)^2}}, \\ x'^2 = x^2, & x'^3 = x^3. \end{cases}$$

(Actually, if it were assumed merely that $L = \sum_{i=1}^{3} (x^i)^2 - c^2 t^2 = 0$ entrains $L' = \sum_{i=1}^{3} (x'^i)^2 - c^2 t'^2 = 0$ by a linear transformation, it is possible to show only that $L' = \kappa(v)L$, with $\kappa(v)$ an arbitrary function of v. However, it is not difficult to show, as is done in more extensive treatments of relativity, that $\kappa(v) \equiv 1$ follows reasonably through plausible inferences of a physical nature.) It is now seen that the Galilean transformation is closely approximated if v/c is small, i.e., if the relative velocity of the two systems is small compared with the speed of light. Thus it is not strange that the Newtonian kinematics was found for so long to be very accurate, since the speeds encountered in practice were in general very much smaller than the speed of light.

The Lorentz transformations (9.84) have consequences which are, at first sight, somewhat strange. Consider, for example, a series of points $x'^1 = n$, with n an integer, as measured in the system S', and at time $t = 0$. From (9.84) it follows that these points are located in the system S at $x^1 = n\sqrt{1 - (v/c)^2}$, i.e., the distances are contracted in the ratio $\sqrt{1 - (v/c)^2}$, called the Fitzgerald contraction. Consider next the origin $x^1 = 0$ in the system S and a clock there which marks the time $t = n$; with n once more an integer. Again it follows from (9.83) that the corresponding time t' in S' are given by $t' = n/\sqrt{1 - (v/c)^2}$; thus there is a dilation in the time, or, as it could also be put, the time runs more slowly as observed from the system S'. Finally, events that occur at the same time t at two different points x_1^1, x_2^1 in S will not occur simultaneously when observed in S', since

$$t_1' = \frac{t - (v/c^2) \, x_1^1}{\sqrt{1 - (v/c)^2}} \quad \text{and} \quad t_2' = \frac{t - (v/c^2) \, x_2^1}{\sqrt{1 - (v/c)^2}}, \quad \text{and} \quad t_1' \neq t_2'$$

unless $x_1^1 = x_2^1$.

Implied in this discussion is a *definition* of what is to be understood by *simultaneity*, as Einstein first pointed out, and it is necessary to analyze carefully what lies behind this notion. In fact, simultaneity in any one system S is very reasonably defined by using the postulated constancy of the speed of light: at any point P at distance r from the origin the local time is synchronized with that at the origin by sending out a light ray at time $t = 0$ from the origin and setting a clock at the point P to read $t = r/c$. Events at two different points are called simultaneous events if the clocks, after synchronization at these points, have the same reading. The same is done in any other system S' (it being implied, of course, that standard clocks and measuring rods of like composition are available for all systems). With this notion of simultaneity—which is clearly reasonable—things fall nicely into order and the new kinematics is not at all paradoxical.

12. Relativistic Dynamics

So far, only the kinematical part of the special theory of relativity has been discussed. It is, however, not difficult to formulate the appropriate new dynamical law that replaces Newton's law for the mechanics of a particle. This was done by Minkowski in a very elegant and quite geometrical way. What is wanted is a generalization of the Newtonian law that holds in a linear vector space of four dimensions. To this end, suppose that a particle moves with velocity $\mathbf{v}$ relative to a system $S: x^1, x^2, x^3, t$. Thus $\mathbf{v} = (dx^1/dt, dx^2/dt, dx^3/dt)$, and write $v^2 = (dx^1/dt)^2 + (dx^2/dt)^2 + (dx^3/dt)^2$ for the square of the speed. Suppose that τ is the time in a coordinate system S' with origin coinciding momentarily with the particle and moving with constant velocity $\mathbf{v}$ relative to S. The motion of the particle at that instant is a translation with this velocity; it is therefore reasonable to regard t as a local time associated with the particle at the instant in question. Hence, in accordance with (9.84)[1] for small times dt and $d\tau$ in the two systems, a time dilatation for the origin of S' is given by the relation

$$(9.85) \qquad \frac{dt}{d\tau} = \left(1 - \frac{v^2}{c^2}\right)^{-\frac{1}{2}},$$

or, as this can also be written,

$$(9.86) \qquad d\tau^2 = \frac{1}{c^2}\left(c^2\,dt^2 - v^2\,dt^2\right)$$

$$= \frac{1}{c^2}\left(c^2\,dt^2 - (dx^1)^2 - (dx^2)^2 - (dx^3)^2\right).$$

Thus the quantity $d\tau$ is a Lorentz invariant. If, then, the position vector of

---

[1] Strictly speaking, the analogue of (9.84) for the general case in which $\mathbf{v}$ does not necessarily fall along the x^1-axis should be used. The result would be the same, however, for the origin of S'.

the particle is supposed given in Minkowski's four-space by $\mathbf{x} = (x^1, x^2, x^3, ict)$, a new vector $\mathbf{q}$ can be defined by differentiation with respect to the invariant time τ, which is a scalar:

$$(9.87) \qquad \mathbf{q} = \left(\frac{dx^1}{d\tau}, \frac{dx^2}{d\tau}, \frac{dx^3}{d\tau}, ic\frac{dt}{d\tau}\right) = (u^1, u^2, u^3, u^4),$$

and it is by definition the *four-velocity* of the particle. Since $dt/d\tau$ is given by (9.85) it follows that

$$(9.88) \qquad u^i = \frac{dx^i}{dt}\cdot\frac{dt}{d\tau} = \frac{v^i}{\sqrt{1 - v^2/c^2}}, \qquad i = 1, 2, 3,$$

when the ordinary velocity $\mathbf{v} = (v^1, v^2, v^3)$ is defined in the customary way by $v^i = dx^i/dt$, as above. In addition the component u^4 is given by

$$(9.89) \qquad u^4 = ic\frac{dt}{d\tau} = \frac{ic}{\sqrt{1 - v^2/c^2}}.$$

With the aid of these formulas, the Newtonian mechanics of a particle can be generalized. The basic laws to be generalized are the law of conservation of momentum

$$(9.90) \qquad m\frac{d\mathbf{v}}{dt} = \mathbf{F}$$

and the law of conservation of energy

$$(9.91) \qquad \frac{d}{dt}\left(\tfrac{1}{2}\,m\mathbf{v}\cdot\mathbf{v}\right) = \mathbf{F}\cdot\mathbf{v},$$

in which the dot means the scalar product. A Lorentz invariant form of these laws in four-space is desired that will tend to them when v/c is made to approach zero.

The generalization of (9.90) to the system S is not far to seek, in view of the above developments: the Newtonian velocity $\mathbf{v}$ is replaced by the four-velocity $\mathbf{q}$ of (9.87) with the result

$$(9.92) \qquad m_0\frac{du_i}{d\tau} = F_i, \qquad i = 1, 2, 3, 4,$$

upon introducing a mass m_0, which is by definition a scalar, an invariant, and differentiating with respect to the invariant time τ associated with the particle. In order to interpret the physical sense of the four-vector $\mathbf{F}$ introduced as the force in this formal way, it is best to reintroduce the time t

associated with the system S with respect to which the particle is moving. This leads, from (9.85), to the relations

$$(9.93) \qquad \frac{d}{dt}\left(\frac{m_0 v^i}{\sqrt{1 - v^2/c^2}}\right) = F_i \left(1 - \frac{v^2}{c^2}\right)^{\frac{1}{2}}, \qquad i = 1, 2, 3.$$

Thus, as $v^2/c^2 \to 0$, these equations become Newton's law and the quantities $F_i \sqrt{1 - v^2/c^2}$ tend therefore to the force components in that law in three-dimensional space. The differential equations (9.93) have led to some of the best experimental evidence for the validity, from the point of view of experiment, of the special theory of relativity, since with their aid the trajectories of various fast-moving elementary particles have been calculated and found to be in accord with observations.

The fourth component of (9.92) leads to something of extraordinary interest, which is best brought out by taking the scalar product of the vectors on both sides of (9.92) with the four-velocity $\mathbf{q} = (u^1, u^2, u^3, u^4)$ to obtain

$$(9.94) \qquad m_0 \frac{1}{2} \frac{d}{d\tau} \sum_{i=1}^{4} (u^i)^2 = \sum_{i=1}^{4} F_i \cdot u^i.$$

However,

$$\sum_{i=1}^{4} (u^i)^2 = \sum_{i=1}^{3} \left(\frac{dx^i}{d\tau}\right)^2 - c^2 \left(\frac{dt}{d\tau}\right)^2,$$

from (9.87), and since

$$\frac{dx^i}{d\tau} = u^i = \frac{v^i}{\sqrt{1 - v^2/c^2}} \quad \text{and} \quad \frac{dt}{d\tau} = \left(1 - \frac{v^2}{c^2}\right)^{-\frac{1}{2}}$$

it is seen that $\sum (u^i)^2 = -c^2$, and hence (9.94) yields

$$(9.95) \qquad \sum_{i=1}^{4} F_i \cdot u^i = 0.$$

Since $u^4 = ic/\sqrt{1 - v^2/c^2}$, (9.95) can be written in the form

$$(9.96) \qquad \frac{ic F_4}{\sqrt{1 - v^2/c^2}} + \sum_{i=1}^{3} \left[F_i \left(1 - \frac{v^2}{c^2}\right)^{\frac{1}{2}}\right] \cdot v^i = 0.$$

Since, finally

$$F_4 \left(1 - \frac{v^2}{c^2}\right)^{\frac{1}{2}} = \frac{d}{dt}(m_0 u_4) = \frac{d}{dt}\left(\frac{i m_0 c}{\sqrt{1 - v^2/c^2}}\right)$$

it follows from (9.96) that

$$(9.97) \qquad \frac{d}{dt}\left(\frac{m_0 c^2}{\sqrt{1 - v^2/c^2}}\right) = \sum_{i=1}^{3} \left[F_i \left(1 - \frac{v^2}{c^2}\right)^{\frac{1}{2}}\right] \cdot v^i.$$

In the limit when v/c tends to zero, i.e., in Newtonian mechanics, the right-hand side represents the rate at which work is done on the particle, and consequently it is a reasonable inference that the left-hand side should represent the time rate of change of energy; as a further inference, it seems reasonable then to say that the quantity $E = m_0 c^2 / \sqrt{1 - v^2/c^2}$ represents, within a purely additive constant, the energy of a particle. If v^2/c^2 is small, this energy is given closely by $m_0 c^2$, which Einstein interpreted as meaning that with the "rest mass" m_0 is associated a definite amount of energy. Since the speed of light c is very large, it turns out that very large amounts of energy are associated with relatively small amounts of mass. This source of energy has been made convertible by fission and other processes into kinetic energy, with the results well known to all.

13.   The General Theory of Relativity

So far only the special theory of relativity has been in question. It is concerned with systems referred to Cartesian coordinates, but subject not to the Euclidean metric but to the Lorentz metric, so that systems moving with constant velocity relative to each other were alone considered. It should be pointed out also that the constancy of the speed of light is assumed only in this special case. In addition, the only physical phenomena treated were purely kinematic phenomena together with the motion of a particle under given mechanical forces. Einstein in his general theory of relativity goes beyond these restrictions by permitting arbitrary motions and arbitrary coordinate systems, and also—what is of great significance—permitting forces arising from the gravitational effects of a distribution of matter in the space. In fact, these two phases of the generalization are inextricably tied together, as will be seen.

In Newtonian mechanics it is not difficult to deal with coordinate systems that move with an acceleration relative to a system in which Newton's laws are assumed to be valid. It suffices for that to make a formal calculation in which the known motion of the accelerated coordinate system is explicitly taken into account. The result of such a calculation is often interpreted in such a way as to leave Newton's law $\mathbf{F} = m\ddot{\mathbf{x}}$ valid in the accelerated system by adding fictitious forces to the given forces that are determined by the known acceleration of that system. Another way to interpret such a procedure is to imagine that there is a *field* of force defined in the accelerated system that has the desired compensating effect.

It was Einstein's idea to do a similar thing in such a way as to generalize the special theory of relativity so that the restriction to coordinate systems moving with constant velocity relative to each other could be removed. In his famous paper of 1916 he succeeded in doing so for the motion of a mass particle and the paths of light rays in a gravitational field.   The object of the new theory was, roughly speaking, to find the appropriate invariant analogues of the Poisson equation

$$\nabla^2\phi = 4\pi\rho$$

for the Newtonian potential ϕ due to a distribution of mass of density ρ, together with the differential equations

$$\frac{d^2\mathbf{x}}{dt^2} = - \text{ grad } \phi$$

appropriate for the motion of a particle in the field.

In devising such a theory Einstein starts from a fact of experience, which had been established with great accuracy by a variety of experimental measurements, that *the field of gravity imparts the same acceleration to all bodies regardless of their composition*.   Thus on the earth the acceleration of gravity g is a number that depends only upon the *location* of the falling body and upon nothing else.   This fact makes it possible to postulate that the gravitational field could be made to disappear in a properly chosen coordinate system if that system were moving with exactly the acceleration determined by the field.   On the earth, for example, a freely falling body has no acceleration relative to a box falling with it.   This fact of experience led Einstein to postulate that *the gravitational field could be transformed away at any point by introducing locally an appropriate moving coordinate system*, and since this field is the only extraneous factor added to those in the special theory of relativity, it seems reasonable to suppose that the special theory is valid in such a coordinate system to a good approximation in a sufficiently small neighborhood of the point.   Thus Einstein makes the assumption that the four-dimensional space in which the phenomena take place is such that locally it is approximately invariant with respect to the Lorentz transformation, and this in turn is taken to mean that the line element at a point can be put into the form

$$ds^2 = d\bar{x}^{1^2} + d\bar{x}^{2^2} + d\bar{x}^{3^2} + d\bar{x}^{4^2}, \qquad d\bar{x}^4 = ic\,dt',$$

with good accuracy by a suitable coordinate transformation.   From the point of view of the Riemannian geometry developed in the preceding sections this means that in the four-dimensional physical space the linear vector tangent space at a point is not given a *Euclidean* metric, but rather a *Lorentz-invariant* metric.   If, now, an arbitrary curvilinear coordinate system x^1, x^2, x^3, x^4 is introduced, the line element of the space, of course, takes the form

$$ds^2 = g_{ik}\,dx^i\,dx^k,$$

with coefficients g_{ik} that depend on the coordinates. Einstein felt called upon to admit *arbitrary* coordinate systems, since there seemed to be no valid reason to single out any special ones. However, he did not at once find the way to his formulation of the invariant laws of motion in a gravitational field.

The insight which seems to have led him to the final form of his theory was that the presence of a gravitational field must lead in the large to a geometry in which measurements of length and time will give rise to laws of configuration which are neither Euclidean, nor invariant under the Lorentz transformation; and that the proper tool to use in overcoming the difficulties was Riemannian geometry, essentially in the form described above and developed long before 1916 by Ricci and Levi-Civita, since this theory deals with the geometry of a manifold of any dimensions in a form that is invariant in an important sense with respect to all coordinate systems. Einstein gave the following example to explain why measurements in the large must lead to conflicts with Euclidean or Minkowskian geometry as regards space measurements. Consider a coordinate system S with respect to which the special theory of relativity holds and another S' such that x'^3 coincides with x^3, but so that S' rotates about the x^3-axis with constant angular velocity. Imagine now a circle in the x'^1, x'^2-plane with its center at the origin. If S' were not rotating about S, the ratio of the circumference to the diameter of the circle as measured in S would be π. If, however, S' is rotating, a different result is obtained, since at a definite time t in S a small measuring rod on the circumference of the circle would experience the Lorentz contraction, but rods placed along a diameter would not. Hence the ratio of circumference to diameter would now be $> \pi$, as estimated from S. Thus the geometry would not be Euclidean. Since the acceleration of S' relative to S is supposed to be mechanically equivalent to the presence of a field, it is to be expected that such a field would in general have the effect that the geometry in the large would not be characterized by invariance under the Lorentz metric. The same thought experiment also leads to the result that a gravitational field should result in time measurements that are in the large not compatible with the Lorentz transformation. Thus light rays would be bent, and the velocity of light would no longer be constant everywhere, but rather would vary from point to point in the four-dimensional space. Thus the introduction of gravitation in general means that the physical laws will no longer be invariant in the large with respect to the Lorentz transformation. In the language of Riemannian geometry this, in turn, means that the line element $g_{ik}\, dx^i\, dx^k$ cannot be transformed in the large, though it can be at a point, so that it takes on the form $\sum_{i=1}^{3} (dx'^i)^2 - (dx'^4)^2$ in which the coefficients g_{ik} are zero for $i \neq k$, and three of them have the value $+1$ and the fourth has the value -1.

This point of view of Einstein thus means, obviously, that the geometry of the four-dimensional space-time continuum is determined by the gravitational field, and this in turn leads to the expectation that it should be uniquely determined by a given distribution of mass in the system. This remarkable and beautiful conception thus would make the problem of determining the "true" geometry of the world a problem of physics—a point of view which had long before been put forward by Riemann [R.3]. The word "true" was put into quotation marks because the laws of physics are always provisional since they depend upon the existing state of knowledge and experimentation; in the present case, in fact, only gravity is to be taken into account, and such effects as those due to electromagnetic wave propagation, atomic processes, etc., have been neglected. Einstein spent the latter part of his life in pursuit of ever more general field theories which should embrace as many as possible of the physical effects as are at present known, but he succeeded only partially in that enterprise.

Naturally, even if gravitational phenomena alone are considered, it is still necessary to devise a way to determine the field, i.e., to define the metric tensor, once the distribution of mass in space and time has been given. A necessarily brief outline of the manner in which Einstein accomplished this task follows.

In the special theory of relativity a particle upon which no force acts moves uniformly in a straight line—that is, along a geodesic in the Minkowski space. In the presence of a gravitational field it is natural to assume that a particle moves along a *geodesic in the metric appropriate to that field* since that is correct in the absence of a gravitational field and the metric is supposedly so defined that it gets rid of that field locally. In addition, such a stipulation has an invariant character, as it must to make sense from Einstein's point of view. This would mean that the differential equation of motion of a particle is given by

$$\frac{d^2x^i}{ds^2} + \Gamma^i_{rk}\frac{dx^r}{ds}\frac{dx^k}{ds} = 0 \quad \text{with} \quad g_{ik}\frac{dx^i}{ds}\frac{dx^k}{ds} = 1,$$

when the Cartan symbols are defined with respect to the metric form. In particular, if no gravitational field is present, it follows that a coordinate system exists in which all of the Cartan symbols vanish, and the motion is that of a straight line given by

$$\frac{d^2x^i}{ds^2} = 0.$$

If the gravitational field does not vanish, it seems then reasonable to regard the Cartan symbols as giving in some sense a measure of the intensity of the gravitational field. But this means that the metric coefficients are also defined by the intensity of the field since they are uniquely defined by these symbols, as was seen in earlier sections.

A similar consideration can now be applied to the propagation of light rays. In the absence of gravity, i.e., in case the space is a Minkowski space, these are straight lines along which the Lorentz invariant has the value zero, as was seen above, and hence in the general case they should be geodetic lines of zero length; the relations

$$\frac{d^2x^i}{d\lambda^2} + \Gamma^i_{rk}\frac{dx^r}{d\lambda}\frac{dx^k}{d\lambda} = 0 \quad \text{with} \quad ds^2 = g_{ik}\,dx^i\,dx^k = 0,$$

should thus hold for them. In general, the light rays will therefore be curved in a fashion that depends on the metric, i.e., on the mass distribution in the system.

One of the first simple consequences deduced by Einstein from these considerations was a useful approximation to the motion of a particle in a *weak* gravitational field when the particle has a velocity small compared with that of light. To do this it is assumed that the coefficients g_{ii} differ by at most first-order terms from the values $+1, +1, +1, -1$ and that the functions g_{ik} for $k \neq i$ vanish to zero order and thus are at least of first order: this is reasonable since the space should be approximately a Minkowski space. If in addition the field is assumed to be stationary (i.e., the functions g_{ik} are independent of t), it is found by a formal expansion that the equations of motion are given, up to first order terms, by

$$\frac{d^2x^i}{dt^2} = \frac{c^2}{2}\frac{\partial g_{44}}{\partial x^i},$$

and if the function Φ is introduced by the relation

$$\Phi = -\frac{c^2}{2}(g_{44} + 1)$$

it turns out that the equations of motion become

$$\frac{d^2x^i}{dt^2} = -\frac{\partial \Phi}{\partial x^i},$$

and Φ therefore plays exactly the same role as the gravitational potential function in Newtonian mechanics. The function Φ was normed in such a way that it vanishes in case there is no gravitational field, that is to say when $g_{44} = -1$. Thus in this case at least the metric coefficients g_{ik} do play the same role as the Newtonian potential, although it is somewhat of an accident that a single scalar function serves the purpose.

However, it still remains to be seen how the metric is to be determined in general once the distribution of mass is given. It seems likely that appropriate differential equations for the unknown functions $g_{ik} = g_{ki}$ should be derived which would be invariant—that is, have tensor character—and by analogy with Poisson's equation it seems reasonable to suppose that the functions g_{ik} should be determined as solutions of second-order partial differential equations that are linear in the second derivatives.

Only one special case is considered here, i.e., that in which a portion of space where the differential equations are desired is free of matter. There are very few tensors that contain second derivatives of the functions g_{ik} and no higher derivatives, and which are linear in the second derivatives. One such is the Riemann tensor. If, however, $R^{\mu}_{\nu\alpha\beta} = 0$[1] were to be chosen to obtain differential equations no progress would be possible since these are the conditions which make it possible to introduce a coordinate system in the large in which the coefficients g_{ik} would all be constants (cf. Section 7) and thus such that the Lorentz transformation holds, and hence in which no gravitational field exists. In any case, again by analogy with the Poisson equation, whose left-hand side is in general coordinates a tensor of second rank, it would seem reasonable to seek appropriate invariant equations among the tensors of second rank, rather than among those of fourth rank. In fact it was shown by H. Weyl[W.1] that under the present circumstances the only tensor of second rank available is the tensor $R_{ij} = R^{n}_{ijn}$ obtained by contracting the Riemann tensor, and since this tensor is symmetric, it follows that the equations

$$R_{ij} = 0$$

form a system of 10 equations—just the right number to determine the 10 functions g_{ij}. It therefore is reasonable to hope that these 10 nonlinear partial differential equations might serve to determine the field in empty space. (Actually, the count is a little more complicated since the functions g_{ik} are determined only within arbitrary changes of local parameters, and this in turn is compensated for because the tensor components R_{ij} satisfy certain conditions.) This expectation has proved well warranted.

For example,[2] consider the interesting and important special case of the field due to a concentrated mass—that is, a particle—assumed fixed in space. The field is therefore assumed to be static and it clearly is appropriate to assume it to be spherically symmetric. In that case the line element can be shown to be expressible in the form

$$ds^2 = \gamma[(dx^1)^2 + (dx^2)^2 + (dx^3)^2]$$
$$+ l(x^1 \, dx^1 + x^2 \, dx^2 + x^3 \, dx^3)^2 + g_{44}(dx^4)^2,$$

[1] It is to be expected that the right-hand sides should be zero, in analogy with the Poisson equation, since no matter is assumed to be present.

[2] All details not strictly necessary for an understanding of the general procedure are omitted. For a complete discussion see Weyl [W.1], Synge [S.11], Pauli [P.2].

with γ, l, and g_{44} functions of $r = \sqrt{(x^1)^2 + (x^2)^2 + (x^3)^2}$ only. A transformation of the form $x'^i = [f(r)/r] \, x^i$, which contains an arbitrary function $f(r)$, is readily seen to leave ds^2 invariant; hence there is no loss of generality in setting $\gamma \equiv 1$. Outside of the origin the field equations $R_{ik} = 0$ hold. If it is assumed that the functions g_{ik} behave at ∞ as though no matter were present anywhere (i.e., the special theory of relativity is assumed to hold far away from the concentrated mass at $r = 0$), the differential equations $R_{ij} = 0$ are found to have the following solution for g_{44}:

$$g_{44} = -1 + \frac{2m}{r},$$

in which m is a constant of integration. From the previous result for weak fields, i.e., $\Phi = -(c^2/2)(g_{44} + 1) = kM/r$, where now M is the mass of the particle at the origin and k is the gravitational constant, it is indicated that m should be assumed to have the value

$$m = \frac{kM}{c^2},$$

thus determining g_{44} in terms of the mass concentrated at $r = 0$, the gravitational constant, and the speed c of light in a space entirely free of matter. It is then found that the function $l(r)$ given by

$$l(r) = \frac{2m}{r^2(r - 2m)},$$

in conjunction with the function g_{44} above furnish a solution of the field equations, and the line element is now completely determined.

Once the gravitational field due to a mass particle has been determined, it becomes possible to calculate both the trajectories of moving mass particles in the field, the form of light rays, and the local speed of light, which is not everywhere constant. In fact, the integration of the ordinary (though non-linear) differential equations which result for the given field can be carried out with no very great difficulty. An interesting case is that in which m is taken to correspond to the mass of the sun, while the moving particle is supposed located relative to the sun at the distance of the planet Mercury. It turns out that the correction to Newtonian mechanics implied in the use of the new theory leads to a correction to the orbit of this planet, which consists of a slow motion of its perihelion that amounts to 42.89 seconds of arc in a century, and this is what is needed to explain the discrepancy between the observed motion and the motion determined by Newtonian mechanics, and which cannot be explained as the result of perturbations caused by the other planets. In fact, the measured discrepancy from the result of calculations based on the Newtonian theory is 41.24 seconds of arc in a century, with an error of ± 2.09 seconds, so that the theoretical result obtained from Einstein's theory lies within the experimental error.

The bending of a light ray passing near the sun can also be made the object of a calculation, and it turns out that a ray of light from a fixed star behind the sun that passes by the sun would be deflected by 1.75 seconds of arc when observed on the earth. Such a deflection is in principle observable at times when there is a total eclipse of the sun, and appropriate measurements have been made at such times. An effect in the right direction has been observed, but there still seem to be some doubts as to whether the experimental errors are not of about the same order of magnitude as the effects to be detected.

Of the validity and usefulness of the special theory of relativity there can be no doubt, and Einstein contributed greatly to its understanding. It is perhaps not certain that the general theory developed by him will prove to be as useful as the special theory, but that its creation was a remarkable and inspired stroke of genius cannot be put in doubt, and it is in any case a very interesting and perfectly valid piece of mathematics quite aside from the question of its relation to the facts of the physical world.

PROBLEMS

1. Find the transformation rule for the symbols Γ_{ij}^k that must hold by virtue of the fact that the formula for covariant differentiation yields a tensor.

2. Show that the formulas for $D\xi^i$ and $D\xi_i$ yield associated vectors if ξ^i and ξ_i stand in that relation.

3. Prove that the tangent space associated with a system of normal coordinates at a point P is an osculating tangent space. *Hint:* Consider all geodesics $u^i(s)$ going out from P, with s as arc length. Develop their equations near $s = 0$, use the differential equation for geodesics, and observe that what results is the parameter transformation desired (for s small enough).

4. Show that the equation $\phi_{x^i} - (d/dt)\,\phi_{\dot{x}^i} = 0$ that arises from the integral in (9.70) yields the differential equations of the geodetic lines.

5. Prove the statement made in Section 10 concerning the existence of a geodesic g from a point P near to g_0, but not on it, that cuts g_0 orthogonally.

6. Show that (9.79) follows in the circumstances prevailing when the two integrals representing $L''(0)$ are compared.

7. Show, in Euclidean space of three dimensions, that two surfaces that are tangent to each other along a straight line have the same Gaussian curvature along it. *Hint:* Use Synge's lemma.

8. Show, by analogy with what was done in Chapter VIII, that a simply connected covering surface of a complete Riemannian manifold can

be constructed with the aid of geodetic polar coordinates if the Riemannian curvature is everywhere nonpositive. *Hint:* Observe from (9.78) that $L''(0) > 0$ holds, i.e., every geodesic furnishes an arc of minimum length (locally, at least).

9. Suppose R is complete and simply connected with negative Riemannian curvature. Show that all geodetic triangles have interior angles that are less than π in value. *Hint:* From one vertex draw all geodetic arcs to the points of the opposite side; use Synge's lemma, and the Gauss-Bonnet formula for the resulting V_2.

10. Same assumptions as in Problem 9, except that a geodetic *quadrilateral* is involved. Show that the sum of the angles (less than π) at the four vertices is less than 2π.

CHAPTER X

The Wedge Product and the Exterior Derivative of

Differential Forms, with Applications to Surface Theory

1. Definitions

In 1862 Grassmann [G.4] introduced an algebra of vectors, based on a definition for products of vectors that has analogies with the ordinary vector product in Euclidean three-space. This algebraic theory has become well known principally through the work of E. Cartan [C.2], who made use of it for a variety of purposes in geometry and mechanics. However, a calculus as well as an algebra is necessary for such purposes, and Cartan discovered that the algebra of Grassmann when applied to differential forms as elements of the algebra could be combined with a special differentiation process, called exterior differentiation, to produce a very compact notation that has proved to be convenient for many purposes. For example, this notation makes it possible to present the various formulas for the transformation of volume integrals into surface integrals (in all dimensions) in a unified and simple way. It is also an especially compact and convenient notation whenever it is important to deal with compatibility conditions for overdetermined systems of partial differential equations—as was necessary on many occasions in earlier chapters of this book. In Riemannian geometry a good deal of the recent literature makes use of Cartan's notation.

In the present chapter of this book limitations of space make it possible to do no more than give a brief introduction to the subject with a few applications. One of the applications to be given is a rederivation of the fundamentals of surface theory in a three-dimensional space (i.e., of the theory treated earlier in Chapters IV and VI) in the new notation. Other applications include proofs of some interesting theorems of differential geometry in the large; these theorems are in the main uniqueness theorems for closed surfaces in three-dimensional space—for example, the theorem that two closed convex surfaces that are in a one-to-one isometric correspondence are either

335

congruent or symmetrical. This implies that the discussion will be confined mostly to two-dimensional manifolds and hence to functions of two variables. It is, however, only slightly more complicated to consider the general case, and some indication of how that is done will be given at the end of this section.

The basic elements for the algebraic part of the theory to be discussed here are (1) the notion of an *invariant linear differential form of first degree* and (2) the definition of the *exterior* or *wedge product* of two such forms. Differential forms of first degree are denoted by small Greek letters and are defined by

$$(10.1) \qquad \omega = a_1 \, du^1 + a_2 \, du^2 = a_i \, du^i$$

with the stipulation that the coefficients $a_i = a_i(u^1, u^2)$ should be the components of a covariant vector, i.e., they should be such that the functions $a_i(u^1, u^2)$ transform according to the rule $a_i = a_j'(\partial u'^j/\partial u^i)$ upon introducing new coordinates (u'^1, u'^2). This ensures the invariance of ω with respect to coordinate transformations. That is, if $u^i = u^i(u_1', u_2')$ the usual rule of the calculus defines the differential by $du^i = (\partial u^i/\partial u'^j) \, du'^j$ and thus $\omega = a_i \, du^i$ is transformed in the new coordinates into the form $\omega = a_i(\partial u^i/\partial u'^j) \, du'^j = a'_j \, du'^j$ with $a'_j = a_i \, \partial u^i/\partial u'^j$. A differential form ω vanishes, and is written $\omega = 0$, only if a_1 and a_2 are both zero.

The *exterior* or *wedge* product, denoted by the symbol " $\wedge$," is defined for two such forms in the usual way, and obeys the usual algebraic rules for products, except that the product of the differentials du^1, du^2 is assumed to be what Grassmann called an alternating product, i.e., *a product obeying the rules* $du^i \wedge du^j = -du^j \wedge du^i$ *for* $i \neq j$ *and* $du^i \wedge du^i = 0$, while $a \wedge du^i = du^i \wedge a = a \, du^i$ for a multiplication with any function $a(u^1, u^2)$ and $du^i \wedge a \, du^j = a \, du^i \wedge du^j$. As a consequence of these stipulations the following formulas result:

$$(10.2) \qquad \omega_1 \wedge \omega_2 = (a_1 \, du^1 + a_2 \, du^2) \wedge (b_1 \, du^1 + b_2 \, du^2)$$

$$= (a_1 b_2 - a_2 b_1) \, du^1 \wedge du^2 = -\omega_2 \wedge \omega_1.$$

The form $\omega_1 \wedge \omega_2$ is now called a *differential form of second degree*; it also turns out to be an invariant under coordinate transformations, as is to be shown.

In fact, if the transformation $x = x(u, v)$, $y = y(u, v)$ is made, it follows that

$$dx = x_u \, du + x_v \, dv, \qquad dy = y_u \, du + y_v \, dv$$

and a formal calculation of $dx \wedge dy$ yields

$$(10.3) \qquad dx \wedge dy = (x_u y_v - x_v y_u) \, du \wedge dv = \frac{\partial(x, y)}{\partial(u, v)} \, du \wedge dv.$$

Thus if

$$\omega_i = a_i \, du + b_i \, dv$$

are transformed into

$$\omega_i' = \alpha_i \, dx + \beta_i \, dy$$

it follows that $a_i = \alpha_i x_u + \beta_i y_u$, $b_i = \alpha_i x_v + \beta_i y_v$ and hence that

$$\omega_1 \wedge \omega_2 = (a_1 b_2 - a_2 b_1) \, du \wedge dv = (\alpha_1 \beta_2 - \alpha_2 \beta_1) \frac{\partial(x, y)}{\partial(u, v)} \, du \wedge dv$$

$$= (\alpha_1 \beta_2 - \alpha_2 \beta_1) \, dx \wedge dy = \omega_1' \wedge \omega_2',$$

in view of (10.3).

One of the advantages of Grassmann's algebra is that it offers a convenient formalism for the study of determinants and questions of linear dependence. In the present particularly simple case of two variables it is seen at once from (10.2) that *the necessary and sufficient condition for the linear dependence of two nonvanishing* differential forms ω_1, ω_2 *is that their wedge product should vanish*, since this means that $a_1/b_1 = a_2/b_2$ and hence $\omega_1 = f\omega_2$ with f some function of u^1 and u^2. Since in what follows no applications are made in cases involving more than two independent variables, these matters of an algebraic character will not be pursued further.

It can now be seen how the wedge product leads automatically to the correct formula for a change of variables in the integral $\iint_D f(x, y) \, dx \, dy$ over a domain D. In fact, if the element of area $dx \, dy$ is interpreted to mean $dx \wedge dy$, the result of the change of variables $x = x(u, v)$, $y = y(u, v)$ is given by (10.3) in terms of du and dv and their wedge product, i.e., the formula

$$\iint_D f(x, y) \, dx \wedge dy = \iint_D f(x(u, v), y(u, v)) \frac{\partial(x, y)}{\partial(u, v)} \, du \wedge dv$$

results—and this formula is correct, as is proved in the calculus, provided that the transformation of variables is one-to-one and the Jacobian $\partial(x, y)/\partial(u, v)$ is positive. This simple observation perhaps served as the starting point for E. Cartan in the development of his notation. Some writers in fact prefer, for example, to characterize $Q \, du \, dv$ as a differential form of second degree by the requirement that $\iint Q \, du \, dv = \iint Q' \, du' \, dv'$ for all changes of variables that preserve the orientation.

The next step to be taken is the introduction of a process of differentiation of differential forms ω. This process, denoted by $d\omega$ and called *exterior differentiation*, is *defined by the formula*

$$(10.4) \qquad\qquad d\omega = d(a_i \, du^i) = da_i \wedge du^i.$$

In this formula the differentials da_i are given as usual by $da_i = (\partial a_i/\partial u^j) \, du^j$

and the products of the differentials du^i are assumed to obey Grassmann's rule. It follows that

$$(10.5) \qquad d\omega = da_i \wedge du^i = \left(\frac{\partial a_2}{\partial u^1} - \frac{\partial a_1}{\partial u^2} \right) du^1 \wedge du^2.$$

The motivation for this definition again comes from a familiar result of integral calculus, i.e., Green's theorem (or Stokes's theorem in the plane):

$$\oint_{\bar{D}} a_1 \, du^1 + a_2 \, du^2 = \iint_D \left(\frac{\partial a_2}{\partial u^1} - \frac{\partial a_1}{\partial u^2} \right) du^1 \, du^2,$$

in which the line integral is taken in the positive sense around the boundary $\bar{D}$ of the domain of integration D of the area integral. Evidently this theorem is given in the new notation in the following compact form

$$(10.6) \qquad \oint_{\bar{D}} \omega = \iint_D d\omega.$$

This implies, of course, that the calculation of the integrals is carried out in the usual way: the wedge product of the differentials has no special significance for the actual calculation of an integral.

It can now be seen that the exterior derivative $d\omega$ is invariant under parameter transformations. This could be shown by a direct calculation, but it is easier to prove it with the aid of (10.6). If new parameters are introduced so that $\omega = a_i \, du^i$ is transformed to $\omega' = a_i' \, du'^i$, then $d\omega' = (\partial a_2'/\partial u'^1 - \partial a_1'/\partial u'^2) \, du'^1 \wedge du'^2$ and it is known from (10.6) that

$$\iint d\omega' = \oint \omega'.$$

Since ω was so defined as to be invariant under parameter transformations it follows that $\oint \omega' = \oint \omega$ and hence that $\iint d\omega' = \iint d\omega$ for arbitrary domains of integration; thus it follows readily that $d\omega' = d\omega$.

The differential forms ω introduced here are what are called Pfaffian forms; they are not in general the differentials of functions $f(u^1, u^2)$, i.e., they are not in general exact differentials However, as is seen from (10.5), *the necessary and sufficient condition that ω should be an exact differential is that*

$$(10.7) \qquad\qquad d\omega = 0,$$

since this implies $\partial a_2/\partial u^1 = \partial a_1/\partial u^2$, and if this condition is satisfied it is known from calculus that $a_1(u^1, u^2)$ and $a_2(u^1, u^2)$ are such that $a_1 \, du^1 + a_2 \, du^2$ is an exact differential.

Finally, it is useful to have the formula for the exterior derivative of a product of a function f with a differential form $\omega = a\,du + b\,dv$. This is obtained by a direct calculation:

$$(10.8) \qquad d(f\omega) = [(fb)_u - (fa)_v]\,du \wedge dv$$

$$= (bf_u - af_v)\,du \wedge dv + (fb_u - fa_v)\,du \wedge dv$$

$$= df \wedge \omega + f\,d\omega.$$

The above definitions and the accompanying eight formulas suffice for most of the specific applications of this chapter. However, it is perhaps worthwhile to indicate briefly, but with omission of many details, how the theory can be generalized to higher dimensions, i.e., for an arbitrary number of variables $u^1, u^2, \ldots, u^n$, and to differential forms $\omega_1 \wedge \omega_2 \wedge \cdots \wedge \omega_n$ of arbitrary degree. A differential form of first degree is, of course, defined by

$$(10.9) \qquad \omega = a_i\,du^i = a_1\,du^1 + a_2\,du^2 + \cdots + a_n\,du^n,$$

with coefficients $a_i(u^1, u^2, \ldots, u^n)$ that form a covariant vector. The wedge product of two such differentials is then given by

$$(10.10) \qquad \omega_1 \wedge \omega_2 = (a_i\,du^i) \wedge (b_j\,du^j),$$

with the understanding that the rule for multiplication of any two terms is the same as was described above. The definition of forms of higher degree is then obtained recursively, starting with the forms of first degree and the basic rule given by (10.10), by the following stipulations:

1. $$\omega_{i_1} \wedge \omega_{i_2} \wedge \cdots \wedge \omega_{i_p} = \omega_{k_1} \wedge \omega_{k_2} \wedge \cdots \wedge \omega_{k_p}$$

or

$$\omega_{i_1} \wedge \omega_{i_2} \wedge \cdots \wedge \omega_{i_p} = -\omega_{k_1} \wedge \omega_{k_2} \wedge \cdots \wedge \omega_{k_p},$$

according to whether $k_1, k_2, \ldots, k_p$ is an even or an odd permutation of the subscripts $i_1, i_2, \ldots, i_p$.

2. $$(\omega_{i_1} \wedge \omega_{i_2} \wedge \cdots \wedge \omega_{i_p}) \wedge (\omega_{k_1} \wedge \omega_{k_2} \wedge \cdots \wedge \omega_{k_q})$$

$$= \omega_{i_1} \wedge \omega_{i_2} \wedge \cdots \wedge \omega_{i_p} \wedge \omega_{k_1} \wedge \omega_{k_2} \wedge \cdots \wedge \omega_{k_q}.$$

3 Addition of such forms and multiplication by a function are to obey the usual algebraic rules.

These forms are all invariant under coordinate transformations. In fact, it is clear that any such form, of degree p say, can be expressed with the aid of some particular coordinate system in the form

$$(10.11) \qquad \omega = \sum_{i_1 < i_2 \cdots < i_v} a_{i_1 i_2 \cdots i_p}\,du^{i_1} \wedge du^{i_2} \wedge \cdots \wedge du^{i_p}.$$

The coefficients are defined only for $i_1 < i_2 < \cdots < i_p$, but the sum can be extended over all indices if additional coefficients are defined antisymmetrically, i.e., so that $a_{j_1 j_2 \cdots j_p} = 0$ if the indices are not all distinct and otherwise equal to $\pm a_{i_1 i_2 \cdots i_p}$, depending on whether $j_1, j_2, \ldots, j_p$ is an even or an odd permutation of the subscripts $i_1, i_2, \ldots, i_p$. If this is done then a form ω of degree p is given by

$$(10.12) \qquad \omega = \frac{1}{p!}\, a_{i_1 i_2 \cdots i_p}\, du^{i_1} \wedge du^{i_2} \wedge \cdots \wedge du^{i_p},$$

the summation convention being used. With these conventions it turns out rather easily that if the coefficients $a_{i_1 i_2 \cdots i_p}$ are the components of an antisymmetric covariant tensor of rank p, then ω is an invariant form. Some writers, in fact, prefer to develop the entire theory of exterior differential forms by using (10.12) as the basic defining relation for a form of degree p by requiring the coefficients to be the components of an antisymmetric covariant tensor of rank p.

The next step is to define the exterior derivative $d\omega$ of a form of any degree. This is done, for example, by employing (10.12) to define ω. The definition for $d\omega$ is then

$$(10.5)_1 \qquad d\omega = \frac{1}{p!}\, da_{i_1 i_2 \cdots i_p} \wedge du^{i_1} \wedge du^{i_2} \wedge \cdots \wedge du^{i_v}.$$

Here the differentials $da_{i_1 i_2 \cdots i_p}$ are computed as usual with respect to the functions $a_{i_1 i_2 \cdots i_p}$ of the variables $u^1, u^2, \ldots, u^n$, and afterward the wedge products of the differentials are taken. It is then shown without much trouble that $d\omega$ is a differential form of degree $p + 1$, and thus is in particular invariant under changes of coordinates.

A generalization of the rule given by (10.8) is given by

$$(10.8)_1 \qquad d(\omega_1 \wedge \omega_2) = d\omega_1 \wedge \omega_2 + (-1)^p\, \omega_1 \wedge d\omega_2,$$

where p *is the degree of the form* ω_1. The rule is easily verified on the basis of (10.12) as definition of ω. [It is reasonable to define as a form of degree 0 any invariant function f. In that case the formula $(10.8)_1$ when applied to $(f\omega)$ leads at once to (10.8) for f in place of ω_1, ω in place of ω_2, and $p = 0$; however, note that $d(\omega f) = d\omega \wedge f - \omega \wedge df$ when $(10.8)_1$ is applied—but with the expected result that $d(f\omega) = d(\omega f)$.]

A more important general formula, which was not derived above even for the special case of two variables, embodies what is called *Poincaré's theorem*, i.e., that the repeated exterior derivative always vanishes:

$$(10.13) \qquad d(d\omega) = 0.$$

This follows almost at once from (10.12) by considering again a single term first, say the term

$$\frac{\partial a_{i_1 i_2 \cdots i_p}}{\partial u^{ij}} du^{i_j} \wedge du^{i_1} \wedge \cdots \wedge du^{i_p}.$$

In taking the second exterior derivative the mixed second derivatives appear in pairs and these terms are either zero or they cancel because of the antisymmetric property of differential forms.

It should perhaps also be mentioned that the remarks made for two dimensions in connection with (10.7) are generally valid. That is, if $d\omega = 0$ holds, then ω is (at least locally) the differential of a function $\alpha(u^1, u^2, \cdots, u^n)$, i.e., $\omega = d\alpha = (\partial \alpha / \partial u^i) du^i$ holds.

The theorem of Stokes and the divergence theorem in three-dimensional Euclidean space are both contained in a generalization of (10.6). Take the theorem of Stokes first, which is expressed in the formula

$$\oint_C P\,dx + Q\,dy + R\,dz = \iint_S (R_y - Q_z)dy\,dz$$
$$+ (P_z - R_x)dz\,dx + (Q_x - P_y)dx\,dy.$$

Here $\omega = P\,dx + Q\,dy + R\,dz$ is meant to be an invariant form (in hydrodynamics, for example, it represents the tangential velocity component of a fluid along the closed curve C), and the surface integral on the right-hand side is then taken over the exterior derivative $d\omega$ of this form, as is seen upon interpreting the products $dy\,dz$, etc., as wedge products. It follows that

$$\oint \omega = \iint_S d\omega.$$

This interpretation of Stokes's theorem brings out once more the fact that the curl of a vector is really an antisymmetric tensor, rather than a vector. In similar fashion the divergence theorem

$$\iint_S P\,dy \wedge dz + Q\,dz \wedge dx + R\,dx \wedge dy$$
$$\iiint_D (P_x + Q_y + R_z)\,dx \wedge dy \wedge dz$$

can be written in the form

$$\iint_S \omega = \iiint_D d\omega$$

with ω a form of second degree, as can be readily verified. Quite generally, such integral formulas hold in all dimensions (of course, with appropriate stipulations about the smoothness of boundary surfaces).

The divergence theorem also gives an example of a general fact, i.e., that *for forms in n variables the form of degree n has essentially only one term.* It was not mentioned earlier that *no forms of degree higher than n that are differ-ent from zero can occur*—the proof is obvious.

2. Vector Differential Forms and Surface Theory

An algebra and calculus of linear *vector differential forms* (and linear tensor forms as well), which are exactly analogous to the calculus of the scalar forms treated above, are possible and useful.   By a linear vector form $\mathbf{\Omega}$ of first degree is meant a form

$$(10.14) \qquad \mathbf{\Omega} = \mathbf{X}_1\, du^1 + \mathbf{X}_2\, du^2$$

in which the coefficients $\mathbf{X}_i(u^1, u^2)$ are differentiable *vector* functions of the coordinates (u^1, u^2).   Here the vectors $\mathbf{X}_1$ and $\mathbf{X}_2$ are assumed to lie in a Euclidean space of three-dimensions, but to be functions of Gaussian para-meters u^1, u^2.   The connection of (10.14) with the preceding section is estab-lished by noting that each of the three scalar components of $\mathbf{\Omega}$ is a linear differential form of the kind treated there.   In fact, the form $\mathbf{\Omega}$ can be ex-pressed in another way by employing unit orthogonal basis vectors $\mathbf{a}_i(x^1, x^2, x^3) \equiv \mathbf{a}_i(u^1, u^2)$ in the three-dimensional space, which, however, need not be the same at all points where the vectors $\mathbf{X}_i(u^1, u^2)$ are defined.   It is also to be noted that the vectors $\mathbf{a}_i$ need have no particular relation to the vectors $\mathbf{X}_1$ and $\mathbf{X}_2$—even in the special case of most interest here, i.e., that in which $\mathbf{X}_1$ and $\mathbf{X}_2$ are the derivatives of a vector function $\mathbf{X}(u^1, u^2)$ defining a surface in three-space.   (It is one of the essential merits of Cartan's formalism that such a freedom of choice of basis vectors is made available.)

It follows that (10.14) may also be expressed in the form

$$(10.15) \qquad \mathbf{\Omega} = \sum_{j=1}^{3} \mathbf{a}_j \sigma_j,$$

in terms of three scalar differential forms of first degree that are immediately seen to be given by

$$(10.16) \qquad \sigma_j = \mathbf{a}_j \cdot \mathbf{\Omega} = \mathbf{a}_j \cdot \mathbf{X}_1\, du^1 + \mathbf{a}_j \cdot \mathbf{X}_2\, du^2, \qquad j = 1, 2, 3,$$

since the basis vectors $\mathbf{a}_j$ are assumed to be mutually orthogonal unit vectors. The notation $\mathbf{a}_j \cdot \mathbf{\Omega}$ of course implies that the ordinary scalar product is to be taken.   The exterior derivative $d\mathbf{\Omega} = d\mathbf{X}_1 \wedge du^1 + d\mathbf{X}_2 \wedge du^2$ can then be put into the following form:

$$(10.17) \qquad d\mathbf{\Omega} = \sum_{j=1}^{3} (d\mathbf{a}_j \wedge \sigma_j + \mathbf{a}_j\, d\sigma_j),$$

with

(10.18) $d\mathbf{a}_j = \dfrac{\partial \mathbf{a}_j}{\partial u^1} du^1 + \dfrac{\partial \mathbf{a}_j}{\partial u^2} du^2 = \displaystyle\sum_{k=1}^{3} \mathbf{a}_k \omega_{jk}, \qquad j = 1, 2, 3,$

as an easy calculation [using (10.16)] shows. The linear differential forms denoted by ω_{jk} can clearly be expressed as scalar products by the formulas

(10.19) $\omega_{jk} = \mathbf{a}_k \cdot d\mathbf{a}_j, \qquad j, k = 1, 2, 3,$

since $\mathbf{a}_i \cdot \mathbf{a}_j = \delta_{ij}$. Thus (10.17) is a vector differential form of second degree in the differentials du^1, du^2. Again because the local coordinate vectors $\mathbf{a}_j$ are assumed to form orthonormal systems so that $\mathbf{a}_j\,\mathbf{a}_k = \delta_{jk}$, it follows that $d\mathbf{a}_j \cdot \mathbf{a}_k + \mathbf{a}_j \cdot d\mathbf{a}_k = 0$ and hence that

(10.20) $\omega_{jk} + \omega_{kj} = 0.$

Thus the linear differential forms ω_{jk} are antisymmetric. (They were employed in similar circumstances in Chapter IX when Euclidean geometry in curvilinear coordinates was discussed as the intuitive basis for the definitions of tensor calculus in Riemannian geometry.)

In the theory of surfaces, to which the remainder of this chapter is devoted, the vector differential form $\boldsymbol{\Omega}$ will in general be the differential $d\mathbf{X}$ of a regular surface defined in E^3 by means of a vector $\mathbf{X}(u^1, u^2)$. In that case $\boldsymbol{\Omega}$ is the exact differential $\boldsymbol{\Omega} = (\partial \mathbf{X}/\partial u^1)\,du^1 + (\partial \mathbf{X}/\partial u^2)\,du^2$ and hence

(10.21) $d\boldsymbol{\Omega} = 0,$

since each scalar component of $\boldsymbol{\Omega}$ is an exact differential, so that (10.7) applies. The form $\boldsymbol{\Omega}$ will be referred to on occasion as a *tangential form*.

It is convenient to choose the local basis vectors $\mathbf{a}_i$ (they are often referred to as Cartan's *repère mobile*) so that $\mathbf{a}_3$ is the normal vector to the surface $\mathbf{X}(u^1, u^2)$, but, as was remarked above, $\mathbf{a}_1$ and $\mathbf{a}_2$ do not in general fall along the vectors $\mathbf{X}_1 = \partial \mathbf{X}/\partial u^1$, $\mathbf{X}_2 = \partial \mathbf{X}/\partial u^2$. (In fact, they cannot do so unless the curvilinear coordinate system on the surface is an orthogonal system.) It follows from (10.16) that σ_3 vanishes when $\boldsymbol{\Omega}$ is a tangential form:

(10.22) $\sigma_3 = 0.$

[Actually, the assumption $\sigma_3 = 0$ is not necessary for the validity of what follows up to equation (10.25)].

With $\boldsymbol{\Omega} = \mathbf{X}_1\,du^1 + \mathbf{X}_2\,du^2$, but expressed in the form (10.15), it follows that

$$d\boldsymbol{\Omega} = \sum_{j=1}^{3} d\mathbf{a}_j \wedge \sigma_j + \sum_{j=1}^{3} \mathbf{a}_j\,d\sigma_j = 0,$$

because of (10.17) and (10.21).   Insertion of $d\mathbf{a}_j$ as given by (10.18) yields

$$\sum_{j,\,k=1}^{3} \mathbf{a}_k \omega_{jk} \wedge \sigma_j + \sum_{k=1}^{3} \mathbf{a}_k \, d\sigma_k = 0.$$

Because of the linear independence of the local basis vectors $\mathbf{a}_j$ it follows that

$$(10.23) \quad d\sigma_k = -\sum_{j=1}^{3} \omega_{jk} \wedge \sigma_j = \sum_{j=1}^{3} \omega_{kj} \wedge \sigma_j, \quad k = 1, 2, 3,$$

the last step following from (10.20).   Evidently, these equations must be closely related to the group of basic partial differential equations due to Gauss and Weingarten (cf. Chapter VI) since they express the differentials of local basis vectors of the surface in terms of these same vectors.

Since the vectors $\mathbf{a}_j$ are given functions of u^1, u^2 it follows that the differentials $d\mathbf{a}_j$ are exact, and hence these forms of first degree have a vanishing exterior derivative; thus the following formulas hold:

$$0 = d(d\mathbf{a}_j) = \sum_{k=1}^{3} d\mathbf{a}_k \wedge \omega_{jk} + \sum_{k=1}^{3} \mathbf{a}_k \, d\omega_{jk}, \quad j = 1, 2, 3.$$

Upon replacing the differentials $d\mathbf{a}_k$ by using (10.18) the result is

$$0 = \sum_{l,\,s=1}^{3} \mathbf{a}_s \omega_{ls} \wedge \omega_{jl} + \sum_{k=1}^{3} \mathbf{a}_k \, d\omega_{jk}$$

$$= \sum_{k,\,s=1}^{3} \mathbf{a}_k \omega_{sk} \wedge \omega_{js} + \sum_{k=1}^{3} \mathbf{a}_k \, d\omega_{jk}, \quad j = 1, 2, 3.$$

Again the fact that the vectors $\mathbf{a}_k$ are linearly independent leads to conditions on the forms ω_{jk}:

$$(10.24) \quad d\omega_{jk} = -\sum_{s=1}^{3} \omega_{sk} \wedge \omega_{js} = \sum_{s=1}^{3} \omega_{js} \wedge \omega_{sk},$$

upon using (10.20).   These equations are equivalent, as will be seen, to the compatibility conditions of surface theory, i.e., to Gauss's *theorema egregium* and the Codazzi-Mainardi equations, as might be expected from the way they were derived.

The assumption that $\sigma_3 = 0$ is now explicitly used to obtain the full formal apparatus of surface theory.   In this case $\mathbf{\Omega}$ has the form

$$(10.25) \quad \mathbf{\Omega} = d\mathbf{X} = \mathbf{a}_1 \sigma_1 + \mathbf{a}_2 \sigma_2, \quad \sigma_i = \mathbf{a}_i \cdot d\mathbf{X},$$

amd equations (10.18) become, since ω_{jk} is skew-symmetric:

$$(10.26) \quad d\mathbf{a}_1 = \mathbf{a}_2 \omega_{12} + \mathbf{a}_3 \omega_{13}, \quad d\mathbf{a}_2 = \mathbf{a}_3 \omega_{23} + \mathbf{a}_1 \omega_{21},$$

$$d\mathbf{a}_3 = \mathbf{a}_1 \omega_{31} + \mathbf{a}_2 \omega_{32}.$$

The six conditions (10.23) and (10.24), when written out in full are

$$(10.27) \quad \begin{cases} d\sigma_1 = \omega_{12} \wedge \sigma_2, \qquad d\sigma_2 = \omega_{21} \wedge \sigma_1, \\ \qquad\qquad\qquad\qquad d\sigma_3 = 0 = \omega_{31} \wedge \sigma_1 + \omega_{32} \wedge \sigma_2, \\ d\omega_{12} = \omega_{13} \wedge \omega_{32}, \qquad d\omega_{23} = \omega_{21} \wedge \omega_{13}, \qquad d\omega_{31} = \omega_{32} \wedge \omega_{21}. \end{cases}$$

The partial differential equations of surface theory are contained in the equations (10.26) and (10.27). Evidently, the present notation furnishes them in a form remarkably more compact than the form employed in Chapter VI. However, some steps must be taken to put the new notation into relation with the details of the geometry of surfaces.

As a first step in this direction consider the form ϕ of second degree defined by $\sigma_1 \wedge \sigma_2$. It is to be shown that the area of a portion of the surface defined over a domain D of the parameter plane is given by

$$(10.28) \qquad A = \int_D \phi = \int_D \sigma_1 \wedge \sigma_2.$$

This definition makes sense in the large (when overlapping coordinate patches are made necessary) *since the differential forms are invariant with respect to parameter transformations*—a remark which will not be repeated in what follows in dealing with other formulas in the large. The proof of (10.28) is easily carried out by a direct calculation, as follows:

$$\phi = \sigma_1 \wedge \sigma_2 = [(\mathbf{X}_u \, du + \mathbf{X}_v \, dv) \cdot \mathbf{a}_1] \wedge [(\mathbf{X}_u \, du + \mathbf{X}_v \, dv) \cdot \mathbf{a}_2)]$$

$$= [(\mathbf{X}_u \cdot \mathbf{a}_1)(\mathbf{X}_v \cdot \mathbf{a}_2) - (\mathbf{X}_u \cdot \mathbf{a}_2)(\mathbf{X}_v \cdot \mathbf{a}_1)] \, du \wedge dv.$$

The last expression, finally, can be written

$$(10.28)_1 \quad \phi = \sigma_1 \wedge \sigma_2 = \mathbf{X}_u \times \mathbf{X}_v \cdot \mathbf{a}_3 \, du \wedge dv = \sqrt{EG - F^2} \, du \wedge dv.$$

This is verified by noting that $\mathbf{X}_u \times \mathbf{X}_v \cdot \mathbf{a}_3 = \mathbf{X}_u \times \mathbf{X}_v \cdot (\mathbf{a}_1 \times \mathbf{a}_2)$ and expanding the latter expression using Lagrange's identity; since $\mathbf{a}_3$ is the unit normal vector $\mathbf{X}_3$ of the surface the area element is $\mathbf{X}_u \times \mathbf{X}_v \cdot \mathbf{a}_3 \, du \, dv$ and this in turn is $\sqrt{EG - F^2} \, du \, dv$ [cf. Chapter IV, equation (4.8)].

Consider next the spherical image of the surface, which is of course determined by the vector $\mathbf{a}_3(u^1, u^2)$, and its area K_T, which measures the Gaussian curvature of the surface. Upon comparing $d\mathbf{X} = \mathbf{a}_1\sigma_1 + \mathbf{a}_2\sigma_2$ with $d\mathbf{a}_3 = \mathbf{a}_1\omega_{31} + \mathbf{a}_2\omega_{32}$ [see (10.26)] it is clear that the area traced out by $\mathbf{a}_3$ is given by the same formula as (10.28) when σ_1, σ_2 are replaced by ω_{31}, ω_{32}; thus for K_T the following formula holds:

$$(10.29) \qquad K_T = \int_D \omega_{31} \wedge \omega_{32}.$$

It is naturally of interest to express the fundamental forms I, II, III of surface theory, as derived in Chapter IV, in the new notation. For the first fundamental form, or line element, the result is

$$(10.30) \qquad I = d\mathbf{X} \cdot d\mathbf{X} = \sum_j \mathbf{a}_j \sigma_j \cdot \sum_k \mathbf{a}_k \sigma_k = \sigma_1^2 + \sigma_2^2,$$

since $\mathbf{a}_i \cdot \mathbf{a}_k = \delta_{ik}$ and $\sigma_3 = 0$. The second fundamental form is given by $II = -d\mathbf{X} \cdot d\mathbf{X}_3 = -d\mathbf{X} \cdot d\mathbf{a}_3$ (cf. Section 17 of Chapter IV), and can be expressed as follows:

$$(10.31) \quad II = -d\mathbf{X} \cdot d\mathbf{a}_3 = -\sum_j \mathbf{a}_j \sigma_j \cdot (\mathbf{a}_1 \omega_{31} + \mathbf{a}_2 \omega_{32}) = \sigma_1 \omega_{31} + \sigma_2 \omega_{32},$$

when $d\mathbf{a}_3$ is taken from (10.26).

Finally, the third fundamental form (again see Section 17 of Chapter IV) is given by

$$(10.32) \qquad III = d\mathbf{a}_3 \cdot d\mathbf{a}_3 = (\mathbf{a}_1 \omega_{31} + \mathbf{a}_2 \omega_{32}) \cdot (\mathbf{a}_1 \omega_{31} + \mathbf{a}_2 \omega_{32})$$

$$= \omega_{31}^2 + \omega_{32}^2.$$

It should be noted that the products of differentials in these three formulas are *ordinary* products and not wedge products. In other words, the three basic differential forms of surface theory are *not* differential forms in the sense of Cartan—in fact, they are all symmetric forms rather than skew-symmetric forms.

One of the ways in which the new formalism shows its power is in the proof of Gauss's *theorema egregium*. The Gaussian curvature K is defined as the ratio of the area elements of the spherical image and the surface (cf. Section 15 of Chapter IV). In view of (10.29) and (10.30) it follows that K satisfies the relation

$$(10.33) \qquad K\sigma_1 \wedge \sigma_2 = \omega_{31} \wedge \omega_{32},$$

and since all of the differentials and their wedge products are invariants under parameter transformations this defines K in an invariant way. In the literature it is a common practice to define K by the formula

$$(10.33)_1 \qquad K = \frac{\omega_{31} \wedge \omega_{32}}{\sigma_1 \wedge \sigma_2}.$$

At first sight it might seem odd to define a function $K(u^1, u^2)$ as a ratio of differentials. However, this is a very special case since the differentials are of second degree and hence have only one term since there are only two independent variables (cf. remarks made at the end of Section 1).

Gauss's theorem states that $K(u, v)$ is the same for two surfaces $\mathbf{X}(u, v)$ and $\mathbf{X}^*(u, v)$ if both are referred to the same parameters and all corresponding curves have the same lengths. In that case $d\mathbf{X} \cdot d\mathbf{X} = d\mathbf{X}^* \cdot d\mathbf{X}^* = ds^2$ for all du, dv and it follows that $\sigma_1^* = \sigma_1$, $\sigma_2^* = \sigma_2$. It is convenient to prove this by choosing the base vectors $\mathbf{a}_1$, $\mathbf{a}_2$ in the tangential planes on both surfaces so that they will make the same angles with the coordinate vectors $\mathbf{X}_1$, $\mathbf{X}_2$ and $\mathbf{X}_1^*$, $\mathbf{X}_2^*$ of both surfaces. This can be done since angles are measured solely in terms of the coefficients E, F, G of the first fundamental form. From (10.25) it follows that

$$\sigma_1 = \mathbf{a}_i \cdot (\mathbf{t}_1 \sqrt{E}\, du^1 + \mathbf{t}_2 \sqrt{G}\, du^2)$$

in terms of unit vectors $\mathbf{t}_i$ along the parameter curves. Since E, F, G are the same for both surfaces it follows that $\sigma_i = \sigma_i^*$, as was to be shown.

The theorem of Gauss says in effect that K as given by (10.33) or (10.33)$_1$ really depends only upon the forms σ_1 and σ_2, and thus is independent of ω_{31} and ω_{32}. Since $\sigma_1 \wedge \sigma_2 \neq 0$ [cf. (10.28)$_1$], it follows that σ_1 and σ_2 are linearly independent. It follows in turn that any scalar linear differential form in du^1 and du^2 can be expressed as a linear combination of σ_1 and σ_2; in particular, functions g_1 and g_2 exist such that

$$\omega_{12} = g_1\sigma_1 + g_2\sigma_2.$$

This follows from (10.27) in the following way:

$$d\sigma_1 = \omega_{12} \wedge \sigma_2 = (g_1\sigma_1 + g_2\sigma_2) \wedge \sigma_2 = g_1\sigma_1 \wedge \sigma_2,$$

$$d\sigma_2 = \omega_{21} \wedge \sigma_1 = g_2\sigma_1 \wedge \sigma_2,$$

thus in effect determining g_1 and g_2 since $d\sigma_1$ and $d\sigma_2$ are known forms of second degree. Thus ω_{12} is given by

$$\omega_{12} = \frac{d\sigma_1}{\sigma_1 \wedge \sigma_2}\sigma_1 + \frac{d\sigma_2}{\sigma_1 \wedge \sigma_2}\sigma_2.$$

In addition, again from (10.27), $d\omega_{12} = \omega_{13} \wedge \omega_{32}$ and hence (10.33)$_1$ takes the form

$$(10.34) \quad K = \frac{-d\omega_{12}}{\sigma_1 \wedge \sigma_2} = \frac{-1}{\sigma_1 \wedge \sigma_2}\left[d\left(\frac{d\sigma_1}{\sigma_1 \wedge \sigma_2}\sigma_1 + \frac{d\sigma_2}{\sigma_1 \wedge \sigma_2}\sigma_2 \right) \right],$$

in which the operator d of course refers to the exterior derivatives. Thus it is seen that the Gaussian curvature does indeed depend only upon σ_1 and σ_2 and therefore only upon the first fundamental form.

In order to compare (10.34) with the classical formula consider $\mathbf{X}(u, v)$ as given in terms of orthogonal parameter curves so that

$$d\mathbf{X} = \mathbf{X}_u\, du + \mathbf{X}_v\, dv = \mathbf{a}_1 \sqrt{E}\, du + \mathbf{a}_2 \sqrt{G}\, dv,$$

by choosing $\mathbf{a}_1$ and $\mathbf{a}_2$ as tangents to the parameter curves. Thus $\sigma_1 = \sqrt{E}$

du and $\sigma_2 = \sqrt{G}\,dv$ [cf. (10.25)] and $\sigma_1 \wedge \sigma_2 = \sqrt{EG}\,du \wedge dv$. The exterior derivatives of the linear forms σ_1 and σ_2 are

$$d\sigma_1 = [(\sqrt{E})_u\,du + (\sqrt{E})_v\,dv] \wedge du = -\frac{1}{2}\frac{E_v}{\sqrt{E}}\,du \wedge dv,$$

$$d\sigma_2 = \frac{1}{2}\frac{G_u}{\sqrt{G}}\,du \wedge dv.$$

Substitution in (10.34) finally gives

(10.35)
$$K = -\frac{1}{2}\frac{d[-(E_v/\sqrt{EG})\,du + (G_u/\sqrt{EG})\,dv]}{\sqrt{EG}\,du \wedge dv}$$

$$= -\frac{1}{2}\frac{1}{\sqrt{EG}}\left[\left(\frac{E_v}{\sqrt{EG}}\right)_v + \left(\frac{G_u}{\sqrt{EG}}\right)_u\right],$$

and this is the well-known formula of Chapter VI [cf. (6.17)] that expresses K in terms of the coefficients of the first fundamental form in case the coordinate curves are orthogonal. However (10.35), unlike (10.34), does not express K in an invariant form.

The forms ω_{13} and ω_{23} can also be expressed as linear combinations of σ_1 and σ_2:

(10.36)
$$\omega_{13} = a\sigma_1 + b\sigma_2,$$
$$\omega_{23} = b\sigma_1 + c\sigma_2.$$

That two of the coefficients must be alike follows from the relation $d\sigma_3 = 0 = \omega_{31} \wedge \sigma_1 + \omega_{32} \wedge \sigma_2$ given in (10.27). The second fundamental forms as given by (10.31) can now be expressed as follows:

(10.37) $\mathrm{II} = (\sigma_1\omega_{31} + \sigma_2\omega_{32}) = a\sigma_1^2 + 2b\sigma_1\sigma_2 + c\sigma_2^2.$

Since $K = (\omega_{31} \wedge \omega_{32})/(\sigma_1 \wedge \sigma_2)$ it is easily seen that

(10.38) $K = ac - b^2.$

In terms of an orthogonal coordinate system and basis vectors tangent to them $\sigma_1 = \sqrt{E}\,du$, $\sigma_2 = \sqrt{G}\,dv$, and it follows that

(10.39) $\mathrm{II} = aE\,du^2 + 2b\sqrt{EG}\,du\,dv + cG\,dv^2,$

and hence that the coefficients L, M, N of II are given by

(10.40) $L = aE, \qquad M = b\sqrt{EG}, \qquad N = cG.$

Once these notations have been introduced it is easy to check that the equations in (10.27) for $d\omega_{23}$ and $d\omega_{31}$ are the Codazzi-Mainardi equations.

3. Scalar and Vector Products of Vector Forms on Surfaces and Their Exterior Derivatives

In the preceding section the elements of surface theory in Euclidean three-space were rederived on the basis of the algebra and calculus of tangential vector differential forms:

$$(10.41) \qquad \Omega = \sum_{i=1}^{2} \mathbf{a}_i \, du^i = \mathbf{a}_i \, du^i.$$

In this section the vector coefficients $\mathbf{a}_i = \mathbf{a}_i(u^1, u^2)$ are again assumed to lie in three-space. The indicated summation is taken over 1 and 2. The vector $\mathbf{a}_3(u^1, u^2)$ is the unit normal vector of the surface.

In the preceding section the scalar product $\mathbf{b} \cdot \Omega = \mathbf{b} \cdot \mathbf{a}_i \, du^i$ was often used to produce a scalar linear differential form of first degree. At that time no need arose for a specific formula expressing the exterior derivative of such a product. The formula, which will be used extensively in the following, is

$$(10.42) \qquad d\alpha = d(\mathbf{b} \cdot \Omega) = d\mathbf{b} \cdot \Omega + \mathbf{b} \cdot d\Omega.$$

The sign d and the dot in the term $d\mathbf{b} \cdot \Omega$ mean that the scalar product of the vectors is to be taken, but the wedge product of the differentials. The formula (10.42) is justified by the following calculation:

$$\begin{aligned}
d\alpha &= d(\mathbf{b} \cdot \mathbf{a}_i \, du^i) = d(\mathbf{b} \cdot \mathbf{a}_1) \wedge du^1 + d(\mathbf{b} \cdot \mathbf{a}_2) \wedge du^2 \\
&= (d\mathbf{b} \cdot \mathbf{a}_1 + \mathbf{b} \cdot d\mathbf{a}_1) \wedge du^1 + (d\mathbf{b} \cdot \mathbf{a}_2 + \mathbf{b} \cdot d\mathbf{a}_2) \wedge du^2 \\
&= d\mathbf{b} \cdot (\mathbf{a}_1 \, du^1 + \mathbf{a}_2 \, du^2) + \mathbf{b} \cdot (d\mathbf{a}_1 \wedge du^1 + d\mathbf{a}_2 \wedge du^2) \\
&= d\mathbf{b} \cdot \Omega + \mathbf{b} \cdot d\Omega.
\end{aligned}$$

(Here, and in what follows in this section, various algebraic operations involving addition and also scalar and vector multiplications of differential forms by vectors will be performed; these operations are permissible and consistent, but they will not be explicitly formulated and discussed.)

It turns out that a variety of problems in the large concerning the embedding of closed surfaces in Euclidean space can be treated very conveniently by introducing still other kinds of products. The first such product is a multiplication of a linear vector form Ω by a *vector* $\mathbf{b}$ in such a way as to produce a *vector form*; this is done simply by using the familiar product discussed in Chapter I:

$$(10.43) \qquad \mathbf{b} \times \Omega = \mathbf{b} \times \mathbf{a}_1 \, du^1 + \mathbf{b} \times \mathbf{a}_2 \, du^2.$$

Evidently, this multiplication does produce a new linear *vector* form of first degree. The second product to be defined is the *vector product* $\Omega_1 \times \Omega_2$ of *two linear vector forms* $\boldsymbol{\Omega}_1 = \mathbf{a}_i \, du^i$, $\boldsymbol{\Omega}_2 = \mathbf{b}_i \, du^i$. The definition chosen is, quite naturally, the following:

$$(10.44) \qquad \boldsymbol{\Omega}_1 \times \boldsymbol{\Omega}_2 = (\mathbf{a}_1 \, du^1 + \mathbf{a}_2 \, du^2) \times (\mathbf{b}_1 \, du^1 + \mathbf{b}_2 \, du^2)$$

$$= (\mathbf{a}_1 \times \mathbf{b}_2 - \mathbf{a}_2 \times \mathbf{b}_1) du^1 \wedge du^2.$$

This means that the ordinary vector product of the vectors, but the wedge product of the differentials, is taken. This product, unlike the wedge product of scalar differential forms, is commutative, since a double reversal of sign occurs when the factors are interchanged, due to the noncommutative character of both of the two types of products involved. Thus the following symmetry relation holds:

$$(10.45) \qquad \boldsymbol{\Omega}_1 \times \boldsymbol{\Omega}_2 = \boldsymbol{\Omega}_2 \times \boldsymbol{\Omega}_1.$$

The vector product of a linear vector form with itself, it should be noted, does *not* vanish in general.

In the preceding section the exterior derivative $d\boldsymbol{\Omega}$ was defined as follows:

$$(10.46) \qquad d\boldsymbol{\Omega} = d\mathbf{a}_i \wedge du^1$$

$$= (-\mathbf{a}_{12} + \mathbf{a}_{21}) \, du^1 \wedge du^2.$$

The symbols $\mathbf{a}_{ij}$ of course refer to the partial derivatives $\mathbf{a}_{ij} = \partial \mathbf{a}_i / \partial u_j$. From this it follows, as was already noted in the preceding section, that $d\boldsymbol{\Omega} = 0$ if, and only if, $\boldsymbol{\Omega}$ is the differential of some vector function $\mathbf{X}(u^1, u^2)$.

The next formula is simply the result of applying the operation d to the vector product defined by (10.43); the result is

$$(10.47) \qquad d(\mathbf{b} \times \boldsymbol{\Omega}) = d\mathbf{b} \times \boldsymbol{\Omega} + \mathbf{b} \times d\boldsymbol{\Omega}.$$

In fact, the application of d to (10.43) yields, in view of (10.46):

$$(d\mathbf{b} \times \mathbf{a}_1 + \mathbf{b} \times d\mathbf{a}_1) \wedge du^1 + (d\mathbf{b} \times \mathbf{a}_2 + \mathbf{b} \times d\mathbf{a}_2) \wedge du^2$$

$$= d\mathbf{b} \times (\mathbf{a}_1 \wedge du^1 + \mathbf{a}_2 \wedge du^2) + \mathbf{b} \times (d\mathbf{a}_1 \wedge du^1 + d\mathbf{a}_2 \wedge du^2),$$

and that is the right-hand side of (10.47).

Finally, it is of interest to introduce a *triple* product analogous to the scalar triple product of ordinary vector algebra. This is a scalar differential form β of first degree that results from the scalar product $\mathbf{b} \cdot \boldsymbol{\Omega} = \beta$ when $\mathbf{b}$ is a vector given by the vector product of two other vectors $\mathbf{A}$ and $\mathbf{B}$ and $\boldsymbol{\Omega}$ is a linear vector form:

$$(10.48) \qquad \beta = (\mathbf{A} \times \mathbf{B}) \cdot \boldsymbol{\Omega}$$

$$= (\mathbf{A} \times \mathbf{B}) \cdot \mathbf{a}_1 \, du^1 + (\mathbf{A} \times \mathbf{B}) \cdot \mathbf{a}_2 \, du^2.$$

The exterior derivative $d\beta$ of forms of this type will occur frequently in this and also in the following section. The formula is

(10.49) $$d\beta = d\mathbf{A} \times \mathbf{B}\cdot\mathbf{\Omega} + \mathbf{A} \times d\mathbf{B}\cdot\mathbf{\Omega} + \mathbf{A} \times \mathbf{B}\cdot d\mathbf{\Omega},$$

in which once more the vector and scalar products of vectors, but the wedge products of differentials, are to be taken; in addition, the symbol $d\mathbf{\Omega}$ of course refers to the exterior derivative of $\mathbf{\Omega}$. The justification of (10.49) is obtained from (10.48) by a straightforward calculation. To do so, it is to be noted first that (10.48) can be written in the form $\beta = \mathbf{A}\cdot(\mathbf{B} \times \mathbf{\Omega})$ since dot and cross may be interchanged without affecting the value of the triple product of vectors. It follows from (10.42) that

$$d\beta = d\mathbf{A}\cdot(\mathbf{B} \times \mathbf{\Omega}) + \mathbf{A}\cdot d(\mathbf{B} \times \mathbf{\Omega})$$
$$= d\mathbf{A}\cdot(\mathbf{B} \times \mathbf{\Omega}) + \mathbf{A}\cdot(d\mathbf{B} \times \mathbf{\Omega} + \mathbf{B} \times d\mathbf{\Omega}),$$

the second line resulting from use of (10.47). Again dot and cross may be interchanged since no change in the triple product of vectors occurs, and also no change in the order of occurrence of the differentials. Since in addition the vector product has meaning only for two vectors it follows that the parentheses may be omitted without causing ambiguities; thus the second line is the same as the right-hand side of (10.49).

4. Some Formulas for Closed Surfaces. Characterizations of the Sphere

With the aid of the notation introduced in the preceding section a variety of formulas for regular surfaces in three-dimensional space, which are useful for the solution of problems in the large, can be obtained. In doing so, the basic differential forms that come into play are provided by the differentials of the vector $\mathbf{X}(u, v)$, which defines the surface, and of its unit normal vector $\mathbf{Y} = (\mathbf{X}_1 \times \mathbf{X}_2)/|\mathbf{X}_1 \times \mathbf{X}_2|$. For these vector differential forms of first degree the following notation is introduced:

(10.50) $$d\mathbf{X} = \mathbf{X}_u\, du + \mathbf{X}_v\, dv, \qquad d\mathbf{Y} = \mathbf{Y}_u\, du + \mathbf{Y}_v\, dv.$$

It is important to observe that $d(d\mathbf{X}) = d(d\mathbf{Y}) = 0$ since both $d\mathbf{X}$ and $d\mathbf{Y}$ are exact differentials.

First of all, the following three formulas (each one of which involves a vector form of second degree) are to be derived:

(10.51) $$d\mathbf{X} \times d\mathbf{X} = 2\mathbf{Y}\, dA,$$

(10.52) $$d\mathbf{X} \times d\mathbf{Y} = -2H\mathbf{Y}\, dA,$$

(10.53) $$d\mathbf{Y} \times d\mathbf{Y} = 2K\mathbf{Y}\, dA.$$

In these formulas dA is the element of area of the surface, H is its mean curvature, and K is its Gaussian curvature. They are verified by calculations that employ the formulas derived in the preceding section:

$$dX \times dX = (X_u\, du + X_v\, dv) \times (X_u\, du + X_v\, dv) = 2X_u \times X_v\, du \wedge dv$$

$$= 2\sqrt{EG - F^2}\, Y\, du \wedge dv = 2Y\, dA,$$

since $|X_u \times X_v| = \sqrt{EG - F^2}$. To verify (10.52) it has some advantage to suppose that the parameter curves at a given point are in orthogonal directions of principal curvature, and this procedure causes no loss of generality provided that the end result, as is the case here, is a relation involving only quantities that are invariant under parameter transformations. In that case the formula of Rodrigues (cf. Section 11 of Chapter IV) gives

$$dX \times dY = (X_u\, du + X_v\, dv) \times (Y_u\, du + Y_v\, dv)$$

$$= (X_u\, du + X_v\, dv) \times (-k_1 X_u\, du - k_2 X_v\, dv)$$

$$= -(k_1 + k_2)X_u \times X_v\, du \wedge dv$$

$$= -2HY\, dA.$$

As for (10.53), no calculation by way of justification is needed, since it is seen to be correct because of (10.51): the unit normal to the unit sphere defined by the vector Y is Y itself and $K\, dA$ is the element of area of that surface (cf. the discussion leading to [10.29]).

The first application to be made of the above formulas is to the calculation of areas of closed orientable surfaces and to the volumes enclosed by them. The starting point is a regular surface given by $X(u^1, u^2)$, with (u^1, u^2) ranging over a domain D, and the set of its *parallel surfaces* given by $X_t = X(u^1, u^2) + tY(u^1, u^2)$. It is easy to show that the surfaces X_t are regular for $|t|$ sufficiently small and that all of them have the same unit normal $Y(u^1, u^2)$, in terms of the parameters employed for $X(u^1, u^2)$. It follows therefore from the basic formulas that

$$2Y\, dA_t = (dX + t\, dY) \times (dX + t\, dY)$$

$$= dX \times dX + 2t\, dX \times dY + t^2\, dY \times dY$$

$$= 2Y(dA - 2tH\, dA + t^2K\, dA).$$

In the calculation the commutative property of the vector product of linear forms was used. By taking the scalar product of both sides of this equation with Y and then integrating over the domain D the area $A(t)$ of the parallel surface is obtained in the following form:

$$(10.54) \qquad A(t) = A(0) - 2t \iint_D H\, dA + t^2 \iint_D K\, dA.$$

In case the surface is a closed surface, the domain D is the entire parameter surface. It is now possible to compute the volume ΔV covered by the points of all of the surfaces obtained in letting t vary (for t sufficiently small), since $dV = t\,dA$ and hence

$$\Delta V = tA(0) - t^2 \iint_D H\,dA + \frac{t^3}{3}\iint_D K\,dA.$$

If the surface is a *simple* closed surface S of genus g it is the boundary of its interior domain, which has a volume denoted by V_0. It follows from the last formula that the volume $V(t)$ enclosed by a surface parallel to S (and close to it) is given by

$$(10.55) \qquad V(t) = V_0 - t\,A(0) + t^2 \iint_S H\,dA - \frac{4\pi}{3}(1 - g)t^3$$

since, as was seen in Chapter VIII, $\iint_S K\,dA = 4\pi(1 - g)$. In this formula the inner normal to S is taken; i.e., the distances t are positive toward the interior. This interesting formula is due to Steiner.

The next formulas to be derived involve what is called the *support function* $p = p(\mathbf{X})$ of a closed orientable surface S given by $\mathbf{X}(u^1, u^2)$; its definition is

$$(10.56) \qquad\qquad p(\mathbf{X}) = -\mathbf{X}\cdot\mathbf{Y}.$$

Thus p is the distance (with a sign) from the origin to the tangent plane of S.

It is convenient to introduce two scalar differential forms of first degree that are defined as triple products:

$$(10.57) \qquad\qquad \alpha = \mathbf{X} \times \mathbf{Y}\cdot d\mathbf{X}, \qquad \beta = \mathbf{X} \times \mathbf{Y}\cdot d\mathbf{Y}.$$

The exterior derivative $d\alpha$ is given by [cf. (10.49)]:

$$d\alpha = d\mathbf{X} \times \mathbf{Y}\cdot d\mathbf{X} + \mathbf{X} \times d\mathbf{Y}\cdot d\mathbf{X}$$

since $d(d\mathbf{X}) = 0$. This can also be written in the form

$$d\alpha = -\mathbf{Y}\cdot(d\mathbf{X} \times d\mathbf{X}) + \mathbf{X}\cdot(d\mathbf{Y} \times d\mathbf{X})$$

since dot and cross may be interchanged but $\mathbf{A} \times \mathbf{B}\cdot C = -\mathbf{B} \times \mathbf{A}\cdot C$ for any three vectors. From (10.51) and (10.52) it is seen that

$$\frac{d\alpha}{2} = -dA + pH\,dA.$$

From this in turn the following formula results since $\iint_S d\alpha = 0$:

$$(10.58) \qquad\qquad \iint_S pH\,dA = \iint_S dA = A.$$

In deriving (10.58) a general theorem was used, i.e., that $\iint_S d\omega = 0$ when ω is any differential form of first degree and S is a closed orientable surface. This follows from the integral identity $\iint_D d\omega = \oint_{\bar{D}} \omega$ when D is any domain and $\bar{D}$ is its boundary curve; the theorem then follows by a familiar argument (employing a triangulation) when D is a closed orientable surface.

Similarly, for $d\beta$ the following calculation is made:

$$d\beta = d\mathbf{X}\cdot\mathbf{Y} \times d\mathbf{Y} + \mathbf{X}\cdot d\mathbf{Y} \times d\mathbf{Y}$$

$$= -\mathbf{Y}\cdot d\mathbf{X} \times d\mathbf{Y} + \mathbf{X}\cdot d\mathbf{Y} \times d\mathbf{Y}$$

$$= 2H\,dA - 2pK\,dA.$$

From this follows another interesting formula:

$$(10.59) \qquad \iint_S pK\,dA = \iint_S H\,dA,$$

when S is a closed orientable surface.

The last two formulas will be used now to prove two theorems in the large for certain simple closed surfaces.

THEOREM I.   *If S is star-shaped, i.e., such that for a suitable choice of origin $p > 0$ holds for all points of S, then S is a sphere if its mean curvature H is a constant c.*   (At the end of Chapter VI a less general theorem was proved, i.e., that S is a sphere if $K > 0$ and $H = $ constant holds; thus such a surface is convex by Hadamard's theorem and consequently star-shaped—but a surface need not be convex to be star-shaped.)

Proof.   It is known that S must have points where K is positive, simply because it is a closed surface, and by choosing the inner normal as positive and equal to $(\mathbf{X}_u \times \mathbf{X}_v)/|\mathbf{X}_u \times \mathbf{X}_v|$ it follows that the principal curvatures are both positive at such points, hence that H is also positive; thus c is positive. By replacing the function $\mathbf{X}(u^1, u^2)$, which defines S, by $\lambda\mathbf{X}$ for $\lambda > 0$, it is seen that H is changed to $(1/\lambda)H$ and K to $(1/\lambda^2)K$.   Thus no loss of generality results if $H = c$ is assumed to have the value one.   From (10.58) and (10.59) it follows that

$$\iint_S p\,dA = \iint_S dA, \qquad \iint_S pK\,dA = \iint_S dA$$

and hence that

$$\iint_S p\,dA = \iint_S pK\,dA,$$

or

$$\iint_S p(1 - K)\,dA = 0.$$

Since

$$H = \frac{k_1 + k_2}{2} = 1 \quad \text{and} \quad H^2 - K = \left(\frac{k_1 - k_2}{2}\right)^2$$

it follows that $H^2 > K$ holds, with equality only if $k_1 = k_2$, i.e., at umbilical points. The origin was chosen so that $p > 0$ holds, and since $K \leq 1$ holds, the integrand in the last formula is non-negative; thus $K = 1$ holds; hence $k_1 = k_2$ everywhere, and the surface must be a sphere since all points are umbilical points.

The second theorem is:

THEOREM II. *If the Gaussian curvature K is a constant over any closed surface it is a sphere.* (A more general theorem including this was proved at the end of Chapter VI.)

Proof. By Hadamard's theorem, which was proved in the last section of Chapter VIII, S is a convex surface since K is of necessity positive because it is positive at some points at least. Again $K \equiv 1$ may be assumed. In this case the formulas (10.58) and (10.59) become

$$\iint_S pH \, dA = \iint_S dA, \qquad \iint_S p \, dA = \iint_S H \, dA.$$

Since now $H \geq 1$ holds if the inner normal is taken as positive, it follows that $\iint_S H \, dA \geq \iint_S dA$ and hence that $\iint_S p \, dA \geq \iint_S pH \, dA$. But $p > 0$ holds if the origin is chosen in the interior of S, and hence $\iint_S p(1 - H) \, dA \geq 0$ leads to $H = 1$, and that again means that all points of S are umbilical points and thus S is a sphere.

There is an important and interesting corollary to Theorem II which in turn foreshadows a more general theorem. The corollary is:

COROLLARY. *If S and $\bar{S}$ are two closed surfaces in one-to-one correspondence in such a way that the lengths of all corresponding curves are the same on both—or, as it is also put, they are isometric—then if one of the two surfaces is a sphere so is the other.*

The proof is immediate since K is the same for both at corresponding points. Thus the only deformations of a sphere into a regular surface which preserve the lengths of all curves on it are rigid motions. It should be noted that the theorem is not true in general without the hypothesis of regularity. For example if a sphere is cut by a plane and one of the two portions into which the plane divides it is reflected in the plane the result is a surface isometric to the sphere, but it is not a sphere. However, a singular curve exists on the deformed surface. Later in this chapter it will be shown that the analogous theorem is valid for all closed surfaces of positive Gaussian curvature, that is for so-called ovaloids.

5. Minimal Surfaces

Formulas (10.51) and (10.52) of the previous section lend themselves well to a characterization of *minimal surfaces*. They are defined as the regular surfaces of stationary area in Euclidean three-space that are spanned by a closed curve C. It is among such surfaces that surfaces of least area spanning the curve would be sought. To find a necessary condition for such surfaces consider a surface $\mathbf{X}(u, v)$ spanned in C and a set of surfaces $\mathbf{X}(\epsilon)$, so-called normal variations of $\mathbf{X}$, that are given by the formula

$$(10.60) \qquad \mathbf{X}(\epsilon) = \mathbf{X} + \epsilon\phi(u, v)\mathbf{Y}.$$

Here $\phi(u, v)$ is a sufficiently smooth function that vanishes on the boundary C and $\mathbf{Y}$ is the unit normal on $\mathbf{X}$. It is to be recalled that $\sigma = \sigma_1 \wedge \sigma_2$ is the differential form that defines the element of area of a surface. The formulas (10.51) and (10.52) then lead to the following equations involving $\sigma(\epsilon)$:

$$dX(\epsilon) \times dX(\epsilon) = 2\mathbf{Y}(\epsilon) \times \sigma(\epsilon)$$

$$= [d\mathbf{X} + \epsilon \, d(\phi\mathbf{Y})] \times [d\mathbf{X} + \epsilon \, d(\phi\mathbf{Y})]$$

$$= d\mathbf{X} \times d\mathbf{X} + 2\epsilon\phi \, d\mathbf{X} \times d\mathbf{Y} + 2\epsilon \, d\mathbf{X} \times \mathbf{Y} \, d\phi + \epsilon^2(\cdots)$$

$$= 2\sigma(0) \, \mathbf{Y}(0) - 4\epsilon\phi \, H(0) \, \sigma(0) \, \mathbf{Y}(0) + \cdots.$$

The dots refer to terms which are either of order ϵ^2, or are such as to involve a vector orthogonal to $\mathbf{Y}(0)$. From (10.60) it is easily checked that $\mathbf{Y}(\epsilon) = \mathbf{Y}(0) + \epsilon\mathbf{Z} + \epsilon^2(\cdots)$ with $\mathbf{Z}$ orthogonal to $\mathbf{Y}(0)$. Upon comparing the first line on the right with the last it is then seen that

$$\sigma(\epsilon) = \sigma(0) - 2\epsilon\phi \, H(0) \, \sigma(0) + \cdots,$$

where the dots refer to a neglected term of order ϵ^2. In order that $\mathbf{X}(0)$ should furnish the surface of minimum area it is necessary that $A(\epsilon) = \iint \sigma(\epsilon)$ should be stationary, i.e., that $dA/d\epsilon|_{\epsilon=0} = 0$, and this in turn requires that $\iint \phi \, H(0) \, \sigma(0)$ should vanish for arbitrary functions ϕ that vanish on C. A familiar argument of the calculus of variations (cf. Courant-Hilbert [C.12], Vol. I, for example) then leads to the condition

$$(10.61) \qquad H = 0$$

at all points as a necessary condition in order that $\mathbf{X}(0)$ should furnish the minimum area. *Any surfaces for which the mean curvature is zero are by definition minimal surfaces.*

The condition (10.61) leads to another condition that also characterizes the minimal surfaces, i.e., that *any such surface is mapped conformally on its own spherical image.* This is seen from the identity between the three fundamental forms: $\text{III} - 2H \ \text{II} + K \ \text{I} = 0$. If $H = 0$ everywhere, then $\text{III}/\text{I} = -K$ and this implies that the mapping is conformal since the line elements of the two surfaces are proportional independent of the ratio of du and dv, which in turn means that pairs of curves that correspond on the surfaces cut each other at equal angles. (It should be observed that $H = 0$ implies $K \leq 0$, so that the form III is not negative.) Since the sphere can be mapped conformally on the plane by a stereographic projection and thus has a line element of the form $\mu^2(du^2 + dv^2)$ in terms of rectangular Cartesian coordinates in the plane, it follows that a minimal surface also has a line element of the form $-(\mu^2/K)(du^2 + dv^2) = \lambda^2(du^2 + dv^2)$. In such a case the parameters, u, v are called *isothermal parameters*.[1] Since the mean curvature is given, in general, by

$$H = \frac{EN - 2FM + GL}{EG - F^2},$$

it follows in the present case with the use of isothermal parameters that $H = 0$ leads to the equation

$$L + N = 0.$$

Since $L = \mathbf{X}_{uu} \cdot \mathbf{Y}$ and $N = \mathbf{X}_{vv} \cdot \mathbf{Y}$ it is at once seen that minimal surfaces satisfy the equation

(10.62) $$\nabla^2 \mathbf{X} \cdot \mathbf{Y} = 0,$$

with ∇^2 the Laplacian $\partial^2/\partial u^2 + \partial^2/\partial v^2$. However, the parameters are such that

$$\mathbf{X}_1 \cdot \mathbf{X}_1 = \mathbf{X}_2 \cdot \mathbf{X}_2, \qquad \mathbf{X}_1 \cdot \mathbf{X}_2 = 0.$$

Differentiation of these relations leads to

$$\nabla^2 \mathbf{X} \cdot \mathbf{X}_1 = \nabla^2 \mathbf{X} \cdot \mathbf{X}_2 = 0,$$

and these together with (10.62) imply

(10.63) $$\nabla^2 \mathbf{X} = 0.$$

In other words, the components of a vector $\mathbf{X}$ that defines a minimal surface are harmonic functions of isothermal parameters on it. It follows, therefore, that any such surface that has continuous second derivatives with respect to isothermal parameters is actually *analytic*, since that is a general property of harmonic functions. This means that any such minimal surfaces can be

---

[1] Every surface with continuous third derivatives can be parametrized, at least locally, in terms of isothermal parameters. The proof belongs rather specifically to the theory of linear elliptic partial differential equations rather than to differential geometry, and will not be given here.

represented in terms of analytic functions of a single complex variable $z = u + iv$, with u, v as isothermal parameters.

It follows that $\mathbf{X}(u, v)$ can be defined as the real part of $\mathbf{Z}(u, v)$, with $\mathbf{Z}(u, v) = \mathbf{X}(u, v) + i\mathbf{Y}(u, v)$ a complex-valued vector each of whose three components $x_i + iy_i$ is an analytic function of the complex variable $z = u + iv$. In order that the real part of $\mathbf{Z}$ should represent a minimal surface with u, v as isothermal parameters it is necessary, as was noted above, that $\mathbf{X}_u^2 = \mathbf{X}_v^2$, $\mathbf{X}_u \cdot \mathbf{X}_v = 0$ should hold; since the Cauchy-Riemann equations are easily seen to lead to the relation $d\mathbf{Z}/dz = \mathbf{X}_u - i\mathbf{X}_v$, it follows that these conditions are equivalent to the condition $(d\mathbf{Z}/dz)^2 = (\mathbf{X}_u - i\mathbf{X}_v)^2 = 0$. Hence all minimal surfaces with a few continuous derivatives are obtained as the real part of complex-valued analytic vector functions $\mathbf{Z}(z)$ such that $(d\mathbf{Z}/dz)^2 = 0$.

6. Uniqueness Theorems for Closed Convex Surfaces

In differential geometry in the large few problems have attracted as much attention as those relating to the existence and uniqueness under a variety of conditions of the closed regular convex surfaces embedded in three-dimensional space. By uniqueness is to be understood uniqueness within rigid motions possibly combined with a reflection in a plane. Some special cases of such problems have already been treated in Section 4 of this chapter, where uniqueness theorems for the sphere were proved.

By convexity is meant here that the surface is the full boundary of a bounded three-dimensional convex set, that it is a regular surface with continuous fourth derivatives (for some of the theorems to be discussed fewer derivatives are needed), and that the Gaussian curvature K is everywhere positive. (It was shown at the end of Chapter VIII that a closed regular surface with $K > 0$ is necessarily the boundary of a convex body.)

There are four classical problems concerning such closed convex surfaces in three-space:

Weyl's problem. The coefficients E, F, G of the line element $ds^2 = E\, du^2 + 2F\, du\, dv + G\, dv^2$ are prescribed functions over the unit sphere $\sum$ such that $K > 0$ holds, and with $\int_\Sigma K\, dA = 4\pi$ as a necessary condition to be satisfied. It is to be shown that a regular surface S exists in three-space that is in one-to-one correspondence with the sphere such that its line element has the prescribed coefficients E, F, G at corresponding points, and that S is uniquely determined within rigid motions and reflections.

Minkowski's problem. A positive point function K is defined on the unit sphere $\sum$. A regular closed surface S is to be found in three-space such that its outer normals correspond to the points of $\sum$ determined by the parallel rays from its center, and such that the Gaussian curvature K of S has the value prescribed at the corresponding point on $\sum$. (Certain obvious neces-sary conditions on K must be satisfied simply because S is a closed surface.) It is to be shown in addition that all such surfaces S differ by at most a translation.

Christoffel's problem. This problem is like Minkowski's except that it is the sum of the principal radii of curvature $1/k_1 + 1/k_2$ that is prescribed as a positive function of the direction of the normal vector, and only the *unique-ness theorem* for a closed convex surface is in question.

Liebmann's problem. A closed convex surface in three-space is given. It is to be shown that the only small deformations of it that preserve the line element within terms of second order in the deformation parameter are small rigid motions. It is customary to say under such circumstances that the surface is *rigid*—in conformity with the analogous circumstances in the statics of framed structures.

The fundamental reason for the difficulty of the first two problems is that their formulations belong basically to the theory of nonlinear boundary value problems for partial differential equations which, being usually of Monge-Ampère type, are not as a rule even quasi-linear. In studying the unique-ness questions, however, it is possible to linearize the second of the above uniqueness problems, and to make some progress with the first, by formulat-ing them in terms of Minkowski's support function (the same function $p(\mathbf{X})$ that was employed in Section 4 above). The third and fourth problems are linear from the outset; in fact the fourth problem is the linearized version of the uniqueness question in Weyl's problem. A treatment of the uniqueness questions with the aid of the support function, together with a history of the problems and references to the literature can be found in Stoker [S.5]. Another way of dealing with uniqueness questions was initiated by Cohn-Vossen [C.11] in 1928 for the case of Weyl's problem. This method reduces the uniqueness problem to the study of the indices of the singularities of certain vector fields defined on the surfaces, and it has been proved useful for a variety of problems, e.g., for Minkowski's problem (cf. Lewy [L.4]).[1]

---

[1] It perhaps has some relevance at this point to mention that existence and uniqueness problems analogous to those under discussion here can be formulated for polyhedra with plane faces in three-space. The best known of these is the uniqueness theorem for closed convex polyhedra proved by Cauchy [C.3] in 1813. This theorem states that two closed convex polyhedra in three-space with faces that are congruent and arranged in the same way are either congruent or symmetric. This is, of course, analogous to the uniqueness theorem in the case of Weyl's problem: the two polyhedra are isometric, but might not be congruent since bending of the faces about their edges is not a priori ruled out. Cauchy's method has been used and extended by A. D. Alexandrov [A.2] in a wide variety of amusing and interesting ways; still other extensions of Cauchy's ideas have been carried out by Stoker [S.4], [S.8], including examples of indeformable nonconvex surfaces of arbitrary genus.

An extended discussion of this method has been given by Voss [V.3]; this includes generalizations that permit $K \geq 0$, whereas only $K > 0$ is permitted here (see also another interesting paper of the same author concerning uniqueness questions [V.2]).

In this section the uniqueness theorems for the first three problems will be proved without reference to a formulation in terms of partial differential equations or in terms of singularities of vector fields. In all of the cases the proofs given here may well be regarded as somewhat tricky, since it is not at all apparent geometrically why such devices should be used; on the other hand these proofs have the virtues of simplicity and elegance. The solutions of all three problems[2] depend upon finding an integral $\iint_S P(u, v)\sigma$ which vanishes over one of the surfaces, while P is an invariant function depending on both surfaces being compared and has a fixed sign, but is such that its vanishing means that the two surfaces are identical (within motions). Of course, the difficulty is to find the appropriate function P in each of the three cases, and for that experience and sound intuition seem to be the only guides. In defense of this somewhat mysterious procedure it might perhaps be noted that the proofs of uniqueness theorems for boundary-value problems involving other partial differential equations also usually require the invention of special tricks and devices, above all if the problems are nonlinear, and such devices commonly involve integrals over the domains in question (e.g., energy integrals in problems having their origin in mathematical physics).

The uniqueness theorems for the first three problems listed above are treated in reverse order.

1. *Christoffel's problem.* Christoffel gave a very intricate proof of his theorem in 1865. Hurwitz gave a much simpler proof in 1902 which nevertheless involves a rather sophisticated theorem, i.e., the theorem on the completeness of the set of all spherical harmonics; in the paper by Stoker cited above a very simple proof is given that hinges only on the theorem that a harmonic function cannot take on a maximum or minimum in the interior of its domain of definition. The following proof was communicated to the author by L. Nirenberg.

Consider two surfaces $\mathbf{X}$ and $\mathbf{X}'$ satisfying the hypotheses of Christoffel's theorem and set $\mathbf{Z} = \mathbf{X} - \mathbf{X}'$. Denote, as usual, the unit normals in $\mathbf{X}$ and $\mathbf{X}'$ by $\mathbf{Y}$ and $\mathbf{Y}'$. From the formulas (10.52) and (10.53) at the beginning of Section 4 the following formulas are obtained:

$$(10.64) \qquad\qquad -d\mathbf{Y} \times d\mathbf{X} = 2H\mathbf{Y}\, dA,$$

$$(10.65) \qquad d\mathbf{Y} \times d\mathbf{Y} = d\mathbf{Y}' \times d\mathbf{Y}' = 2K\mathbf{Y}\, dA = 2K'\mathbf{Y}\, dA'.$$

[2] The fourth problem, Liebmann's problem, can also be treated in the same way, as was shown by Blaschke [B.6] in 1912.

In the second of these equations the important fact that parameters (u, v) exist in terms of which $\mathbf{Y}(u, v) \equiv \mathbf{Y}'(u, v)$ holds has been used. Also valid for the same reason are the identities

$$(10.66) \qquad -d\mathbf{Y}' \times d\mathbf{X}' = -d\mathbf{Y} \times d\mathbf{X}' = 2H'\mathbf{Y}\,dA'.$$

By the basic assumption of the theorem $\frac{1}{2}(1/k_1 + 1/k_2) = H/K$ is the same at corresponding points of the two surfaces; i.e., $H/K = H'/K'$. From (10.65) it follows that $K\,dA = K'\,dA'$, and hence that $H'\,dA' = H\,dA$ at corresponding points. In that case (10.64) and (10.66) lead to

$$(10.67) \quad d\mathbf{Y} \times d\mathbf{X}' = d\mathbf{Y} \times d\mathbf{X} \quad \text{or} \quad d\mathbf{Y} \times d(\mathbf{X}' - \mathbf{X}) = d\mathbf{Y} \times d\mathbf{Z} = 0.$$

A scalar differential form α of first degree is introduced by means of the following triple product (cf. Section 3, especially equations (10.48), (10.49)]:

$$\alpha = \mathbf{Z} \cdot \mathbf{Y} \times d\mathbf{Z}.$$

The exterior derivative of this form is given by

$$d\alpha = d\mathbf{Z} \cdot \mathbf{Y} \times d\mathbf{Z} + \mathbf{Z} \cdot d\mathbf{Y} \times d\mathbf{Z}$$
$$= -\mathbf{Y} \cdot d\mathbf{Z} \times d\mathbf{Z},$$

since $d(d\mathbf{Z}) = 0$, and $d\mathbf{Y} \times d\mathbf{Z} = 0$ from (10.67); the order of two factors in a triple product was also changed. Since $d\alpha$ is a differential form of second order, it follows that $\iint_S d\alpha = 0$, since S is a closed surface; hence

$$(10.68) \qquad \iint_S \mathbf{Y} \cdot d\mathbf{Z} \times d\mathbf{Z} \times 0.$$

The proof of the theorem of Christoffel is completed by showing that the integral identity holds only if $d\mathbf{Z}$ vanishes everywhere, and hence $\mathbf{X}'$ and $\mathbf{X}$ would differ only by a constant vector and thus the two surfaces would differ only by a translation. The proof of the statement requires a calculation that starts from the formula $d\mathbf{Z} \cdot \mathbf{Y} = 0$ (which holds simply because $\mathbf{Y}$ is the normal to S and S') from which the following further formula is derived:

$$(10.69) \qquad d(d\mathbf{Z} \cdot \mathbf{Y}) = d\mathbf{Z} \cdot d\mathbf{Y} = 0.$$

The spherical image surface $\mathbf{Y}(u, v)$ is a regular surface, and hence $\mathbf{Y}_u \times \mathbf{Y}_v \neq 0$, since the curvature K does not vanish because the surface is assumed to be convex.[1] Choose local coordinates in such a way that $\mathbf{Y}_u$ and $\mathbf{Y}_v$ are ortho-

---

[1] A reasonable existence theorem for Christoffel's problem analogous to those for Minkowski's and Weyl's problems is hardly possible because the prescription of the sum of the principal radii of curvature does not of necessity lead to a regular surface with $K > 0$, and that causes difficulties in dealing with the partial differential equations. On the other hand Weyl [W.4] has treated the existence question corresponding to the uniqueness theorem of Liebmann.

gonal (which causes no loss of generality since all relations are between quantities invariant with respect to changes of parameters), and set

$$(10.70) \qquad \begin{cases} \mathbf{Z}_u = a\mathbf{Y}_u + \dfrac{b}{|\mathbf{Y}_v|^2}\,\mathbf{Y}_v, \\[2mm] \mathbf{Z}_v = c\mathbf{Y}_u + d\mathbf{Y}_v, \end{cases}$$

in which a, b, c, d are certain scalar functions; such relations hold since $\mathbf{Y}_u$ and $\mathbf{Y}_v$, as derivatives of the unit normal, lie in the tangent planes of $\mathbf{X}$ and $\mathbf{X}'$. From $d\mathbf{Y} \times d\mathbf{Z} = 0$ [cf. (10.67)] and (10.69) it follows that

$$(10.71) \qquad \mathbf{Z}_u \cdot \mathbf{Y}_v = \mathbf{Z}_v \cdot \mathbf{Y}_u, \qquad \mathbf{Z}_u \times \mathbf{Y}_v = \mathbf{Z}_v \times \mathbf{Y}_u,$$

when the fact that the symbol d refers to exterior differentiation is noted and that the $\cdot$ and $\times$ imply that the differentials of u and v are subject to the Grassmann rule. The orthogonality of $\mathbf{Y}_u$ and $\mathbf{Y}_v$ in combination with (10.71), when used in conjunction with (10.70), lead by an easy calculation to the following conditions on the coefficients a, b, c, d:

$$c|\mathbf{Y}_u|^2 = b, \qquad d = -a.$$

Insertion of these in (10.70) leads to

$$d\mathbf{Z} \times d\mathbf{Z} = -2\left(a^2 + \frac{b^2}{|\mathbf{Y}_u|^2|\mathbf{Y}_v|^2}\right)\mathbf{Y}_u \times \mathbf{Y}_v \, du \wedge dv.$$

The scalar product of this with $\mathbf{Y}$ is everywhere of one sign since $\mathbf{Y}(u, v)$ represents the unit sphere in a regular parameter representation. Hence from (10.68) $\mathbf{Y} \cdot d\mathbf{Z} \times d\mathbf{Z}$ is everywhere zero, and this in turn is possible only if $a = b = 0$ since $\mathbf{Y} \cdot \mathbf{Y}_u \times \mathbf{Y}_v$ does not vanish; hence $c = d = 0$ and from (10.70) $\mathbf{Z}_u = \mathbf{Z}_v = 0$ everywhere. This completes the proof of Christoffel's theorem.

2. *Uniqueness theorem for Minkowski's problem.* In this problem two closed convex surfaces S and S' are given such that the Gaussian curvature K is positive and has the same value at points with parallel outer normal vectors. Again parameters for both surfaces can be chosen so that they are given by $\mathbf{X}(u, v)$, $\mathbf{X}'(u, v)$ and $\mathbf{Y}(u, v) = \mathbf{Y}'(u, v)$, $K(u, v) = K'(u, v)$. Minkowski [M.3] gave the first proof that S and S' are identical in 1903, but this proof makes essential use of the rather difficult Brunn-Minkowski inequality for the so-called mixed volumes of convex bodies. H. Lewy [L.4] gave a proof in 1938 valid for analytic surfaces, and Stoker [S.5] in 1950 an elementary proof not requiring analyticity. The proof that follows is essentially due to Chern [C.4], who devised it probably by analogy with the proof of Herglotz [H.7] in 1942 for the uniqueness theorem in the case of Weyl's problem. The

first uniqueness proof for the solution of Weyl's problem was given by Cohn-Vossen [C.11] in 1927 and it makes essential use of the assumption that the surfaces are analytic. Thus it is seen that proofs of an elementary character for these uniqueness theorems have been found only rather recently.

The proof of Minkowski's theorem begins with the formula [cf. (10.53)]

$$(10.72) \qquad d\mathbf{Y} \times d\mathbf{Y} = d\mathbf{Y}' \times d\mathbf{Y}' = 2K\mathbf{Y}\,dA = 2K'\mathbf{Y}\,dA',$$

in which $\mathbf{Y} \equiv \mathbf{Y}'$ has been used. It follows from (10.72) that $K\,dA = K'\,dA' = K\,dA'$, and hence $dA \equiv dA'$ since $K \equiv K'$ is one of the basic hypotheses of the theorem. As in the proof of the previous theorem there are advantages to be gained in assuming a local coordinate system such that $\mathbf{Y}_u$ and $\mathbf{Y}_v$ are orthogonal, and to express the tangent vectors $\mathbf{X}_u$ and $\mathbf{X}_v$ of S (and later also S') in terms of them:

$$(10.73) \qquad \begin{cases} \mathbf{X}_u = \dfrac{a}{|\mathbf{Y}_u|^2}\,\mathbf{Y}_u + \dfrac{b}{|\mathbf{Y}_v|^2}\,\mathbf{Y}_v, \\[3mm] \mathbf{X}_v = \dfrac{b}{|\mathbf{Y}_u|^2}\,\mathbf{Y}_u + \dfrac{c}{|\mathbf{Y}_v|^2|}\,\mathbf{Y}_v. \end{cases}$$

In these formulas the scalar coefficient $b(u, v)$ occurs twice. This is due to the fact that $d\mathbf{X}\cdot d\mathbf{Y} = 0$ holds because $d\mathbf{X}\cdot \mathbf{Y} = 0$, from which $\mathbf{X}_u\cdot\mathbf{Y}_v = \mathbf{X}_v\cdot\mathbf{Y}_u$ follows. Since, in addition, $\mathbf{Y}_u$ and $\mathbf{Y}_v$ were chosen to be orthogonal, the statement is seen to hold. For $\mathbf{X}_u'$ and $\mathbf{X}_v'$ the same formulas hold, since $\mathbf{Y} = \mathbf{Y}'$, except that a, b, c become certain other coefficients a', b', c'. From (10.73) the following relations are obtained:

$$(10.74)$$

$$\mathbf{X}_u \times \mathbf{X}_v = \frac{ac - b^2}{|\mathbf{Y}_u|^2|\mathbf{Y}_v|^2}\,\mathbf{Y}_u \times \mathbf{Y}_v, \qquad \mathbf{X}_u' \times \mathbf{X}_v' = \frac{a'c' - b'^2}{|\mathbf{Y}_u|^2|\mathbf{Y}_v|^2}\,\mathbf{Y}_u \times \mathbf{Y}_v.$$

From (10.51) in the present case the following identities are valid:

$$(10.75) \qquad d\mathbf{X} \times d\mathbf{X} = 2\mathbf{Y}\,dA = 2\mathbf{Y}'\,dA' = d\mathbf{X}' \times d\mathbf{X}',$$

since $\mathbf{Y} \equiv \mathbf{Y}'$, and $dA \equiv dA'$ was shown to hold; it follows from this that $\mathbf{X}_u \times \mathbf{X}_v\,du \wedge dv = \mathbf{X}_u' \times \mathbf{X}_v'\,du \wedge dv$ and hence that $\mathbf{X}_u \times \mathbf{X}_v = \mathbf{X}_u' \times \mathbf{X}_v'$. The equations (10.74) then yield

$$(10.76) \qquad\qquad ac - b^2 = a'c' - b'^2 \neq 0,$$

since all surfaces, including $\mathbf{Y}(u, v)$, are regular. Since K is positive, it is clear from (10.74) that $ac - b^2 > 0$ holds, hence $a'c' - b'^2$ is also positive.

The crucial step in the discussion is the introduction of the appropriate invariant form of first degree; in the present case that is the form $\alpha =$

$\mathbf{X} \cdot \mathbf{X}' \times d\mathbf{X}'$. The exterior derivative of this form is expressed as follows:
(10.77)

$$d\alpha = d\mathbf{X} \cdot \mathbf{X}' \times d\mathbf{X}' + \mathbf{X} \cdot d\mathbf{X}' \times d\mathbf{X}'$$

$$= -\mathbf{X}' \cdot d\mathbf{X} \times d\mathbf{X}' + \mathbf{X} \cdot d\mathbf{X}' \times d\mathbf{X}'$$

$$= \mathbf{X} \cdot d\mathbf{X}' \times d\mathbf{X}' + \tfrac{1}{2}\mathbf{X}' \cdot (d\mathbf{X}' - d\mathbf{X}) \times (d\mathbf{X}' - d\mathbf{X}) - \mathbf{X}' \cdot d\mathbf{X}' \times d\mathbf{X}'$$

$$= (\mathbf{X} - \mathbf{X}') \cdot d\mathbf{X}' \times d\mathbf{X}' + \tfrac{1}{2}\mathbf{X}' \cdot (d\mathbf{X}' - d\mathbf{X}) \times (d\mathbf{X}' - d\mathbf{X}).$$

In this calculation (10.75) was used and also the fact that $d\mathbf{X}' \times d\mathbf{X} = d\mathbf{X} \times d\mathbf{X}'$ [cf. (10.45)]. Since the integral of $d\alpha$ over S is zero, it follows that

$$(10.78) \quad \iint_S (\mathbf{X} - \mathbf{X}') \cdot d\mathbf{X}' \times d\mathbf{X}' = -\frac{1}{2} \iint_S \mathbf{X}' \cdot (d\mathbf{X}' - d\mathbf{X}) \times (d\mathbf{X}' - d\mathbf{X}).$$

It is now to be proved that the integrand on the right-hand side of this equation is of one sign, and is zero only where $d(\mathbf{X}' - \mathbf{X}) = 0$. It will also be shown that an interchange of $\mathbf{X}$ and $\mathbf{X}'$ in the integrand leaves its sign unchanged, and that $d(\mathbf{X}' - \mathbf{X}) = 0$ is again necessary at points where it vanishes. Thus the integral I on the left-hand side of (10.78) has the same sign when $\mathbf{X}$ and $\mathbf{X}'$ are interchanged; its value must therefore be zero since $d\mathbf{X}' \times d\mathbf{X}' = d\mathbf{X} \times d\mathbf{X}$ from (10.75) so that I and $-I$ have the same sign. Thus the integral on the right-hand side vanishes, its integrand therefore vanishes everywhere, and hence $d(\mathbf{X}' - \mathbf{X}) = 0$ at all points of S. Consequently $\mathbf{X}_u = \mathbf{X}'_u$, $\mathbf{X}_v = \mathbf{X}'_v$ everywhere and $\mathbf{X}$ and $\mathbf{X}'$ therefore differ by a constant vector. The theorem would thus be proved.

It remains therefore to be shown that the right-hand side of (10.78) is of one sign and vanishes only if $d(\mathbf{X}' - \mathbf{X}) = 0$ everywhere, and that this holds upon interchanging $\mathbf{X}$ and $\mathbf{X}'$. This follows from the following algebraic lemma:

LEMMA. If

$$\begin{pmatrix} a & b \\ b & c \end{pmatrix} \quad \text{and} \quad \begin{pmatrix} a' & b' \\ b' & c' \end{pmatrix}$$

are matrices of positive definite quadratic forms with equal determinants, i.e., $ac - b^2 = a'c' - b'^2$, then the matrix

$$\begin{pmatrix} a - a' & b - b' \\ b - b' & c - c' \end{pmatrix}$$

has a determinant which is nonpositive and is zero only if the matrices are identical.

Assuming the correctness of the lemma, it is easily seen that the result desired follows, simply by calculating $d\mathbf{X} - d\mathbf{X}'$ by using (10.73), the discussion following (10.76), and their counterparts for $\mathbf{X}'$; in this way the following relation is easily found:

$$(10.79)\quad (d\mathbf{X} - d\mathbf{X}') \times (d\mathbf{X} - d\mathbf{X}')$$
$$= [(a - a')(c - c') - (b - b')^2]\mathbf{Y}_1 \times \mathbf{Y}_2 \, du \wedge dv,$$

when $\mathbf{Y}_1$ and $\mathbf{Y}_2$ are introduced for $(|\mathbf{Y}_u|^2)^{-1}\,\mathbf{Y}_u$ and $(|\mathbf{Y}_v|^2)^{-1}\,\mathbf{Y}_v$. Since it is always possible to place S and S' so that the origin of coordinates lies inside of both of them, and since $\mathbf{X} \cdot \mathbf{Y}$ and $\mathbf{X}' \cdot \mathbf{Y}$ have then a fixed sign throughout because S and S' are convex, it follows from the algebraic lemma that the scalar product of (10.79) with $\mathbf{X}'$ has indeed a fixed sign everywhere, and that it vanishes only if $a = a', b = b', c = c'$. It is also clear that an interchange of the roles of $\mathbf{X}$ and $\mathbf{X}'$ gives the same result, including the fact that the right-hand side of (10.79) retains its sign.

The algebraic lemma remains to be proved. It is possible to give a geometric proof of the lemma (this is left as an exercise at the end of the chapter), but a simple algebraic proof will be given instead. Consider the positive definite quadratic forms $ax^2 + 2bxy + cy^2$ and $a'x^2 + 2b'xy + c'y^2$. By a well-known theorem in linear algebra it is possible to transform both of these forms simultaneously with the use of a nonsingular linear transformation to new coordinates ξ, η in such a way that the first goes into $\xi^2 + \eta^2$ and the second into $\lambda_1 \xi^2 + \lambda_2 \eta^2$. The positive numbers λ_1, λ_2 are the roots of

$$\begin{vmatrix} a - \lambda a' & b - \lambda b' \\ b - \lambda b' & c - \lambda c' \end{vmatrix} = 0.$$

By hypothesis $\Delta = ac - b^2 = a'c' - b'^2 > 0$. Clearly, $\Delta = 1$ may be assumed without loss of generality. The quadratic equation for λ is then $\lambda^2 - \lambda(ac' - 2bb' + ca') + 1 = 0$, and since it has two real roots it follows that $(ac' - 2bb' + ca')^2 - 4 \geq 0$, and hence $ac' - 2bb' + ca' \geq 2$ holds, since $\lambda_1 + \lambda_2 = ac' - 2bb' + ca'$ is positive. The determinant of the matrix that figures in the lemma can be put in the form $(a - a')(c - c') - (b - b')^2 = 2 - (ac' - 2bb' + ca')$ and therefore vanishes only if $\lambda_1 + \lambda_2 = 2$, and since $\lambda_1 \lambda_2 = 1$, only $\lambda_1 = \lambda_2 = 1$ is possible. But this clearly means that the two quadratic forms are identical. The lemma is thus proved, and with it Minkowski's uniqueness theorem.

3. *Uniqueness theorem for Weyl's problem.* Finally the proof of uniqueness for the case of Weyl's problem will be given by using what is essentially the method of Herglotz [H.7], but in the notation of this chapter. To be shown is that *two closed convex surfaces S and S' in one-to-one isometric correspondence are identical.* This is by far the most interesting of the three uniqueness theorems treated in this section, perhaps because of its striking

geometrical interpretation, i.e., that a closed convex surface cannot be mapped in a one-to-one way on another such surface that is not congruent to it while preserving the lengths of all corresponding curves. The theorem is not true in general for convex surfaces with holes. How the matter stands in general for simple closed surfaces that are not convex, as is the case for all

Fig. 10.1   Two closed noncongruent isometric surfaces.

that are of a genus different from zero, is not known and offers a challenging problem; however, it is known that an extension of the uniqueness theorem to *all* such surfaces is not possible. A simple counterexample is shown in Fig. 10.1. The surface shown has a bulge that goes out from a plane convex curve, the plane of which is tangent to the surface (and thus cuts it) at all points of the curve. An isometric surface is defined by reflecting the bulge in the plane, as indicated by the dotted curve in the figure. Evidently the construction could be made so that the surface would be differentiable of as high an order as might be wished.

Sometimes this property of closed convex surfaces is referred to as the impossibility of deforming them without inducing strains, or as the property of *rigidity*. However, this latter way of looking at the matter implies that the surface is regarded as being subjected to a continuous isometric deformation and that is not what is to be proved here. In fact if the problem is formulated with respect to a family of surfaces given by $\mathbf{X}(u, v; t)$ depending differentiably on a parameter t with the requirement that all remain isometric when t is varied, the question of uniqueness is much easier to settle. This was done by Weyl[W.2] and Blaschke [B.5] using a method that is much like that of Herglotz. This property of remaining congruent under continuous isometric deformations can be studied by introducing the notion of an infinitesimal deformation and requiring that the term "congruent" should allow discrepancies of second and higher order in the surface coordinates and in the lengths of curves. If a surface remains "congruent" under such a deformation it is often called infinitesimally rigid. As was already stated, the closed convex surfaces have this property: this is the content of the theorem of Liebmann. It is tempting to believe that infinitesimal rigidity should always hold in cases where the nonexistence of isometric mappings that are other than congruent mappings can be proved. A very simple example (among many others of interest in the theory of framed structures in statics) proves the contrary.   It is indicated in Fig. 10.2 and consists of two rigid rods jointed together, and attached by smooth pins to rigid walls in such a way that they both fall along the same straight line.   The configuration is obviously uniquely determined if the rods retain their lengths, but it

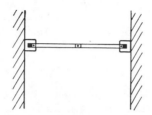

Fig. 10.2   A rigid but not infinitesimally rigid configuration.

is easily seen that a vertical displacement of the hinge in the middle that is small of first order gives rise to changes in length that are small of second order, and hence the configuration is *not* infinitesimally rigid in the sense defined here.

The proof of the theorem concerning what might be called rigidity in the finite sense is now given.   Again two closed convex surfaces S and S' represented by vectors $\mathbf{X}(u, v)$, $\mathbf{X}'(u, v)$ are given with parameters in common and such that at corresponding points the line elements are the same.   It is assumed also that S and S' have been oriented alike—as could be achieved if necessary by a reflection in a plane.   The *repère mobile* of Cartan is supposed given by unit vectors $\mathbf{a}_i(u, v)$, $i = 1, 2, 3$ on S and $\mathbf{a}_i'(u, v)$ on S', both being such that $\mathbf{a}_i$ and $\mathbf{a}_i'$ for $i = 1, 2$ are tangents to the parameter curves, while a_3 and a_3' are the unit normals.   Since S and S' are isometric, it follows [cf. (10.25)] from these stipulations that

(10.80) $$\sigma_1 = \sigma_1', \qquad \sigma_2 = \sigma_2',$$

and hence that the elements of area σ, σ' are also equal:

(10.81) $$\sigma = \sigma_1 \wedge \sigma_2 = \sigma' = \sigma_1' \wedge \sigma_2'.$$

Since the Gaussian curvature K is preserved, it follows from the equation preceding (10.34) that

(10.82) $$\omega_{12} = \omega_{12}'.$$

The object of the proof of Herglotz is to show that the remaining forms ω_{jk} and ω_{jk}' [cf. (10.18) and (10.19)] are equal:

(10.83) $$\omega_{13} = \omega_{13}', \qquad \omega_{23} = \omega_{23}',$$

since this will be seen equivalent to showing that *the second fundamental forms of S and S' are identical*, and hence that S and S' are congruent because their first fundamental forms are identical by assumption.

One step in the desired direction is furnished by (10.33), which under the present circumstances states that

$$(10.84) \qquad \omega_{13} \wedge \omega_{23} = \omega'_{13} \wedge \omega'_{23},$$

since $K = K'$ and $\sigma_1 \wedge \sigma_2 = \sigma'_1 \wedge \sigma'_2$. The integral identity upon which the uniqueness proof depends is obtained in terms of the following linear vector form:

$$(10.85) \qquad \boldsymbol{\Omega} = \mathbf{a}_1 \omega'_{31} + \mathbf{a}_2 \omega'_{32}.$$

This is a "mixed" form involving the basis vectors of S, and the differential forms ω'_{31} and ω'_{32} of S'. The unit normal $\mathbf{Y}$ is denoted in what follows by $\mathbf{a}_3$ and hence [cf. the third equation of (10.26)]

$$(10.86) \qquad d\mathbf{a}_3 = \mathbf{a}_1 \omega_{31} + \mathbf{a}_2 \omega_{32}.$$

The appropriate linear differential form for present purposes is $\alpha = \mathbf{X} \cdot \mathbf{a}_3 \times \boldsymbol{\Omega}$, together with its exterior derivative $d\alpha$. As a preliminary step in the calculation of $d\alpha$ it is shown that $\mathbf{a}_3 \times d\boldsymbol{\Omega}$ is zero. This follows from a separate calculation of $d\boldsymbol{\Omega}$:

$$d\boldsymbol{\Omega} = \mathbf{a}_1 d\omega'_{31} + d\mathbf{a}_1 \wedge \omega'_{31} + \mathbf{a}_2 d\omega'_{32} + d\mathbf{a}_2 \wedge \omega'_{32}$$

$$= \mathbf{a}_1 (\omega'_{32} \wedge \omega'_{21}) + (\mathbf{a}_2 \omega_{12} + \mathbf{a}_3 \omega_{13}) \wedge \omega'_{31}$$

$$+ \mathbf{a}_2 (-\omega'_{21} \wedge \omega'_{13}) + (\mathbf{a}_3 \omega_{23} + \mathbf{a}_1 \omega_{21}) \wedge \omega'_{32}$$

upon using (10.26) and (10.27). However, $\omega_{21} = \omega'_{21}$ from (10.82) and it is then readily seen that all terms involving $\mathbf{a}_1$ and $\mathbf{a}_2$ cancel out. Thus $d\boldsymbol{\Omega}$ is in the direction of $\mathbf{a}_3$ and consequently $\mathbf{a}_3 \times d\boldsymbol{\Omega}$ is zero. The exterior derivative $d\alpha$ is therefore given by

$$(10.87) \qquad d\alpha = d\mathbf{X} \cdot \mathbf{a}_3 \times \boldsymbol{\Omega} + \mathbf{X} \cdot d\mathbf{a}_3 \times \boldsymbol{\Omega}.$$

Upon integrating $d\alpha$ over S the result is of course zero, and therefore

$$0 = \iint_S (\mathbf{a}_1 \sigma_1 + \mathbf{a}_2 \sigma_2) \cdot \mathbf{a}_3 \times (\mathbf{a}_1 \omega'_{31} + \mathbf{a}_2 \omega'_{32})$$

$$+ \iint_S \mathbf{X} \cdot (\mathbf{a}_1 \omega_{31} + \mathbf{a}_2 \omega_{32}) \times (\mathbf{a}_1 \omega'_{31} + \mathbf{a}_2 \omega'_{32})$$

upon using (10.85), (10.86) and $d\mathbf{X} = \mathbf{a}_1 \sigma_1 + \mathbf{a}_2 \sigma_2$. The multiplications can be carried out to yield the following:

$$(10.88) \quad 0 = \iint_S (-\sigma_1 \wedge \omega'_{32} + \sigma_2 \wedge \omega'_{31}) + \mathbf{X} \cdot \mathbf{a}_3 (\omega_{31} \wedge \omega'_{32} - \omega_{32} \wedge \omega'_{31}),$$

when $\mathbf{a}_1 \cdot \mathbf{a}_2 \times \mathbf{a}_3 = 1$ is used. (This means that a local orthogonal parameter system is used—and that is always legitimate.) If the same calculations were to be made in replacing ω'_{31} and ω'_{32} in (10.85) by ω_{31} and ω_{32}, i.e., by working with S instead of S', the result would be, clearly, the following:

$$0 = \iint_S (-\sigma_1 \wedge \omega_{32} + \sigma_2 \wedge \omega_{31}) + \mathbf{X} \cdot \mathbf{a}_3(\omega_{31} \wedge \omega_{32} - \omega_{32} \wedge \omega_{31}).$$

In this last equation the term $\omega_{32} \wedge \omega_{31}$ can be replaced by $-\omega'_{31} \wedge \omega'_{32}$, because of (10.84), and the resulting equation can be subtracted from (10.88) to obtain

(10.89) $$\iint_S \mathbf{X} \cdot \mathbf{a}_3(\omega_{31} - \omega'_{31}) \wedge (\omega_{32} - \omega'_{32})$$

$$= \iint_S \sigma_1 \wedge (\omega'_{32} - \omega_{32}) + \sigma_2 \wedge (\omega'_{31} - \omega_{31}).$$

By choosing the origin in the interior of S, the factor $\mathbf{X} \cdot \mathbf{a}_3$ can be made everywhere positive by choosing the appropriate orientation of S and (S'). It will be shown below that $(\omega_{31} - \omega'_{31}) \wedge (\omega_{32} - \omega'_{32})$ is negative or zero, and is zero only if $\omega_{31} = \omega'_{31}$, $\omega_{32} = \omega'_{32}$. Once this is shown it follows that the right-hand side of (10.89) is also negative or zero, and hence of necessity zero since it changes sign when S and S' are interchanged [which must leave (10.89) still valid], while the left-hand side does not change sign. Hence the integrand on the left-hand side must vanish everywhere and $\omega_{31} = \omega'_{31}$, $\omega_{32} = \omega'_{32}$ holds everywhere, thus proving the theorem.

It remains to be shown that the wedge product occurring in the integrand on the left-hand side of (10.89) has the required property. To prove it, ω_{31} and ω_{32} are expressed, as in (10.36), in terms of σ_1 and σ_2:

$$\omega_{31} = a\sigma_1 + b\sigma_2,$$

$$\omega_{32} = b\sigma_1 + c\sigma_2,$$

and similarly for ω'_{31} and ω'_{32}, though with other coefficients a', b', c'. The wedge product to be investigated is

$$(\omega_{31} - \omega'_{31}) \wedge (\omega_{32} - \omega'_{32}) = [(a - a')(c - c') - (b - b')^2]\sigma_1 \wedge \sigma_2.$$

However, the second fundamental form is given, from (10.37) by

$$\mathrm{II} = a\sigma_1^2 + 2b\sigma_1\sigma_2 + c\sigma_2^2$$

and $K = ac - b^2 = K' = a'c' - b'^2 > 0$ since S and S' are isometric. The two matrices

$$\begin{pmatrix} a & b \\ b & c \end{pmatrix} \quad \text{and} \quad \begin{pmatrix} a' & b' \\ b' & c' \end{pmatrix}$$

are both *positive* definite since S and S' are oriented alike.   The algebraic lemma proved above for use in the discussion of the uniqueness theorem for Minkowski's problem is therefore applicable in the present case, since $\sigma = \sigma_1 \wedge \sigma_2$ does not vanish, and gives the desired result.   It is therefore proved that S and S' are congruent or symmetric.

PROBLEMS

1. Show that the equations for $d\omega_{23}$ and $d\omega_{31}$ in (10.27) are the Codazzi-Mainardi equations.

2. Show that $H = \frac{1}{2}(a + c)$, when a and c are the quantities defined in (10.36).

3. Explain why equations (10.58) and (10.59) are the same independent of the location of the origin.

4. On the basis of the formulas (10.54) and (10.55) consider the problem analogous to a problem formulated with respect to simple closed plane curves at the end of Chapter II, i.e., to give a means of deciding which side of a simple closed surface in E^3 is the inside, say, when measurements are possible only in a neighborhood of the surface.   Note that if the sign of dA/dt is used for the purpose, then the sign of $\iint_S H \, dA$ comes into question.   Show that this integral is always positive for a convex surface when the orientation is chosen so that the surface normal is the inner normal.   Is $\iint_S H \, dA$ always positive for a surface of genus zero?   (The author doubts it, but does not know the answer.)   Show that $\iint_S H \, dA$ is positive for the ordinary torus obtained by rotating a circle about an axis.   Show that if the genus g is large enough the integral $\iint_S H \, dA$ can be made negative if the surfaces have an appropriate form.   Show that the interior region of any simple closed surface of known genus can always be made manifest by measuring volumes and then determining d^3V/dt^3, in connection with the known value of $\iint_S K \, dA$.

APPENDIX A

Tensor Algebra in Affine, Euclidean, and Minkowski Spaces

1. Introduction

The purpose of this appendix is to summarize briefly those algebraic ideas and facts concerning affine, Euclidean, and Minkowski spaces that are of service in differential geometry, without attempting to develop them in detail. (Useful references are the books of Gelfand [G.2], Chapter IV, and Efimov [E.2], Chapter VII.) These ideas are very important for some parts of differential geometry because many of its concepts are introduced by analogy with these simpler geometries, and many of the basic formal tools are carried over from them. These matters are treated here in order to provide convenient references to specific formulas without interrupting the main course of the developments in differential geometry. Also, even though these geometries belong, on the formal side, essentially to linear algebra (a knowledge of which is taken for granted in this book), it has some point to underline here a few of the geometrical concepts that are often passed over rather lightly in books dealing only with linear algebra.

2. Geometry in an Affine Space

The starting point of the discussion is an n-dimensional linear vector space V of elements, or vectors, $\mathbf{x}$; it is assumed here that the scalars are the real numbers. In such a space bases of linearly independent vectors $\mathbf{v}_i$, $i = 1, 2, \ldots, n$, are assumed to exist such that every vector is a linear combination of the basis vectors:

$$(1) \qquad \mathbf{x} = \xi^i \mathbf{v}_i = \sum_{i=1}^{n} \xi^i \mathbf{v}_i,$$

with ξ^i uniquely defined scalars, and conversely all possible vectors are obtained in allowing the "components" ξ^i to range over all real values. (Here, and in what follows, the summation convention is used, i.e., if a subscript and a superscript are repeated in a term, a summation over n is implied, as indicated.) The numbers ξ^i are called the components of the vector. A change to a different set of basis vectors $\mathbf{v}_i'$ must then lead to the following n relations with the original basis vectors:

$$(2) \qquad\qquad \mathbf{v}_i' = c_i^k \mathbf{v}_k, \qquad i = 1, 2, \ldots, n,$$

and the determinant of the matrix (c_i^k) must be different from zero since the vectors $\mathbf{v}_i'$ are supposed linearly independent. The linear vector space is by definition the same, no matter what basis vectors might be used to describe it.

The vectors in linear algebra are in general regarded as based on a certain fixed point O, the origin, and it is customary to say that other points of the space are fixed by the end points of the vectors issuing from the origin. In *affine geometry* it is this space of points that is in question, but it is not in general desirable to distinguish the point O from others. In fact, it is assumed that the whole space can be described by making use of the same set $\mathbf{v}_i$ of basis vectors at any point, and the basis vectors at O and O' are said to form *parallel* systems. If the same point P is located by the vector $\mathbf{x}$ with respect to point O and by $\mathbf{x}'$ with respect to O' then, by definition, $\mathbf{x} = \mathbf{x}' + \mathbf{a}$ with $\mathbf{a}$ the vector locating point O' relative to O. Thus the space can be regarded as homogeneous in the sense that two objects, one described relative to O, the other to O', can be defined as essentially the same if pairs of corresponding points defined by the vectors $\mathbf{x}_\alpha$ and $\mathbf{x}_\alpha'$ all differ by the same constant vector $\mathbf{a}$, i.e., if $\xi_\alpha^i = \xi_\alpha'^i + a^i$, with a^i independent of the pairs of points.

In effect, a notion of parallel displacement has been introduced in the space which makes it possible to compare vectors not based on the same point and to say whether they are parallel or not: it suffices for that to see whether the two vectors have the same components relative to systems of parallel basis vectors, which exist by assumption. This seemingly rather obvious statement is, however, particularly important in relation to much of this book because it is necessary to deal with spaces in which it is very desirable to have a notion of parallelism, but which is not so readily introduced because, for one thing, the whole space cannot be described by a single coordinate system based at one point, as is the case here.

One of the reasons why it is important to have a notion of parallelism available is that with its aid it is possible to compare vectors at different points, and in particular to define differentiation processes with respect to *fields* of vectors defined over regions of the space. In earlier chapters of this book considerable emphasis was placed on the fact that vectors can be combined algebraically in general only when they are based on the same point, simply because in most situations in mechanics and physics it is without meaning to do otherwise. For example, consider the velocity vectors of different particles of a mechanical system—a region occupied by a fluid, say; clearly, it makes no sense to add vectorially the velocities of different mass points. Or, consider an elastic solid subjected to force vectors at two different points; again it would be inappropriate to replace the two forces by their vector sum placed at some point since such a procedure would in general lead to a different distribution of stress in the body and thus to a physical situation different from the original. The instances in which vectors at different points produce the same physical effects, and thus could be regarded as the same, are very rare indeed: they consist of parallel displacement of rigid bodies in Euclidean spaces, and of certain very special entities, such as the couple vector, in rigid-body dynamics.

Nevertheless, in geometry and continuum mechanics it is frequently vital to be able to compare vectors at different points, although usually at points arbitrarily near to some given point. The reason for that is that physical and geometrical laws refer in the great majority of cases to quantities specified in terms of vector *fields* over a region, and these quantities obey laws formulated in terms of differential equations that require derivatives of vectors to be taken when the base points of the vectors vary. For example, Newton's second law for the motion of a single particle has in it the acceleration vector $d\mathbf{v}/dt$, which is obtained by comparing the velocity vectors $\mathbf{v}(t + \Delta t)$ and $\mathbf{v}(t)$ at positions of the particle fixed by times t and $t + \Delta t$. This can be done, *as a matter of convention*, by using the fact that in Euclidean space (a special kind of affine space) a well-defined concept of parallelism of vectors at different points exists. The acceleration vector $d\mathbf{v}/dt$, which figures in Newton's law, is obtained by regarding the vectors $\mathbf{v}(t + \Delta t)$ as having been moved parallel to themselves to the point corresponding to the time t; this set of vectors then can be differentiated with respect to the scalar parameter t on which it depends. The essential point in the present context is that such a process is not a matter of course, but rather a matter of definition; whether it is reasonable or not then hinges on whether an interesting and logically consistent body of geometrical or physical doctrine results. The fact that the process of parallel transport offers itself so naturally in Euclidean spaces as a convenient way to *define* such differentiation processes should not be interpreted as mitigating in any way the firm insistence in this book that vectors should not be combined unless based on the same point. Instead, the device of a preliminary parallel displacement is introduced by definition as a means of bringing them together so that they can be so combined. In spaces other than affine spaces—and a good part of this book is concerned with such spaces—one of the first tasks to be performed in introducing appropriate geometric concepts is that of defining a concept of local parallelism of vectors in order that differentiation processes that yield vectors, or tensors, from vector fields can be defined. The possibility of comparing vectors at distant points in such spaces becomes thus a *problem*, which has not the obvious solution available in affine and Euclidean spaces.

The discussion of geometry in an affine space is now continued. In such a space a *straight line* through a point O is given by the end points of all vectors $\mathbf{x}(\xi)$ of the form $\xi\mathbf{v}$, with $\mathbf{v}$ any nonzero vector at O, when ξ takes on all real values. If $\mathbf{v}_1$ and $\mathbf{v}_2$ are linearly independent vectors at O, then a *plane* is defined by the set of all points given by vectors of the form $\xi_1\mathbf{v}_1 + \xi_2\mathbf{v}_2$ when ξ_1 and ξ_2 range independently over all real values. It is reasonable to say that the plane results by parallel transport of the line given by $\xi_1\mathbf{v}_1$ to all points of the line defined by $\xi_2\mathbf{v}_2$. In similar fashion linear subspaces of all dimensions could be generated. Two such subspaces are defined as parallel if they are generated at different points by systems of vectors that result by parallel displacement.

In affine geometry there are neither distinguished points nor distinguished coordinate systems. Thus if $(O; \mathbf{v}_i)$ and $(O'; \mathbf{v}_i')$ are two points together with a coordinate system defined at each, it follows that any point of the space can be represented by a vector $\mathbf{x} = \xi^i\mathbf{v}_i$ at O or by a vector $\mathbf{x}' = \xi'^i\mathbf{v}_i'$ at O'. Suppose for the moment that O and O' coincide. Since any two systems of basis vectors are related by the equations (2), with $\det(c_i^k) \neq 0$, it follows that the components of an arbitrary vector $\mathbf{x} = \xi^i\mathbf{v}_i$ are related by the formulas

(3) $$\xi^i = c_k^i \xi'^k,$$

as can be seen from the assumed identity $\xi^i\mathbf{v}_i = \xi'^i\mathbf{v}_i'$ and the linear independence of basis vectors. If the points O and O' do not coincide, but rather O' is located relative to O by the vector $\alpha^i\mathbf{v}_i$, it follows that the relation between the coordinates ξ^i and ξ'^i of the same point are related by the equations

(4) $$\xi^i = c_k^i \xi'^k + \alpha^i.$$

Affine geometry can now be treated analytically in representing points by the coordinates of the vectors representing them; and configurations of points in the space which qualify as the objects of affine geometry must be such as to be preserved when the points are subjected to arbitrary transformations of the form (4).

The formulas (4) can be interpreted in a different fashion by regarding them as defining a *mapping* of the points ξ^i into points ξ'^i, or of vectors $\mathbf{x}$ into vectors $\mathbf{x}'$, in employing a *fixed* coordinate system. Two configurations which are transformed into each other by any such transformation are then regarded as identical. Among such configurations are the following: systems of linearly independent vectors, linear subspaces of arbitrary dimension, a coordinate system $(O; \mathbf{v}_i)$ into another $(O'; \mathbf{v}_i')$, parallel subspaces of all dimensions. This is all exactly analogous to the role played by the concept of *congruence* in *Euclidean geometry*, in which configurations in an affine space are regarded as identical when the transformations are restricted to the *orthogonal transformations*, which are such that the inverse of the matrix (c_k^i) is its transpose (c_i^k). Such transformations come about through introduction of the notion of *length* of a vector—a notion that is alien to affine geometry— and the orthogonal transformations are just those that preserve the lengths of all vectors. (How this is done will be explained briefly a little later.) Thus two configurations which go into each other by an orthogonal transformation would be such that vectors drawn between pairs of corresponding points would have the same length—and thus the two configurations would be congruent in the same sense as in the familiar Euclidean geometry of the plane.

In addition to the vectors in affine spaces—which are the prototypes of all invariant objects studied in this book—there exists a whole hierarchy of more complicated invariant objects called *tensors*, and it is important to recall their definitions and properties here, since similar invariants are defined by analogy in n-dimensional manifolds (cf. Chapter IX). For example, it turns out in a natural way that the derivative of a field of vectors defined in a manifold is not in general a vector, but rather a tensor of higher rank, and thus an essentially more complicated thing.

Tensor algebra is carried out in the first instance in an affine space—or rather in what is sometimes called a *centered affine space* in which all vectors considered emanate from the same point O—rather than in a linear space with a Euclidean metric. The same course is followed here, since experience has shown that the geometry of more general manifolds can also be carried out in a useful way without first introducing a metric of any kind. Afterward, the introduction of a metric leads in the geometry of manifolds to a much richer geometrical theory, just as is true in the passage from affine geometry to Euclidean geometry through the introduction of the appropriate metric.

3. Tensor Algebra in Centered Affine Spaces

The algebra of vectors is discussed here in a form that is convenient for the purposes of differential geometry.

Experience has shown that there are advantages to be gained by basing the treatment of vectors (and of the more complicated invariant objects called tensors to be introduced shortly) in terms of the algebra of vectors in a linear vector space V together with what is called its dual space $\bar{V}$. (In what follows the presentation of tensor algebra given by Gelfand [G.2] was followed in the main. This book is recommended as a reference for linear algebra in general.) The definition of the dual space $\bar{V}$ associated with V is based on the notion of a *real-valued linear function* $f(\mathbf{x})$ of a vector $\mathbf{x}$ in V. The requirement of linearity is achieved as would be expected by imposing the following conditions:

1.
$$f(\mathbf{x} + \mathbf{y}) = f(\mathbf{x}) + f(\mathbf{y}),$$

2.
$$f(\lambda\mathbf{x}) = \lambda f(\mathbf{x}),$$

when $\mathbf{x}$ and $\mathbf{y}$ are arbitrary vectors in V (which implies that f is of necessity defined for all $\mathbf{x}$ in V), and λ is an arbitrary real number. If $f(\mathbf{x})$ and $g(\mathbf{x})$ denote two linear functions their sum $f + g$ evaluated for $\mathbf{x}$ is defined as the real number that results upon adding $f(\mathbf{x})$ and $g(\mathbf{x})$, and the product αf is defined as the number $\alpha f(\mathbf{x})$. It is clear that such sums and products are again linear functions. The dual space $\overline{V}$ is defined as follows:

DEFINITION 1. *The dual space $\overline{V}$ of V is the space whose elements are the linear functions $f(\mathbf{x})$, with addition and scalar multiplication in $\overline{V}$ defined in accordance with the rules for addition and for multiplication by scalars of the linear functions.*

It is clear that $\overline{V}$ is a vector space. It is to be checked that its dimension is n, i.e., that its dimension is the same as that of V. This, however, is readily done by making use of a basis $\mathbf{v}_1, \mathbf{v}_2, \ldots, v_n$ in the space V in terms of which an arbitrary vector $\mathbf{x}$ is given by

$$\mathbf{x} = \xi^i \mathbf{v}_i.$$

If, then, $f(\mathbf{x})$ is any linear function it can be expressed in the form

(5)
$$f(\mathbf{x}) = f(\xi^1\mathbf{v}_1 + \xi^2\mathbf{v}_2 + \ldots + \xi^n\mathbf{v}_n)$$

$$= \xi^1 f(\mathbf{v}_1) + \xi^2 f(\mathbf{v}_2) + \ldots + \xi^n f(\mathbf{v}_n)$$

$$= a_i \xi^i,$$

when the rules of linearity are used, and the numbers a_i are defined as follows:

(6)
$$a_i = f(\mathbf{v}_i).$$

Conversely, any set of numbers a_i gives rise through (6) to a uniquely defined linear function $f(\mathbf{x})$—as is readily seen. It is therefore clear that from a given basis $\mathbf{v}_i$ a linear function is determined in a unique way by an ordered n-tuple a_i of real numbers, which give the values $f(\mathbf{v}_i)$ of f with respect to a basis $\mathbf{v}_i$ in V. Also, if f is determined in this way by a_i and g by b_i, then $f + g$ is clearly determined by $a_i + b_i$ and αf by αa_i. Since the vectors f in $\overline{V}$ are determined uniquely by the n-tuples $a_1, a_2, \ldots, a_n$, it follows that *the dimension of the space $\overline{V}$ is the same as that of V*. Thus $\overline{V}$ is a linear space spanned by n vectors that are denoted by f^i. The vectors in $\overline{V}$ are *not* printed in bold-faced type as is the general custom, and also the vectors forming a basis are denoted by *superscripts* rather than by subscripts. On the other hand, the numbers a_i that fix a vector f in $\bar{V}$—and hence serve as coordinates of it—are distinguished by subscripts while the vectors $\mathbf{x}$ in V are fixed by the coordinates ξ^i, which are distinguished by superscripts. The following definition is now introduced:

DEFINITION 2. *Vectors in V are called contravariant vectors and vectors in the dual space $\bar{V}$ are called covariant vectors.*

Thus the two types of vectors are defined in different (though related) spaces, and hence are basically different entities. The distinction is, as it were, underlined by the difference in notation.

It is convenient, and also rather natural, to associate with a given set of basis vectors $\mathbf{v}_i$ in V a special and uniquely determined basis f^i in $\bar{V}$, called the dual basis and defined as follows. The n vectors f^i are defined by requiring the corresponding linear functions $f^i(\mathbf{x})$ to have the values $f^i(\mathbf{x}) = \xi^i$, i.e., the value of the function $f^i(\mathbf{x})$ is defined to be the value of the ith coordinate of $\mathbf{x}$ relative to a basis $\mathbf{v}_k$. Since the following formula holds in general:

$$f^i(\xi^k \mathbf{v}_k) = \xi^1 f^i(\mathbf{v}_1) + \xi^2 f^i(\mathbf{v}_2) + \ldots + \xi^n f^i(\mathbf{v}_n),$$

it follows that

$$(7) \qquad\qquad f^i(\mathbf{v}_k) = \delta_k^i$$

in the special case in which $\mathbf{x}$ is a basis vector $\mathbf{v}_k$. That such a real-valued function defines also a linear function for all $\mathbf{x}$ in V is easily seen; hence the function f^i determines a covariant vector. It remains to be shown that the n vectors f^i so defined form a basis in $\bar{V}$. Since each covariant vector f^i is determined by the n-tuple of real numbers a_k^i defined by $a_k^i = f^i(\mathbf{v}_k)$ [cf. (6)] and since $f^i(\mathbf{v}_k) = \delta_k^i$ in view of the definition of the vectors f^i, it is clear that the covariant vectors f^i are linearly independent, since the corresponding n-tuples of real numbers $a_k^i = \delta_k^i$ form a linearly independent set.

Thus once dual bases $\mathbf{v}_i$ and f^k have been introduced, the following formulas hold with respect to vectors in V:

$$f^i(\mathbf{v}_k) = \delta_k^i,$$
$$(8) \qquad\qquad f^i(\mathbf{x}) = f^i(\xi^k \mathbf{v}_k) = \xi^k \delta_k^i = \xi^i,$$
$$f(\mathbf{x}) = f(\xi^k \mathbf{v}_k) = \xi^k f(\mathbf{v}_k).$$

It is of considerable interest for the sake of later more general developments to make an observation in relation to the notion of the space of covariant vectors. It was defined by making use of a linear space V of contravariant vectors, and resulted as the space $\bar{V}$ of linear functions of those vectors. But *V can also be regarded as essentially the dual $\overline{\overline{V}}$ of V*, as will be shown in a moment. In that case it is clear that the primitive notion from which these developments originate is the notion of an invariant linear function of vectors in a linear vector space. It is by generalizing to invariant *multilinear* functions of vectors in both V and $\overline{V}$ that more general invariant objects are defined.

To see, then, that V and $\overline{V}$ are thus in a certain sense interchangeable, it is shown that if $\phi(f)$ is a linear function of the vectors f in $\overline{V}$ that a unique vector $\mathbf{x}$ in V exists such that $\phi(f) = f(\mathbf{x})$ for each vector f in $\overline{V}$, and that the vectors $\mathbf{x}$ thus found fill the space V. To this end choose a basis $\mathbf{c}_i$ in V and its dual f^i in $\overline{V}$. If the coordinates of f are η_i it follows that $\phi(f) = a^i \eta_i$, say, where $a^i = \phi(f^i)$. Consider the vector $\mathbf{x} = a^i \mathbf{c}_i$, for which $f(\mathbf{x}) = a^i \eta_i$, in view of the definition of the dual basis. In that case $\phi(f) = f(\mathbf{x})$, as was to be shown. Clearly each vector $\mathbf{x}$ in V would result from some vector f in $\overline{V}$.

4. Effect of a Change of Basis

It is of fundamental importance to study transformations from one set of basis vectors to another in V and $\overline{V}$. To this end a new basis $\mathbf{v}'_i$ is introduced in V by setting

$$(9) \qquad \mathbf{v}'_i = c_i^k \mathbf{v}_k$$

with (c_i^k) a nonsingular matrix, so that the vectors $\mathbf{v}'_i$ do indeed form a basis. Along with (9) the following inverse transformation exists:

$$(10) \qquad \mathbf{v}_j = \gamma_j^k \mathbf{v}'_k,$$

and if each of the formulas (9) and (10) in turn is substituted in the other, it is found (because of the linear independence of the basis vectors) that

$$(11) \qquad \gamma_j^k c_k^i = \delta_j^i, \qquad c_k^i \gamma_i^j = \delta_k^j,$$

is the expression of the fact that the matrices (c_i^k) and (γ_i^k) are inverse to one another. A fixed vector $\mathbf{x}$ can be represented in both, as follows:

$$\mathbf{x} = \xi^i \mathbf{v}_i = \xi'^i \mathbf{v}'_i = \xi'^i c_i^k \mathbf{v}_k = \xi^i \gamma_i^k \mathbf{v}'_k,$$

from which

$$(12) \qquad \xi'^k = \gamma_j^k \xi^j, \qquad \xi^i = c_k^i \xi'^k$$

result. Thus, upon comparing (9) and (10) with (12), it is seen that the coordinates do not transform in the same way as the basis vectors, but rather the transformed components ξ'^k could be written in matrix notation as follows:

$$\xi' = \xi(C^{-1}), \quad \text{where} \quad \xi' \equiv (\xi'^1, \xi'^2, \ldots, \xi'^n), \quad \xi \equiv (\xi^1, \xi^2, \ldots, \xi^n),$$

in which C is the matrix (c_i^k) of the basis transformation. Such transformations are said to be *contragredient* to one another.

Suppose now that f^i is the basis in $\overline{V}$ which is dual to the basis v_k in V; the matrix is now to be found which expresses the change from the basis f^i to f'^i when f'^i is the basis dual to v_i'—such a nonsingular matrix exists since the dual basis itself always exists. The inverse transform is first written down:

$$(13) \qquad\qquad f^k = u_{i}^{k} f'^i,$$

and $f^k(v_i')$ is computed in two ways as follows:

$$f^k(v_i') = f^k(c_i^\alpha v_\alpha) = c_i^\alpha f^k(v_\alpha) = c_i^\alpha \delta_\alpha^k = c_i^k,$$
$$f^k(v_i') = u_j^k f'^j(v_i') = u_j^k \delta_i^j = u_i^k,$$

when use is made of the relation (7) which characterizes a dual basis. It follows therefore that $c_i^k = u_i^k$, and hence that the matrices (c_i^k) and (u_i^k) in (9) and (13) are transposed to each other since the right-hand sides of these equations are summed on different indices. It follows therefore from (13) that the matrix (γ_i^k) in the relation

$$(14) \qquad\qquad f'^k = \gamma_i^k f^i$$

is the inverse of the matrix (c_i^k) *in* (9), *and this is correctly designated as the matrix* (γ_i^k) which figures in (12).

It is of importance next to discuss the effect of a change of basis on the coordinates of vectors in $\overline{V}$ as well as in V. Suppose that

$$\mathbf{x} = \xi^i v_i = \xi'^i v_i';$$

it follows that

$$f^i(\mathbf{x}) = f^i(\xi^k v_k) = \xi^i,$$
$$f'^i(\mathbf{x}) = f'^i(\xi'^k v_k') = \xi'^i,$$

by the stipulations which fix a dual basis f^i in $\overline{V}$. On the other hand from (14) it follows that

$$(15) \qquad\qquad \xi'^i = f'^i(\mathbf{x}) = \gamma_k^i f^k(\mathbf{x}) = \gamma_k^i \xi^k;$$

and hence [cf. (12)] *the coordinates* of vectors in V transform like the *vectors* of the dual basis in $\overline{V}$. In the same manner it is shown that the *coordinates* of vectors in $\overline{V}$ transform like the *basis vectors* in V, i.e., that

$$(16) \qquad\qquad a_i' = c_i^k a_k.$$

It is now seen that quantities characterized by upper indices do not transform in the same way as those with lower indices. In particular the coordinates of vectors in the dual space $\bar{V}$ transform like the basis vectors in V, or, as it is customary to say, *cogrediently*; hence it seems fitting to call these vectors *covariant vectors*. This use of upper and lower indices to distinguish the two different types of vectors facilitates the systematic use of the summation convention.

5. Definition of Tensors

The development of tensor algebra proceeds by making use of the spaces V and $\bar{V}$ in order to define real-valued *multilinear functions* $l(\mathbf{x}, \mathbf{y}, \ldots;$ $f, g, \ldots)$ in which the arguments of the type $\mathbf{x}, \mathbf{y}, \ldots$ are vectors which range independently over the whole of the space V, while those of type $f, g, \ldots$ range over its dual $\bar{V}$. The term multilinear function means exactly what would be expected, i.e., that the function is linear in each of its arguments for arbitrary but fixed values of all of the other arguments, and thus in accordance with the conditions 1 and 2 given above for the special case $l(\mathbf{x})$. If, for example, all vectors but f are kept fixed, then

$$l(\mathbf{x}, \mathbf{y}, \ldots \alpha f + \beta f', g, \ldots)$$
$$= \alpha l(\mathbf{x}, \mathbf{y}, \ldots; f, g, \ldots) + \beta l(\mathbf{x}, \mathbf{y}, \ldots; f', g, \ldots),$$

is assumed to hold when α, β are arbitrary real numbers. It has some point to say that l is of type (p, q) if it has p arguments in V and q arguments in $\bar{V}$.

As before, it is of great interest to investigate the representations of a multilinear function—which it need hardly be said is by definition an invariant, or a *scalar*—in terms of a set of basis vectors in V and in $\bar{V}$. It is convenient to do that in the special case in which the basis in $\bar{V}$ is the dual basis to one in V. Thus, if $\mathbf{v}_i$ and f^i are two such bases, and hence

(17) $$\mathbf{x} = \xi^i \mathbf{v}_i, \qquad \mathbf{y} = \eta^j \mathbf{v}_j, \qquad f = a_k f^k$$

are representations of arbitrary vectors in V and $\bar{V}$, a multilinear function of the type $l(\mathbf{x}, \mathbf{y}; f)$—to take a typical example—would have the representation

(18) $$l(\mathbf{x}, \mathbf{y}; f) = l(\xi^i \mathbf{v}_i, \eta^j \mathbf{v}_j; a_k f^k)$$
$$= \xi^i \eta^j a_k l(\mathbf{v}_i, \mathbf{v}_j; f^k);$$

or, upon setting

(19) $$A_{ij}^k = l(\mathbf{v}_i, \mathbf{v}_j; f^k),$$

the result would be

(20) $$l(\mathbf{x}, \mathbf{y}; f) = A_{ij}^k \xi^i \eta^j a_k.$$

Thus (19) shows how the coefficients A^k_{ij}, which evidently define l in terms of bases in V and $\bar{V}$, are determined.   In general such representations have the form

(21) $$l(\mathbf{x}, \mathbf{y}, \ldots; f, g, \ldots) = A^{rs\cdots}_{ij\cdots} \xi^i \eta^j \cdots a_r b_s \cdots,$$

with

(22) $$A^{rs\cdots}_{ij\cdots} = l(\mathbf{v}_i, \mathbf{v}_j, \ldots; f^r, f^s, \ldots).$$

It is now of interest to investigate how the coefficients $A^{rs\cdots}_{ij\cdots}$ change when the basis vectors $\mathbf{v}_i$, f^i (with f^i dual to $\mathbf{v}_i$) are changed in accordance with the basic rules:

(23) $$\mathbf{v}'_\alpha = c^\beta_\alpha \mathbf{v}_\beta, \qquad f'^\beta = \gamma^\beta_\alpha f^\alpha,$$

so that the coordinates of vectors change according to the formulas

$$a'_i = c^k_i a_k, \qquad \xi'^j = \gamma^j_k \xi^k,$$

from which

(24) $$\gamma^i_r a'_i = a_r, \qquad c^i_j \xi'^j = \xi^i$$

in conformity with the developments above in which (γ^β_α) is the inverse of (c^β_α).   To compute the coefficients $A'^{rs\cdots}_{ij\cdots}$, which determine the multilinear function l in the new bases, it is merely necessary to insert the formulas (24) in the equation (21), and then to compare the coefficients of like terms in the equation analogous to (21):

$$l(\mathbf{x}, \mathbf{y}, \ldots; f, g, \ldots) = A'^{rs\cdots}_{ij\cdots} \xi'^i \eta'^j \cdots a'_r b'_s \cdots.$$

The result is

(25) $$A'^{rs\cdots}_{ij\cdots} = c^\alpha_i c^\beta_j \cdots \gamma^r_\sigma \gamma^s_\tau \cdots A^{\sigma\tau\cdots}_{\alpha\beta\cdots}.$$

Thus each lower index in $A'^{rs\cdots}_{ij\cdots}$ gives rise to a multiplying factor of the type c^α_i with summation on the upper index, and each upper index to a factor of the type γ^r_σ with summation on the lower index.

A multilinear function is by its very nature an invariant, since its value is determined for a given set of vectors from V and $\bar{V}$ independent of a special coordinate representation.   It might be thought therefore that discussions of invariants should be conducted without reference to a special basis in either V or $\bar{V}$.   It is possible to do so, but for concrete applications to geometry and physics it is much more practical to make use of a representation in terms of a basis,[1] but then to understand clearly that quite specific transformation

---

[1] In geometry and physics it is often the components of vectors and tensors themselves which have the most direct geometrical or physical significance, and also they are the quantities with which it is usually necessary to deal in coming to grips with concrete problems.   For example, in elasticity the state of stress at a point is described by a tensor σ_{ik} which is covariant of order two, and the physical significance of its individual components is very important; the associated scalar bilinear form is, on the other hand, of minor importance from the physical point of view.

laws for a change of basis must hold. Consequently the following definition of the concept of a tensor is introduced (though a "coordinate-free" definition could be given).

DEFINITION 3. *A tensor is defined by a set of n^{p+q} real numbers $A_{ij\cdots}^{rs\cdots}$ with p subscripts and q superscripts assigned to a definite basis in the n-dimensional vector space V. Under a change of basis defined by a matrix (c_i^j) these numbers are required to transform to numbers $A'^{ij\cdots}_{rs\cdots}$ according to the equations (25), in which the numbers γ_i^j are the elements of the inverse of the matrix (c_i^j). The tensor is said to be p times covariant and q times contravariant, and to be of rank $p+q$.*

Although it is somewhat ungrammatical to do so, nevertheless it will often be said for the sake of brevity that the numbers $A_{ij\cdots}^{rs\cdots}$ *are* a tensor.

In effect, tensors are by the above definition nothing but sets of coefficients of invariant multilinear forms. Some writers on tensor algebra prefer to define them as the elements of a linear vector space of dimension n^{p+q} which is obtained from V and $\bar{V}$ by the construction of what is called a tensor product formed from these spaces. In that formulation, the invariant character of tensors of any order is basically the same as that of a vector, though not of a vector in the original spaces.

A few conclusions of importance concerning tensors can be drawn at once directly from the definition. Clearly, *two tensors of the same type will be identical if they have the same components with respect to any one coordinate system*, since then they would have components which are the same in all coordinate systems. Also, *a tensor which has all components zero in any coordinate system* has components which are zero in all coordinate systems, since the transformed components are given by linear homogeneous functions of the original components.

Two concrete examples of tensors have already been given. Relative to a given basis in V a vector $\mathbf{x}$ is determined by n numbers ξ^i which transform according to the rule

$$\xi'^i = \gamma_j^i \xi^j,$$

so that the set of numbers ξ^i defines a tensor that is of rank one and contravariant; thus it is natural to call it a *contravariant vector*. In the same way the set of numbers a_i that transforms according to the rule

$$a_i' = c_i^k a_k$$

defines a tensor of rank one called a *covariant vector*.

An interesting example of a tensor of higher rank results from consideration of a linear transformation in V and the matrices that can be used to define it with respect to all possible bases in V. In fact, if (a_i^k) is such a matrix of a linear transformation L_i, then $L\mathbf{v}_i = a_i^k \mathbf{v}_k$, and if a new basis is introduced by setting $v_i' = c_i^\alpha v_\alpha$, it is easily seen that the matrix $(a_i'^k)$ of L relative to this basis is given by

$$a_i'^k = c_i^\alpha \gamma_\beta^k a_\alpha^\beta,$$

and this means that the numbers a_i^k obey the law of transformation which makes them represent a tensor of rank two, once covariant and once contravariant. An important special case is that of the matrix of the identity transformation I which, relative to any basis, has the matrix δ_i^k. (The fact that the set of numbers δ_i^k does indeed define such a tensor with components which have the same values in all coordinate systems should be verified directly from the general definition.)

There are various algebraic operations to be defined for tensors which result in new tensors, and which are often useful. First of all let

$$A_{ij\cdots}^{rs\cdots} \quad \text{and} \quad B_{ij\cdots}^{rs\cdots}$$

be two tensors of the same rank and type (i.e., with the same number of covariant and of contravariant indices). Then *the sum*

$$C_{ij\cdots}^{rs\cdots} = A_{ij\cdots}^{rs\cdots} + B_{ij\cdots}^{rs\cdots}$$

is readily seen, on the basis of the definition, to be a tensor. A *product* of tensors can also be defined, and be readily seen to be a tensor; such a product is defined by

$$C_{ij\cdots kl}^{rs\cdots tu} = A_{ij\cdots}^{rs\cdots} \cdot B_{kl\cdots}^{tu\cdots};$$

it is a tensor with a rank equal to the sum of the ranks of the factors. This affords a means of defining tensors of any rank: they can be obtained by multiplying any number of covariant and contravariant vectors, e.g.,

$$C_{ij\cdots}^{rs\cdots} = A_i B_j \cdots L^r \cdot M^{s\cdots}.$$

Multiplication by a scalar, i.e., by a tensor of order zero, also yields a tensor. (However, not every tensor can be obtained as such a product of tensors of lower order.)

A different type of operation with tensors which is perhaps of more frequent occurrence than addition and multiplication is that called *contraction*. This means that in a tensor $A_{ij\cdots}^{rs\cdots}$, of rank n say, an upper index is made the same as a lower index, which in turn means of course that a summation (yielding n terms) is implied. It will be shown now that the result is a new tensor, which then evidently would be of rank two less than the original. Consider, for example, the tensor A_{rst}^{ij} with the contraction A_{rsj}^{ij}. From the definition of a tensor the components in a new coordinate system are given by

$$A_{\mu\nu\beta}'^{\alpha\beta} = c_\mu^r c_\nu^s c_\beta^t \gamma_i^\alpha \gamma_j^\beta A_{rst}^{ij},$$

and since $c_\beta^t \gamma_j^\beta = \delta_j^t$ [cf. (11)] it follows that

$$A_{\mu\nu\beta}^{\prime\alpha\beta} = c_\mu^r c_\nu^s \gamma_i^\alpha A_{rsj}^{ij},$$

and this means that A_{rsj}^{ij} is indeed a tensor. Clearly, the general case could be dealt with in the same fashion. It is to be noted also—a fact of importance—that this proof would not hold in general if a contraction on two indices of the same type were to be attempted.

In practice the last two processes are often combined. Consider, for example, a familiar case, i.e. that of the first fundamental form of a surface with coefficients g_{ik}. Suppose it is known that the functions g_{ik} form a covariant tensor of rank two, and that the differentials du^i define a contravariant tensor of rank one (and thus called here a vector). It would follow that $g_{\alpha\beta}\, du^i\, du^k$ would be a tensor, and if it is contracted twice to obtain $g_{ik}\, du^i\, du^k$ the result is an invariant, a scalar, which has a value independent of all coordinate systems. In fact, *any tensor with the same number of covariant and contravariant indices leads to a scalar by contraction on all of its indices*, and this is a very important matter indeed.

It happens frequently that a set of functions can be identified as the components of a tensor without the necessity for an explicit verification that the transformation law holds. A particular process by which that can be done is best illustrated by a concrete example, which however indicates clearly the nature of the general process. Suppose that the quantities A_{ijk} form a set of n^3 numbers depending on bases in V and $\overline{V}$, and that the numbers B^s are the components of a contravariant tensor. Furthermore suppose that the sum $A_{\alpha jk}B^\alpha$ is a tensor of type C_j^k when B^α is any *arbitrarily chosen tensor*. It follows—and that is the content of what is called the *quotient law* for tensors—that the *quantities A_{ijk} are the components of a tensor of type A_{ij}^k*. The proof is straightforward. Since C_j^k is a tensor, it follows that its components $C_j'^k$ in a new coordinate system are given [cf. (25)] by

$$C_j'^k = c_j^\alpha \gamma_\sigma^k C_\alpha^\sigma;$$

and hence

$$A_{rjk}'B'^r = c_j^\alpha \gamma_\sigma^k C_\alpha^\sigma = c_j^\alpha \gamma_\sigma^k A_{r\alpha\sigma}B^r$$

$$= c_j^\alpha \gamma_\sigma^k c_\tau^r A_{r\alpha\sigma}B'^\tau,$$

and this in turn can be written

$$(A_{\tau jk}' - c_\tau^r c_j^\alpha \gamma_\sigma^k A_{r\alpha\sigma})B'^\tau = 0.$$

Since the components of the tensor B'^τ can be chosen arbitrarily (since that was the hypothesis with respect to B^α) it follows that the quantity in the parentheses vanishes; but that furnishes exactly the relations needed to identify the quantities $A_{\tau\alpha\sigma}$ as defining tensor of the type $A_{\tau j}^k$.

6. Tensor Algebra in Euclidean Spaces

It is of interest for later purposes to state here a few facts about Euclidean spaces; these are linear vector spaces in which a positive definite metric form is defined. Such a metric form can be introduced in an abstract way (see, for example, the book of Gelfand [G.2]), but all purposes of this book are served by defining the *metric form as a positive definite quadratic form* $(\mathbf{x}, \mathbf{x}) = g_{ik}\xi^i\xi^k$ in the contravariant components ξ^i of $\mathbf{x}$, with coefficients g_{ik}, which are the components of a covariant tensor of rank two, which is in addition assumed to be *symmetric*, i.e., $g_{ik} = g_{ki}$. (It is left as an exercise to show that $g'_{ik} = g'_{ki}$ continues to hold upon transforming to any new coordinates.) Consequently, the quadratic form is an invariant, as is known from the discussion above. It serves as a means to define angles and distances in the Euclidean space. Angles are defined by making use of the *associated bilinear form* $(\mathbf{x}, \mathbf{y}) = g_{ik}\xi^i\eta^k$—also called the *scalar product* of $\mathbf{x}$ and $\mathbf{y}$—in two vectors $\mathbf{x} = \xi^i\mathbf{v}_i$, $\mathbf{y} = \eta^i\mathbf{v}_i$; the angle ϕ between $\mathbf{x}$ and $\mathbf{y}$ is defined in an invariant form by the equation

$$(26) \qquad \cos\phi = \frac{(\mathbf{x}, \mathbf{y})}{\sqrt{(\mathbf{x}, \mathbf{x})}\,\sqrt{(\mathbf{y}, \mathbf{y})}}, \qquad 0 \le \phi \le \pi,$$

where the positive square root is always intended. This definition makes sense, since $(\mathbf{x}, \mathbf{x})$ is positive definite, since then $\cos\phi$ is always a real number, which is numerically less than or equal to one, because of the *Schwarz inequality*:

$$(27) \qquad (\mathbf{x}, \mathbf{y})^2 \le (\mathbf{x}, \mathbf{x})\cdot(\mathbf{y}, \mathbf{y}).$$

Two vectors are said to be *orthogonal* if $\cos\phi = 0$, which requires that $(\mathbf{x}, \mathbf{y}) = 0$. The *length of a vector* $\mathbf{x}$ is defined as the number $\sqrt{(\mathbf{x}, \mathbf{x})}$.

It is of interest to calculate the scalar product of any pair of basis vectors $\mathbf{v}_i$, $\mathbf{v}_k$ in the Euclidean space; it is given by

$$(28) \qquad (\mathbf{v}_i, \mathbf{v}_k) = g_{ik},$$

as can be readily seen. It is a well-known property of Euclidean spaces that they possess orthonormal bases, i.e., bases such that the vectors $\mathbf{v}_i$ are mutually orthogonal and all have the length one; in that case (28) yields

$$(29) \qquad (\mathbf{v}_i, \mathbf{v}_k) = \delta_{ik}.$$

If such a basis is used the quadratic form $(\mathbf{x}, \mathbf{x})$ becomes $\delta_{ik}\xi^i\xi^k$, which means that it is the sum of the squares of the components of $\mathbf{x}$; similarly, the scalar product $(\mathbf{x}, \mathbf{y})$ is given by $\delta_{ik}\xi^i\eta^k$.

The introduction of the metric form, or scalar product, on the basis of the tensor g_{ik}—often called the fundamental tensor—brings with it the possibility of defining quite explicitly the space $\overline{V}$ dual to V in terms of the scalar product. In fact the linear functions $f(\mathbf{x})$ which determine $\overline{V}$ can be put into a one-to-one correspondence with the vectors $\mathbf{y}$ of V by setting $f(\mathbf{x}) = (\mathbf{x}, \mathbf{y})$ for every vector $\mathbf{x}$ in V. That every vector $\mathbf{y}$ determines a linear function f in this way is obvious, since $(\mathbf{x}, \mathbf{y})$ is a bilinear form. To show that $\mathbf{y}$ is uniquely determined by $f(\mathbf{x})$, set $\mathbf{x} = \xi^i \mathbf{v}_i$ in terms of an orthonormal basis, for the sake of convenience, so that $f(\mathbf{x}) = a_i \xi^i$ [cf. (8)], and $a_i = f(\mathbf{v}_i)$. Define $\mathbf{y}$ as the vector whose coordinates in relation to $\mathbf{v}_i$ are the numbers a_i. Since the basis $\mathbf{v}_i$ is orthonormal it follows that $(\mathbf{x}, \mathbf{y}) = a_i \xi^i$, as was seen above. Thus $\mathbf{y}$ is a vector for which $(\mathbf{x}, \mathbf{y}) = f(\mathbf{x})$. If there were two vectors $\mathbf{y}_1, \mathbf{y}_2$ such that $(\mathbf{x}, \mathbf{y}_1) = (\mathbf{x}, \mathbf{y}_2)$, then $(\mathbf{x}, \mathbf{y}_1 - \mathbf{y}_2) = 0$ would hold for all vectors $\mathbf{x}$ and in particular for $\mathbf{x} = \mathbf{y}_1 - \mathbf{y}_2$; thus $\mathbf{y}_1 - \mathbf{y}_2 = 0$ results since the metric form (x, x) is assumed to be positive definite and hence vanishes only for $x = 0$. When $\mathbf{x}$ ranges over the whole of V it is clear that the vector $\mathbf{y}$ also ranges over the whole of V.

This discussion makes it clear that it is possible in a Euclidean space to regard the dual space as the same space as the original space, with the vectors $\mathbf{f}$ in $\overline{V}$ regarded as the uniquely determined set of vectors $\mathbf{y}$ in V obtained as above with the aid of the scalar product. It is, however, convenient to retain the notation f for the vectors of the dual space $\overline{V}$ even though it is now regarded as the same space as V, and to write $f = a_i f^i$ in terms of a basis f^i for an arbitrary vector of the dual space. On the other hand it is evidently also possible to represent the vectors f in terms of a basis $\mathbf{v}_i$ in V. To this end consider a basis vector f^i in $\overline{V}$ and therefore such that $f^i(\mathbf{x}) = (\mathbf{y}^i, \mathbf{x})$, defines a vector $\mathbf{y}^i$ in V, and identify f^i with $\mathbf{y}^i$. In terms of a basis $\mathbf{v}_\alpha$ the vectors $\mathbf{y}^i$ are given by $\mathbf{y}^i = \xi^{i\alpha} \mathbf{v}_\alpha$, and hence $\mathbf{v}_\alpha = \xi_{\alpha i} \mathbf{y}^i$ in terms of the matrix $\xi_{\alpha i}$ inverse to $\xi^{i\alpha}$; such an inverse exists since $\mathbf{y}^i$ and $\mathbf{v}_\alpha$ are two sets each of n linearly independent vectors in V. Since $g_{ik} = (\mathbf{v}_i, \mathbf{v}_k)$, by (28), and $(\mathbf{v}_i, \mathbf{v}_k) = (\xi_{ij} \mathbf{y}^j, \mathbf{v}_k) = \xi_{ij}(\mathbf{y}^j, \mathbf{v}_k) = \xi_{ij}(f^j, \mathbf{v}_k) = \xi_{ij} \delta^j_k = \xi_{ik}$, it follows that $g_{ik} = \xi_{ik}$. As a consequence the following formulas hold:

$$(30) \qquad\qquad \mathbf{v}_i = g_{ik} f^k, \qquad f^k = g^{ik} \mathbf{v}_i,$$

when (g^{ik}) is the matrix inverse to (g_{ik}). The operation carried out in (30) defines a covariant vector in terms of a contravariant vector with the aid of a multiplication by the fundamental tensor followed by a contraction, and also an analogous operation in the reverse sense. (It is left as an exercise to show that the quantities g^{ik} defined as the elements of the matrix inverse to (g_{ik}) are indeed the components of a contravariant tensor of second rank. It is also left as an exercise to show that the scalar product (f^i, f^k) is given by g^{ik}.)

Suppose next that a tensor of the type a^r_{ij} is given. By setting $b_{ijk} =$

$g_{i\alpha}a^{\alpha}_{jk}$ it is clear that a new tensor b_{ijk} of the same order has been defined (by multiplication and contraction of two tensors), but with a lowering of an index. Evidently this process could be extended, e.g., to the case $b^{ij\cdots rs\cdots} = g^{\alpha r}g^{\beta s}\cdots a^{ij\cdots}_{\alpha\beta\cdots}$. There are many occasions in this book when use of such possibilities is made. Thus it is seen that in a Euclidean space a tensor of a given type can always be replaced if desired by another tensor of the same rank but arbitrary type with the aid of the tensors g_{ik} and g^{ik}, and the relations between the two are precisely given.

Finally, a few remarks might be added here concerning facts of some importance for the development of the special theory of relativity. This theory is concerned with a four-dimensional linear space, called a Minkowski space, in which an indefinite metric form occurs such that with respect to a certain set of basis vectors the form reduces to a sum of squares with three positive and one negative coefficient (reference is again made to the book of Efimov [E.2], Chapter VII, which contains an excellent treatment of this subject). The space is thus a linear metric space which is not Euclidean, but which has a three-dimensional Euclidean subspace. The facts of importance about such spaces that are needed later are that (1) every quadratic form in any linear space can be reduced to a sum of squares with coefficients $1, 0, -1$ by an appropriate choice of basis vectors, and (2) the number of positive and negative coefficients in any such reduction is the same independent of the set of basis vectors which yield it. The second fact is sometimes called the *law of inertia* for quadratic forms.

APPENDIX B

Differential Equations

Ordinary differential equations have a place of central importance in differential geometry. This appendix summarizes briefly the basic theorems that are most frequently used. In Section 1 these theorems are formulated for a single equation of first order, then extended to systems. In Section 2 overdetermined systems of first-order *partial differential equations*, plus compatibility conditions, are treated. Such systems are of particular importance in differential geometry—so much so that proofs of the basic existence and uniqueness theorems are given.

1. Theorems on Ordinary Differential Equations[1]

The starting point of the discussion is the differential equation

$$(1) \qquad \frac{dy}{dx} = f(x, y),$$

in which $f(x, y)$ is a continuous function defined in a convex domain D of the x,y-plane. In D the function $f(x, y)$ is assumed to be bounded:

$$(2) \qquad |f(x, y)| < M_1,$$

and, in addition, it is required to satisfy the following Lipschitz condition:

$$(3) \qquad |f(x, y_1) - f(x, y_2)| \leq M_2|y_1 - y_2|,$$

when (x, y_1), (x, y_2) are any points in D. M_1 and M_2 are, of course, suitable constants.

The main interest here is in the initial value problem associated with (1); that is, a solution of (1) is wanted that passes through the point (x_0, y_0) in D. By a solution of (1) is meant a function $y = y(x)$, which has a continuous

---

[1] For detailed proofs of these and other theorems in this chapter the books of Bierberbach [B-4], Coddington and Levinson [C.9], and the lecture notes of Friedrichs [F.2] are recommended.

derivative $y'(x)$, such that the differential equation (1) becomes an identity in x:

(4) $$y'(x) \equiv f(x, y(x)).$$

The initial condition of course requires $y(x)$ to satisfy the equation

(5) $$y(x_0) = y_0.$$

Two positive constants a and b are introduced subject to the condition

(6) $$b > aM_1,$$

and also to the condition that the rectangle R,

(7) $$|x - x_0| < a, \qquad |y - y_0| < b,$$

lies in D. The basic existence and uniqueness theorem then states:

THEOREM I. *A uniquely determined function $y = y(x)$ with a continuous derivative exists over the interval $|x - x_0| < a$ that satisfies (4) and (5), provided that the conditions (2), (3), (6), and (7) are satisfied. In addition, the solution curve $y = y(x)$ lies in R.*

It should be remarked that the Lipschitz condition (3) will always be fulfilled if $f(x, y)$ has a continuous partial derivative f_y in D which is bounded by the constant M_2, since (3) then follows from the mean value theorem.

This important and basic theorem is a theorem in the small, but it has a quantitative formulation since an estimate for the size of the domain within which it holds is contained in it. This in turn is often used in this book to yield a *uniform* estimate in a compact domain as one of the steps in a variety of discussions of a global nature.

Because of its importance in a wide variety of cases in this book, it is worthwhile to make use of Theorem I in order to prove the classical implicit function theorem in a quantitative formulation. This theorem refers to a function $f(x, y)$ defined in an open domain D of the plane. It is assumed here to have continuous second derivatives,[1] to take on the value 0 at some point (x_0, y_0) in D; and to be such that $f_y(x_0, y_0)$ is not zero. To be shown is that *a uniquely determined differentiable curve $y = g(x)$, with $y_0 = g(x_0)$, exists in a certain interval $x_0 - \delta < x < x_0 + \delta$, with $\delta > 0$ a number determined solely by bounds on $f(x, y)$ and its derivatives, and such that $f(x, g(x))$ vanishes along the curve.* Theorem I is used to prove this statement by considering the differential equation

$$\frac{dy}{dx} = -\frac{f_x(x, y)}{f_y(x, y)},$$

---

[1] This is a more stringent requirement than is necessary for the validity of the theorem. A proof in greater generality is given in the book of Goursat [G.3], also with a quantitative formulation.

to which the theorem applies in a neighborhood of (x_0, y_0) since f_y does not vanish there and the right-hand side has continuous first derivatives. It follows that a uniquely determined differentiable solution $y = g(x)$ through (x_0, y_0) exists in an interval about x_0 that depends only on bounds on $f(x, y)$ and its derivatives. This solution is evidently such that $f_x + f_y(dy/dx) = 0$ holds along it, and consequently $f(x, g(x)) = $ const. also holds along it; but since $f(x_0, g(x_0)) = 0$, the constant vanishes. This completes the proof of the statement. Once this result is obtained, the general implicit function theorem is obtained in the general case, i.e., for sets of functions depending on several variables, by an induction process that is based on the special case just treated.

A generalization of the theorem of great importance generally in mathematics, and particularly in differential geometry, results when the function f in (1) depends upon a parameter λ as well as upon x and $y: f = f(x, y, \lambda)$. If in addition to the conditions of Theorem I, it is supposed that f is continuous in λ in an interval $|\lambda - \lambda_0| \leq c$ and uniformly for x and y in D, then the following theorem holds:

THEOREM II. *Any solution $y = y(x, \lambda)$ of the initial value problem in D is continuous in λ in $|\lambda - \lambda_0| \leq c$.*

In case the function $f(x, y, \lambda)$ is restricted to have still further regularity properties, it turns out that the solutions will also be smoother with respect to x and λ. For example, if f is continuous in all three arguments and also has continuous partial derivatives with respect to y and λ, then the following holds:

THEOREM III. *Any solution $y = y(x, \lambda)$ of the differential equation in D has a derivative with respect to λ which is continuous in x and λ.*

So far only a single first-order differential equation was considered. However, the extension of the theorems to *systems* of first-order differential equations is more or less immediate. Consider, for example, a system of the form

$$(8) \qquad \frac{dy_r}{dx} = f_r(x; y_1, y_2, \ldots, y_n; \lambda_1, \lambda_2, \ldots, \lambda_m), \qquad r = 1, 2, \ldots, n.$$

The functions f_r are now assumed to have continuous partial derivatives of order k in all of their arguments, i.e., in the independent variable x, the dependent variables y_r, and the parameters λ_s. A solution $y_r = y_r(x; \lambda_1, \lambda_2, \ldots, \lambda_m)$ of (8), which satisfying the initial conditions

$$(9) \qquad\qquad\qquad y_r(x_0; \lambda_1, \lambda_2, \ldots, \lambda_m) = y_{r0},$$

is in question, naturally at a point $(x_0; y_{10}, \ldots, y_{n0})$ in the domain of definition of the functions f_r. The following theorem then holds:

THEOREM IV. *A uniquely determined solution $y_r(x; \lambda_1, \lambda_2, \ldots, \lambda_m)$ of (8) satisfying (9) exists, and it has continuous kth derivatives with respect to the parameters λ_s and a continuous $(k + 1)$th derivative with respect to x.*

Theorem IV has two corollaries which are of great importance, not only for differential geometry, but also for many other fields in analysis and mathematical physics. The first corollary refers to a single differential equation of nth order of the form

$$(10) \qquad \frac{d^n y}{dx^n} = f\left(x; \frac{dy}{dx}, \frac{d^2 y}{dx^2}, \ldots, \frac{d^{n-1} y}{dx^{n-1}}; \lambda_1, \lambda_2, \ldots, \lambda_m\right),$$

By writing $y = y_1, dy_1/dx = y_2, \ldots, dy_{n-1}/dx = y_n$ the equation (10) is seen to be equivalent to the system

$$\frac{dy_1}{dx} = y_2, \frac{dy_2}{dx} = y_3, \ldots, \frac{dy_{n-1}}{dx} = y_n,$$

$$\frac{dy_n}{dx} = f(x; y_1, y_2, \ldots, y_n),$$

which is of the same type as (8). Theorem IV thus applies, with obvious modifications, if the *initial conditions* for (10) *are prescribed as the values of y and its first $(n - 1)$-derivatives at a certain point x_0.*

Another very important corollary concerns the behavior of the solution of the initial value problem for (8) or (10) when the initial conditions themselves are varied. Suppose that $x = x_0$, $y_r(x_0) = y_{r0}$ fix the initial conditions. By setting $\xi = x - x_0$, $u_r = y_r(x) - y_{r0}$ an equivalent system of equations exactly like (8) is obtained for the functions $u_r(\xi)$, but in which the quantities x_0, y_{r0} figure as parameters in the new functions F_r on the right-hand sides. Thus Theorem IV applies and tells us that *the solution $y_r = y_r(x, x_0, y_{10}, \ldots, y_{r0})$ is differentiable a certain number of times with respect to the coordinates fixing the initial condition.*

The theorems discussed so far apply *whether the differential equations are linear or not*, i.e., whether the dependent variables and their derivatives do or do not occur linearly. If the differential equations are linear their solutions have many special properties of great interest, and there are many situations in which such types of differential equations are extremely important. In differential geometry, for example, the basic partial differential equations of surface theory are linear, as are also the ordinary differential equations which serve to determine a parallel field of vectors along a given curve. A system of differential equations of the form

$$(11) \quad \frac{dy_k}{dx} = \sum_{i=1}^{n} a_{ir}(x; \lambda_1, \lambda_2, \ldots, \lambda_m)y_i + b_r(x; \lambda_1, \lambda_2, \ldots, \lambda_m),$$

$$r = 1, 2, \ldots, n,$$

is considered, in which the so-called *coefficients* a_{ir} and b_{ir} of the linear system may depend on parameters λ_s as well as upon x. For this system Theorem I takes a modified form as follows:

THEOREM I$_l$. *If the functions a_{ir} and b_{ir} are continuous in an interval $a \leq x \leq b$, a unique differentiable solution $y_r(x)$ of (11) exists over the same interval and for initial values $y_r(x_0)$ for any x_0 in $a \leq x \leq b$, when the initial values are chosen arbitrarily.*

The main differences here are that the solutions exist over the whole interval in which the coefficients are defined,[1] and the initial values of the dependent quantities can be chosen arbitrarily.

2. Overdetermined Systems of Partial Differential Equations

The system of partial differential equations to be considered is

$$(12) \qquad \frac{\partial x_k}{\partial u^\alpha} = U_\alpha^k(u^1, u^2, \ldots, u^m; x^1, x^2, \ldots, x^n),$$

$$k = 1, 2, \ldots, n, \ \alpha = 1, 2, \ldots, m.$$

from which n functions $x_k(u^1, u^2, \ldots, u^m)$ of m independent variables are to be determined. Evidently, there are more equations than there are unknown functions. Hence it is not to be expected that solutions of the differential equations will exist unless the functions U_α^k satisfy some appropriate conditions. Since the only solutions $x^k(u^\alpha)$ of interest here are those which have continuous second derivatives, it is clear in particular that the functions U_α^k must be such that the mixed second derivatives $\partial^2 x_k / \partial u^\alpha \, \partial u^\beta$ and $\partial^2 x^k / \partial u^\beta \, \partial u^\alpha$ are identical; thus *necessary* conditions for the existence of such a solution of (12) would be

$$(13) \qquad U_{\alpha,\beta}^k + U_{\alpha,j}^k x_{,\beta}^j = U_{\beta,\alpha}^k + U_{\beta,j}^k x_{,\alpha}^j,$$

with $,\alpha$ and $,j$ meaning derivatives with respect to u^α and x^j, respectively. The summation convention is of course used here. It turns out, however, that these so-called *compatibility conditions* are also *sufficient* conditions.[2]

---

[1] That this is not true in general for nonlinear differential equations is readily seen on the basis of simple counterexamples. Take, for instance, the differential equation $dy/dx = 1 + y^2$, to which all of the theorems in the small apply when x and y range over the entire x,y-plane. But an explicit integration of it shows that no solution exists over an x-interval of length larger than π.

[2] This is a generalization of the well-known situation in calculus in which it is desired to determine a function $u = u(x, y)$ when its derivatives $u_x = f(x, y)$, $u_y = g(x, y)$ are given; it is easily shown that the solution exists and is uniquely determined in a neighborhood of a point (x_0, y_0) if the compatibility condition $f_y = g_x$ is satisfied and $u(x_0, y_0)$ is prescribed. Similarly $u = u(x, y, z)$ is determined when grad $u = (u_x, u_y, u_z)$ is prescribed if curl $u = 0$—thus three compatibility conditions are needed.

In fact, the following theorem will be proved:

THEOREM V.   *If the functions U_α^k have continuous derivatives of second order in all arguments, and if they satisfy the compatibility conditions* (13), *a uniquely determined solution of* (12) *exists as soon as values of the functions $x_k(u)$ are prescribed at some point* (u_0).

Instead of considering the case of m independent variables directly a proof of the theorem will now be given by an induction on m, and therefore beginning with the case $m = 2$. Thus the following system of partial differential equations for n functions x_k of two independent variables u, v is considered:

$$\frac{\partial x_k}{\partial u} = U_k(u, v, x),$$

(12)₁
$$(k = 1, 2, \ldots, n),$$

$$\frac{\partial x_k}{\partial v} = V_k(u, v, x).$$

The compatibility conditions for this case are

(13)₁ $$\frac{\partial U_k}{\partial v} + \sum_i \frac{\partial U_k}{\partial x_i} V_i = \frac{\partial V_k}{\partial u} + \sum_i \frac{\partial V_k}{\partial x_i} U_i, \qquad (k = 1, 2, \ldots, n);$$

they are to be satisfied identically in the variables x_k, u and v.

To prove the Theorem V in this case the first step is to begin by determining n functions $y_k(u, v_0)$ which satisfy the following system of ordinary differential equations:

(14) $$\frac{dy_k}{du} = U_k(u, v_0, y)$$

and the initial conditions

(15) $$y_k(u_0, v_0) = x_{k0}, \qquad (k = 1, 2, \ldots, n);$$

here the initial values x_{k0} are the same as those postulated in Theorem V. From the theory of ordinary differential equations it is known that the functions y_k are uniquely determined and have continuous second derivatives with respect to u. The next step is to determine functions $x_k(u, v)$ which satisfy the ordinary differential equations

(16) $$\frac{dx_k}{dv} = V_k(u, v, x)$$

and the initial conditions

(17) $$x_k(u, v_0) = y_k(u, v_0),$$

when the functions $y_k(u, v_0)$ are chosen as the solutions of (14) determined above. The functions $x_k(u, v)$ thus determined are unique, have continuous second derivatives with respect to both u and v [since the initial conditions (17) have continuous second derivatives with respect to u], and satisfy the conditions $x_k(u_0, v_0) = y_k(u_0, v_0) = x_{k0}$.

The proof of the theorem with respect to the system $(12)_1$ will be complete once it is shown that the functions $x_k(u, v)$ thus determined satisfy the differential equations $(12)_1$. It is clear that

$$(18) \qquad \frac{\partial x_k}{\partial v} = V_k$$

and that

$$\left(\frac{\partial x_k}{\partial u} \right)_{v = v_0} = U_k(u, v_0, x).$$

The functions $\partial x_k / \partial u$ obtained as solutions of (16), with initial conditions (17) are written as follows:

$$(19) \qquad \frac{\partial x_k}{\partial u} = U_k(u, v, x) + \bar{U}_k(u, v),$$

and it is noted that

$$\bar{U}_k(u, v_0) = 0.$$

The proof of the theorem will be carried out by showing that the functions $\bar{U}_k$ vanish identically in a neighborhood of the point (u_0, v_0). Differentiating (19) with respect to v leads to

$$(20) \qquad \frac{\partial \bar{U}_k}{\partial v} = \frac{\partial^2 x_k}{\partial u \, \partial v} - \frac{\partial U_k}{\partial v} - \sum_i \frac{\partial U_k}{\partial x_i} V_i$$

upon using (18). However, $\partial^2 x_k / \partial u \, \partial v = \partial^2 x_k / \partial v \, \partial u$, and this, through use of (18) and (19) leads to

$$(21) \qquad \frac{\partial^2 x_k}{\partial v \, \partial u} = \frac{\partial^2 x_k}{\partial u \, \partial v} = \frac{\partial V_k}{\partial u} + \sum_i \frac{\partial V_k}{\partial x_i} \frac{\partial x_i}{\partial u}$$

$$= \frac{\partial V_k}{\partial u} + \sum_i \frac{\partial V_k}{\partial x_i} (U_i + \bar{U}_i)$$

$$= \frac{\partial U_k}{\partial v} + \sum_i \frac{\partial U_k}{\partial x_i} V_i + \sum_i \frac{\partial V_k}{\partial x_i} \bar{U}_i,$$

the last step resulting by virtue of the compatibility conditions $(13)_1$. Upon substituting the relation (21) in (20) the equations

$$(22) \qquad \frac{\partial \bar{U}_k}{\partial v} = \sum_i \frac{\partial V_k}{\partial x_i} \bar{U}$$

are obtained.   The functions $\overline{U}_k$ satisfy the initial conditions

$$\overline{U}_k(u, v_0) = 0.$$

But (22) is a system of ordinary differential equations (for any fixed value of u) which has a unique solution for prescribed initial conditions (i.e., conditions at $v = v_0$).   Obviously the functions

$$\overline{U}_k(u, v) \equiv 0$$

yield a solution of (22) which satisfies the condition $\overline{U}_k(u, v_0) = 0$.   Thus $\overline{U}_k \equiv 0$ are the only solutions, so that the functions x_k obtained previously satisfy the differential equations (12), and this completes the proof of the theorem for two independent variables.   It is, however, quite clear that this proof could be extended by induction to the case of any number of independent variables.

It is well known that partial differential equations usually have a manifold of solutions that depends on arbitrary *functions* rather than on arbitrary *constants*, which turned out to be the case here.   The distinguishing feature of the present case is that the addition of the compatibility conditions makes the underlying integration theory essentially the same as that for ordinary differential equations.

BIBLIOGRAPHY

[A.1] Ahlfors, L. V., *Complex Analysis*, McGraw-Hill, New York, 1953.

[A.2] Alexandrov, A. D., *Konvexe Polyeder*, Akad. Verlag, Berlin, 1958.

[A.3] Alexandrov, A. D., *Die innerie Geometrie der konvexen Flächen*, Akad. Verlag, Berlin, 1955. (Translated from the Russian.)

[A.4] Amsler, Marc-Henri, *Des surfaces à courbure négative constante dans l'espace à trois dimensions et de leurs singularités*, Brünsche Universitätsdruckerei Giessen, 1955, Prom. No. 2383.

[A.5] Auslander, L., and MacKenzie, R., *Introduction to Differentiable Manifolds*, McGraw-Hill, New York, 1963.

[B.1] Bieberbach, L., *Differentialgeometrie*, Teubner, Leipzig and Berlin, 1932.

[B.2] Bieberbach, L., *Hilbert's Satz über Flächen konstanter negativer Krümmung*, Acta Math., 48, 1926.

[B.3] Bieberbach, L., *Über Techebyscefsche Netze auf Flächen negativer Krümmung, sowie auf einigen weiteren Flächenarten*, Sitzungsberichte der Preussischen Akademie der Wissenschaften, Vol. XXIII, 1926.

[B.4] Bieberbach, L., *Differentialgleichungen*, Springer, Berlin, 1923.

[B.5] Blaschke, W., *Vorlesungen über Differentialgeometrie*, Springer, Berlin, 1924.

[B.6] Blaschke, W., *Ein Beweis für die Unverbiegbarkeit geschlossener knovexer Flächen*, Nachr. Ges. Wiss., Göttingen, 1912, S. 607–616.

[B.7] Bliss, G. A., *Caculus of Variations*, Open Court, La Salle, Ill., 1925.

[B.8] Bonnesen, T., u. Fenchel, W., *Theorie der konvexen Körper*, Ergeb. der Math., Springer, Berlin, 1934.

[B.9] Buseman, H., *Convex Surfaces*, Wiley-Interscience, New York, 1958.

[B.10] Buseman, H., *The Geometry of Geodesics*, Academic Press, New York, 1955.

[C.1] Caratheodory, C., *Calculus of Variations and Partial Differential Equations*, Holden-Day, San Francisco, 1965.

[C.2] Cartan, E., *Leçons sur la géometrie des espaces de Riemann*, Gauthier-Villars, Paris, 1928, 2nd ed., 1946.

[C.3] Cauchy, A., *Sur les polygones et polyèdres*, Journ. de l'école polytechnique, 9, 1813.

[C.4] Chern, S., *A proof of the uniqueness of Minkowski's problem for convex surfaces*, Am. J., Math., LXXIX, 1957.

[C.5] Chern, S., *Some new characterizations of the Euclidean sphere*, Duke Math. Journ., 12, 1945.

[C.6] Chern, S. S., and Lashof, R. K., *On the total curvature of immersed manifolds*, Mich. Math. Journ., 5, 1958.

[C.7] Chevalley, C., *Theory of Lie Groups*, Vol. I, Princeton Univ. Press, Princeton, New Jersey, 1946.

[C.8] Christoffel, E., *Über die Bestimmung der Gestalt einer krümmen Oberfläche durch lokale Messungen auf derselben*, Journal für die reine und angewandte Mathematik, 64, 1865.

[C.9] Coddington, E. A., and Levinson, N., *Theory of Ordinary Differential Equations*, McGraw-Hill, New York, 1955.

[C.10] Cohn-Vossen, S., *Singularitäten konvexer Flächen*, Math. Ann., 97, 1927.

[C.11] Cohn-Vossen, S., *Zwei Sätze über die Starrheir der Eiflächen*, Nachr. Ges. Wiss., Göttingen, 1927, S. 125–134.

[C.12] Courant, R., and Hilbert, D., *Methoden d. Math. Physik*, Springer, Berlin, Bd. I, 1931; Bd. II, 1937.

[C.13] Courant, R., and Robbins, H., *What Is Mathematics?*, Oxford Univ. Press, New York, 1941.

[D.1] Darboux, G., *Théorie des surfaces*, Gauthier-Villars, Paris, 1896.

[D.2] DeRham, G., *Variétés differentiables*, Hermann, Paris, 1955.

[E.1] Efimov, N. V., *Generation of singularities of surfaces of negative curvature*, Matem. Sborn, N.S., 64, 286–320, 1964.

[E.2] Efimov, N. V., *Höhere Geometrie*, VEB Deutsche Verlag der Wiss., Berlin, 1960. (Translated from the Russian.)

[E.3] Efimov, N. V. *Flächenverbiegung im Grossen*, Akad. Verlag, Berlin, 1957. (Translated from the Russian.)

[E.4] Einstein, A., *Zür Elektrodynamik bewegter Körper*, Ann. d. Physik, 17, 1905.

[E.5] Einstein, A., *Die Grundlage der allgemeinen Relativitätstheorie*, Ann. d. Physik, 49, 1916.

[E.6] Eisenhart, L. P., *Riemannian Geometry*, Princeton Univ. Press, Princeton, N.J., 1926.

[F.1] Flanders, H., *Differential Forms*, Academic Press, New York, 1963.

[F.2] Friedrichs, K. O., *Advanced Ordinary Differential Equations*, N.Y.U. Lecture Notes, 1947–48.

[G.1] Gauss, C. F., *Disquisitiones generales circa superficies curvas*, Comm. Soc. Göttingen, Bd. 6, 1923–27. (Translated recently as General Investigations of Curved Surfaces, New York, 1965.)

[G.2] Gelfand, I. M., *Lectures on Linear Algebra*, Wiley-Interscience, New York, 1961.

[G.3] Goursat, E., *Cours d'analyse mathématique*, Gauthier-Villars, Paris, 1942.

[G.4] Grassmann, H., *Lineale Ausdehnungslehre*, Leipzig, 1844. In: *Gesammelte Werke*, Teubner, 1894.

[G.5] Graustein, W. C., *Differential Geometry*, Macmillan, New York, 1935.

[H.1] Hadamard, J., *Leçons de géométrie élémentaire*, Vol. II, 7th ed. 1931.

[H.2] Hadamard, J., *Sur certaines propriétés des trajectoires en dynamique*, Journ. de Math. (5), T. 3, 331–387, 1897.

[H.3] Halmos, P. R., *Finite-dimensional Vector Spaces*, Van Nostrand, Princeton, N.J., 1958.

[H.4] Hartman, P., and Nirenberg, L., *On spherical image maps whose Jacobians do not change sign*, Am. J. Math., 81, 1959.

[H.5] Hartman, P., and Wintner, A., *On the asymptotic curves of a surface*, Amer. J. Math., LXXIII, 1951.

[H.6] Hazzidakis, J. N., *Über einige Eigenschaften der Flächen mit konstantem Krümmungsmass*, Crelles J., 88, 1880.

[H.7] Herglotz, G., *Über die Starrheit der Eiflächen*, Abhandlungen aus dem mathematischen Seminar der Hansischen Universität, Vol. 15, 1943.

[H.8] Hicks, Noel J., *Notes on Differential Geometry*, Van Nostrand Math. Studies, No. 3, 1964, Princeton, N.J.

[H.9] Hilbert, D., Appendix of the fifth edition of the *Grundlagen der Geometrie*, Leipzig and Berlin, 1909.

[H.10] Hilbert, D., and Cohn-Vossen, S., *Geometry and the Imagination*, Chelsea, New York, 1952.

[H.11] Holmgren, E., *Sur les surfaces à courbure constante négative*, C. R. Acad. Sci. (Paris), 134, 1902, 740–743.

[H.12] Hopf, H., *Selected topics in geometry*, N.Y.U., 1946.

[H.13] Hopf, H., *Über die Drehung der Tangenten und Sehnen ebener Kurven*, Comp. Math., 2, S. 50–62, 1935.

[H.14] Hopf, H., *Lectures on differential geometry in the large*, Stanford, California.

[H.15] Hopf, H., *Selected topics in differential geometry in the large*, N.Y.U., 1955.

[H.16] Hopf, H., u, Rinow, W., *Über den Begriff der vollständigen differentialgeometrischen Fläche*, Comm. Math. Helvetici, 3, S. 209–225, 1931.

[H.17] Hopf, H., and Samelson, H., *Zum Beweis des Kongruenzatzes für Eiflächen*, Math. Zeit., 43, 1938, S. 749–766.

[K.1] Karcher, H., *Konvexe Kurven auf zweidimensionalen Riemannschen Mannigfaltigkeiten*, Ph.D. Thesis, Tech. Univ., Berlin, 1966.

[K.2] Klingenberg, W., *Riemannsche Geometrie im Grossen*, Math. Inst. Univ. Bonn, 1962.

[K.3] Knopp, K., *Theory of Functions*, Dover Publications, New York, 1945.

[K.4] Kobayashi, S., and Nomizu, K., *Foundations of Differential Geometry*, Wiley, New York, 1963.

[K.5] Kuiper, N. H., *Ausgewählte Kapitel der Riemannschen Geometrie*, Vorles. d. Univ., Bonn, 1957.

[L.1] Levi-Civita, T., *Lezioni di meccanica razionale*, Zanichelli, 1923.

[L.2] Levi-Civita, T., *Nozione di parallelismo in una varietà qualunque*, Rend. Circ. Matem., Palermo, 42, 1917.

[L.3] Lewy, H., *On the existence of a closed convex surface realizing a given Riemannian metric*, Proceedings of the National Academy of Sciences, 24, No. 1938.

[L.4] Lewy, H., *On differential geometry in the large, I* (Minkowski's problem), Transactions of the American Mathematical Society, 43, 1938.

[L.5] Liebmann, H., *Über die Verbiegung der geschlossenen Flächen positiver Krümmung*, Math. Ann., Bd. 53, S. 81–112, 1900.

[L.6] Liebmann, H., *Eine neue Eigenschaft der Kugel*, Göttingen Nach. 1899.

[M.1] Massey, W. S., *Surfaces of Gaussian curvature zero in Euclidean 3 space*, Tohoku Math. J., 14, 1962.

[M.2] Minkowski, H., *Space and Time* (a translation of an address of 1908), in The Principle of Relativity, Dover, New York, 1923.

[M.3] Minkowski, H., *Volumen und Oberfläche*, Math. Ann., 57, 1903.

[O.1] O'Neill, B., *Elementary Differential Geometry*, Academic Press, New York, 1966.

[O.2] Orland, G. M., *On non-convex polyhedral surfaces*, Pacific Journ. Math., 16, 1965.

[P.1] Patterson, E. M., *Topology*, Oliver and Boyd, Edinburgh and London, 1956.

[P.2] Pauli, W., *Relativitätstheorie*, Encykl. Math. Wiss., Leipzig, 1921.

[P.3] Poincaré, P., *Œuvres*, Vol. 1, p. 3 (singularities of first order differential equations), p. 121 (sum of indices on a closed surface).

[P.4] Poincaré, P., *Théorie des groupes Fuchsiennes*, Acta Math., 1, 1882.

[P.5] Pogorelov, A. V., *Differential Geometry*, Noordhoff, Groningen, 1966.

[P.6] Pontrjagin, L. S., *Topological Groups*, Princeton Univ. Press, Princeton, N.J.

[P.7] Preissmann, A., *Quelques propriétés globales des espaces de Riemann*, Comm. Mat. Helv., 15, 1943.

[R.1] Rembs, E., *Unverbiegbare offene Flächen*, S.-B. preuss. Akad. Wiss., 1930, S. 123–133.

[R.2] Ricci, G., and Levi-Civita, T., *Méthodes de calcul différentiel absolu et leurs applications*, Math. Ann., 54, 1901.

[R.3] Riemann, B., *Über die Hypothesen welche der Geometrie zu grundeliegen*, Göttinger Abh., 1868; Ges. Werke, 1892.

[S.1] Sacksteder, R., *On hypersurfaces with no negative sectional curvatures*, Am. J. Math., 1960.

[S.2] Steinitz, E., and Rademacher, H., *Vorlesungen über die Theorie der Polyeder*, Springer, Berlin, 1934.

[S.3] Sternberg, S., *Lectures on Differential Geometry*, Prentice-Hall, Englewood Cliffs, N.J., 1964.

[S.4] Stoker, J. J., *Geometrical problems concerning polyhedra in the large*, Comm. Pure Appl. Math., 21, No. 2, 1968.

[S.5] Stoker, J. J., *On the uniqueness theorems for the embedding of convex surfaces in three-dimensional space*, Comm. Pure Appl. Math., III, No. 3, September 1950.

[S.6] Stoker, J. J., *Open convex surfaces which are rigid*, Courant Anniversary Volume, 1948.

[S.7] Stoker, J. J., *Unbounded convex point sets*, Am. J. Math., LXII, 1940.

[S.8] Stoker, J. J., *Uniqueness theorems for polyhedra*, Proc. Nat. Acad. of Sci., 55, No. 6, 1967.

[S.9] Stoker, J. J., *Über die Gestalt der positiv gekrümmten offenen Flächen im dreidimensionalen Raume*, Comp. Math. 3, S. 55–89, 1936.

[S.10] Stoker, J. J., *Developable surfaces in the large*, Comm. Pure Appl. Math., 14, 1961.

[S.11] Synge, J. L., *Relativity: the general theory*, Wiley-Interscience, New York, 1960.

[S.12] Synge, J. L., *On the connectivity of spaces of positive curvatures*, Quart. Journ. Math., 1936.

[S.13] Synge, J. L., *The first and second variations of length in Riemannian space*, Proc. London Math. Soc., 25, 1926, 247–264.

[T.1] Tchebychef, P. L., *Sur le coupure des vêtements*, Œuvres, Bd. II, S. 708, 1878.

[V.1] van Heijenoort, J., *On locally convex manifolds*, Comm. Pure Appl. Math., 5, 1952.

[V.2] Voss, K., *Differentialgeometrie geschlossener Flächen im Euklidischen Raum*, I, Jahresber, D. Math. Ver., Bd. 63, 1960.

[V.3] Voss, K., *Einige differentialgeometrischer Kongruenzsätze f. geschlossene Flächen u. Hyperflächen*, Math. Ann., Bd. 131, 1956.

[W.1] Weyl, H., *Raum, Zeit, Materie*, Springer, Berlin, 1923.

[W.2] Weyl, H., *Über die Starrheit der Eiflächen und convexen Polyeder*, S.-B. preuss. Akad. Wiss., S. 250–266, 1917.

[W.3] Weyl, H., *Über die Bestimmung einer geschlossenen konvexen Fläche durch ihr Linienelement*, Vierteljahrsschrift det Naturforschenden Gesellschaft in Zürich, 61, 1916.

[W.4] Weyl, H., *Die Idee d. Riemannschen Fläche*, Chelsea, N.Y., 1947.

[W.5] Whitehead, J. H. C., *Convex regions in the geometry of paths*, Quart. J. Math., 3, 1932, 33–42.

[W.6] Whitney, H., *Differentiable Manifolds*, Ann. Math., 37, 1936.

[W.7] Willmore, T. J., *An Introduction to Differential Geometry*, Oxford, 1959.

[Y.1] Yaglom, I. M., and Boltyanskii, V. G., *Convex Figures*, Holt, Rinehart and Winston, New York, 1961.

INDEX